WITHDRAWN
FROM
COLLECTION

FORDHAM
UNIVERSITY
LIBRARIES

The Natural Philosophy of Galileo

The MIT Press Cambridge, Massachusetts, and London, England

The Natural Philosophy of Galileo

Essay on the Origins and Formation of Classical Mechanics

Maurice Clavelin

translated by A. J. Pomerans

B
785
.G24C5
Cp. 3

Fordham University
LIBRARY
AT
LINCOLN CENTER
New York, N. Y.

Copyright © 1968 Librairie Armand Colin
English translation
Copyright © 1974 by
The Massachusetts Institute of Technology

All rights reserved. No part of this book may be reproduced in any form or by any means, electronic or mechanical, including photocopying, recording, or by any information storage and retrieval system, without permission in writing from the publisher.

This book was set in Monotype Baskerville
by The Eastern Typesetting Co., Inc.,
printed on Fernwood Opaque,
and bound in G.S.B. S/535/83 "Lime"
by Halliday Lithograph Corp.
in the United States of America.

Library of Congress Cataloging in Publication Data

Clavelin, Maurice.
The natural philosophy of Galileo.

Translation of La philosophie naturelle de Galilée.
Bibliography: p.
1. Galilei, Galileo, 1564–1642. 2. Philosophy of nature. I. Title.
B785.G24C513 195 74–978
ISBN 0–262–03050–0

To the memory of my father and to all my family

Translator's Note ix

Preface xi

Chronology xiii

I The Tradition and Early Writings

1 Aristotle and Local Motion 3

2 The Fourteenth-Century Tradition 61

3 The Preparatory Years (1589–1602) 118

II The Copernican Doctrine and the Science of Motion

4 The Construction of a Copernican Cosmology 183

5 The Problem of the Earth's Diurnal Motion 224

III The Birth of Classical Mechanics

6 The Geometrization of the Motion of Heavy Bodies (Part 1) 277

7 The Geometrization of the Motion of Heavy Bodies (Part 2) 324

IV

8 Reason and Reality 383

Conclusion 457

Appendix 1
Additional Remarks on Oresme's Theory of
Configurationes Qualitatum 465

Appendix 2
Bradwardine's Law 468

Appendix 3
On the Force of Percussion 472

Appendix 4
Did Galileo Consider the Moving Earth an Inertial System? 476

Appendix 5
Did Galileo Continue to Hold that Forces Are Proportional to Speeds? 482

Appendix 6
The Determination of the *Impeto* (or Speed) of a Projectile at Every Point on Its Trajectory 484

Bibliography 487

Index 493

Translator's Note

In quoting from the works of Aristotle and Galileo, I have drawn freely on the following outstanding English versions, to whose authors I hereby acknowledge my great debt.

Aristotle:
Physics, translated by Richard Hope, University of Nebraska Press, Lincoln, 1961.
On the Heavens, translated by W. K. C. Guthrie, Loeb Classical Library, London, 1939.

Galileo:
On Motion and on Mechanics, translated by I. E. Drabkin and Stillman Drake, the University of Wisconsin Press, Madison, 1960.
Discoveries and Opinions of Galileo (including *The Sidereal Messenger*, *Letters on Sunspots*, and excerpts from *The Assayer*), translated by Stillman Drake, Doubleday, New York, 1957.
A Dialogue Concerning Two Chief World Systems, translated by Stillman Drake, University of California Press, Berkeley and Los Angeles, 1962.
Dialogues Concerning Two New Sciences (the *Discourses*), translated by Henry Crew and Alfonso de Salvio, Northwestern University Press, 1914; reprinted as a McGraw-Hill paperback, 1968.

Because the different translators have not used the same terminology, we shall, in following them, render *movimento* variously by "motion" or "movement," *velocità* by "velocity" or "speed," and *gravità* by "gravity" or "weight." Unless otherwise indicated, all these pairs of terms are used as synonyms; none serve to represent vector quantities. Moreover, for the sake of consistency, the translations have been modified in certain minor respects.

Preface

This book tries to assess Galileo's work in its historical singularity. It is constructed around a precise question: How did Galileo create the modern science of motion? Starting from this question, I shall go on to determine as accurately as I can what concepts and methods helped classical mechanics to take shape.

I have not attempted a systematic reconstruction of the intellectual and social context in which Galilean science evolved. Of course, I am not suggesting that this type of reconstruction would have been pointless. In many respects Galileo's work was a natural consequence of the Italian Renaissance, and even more directly of the return to the Greek sources which began to be felt so strongly during the middle of the sixteenth century. It is also highly probable that the new world view imposed by the great maritime discoveries and the increasing contacts between European countries greatly encouraged the maturation of a mode of thought that utterly repudiated the finite approach of traditional philosophy. However, I do not think that Galileo's essential contribution can be characterized in this way. Galilean science was first of all a transition from one conceptual framework to another, the replacement of one explanatory ideal with another and an unprecedented fusion of reason and reality; to appreciate this tremendous revolution, to evaluate its scope and full significance, we can follow one path only: we must try to view Galileo's work from within. This is the only way of reconstructing his work in its full complexity but also in all its shortcomings and contradictions. Galilean science proceeded by a series of intellectual steps that were so many beginnings and that I have tried to retrace faithfully, eschewing all recourse to extraneous factors.

The same desire to grasp Galileo's contribution in all its originality explains why I have dwelled at such length on the traditional analysis of motion. Its systematic reconstruction struck me as being inescapable for two reasons. If Galilean science did indeed mark the advent of a different conceptual universe, there is no better way of proving this transformation than by reconstructing the very universe it brought tumbling down. Moreover, the further I pursued my studies, the more obvious it became that the traditional physics to which Galileo appealed was not limited to the letter of Aristotle's doctrine, but that some of the themes introduced during the fourteenth century in both Oxford and

Paris also played a considerable part in the formation of the new mechanics. To offer a fragmentary summary of these themes, instead of reconstructing the general framework from which they are inseparable, would inevitably have prevented a complete characterization of the transformation by which Galileo changed the face of science in the space of fifty years.

All these factors have forced me to follow a definite plan. The first two chapters deal with the traditional science of motion, first in its Aristotelian form and then in the renewed form in which fourteenth-century scientists presented it; Chapter 1 discusses the concepts and intellectual norms on which physical speculation relied for twenty centuries; Chapter 2 deals with the medieval contribution and tries to establish its importance as well as its limitations. Terminating the first part, Chapter 3 shows what earlier influences and what original ideas helped Galileo in breaking with the traditional system.

These preliminaries out of the way, the book goes on to analyze Galileo's own science of motion. For reasons that will be discussed at some length, that science will be examined first on the cosmological plane (Galileo's justification of the Copernican doctrine, in Part Two); next on the plane of "rational" mechanics (the geometrized theory of the motion of heavy bodies, in Part Three). A chapter devoted to the synthesis of Galileo's method and explanatory ideal (Part Four) completes the book.

Maurice Clavelin

Chronology

In this summary account of the chief dates in the life and work of Galileo, I have followed Favaro's excellent *Cronologia galileiana* (Padua, 1892). For the events of 1633, which I have mentioned only in passing, the reader is referred to G. de Santillana: *The Crime of Galileo* (University of Chicago Press, 1955).

1564
Galileo Galilei born in Pisa on February 15.

1581
Enters the University of Pisa to read medicine and philosophy.

1583
Discovers the isochrony of small pendular oscillations and applies it to the construction of an original pulsilogium.

1584
Begins to study geometry.
Becomes acquainted with traditional science and writes a series of long notes, first published in 1890 under the title of *Juvenilia*.

1585
Returns to Florence. Proves several propositions "on the center of gravity of various solids."

1586
Studies the works of Archimedes, especially the *De aequiponderantibus* and the *De his quae vehuntur in acqua*.
Constructs a hydrostatic balance (*la bilancetta*) and describes its workings.
Studies and annotates Archimedes' *De Sphaera et Cylindro*.

1587
Visits Rome.

1588
Lectures in Florence on Dante's *Inferno*.
Informs Guido Ubaldo del Monte of the result of his work on centers of gravity.

1589
Offered the chair of mathematics at the University of Pisa in July.

1590–1591

Invents the cycloid and recommends its use in the construction of bridges.
Experiments at Pisa on the fall of heavy bodies.
Writes the *De Motu.*
Writes commentaries on Ptolemy's *Almagest.*

1592

Offered the chair of mathematics at the University of Padua (September). His first lecture on December 7 is greeted with acclaim.

1593

Publishes a treatise on fortifications.
Reads Euclid.
Invents a machine for raising water.

1593–1594

Writes *Le Mecaniche* (theory of simple machines) for his students.
Lectures on the Fifth Book of Euclid.

1597

Constructs a geometrical and military compass.
May 30: Declares his adherence to the Copernican doctrine in a letter to Jacopo Mazzoni.
August 4: In a letter to Kepler, states his long-standing adherence to the Copernican doctrine.
Publishes the *De Sphaera*, a cosmographical treatise on Ptolemaic lines.

1598

Probable date of completion of the first *Discourse on the Force of Percussion* (*Discorso primo et antico della percossa*).

1599

Reedits and improves his treatise on simple machines.

1600

Birth of his daughter Virginia.

1601

Birth of his daughter Livia.

1602

Early work on magnets.

November: First mention of the isochrony of pendular oscillations and the law of vibrating strings.

1603

Early experiments leading to the construction of a thermoscope in 1606.

1604

October: First mention of the law governing the distances covered by freely falling bodies.
Observes a nova in the constellation of Sagittarius.
December: Delivers three lectures on the nova observed in October.

1605

January: Publishes the *Discorso intorno alla nuova stella.*
February: Polemics with Baldassare Capra and Antonio Lorenzi on the nova of 1604.
August: Admitted to the Accademia della Crusca.

1606

June: Prints sixty copies of his *Operazioni del compasso geometrico e militare.*
Birth of his son Vicenzio.

1607

Sues Baldassare Capra for falsely claiming the invention of the geometric and military compass; wins his case in May.
August: Publishes the *Difesa contra alle calunnie et imposture di Baldassare Capra.*
November: Devotes several months to the study of magnetism.

1609

Writes to Antonio de' Medici about his work on mechanics, especially on projectile motion and the resistance of materials to fracture.
Claims to have discovered several important properties of water and other liquids.
June: Informs Luca Valerio of his discovery of two mechanical principles. (Several important theorems on naturally accelerated motion had by then become part and parcel of Galilean science.)
July–August: Construction of the astronomical telescope and demonstration of its capabilities before the Doge and the Senate of Venice on top of the campanile of St. Mark's. Confirmed in his chair at Padua for life.
Autumn: First great astronomical discoveries.

1610

January: Discovery of the satellites of Jupiter (the so-called Medicean planets).

January 30: Sends the *Siderius Nuncius* (*The Sidereal Messenger*) to the printer.

March: Publication of 500 copies of the *Siderius Nuncius.*

April: Kepler publishes the *Dissertatio cum Nuncio Sidereo*, in which he fully endorses Galileo's conclusions.

June: Horky publishes an attack on Galileo.

July: Appointed chief mathematician and philosopher to the Grand Duke of Tuscany.

Observes Saturn, which, he believes, consists of three distinct bodies.

Begins his observations of sunspots.

August: Kepler publishes the *Narratio*, in which he confirms the existence of the satellites of Jupiter.

September: Galileo leaves Padua for good and returns to Florence.

Observation of the phases of Venus.

October–November: Correspondence with G. G. Breugger concerning the mountains of the moon.

December: Asks Giuliano de' Medici to act as a witness for his discovery of the phases of Venus.

Observations to determine the periods of the four satellites of Jupiter.

Writes the *Teorica speculi concavi sphaerici.*

Reads and annotates Lodovica delle Colombe's work denying the motion of the earth.

1611

January–March: Continues to observe Jupiter's satellites.

March–June: Visit to Rome, where the mathematicians in the Jesuit Roman College (notably Father Clavius) acclaim his astronomical discoveries.

Elected a member of the Accademia dei Lincei.

April: Establishes the periods of the revolution of Jupiter's satellites.

Resumes his observation of sunspots.

Summer: Back in Florence, gives public demonstrations on condensation and rarefaction and the behavior of floating bodies.

October: First mention of the sun's revolution.

1612

January: Receives from Marcus Welser copies of Father Christopher Scheiner's three letters on sunspots.

Fits a rudimentary type of micrometer to his telescope and measures the apparent elongations of Jupiter's satellites.

March–April: Publishes the *Discorso intorno alle cose che stanno in su l'acqua* (*Discourse on Bodies in Water*).

May–December: Replies to Scheiner's arguments with his *Letters on Sunspots.*

June: Announces that he has discovered the periods of revolution of Jupiter's satellites and constructs the appropriate tables.

November: Observation of Saturn.

Niccolo Lorini delivers a sermon at St. Mark's, Florence, opposing the motion of the earth.

1613

March: Publication of the *Letters on Sunspots.*

December 21: Letter to Castelli on the relationship between science and faith.

1614

March: Describes a method of weighing air.

Death of Filippo Salviati, a Florentine friend, who was to become one of the interlocutors of the *Dialogue* and the *Discourses.*

November: Receives a visit from Louis Salignac, Bishop of Sarlat.

December: Tomasso Caccini, a Dominican friar, attacks Galileo from the pulpit of Santa Maria Novella.

1615

February: Niccolo Lorini denounces Galileo to the Holy Office for his Copernican views. Galileo writes to Mgr. Dini in defense of Copernicus.

March: Second letter to Mgr. Dini.

April: Letter from Cardinal Bellarmine to Father Paulo Antonio Foscarini on the interpretation of the Copernican system.

May: Third letter to Mgr. Dini.

Replies, in Castelli's name, to Aristotelian attacks on his *Discourse on Bodies in Water.*

June: Letter to the Grand Duchess Christina on the possibility of reconciling the Scriptures with the new astronomy.

November: Visit to Rome to defend himself against accusations of heresy and to have the condemnation of Copernicanism quashed.

1616

January: Writes the *Discorso del flusso e reflusso del mare* (which was to become the Fourth Day of the *Dialogue*), in which he attempts to show that only the motion of the earth can explain the tides.
February: Writes the *Reflections on the Copernican View.* Accepting the propositions of the Theological Fathers, the Congregation of the Roman Inquisition condemns the Copernican doctrine. The condemnation is ratified by the Pope. Cardinal Bellarmine enjoins Galileo to abandon the Copernican doctrine.
March: The decree condemning the Copernican doctrine is published, and the work of Copernicus is placed on the Index.
Pope Paul V receives Galileo.
Publication of the *Disputatio de situ et quiete Terrae contra Copernici systema,* by Father F. Ingoli.
Tomasso Campanella publishes an *Apologia* in support of Galileo.
June: Returns to Florence.
Spain turns to Galileo for help in determining longitudes at sea.
August: Observes Saturn.
November: Makes direct contact with the Spanish government.

1618

June: Pilgrimage to Loretto.

1619

May: Mario Guiducci reads the *Discorso delle comete* (written mainly by Galileo) before the Florentine Academy. (The work was published six months later.)
October: Using the pseudonym of Lothario Sarsi, Father Grassi publishes his *Libra astronomica ac philosophica,* a work to which Galileo's *Saggiatore* would be the reply.

1620

January: New negotiations with Spain on the determination of longitudes at sea.
March: Death of Francesco Sagredo, a Venetian friend who, like Salviati, became one of the interlocutors of the *Dialogue* and the *Discourses.*
June: Letter from M. Guiducci to Father T. Galuzzi on the controversy of the comets.

August: Cardinal Maffeo Barberini (later Pope Urban VIII) presents Galileo with the *Adulatio perniciosa*, a poem written in his honor.

1621

January: Galileo elected a member of the Accademia fiorentina.
November: Works on his reply to Father Grassi's *Libra*.

1622

October: Sends the manuscript of his *Saggiatore* (the *Assayer*) to the Accademia dei Lincei.
December: Informs Ciampoli that he has improved his *Discorso del flusso e reflusso* and, for the first time since the condemnation of 1616, mentions his intention to resume his studies of the earth's motion.

1623

February: The *Saggiatore* is granted the imprimatur.
August: Cardinal Maffeo Berberini is proclaimed Pope Urban VIII.
October: Publication of the *Saggiatore*, dedicated to the new Pope.

1624

April–June: Fourth visit to Rome. Received by Urban VIII but given no further encouragement.
July–August: Improves the compound microscope.
September–October: Completes his reply to Ingoli's *Disputatio* and sends it to Mario Guiducci, who makes it widely known in Rome.
December: Informs Cesare Marsili that he is continuing his work on the problem of the tides.

1625

March: Works on the manuscript that was to become his *Dialogue Concerning the Two Chief World Systems* (*Dialogo sopra i due massimi sistemi del mondo*).
November–December: Studies hydraulic problems.

1626

April–June: Resumes his work on magnetism.
September: Postpones the writing of the *Dialogue* for several years.

1627

August: Informs Castelli of his most recent studies of the orbits of Jupiter's satellites.

1628
March: Falls seriously ill.

1629
Resumes the writing of the *Dialogue.*
November: Fresh negotiations with Spain on the determinations of longitudes at sea.

1630
January: The *Dialogue* is completed.
March: Urban VIII declares that he has never felt hostile to Copernicus.
April–June: Visits Rome to obtain permission for printing the *Dialogue*, now dedicated to the Grand Duke of Tuscany.
September: Obtains permission to have the *Dialogue* printed in Florence.
October: Sends Niccolo Aggiunti the results of his mechanical studies.

1631
January: Studies the possibility of harnessing the River Bisenzio.
January–May: Continued negotiations on the publication of the *Dialogue.*
July: Reports on a canalization scheme for the River Arno.
December: The Republic asks Galileo to publish his *Dialogue* in Venice, and invites him to return to his chair in Padua.

1632
February: Printing of the *Dialogue* completed.
March–April: Galileo contracts a serious inflammation of the eye.
May: Tries to resume negotiations with Spain.
August: Sale of the *Dialogue* prohibited.
September: Declares his intention of publishing a work on projectile motion.
October: Ordered to appear before the Inquisition in Rome; for reasons of health he is given until December, after which he would be brought to Rome forcibly and in chains (*carceratum et ligatum cum ferris*).

1633
February: Having been granted a further postponement, Galileo arrives in Rome.
April–May: Appears before the Commissary General of the Inquisition.
June 21: Further interrogation on the Pope's orders with threat of torture.

June 22: In the Dominican Convent of Santa Maria Sopra la Minerva, Galileo is told that his *Dialogue* has been placed on the Index and he himself sentenced to prison at the pleasure of the Holy Office. He recites the prepared formula of abjuration on his knees.
July: Leaves Rome for Siena, where he is the guest of Archbishop Ascanio Piccolomini.
August: Diodati, with whom Galileo had been in touch for several years, sends the *Dialogue* to Bernegger for Latin translation.
Autumn: Works in Siena on the resistance of materials to fracture.
December: Returns to his home in Arcetri, where he remains under the surveillance of the Inquisition.
Receives a visit from the Grand Duke.

1634
February: Asks for permission to receive medical treatment in Florence; the Pope refuses and lets it be known that if Galileo urges any more petitions for release, he will be confined in the prisons of the Holy Office.
September: Mersenne publishes a French translation of the *Mecaniche* under the title of *Les mécaniques de Galilée.*
November: Informs Micanzio that he is working on an entirely new treatise on motion.
December: The French Ambassador François de Noailles and the French astronomer Nicolas Peiresc petition the Pope to free Galileo.

1635
January: Galileo begins the writing of his *Mathematical Discourses and Demonstrations Concerning Two New Sciences* (*Discorsi e dimostrazioni matematiche intorno a due nuove scienze*).
April: Publication of the Latin translation of the *Dialogue* in Strasbourg.
June: The first Two Days of the *Discourses* are completed, and Galileo entertains hopes of having the work published in Holland.
September: Sustermans paints Galileo's portrait.

1636
April: Bernegger publishes the Italian text and the Latin translation of Galileo's *Letter to the Grand Duchess Christina.*
Summer: Galileo negotiates with Louis Elzevir of Leyden about the printing of the *Discourses* and provides him with a copy of the finished parts of the manuscript.

August: Offers his services to Holland for the determination of longitudes at sea.
October: Meets the Count of Noailles, the French Ambassador, to whom the *Discourses* were to be dedicated.
November: The Dutch States-General examine Galileo's proposal.

1637

January: Works on the Fourth Day of the *Discourses*, dealing with projectile motion.
April: The Dutch States-General decide to present Galileo with a collar of gold.
June: Letter to Carcavy dealing with various mechanical problems discussed in the *Dialogue*.
July: Loses the sight in his right eye.
October–November: New observations of the moon.
November–December: Works on the text of his *Operazioni astronomiche*.
December: Becomes totally blind.

1638

February: Letter to Alfonso Antonini on the "Libration of the Moon" (*titubazione lunare*).
Retires permanently to his house in Arcetri.
April: Works on the force of percussion.
June: The Holy Office orders him not to receive an envoy from Holland.
July: Publication of the *Discourses* in Leyden.
August: On orders of the Inquisition, refuses the Dutch present of a collar of gold.
September: Receives a visit from Grand Duke Ferdinand II and from Milton.
October: Tells Niccolo Arrighetti of a very accurate means of measuring the diameter of the fixed stars.
November: Discovers the proof of the proposition presented during the Third Day of the *Dialogue* as the "single and unique" principle of the science of naturally accelerated motion.
December: Dictates the beginning of a new Day for the *Discourses* (referred to as the Sixth Day), dealing with the force of percussion.

1639

Spring: Viviani takes up residence near Galileo.
June: At long last receives a few copies of the *Discourses*.

September: In a letter to Baliani explains how time can be measured with the help of a pendulum.
December: Sends Castelli the proof of the principle introduced during the Third Day of the *Discourses*, which he had discovered a year earlier and dictated to Viviani.
Tries vainly to resume negotiations with Holland.
Dictates to Viviani, in the form of a dialogue, various reflections on Euclid's theory of proportions. (This text is sometimes called the Fifth Day of the *Discourses.*)

1640
March: Dictates a letter to Prince Leopold de' Medici on the moon's earthlight.
September–October: Revises his letter to Prince Leopold and addresses it to Liceti.

1641
April: Advocates the application of the pendulum to clocks of different design.
October: Torricelli is invited to Galileo's home.
Continues to work on Euclid's theory of proportions.

1642
January 9: Galileo dies at 4 A.M.

The Natural Philosophy of Galileo

I The Tradition and Early Writings

1 Aristotle and Local Motion

Aristotle approached change and movement in two distinct ways: thus, while he repeatedly asserted that all the evidence of our senses points to the reality of change,[1] he also viewed change and motion as ontological problems. To him, the evidence of the senses, though incontrovertible, was merely a starting point: only within the frame of a general conception of reality could such evidence pave the way for natural science. Here, despite Plato, Parmenides' challenge still held good: to reconcile a rigorous definition of "being" with the idea of change and movement. For, to Parmenides, being was eternal and immutable. Now one of the great merits of Aristotelian physics was precisely that it tried to show that the idea of change could be made meaningful without any sacrifice of logical rigor. Thus, though he is remembered as the champion of mechanical thought naïvely attached to sense perception, Aristotle was, in fact, the first to attempt to construct a science of change and motion and to treat them as something other than superficial states or irrational processes. How and with the help of what concepts did he succeed in this task?

Unquestionably, the crux of Aristotle's theory lay in the idea that being, by its very nature, does not lend itself to unequivocal definition. This conviction was clearly expressed in his doctrine of the categories; not only did he interpret being in several distinct senses,[2] but these, he claimed, could not even be reduced to a common denominator. "Being is not a genus," he stated in the *Posterior Analytics;* this was tantamount to asserting that the unity of the categories is simply one of analogy.[3] Nor could an unambiguous definition of being be culled from physics: it is possible to attribute two distinct modes of being to any concrete substance or subject. Thus bronze, though a metal, has special properties by which it can be distinguished from all other metals; similarly, living beings have special characteristics and aptitudes by which they can be distinguished from all other living beings. Because they are fully realized, these properties or characteristics thus determine what this

[1]*Physics* I, 2, 185 a 12-14; VIII, 3, 253 a 32-b 1 and 254 a 22-33. (The English quotations are based largely, though not exclusively, on the excellent modern translation by Richard Hope, University of Nebraska Press, Lincoln, 1961.)

[2]Being is, in fact, *πολλαχῶσ λεγόμενον*: *Physics* I, 2, 185 a 21; and *Categories* IV, 1 b 25–2 a 4.

[3]*Posterior Analytics* II, 7, 92 b 14.

lump of bronze or that living creature is at any given moment: they define its actual being. However, no subject can be defined exclusively in terms of its actual existence, not only because of the difficulty or impossibility of distinguishing between different states in one and the same subject—for example, a man asleep or awake, an architect building a house or taking a rest, a doctor treating a patient or off duty[4]—but above all because every substance, every concrete being, is at any given moment capable of acquiring a virtually infinite number of new qualities. Thus a lump of bronze may be turned into a statue; a slab of wood may be fashioned into a bed, and an uneducated child may become a scholar. In addition to seeing every subject as an actual being—the totality of his actual qualities—we must therefore also treat him as a potential being, that is, consider all the qualities he is capable of expressing or acquiring if and when the need arises.[5]

This distinction, which agrees with observation, disposes of the false dilemma by which the Eleatic philosophers tried to demonstrate that being is incompatible with change. According to them, Aristotle pointed out, "nothing comes into being or passes out of being, because whatever comes into being would have to come from what is or from what is-not, two equally impossible alternatives: what is does not become anything because it already is, and nothing can come from what is-not because it must proceed from something."[6] The notion of potential being, on the other hand, makes it possible to introduce an element of being into nonbeing, and this without denying the axiom that "anything either is or is-not."[7] Compared with actual being, potential being is, in fact, nonbeing since, after all, whatever qualities the lump of brass may acquire do not, strictly speaking, come into being until they are actualized. However, because the new qualities can be actualized, that is, because their absence is not incompatible with their subsequent presence, it would be wrong to speak of potential being as *absolute* nonbeing. All change implies a degree of nonbeing, and Aristotle accepts that this should be so: thanks to potential being, nonbeing ceases to be absolute and hence becomes relative to being.[8] In his *De generatione et corruptione* he says, "In one sense, generation is rooted in absolute nonbeing, but in another sense, it always derives from 'what is.' For coming-to-be

---

[4]*Metaphysics* IX, 1048 a 25-b 4.
[5]*Physics* I, 8, 191 b 27: cf. *Metaphysics* IX.
[6]*Physics* I, 8, 191 a 27–31.
[7]Ibid., 191 b 26–27.
[8]Ibid., 191 a 34–b 27.

necessarily implies the preexistence of something which *potentially* 'is' but *actually* 'is-not'; and this something is spoken of both as 'being' and as 'not being.'[9]

But though the distinction between actual and potential being helps to make a sense of change, it does not fully explain it. What is needed, in addition, is an ontological analysis of the principles on which both actual and potential being are based. Let us take actual being first. Based as it is on the properties a subject possesses at any one moment, actual being can be defined only by recourse to a clearly determined explanatory principle. This is the principle of form (εἶδοσ); it accounts for all the qualities destined to appear in the concrete individual or substance (οὐσία).[10] It is its form that explains why a substance is what it is and that also ensures its unity in the present and its continuity in time. It is a principle of organization no less than of development, and hence immutable and perfect. It is thanks to its actual form, that is, to those of its properties that have already been realized, that a being exists in actuality; and it is by reference to its possible form (as bronze is to the statue) or to its unrealized possibilities (as childhood is to maturity) that a being owes its potential existence. Does this mean that form is the sole ground of potential existence? There are two reasons, according to Aristotle, for not treating it as such. In the first place, form cannot exist independently: it presupposes the existence of a support or subject (ὑποκείμενον).[11] In the second place, while form by its very nature is general, individuals alone can enjoy real existence.[12] Hence, another principle has to be added to form, and this principle is matter. Now, matter cannot be known directly: since all material objects have a form, matter as a principle (πρώτη ὕλη) can be grasped only by analogy.[13] Take the case of a bronze statue or a wooden bed: according to Aristotle, the bronze is to the statue and the wood to the bed as something relatively unformed is to something having a certain form.[14] In other words, matter may be called the indeterminate receptacle (δέκτικου) from which spring all determinate and concrete beings, once a form has been imposed upon it, or—which amounts to the same thing —matter, while being the first persistent thing out of which anything

---

[9]*De generatione et corruptione* I, 3, 317 b 14–17.

[10]See *Physics* II, 3, 194 b 27; and *Metaphysics* V, 2, 1013 a 27.

[11]*Physics* I, 7.

[12]It is matter, and matter alone, which ensures individuation; see *Metaphysics* VII and X, passim.

[13]*Physics* I, 7, 191 a 7 ff.

[14]Ibid., 191 a 10–11.

arises, is itself lacking in essence.[15] We can now see why every inanimate object no less than every living being must enjoy actual as well as potential existence. Because of its form, a block of marble undoubtedly has an actual existence and specific properties. However, these properties cannot determine its material aspects exhaustively, because, in addition, it is capable of assuming a host of forms, so that even while it is actual marble, it is also a potential statue. The same applies, mutatis mutandis, to a living being. As he is bound up with matter lacking in quality, his form must express each one of his material attributes in turn. At any given moment, a greater or lesser number of these attributes will have no more than potential existence, either because, having already been realized, they have been allowed to fall into temporary abeyance or else because they have not yet been actualized. Conversely, should the opportunity arise, matter will invariably actualize its potential qualities. The distinction between actuality and potentiality thus enables the philosopher to dispose of the irreconcilable distinction, based on seemingly logical grounds, between being and movement: once all substances are defined as irreducible amalgamations of matter and form, being and movement, far from being incompatible, become ontologically fused.

We are now able to appreciate the exact role of change. Thanks to the potential qualities of its matter, a given block of marble is capable of assuming new attributes, say, the shape of Hermes. It can also be subjected to other manipulations, clearly alien to its essence. Thus, being a heavy body, it will tend to move as closely as possible to the earth's center. Yet if it is lifted up by force, this property of its form can be temporarily suspended. In either case, material substance involves a measure of "nonbeing": "natural" nonbeing resulting from its capacity to assume new forms, and "artificial" nonbeing resulting from the negation of an existing property. Similarly, a man, however educated and whatever his other attributes, is unlikely to have realized his full human potential, so that, in addition to his actual being, he also partakes of a degree of nonbeing, the sum of all the qualities he might possess but does not. Now, one of Aristotle's most important contributions is the idea that the nonbeing embodied in every substance is experienced by

[15]Ibid., I, 9, 192 a 33. This, by the way, is why matter is eternal: "If matter were produced, some first constituent would have to be present out of which matter would arise, but to be such a constituent is matter's own nature. Were matter produced, it would therefore have been before it arose, which is absurd." Ibid., 192 a 25 ff.

it as *privation* (στέρησισ):[16] the substance is deprived of qualities it is capable of acquiring, or of properties it has already acquired but which for one reason or another have been allowed to fall into desuetude. Now, by defining potentiality as privation, Aristotle was simply asserting that every concrete being is constantly impelled, from the inside as it were, to make all sorts of changes, and so acquires a sense of expectation. As a result, he constantly anticipates the future in the very core of the present. Far from being incompatible with being, it is in being itself that the origins of change must ultimately be sought: as soon as the environment allows, either by removing an obstacle or else by supplying an agent capable of actualizing potential qualities, privation ensures that a change be brought about.[17] Privation must be clearly distinguished from both matter and form; in itself it is nothing or, as Aristotle puts it more precisely, it is only a way of describing what potential existence means for every substance.[18] The very fact that Aristotle treats privation, like matter and form, as a principle (ἀρχή)[19] shows to what extent he considered change as coextensive with being, and explains why it played so important a part in his philosophy. For, to the extent that potential existence is privation, change remains the only way by which properties absent from, yet virtually contained in, substances can be actualized. "Movement," he says in a famous passage, "is the functioning of what is potential as potential," which can mean only that it is the natural nexus between potential and actual being.[20] Aristotle's treatment of privation as a principle not only ensures the reality of movement and change but also endows them with an indispensable function: it is thanks to them, and to them alone, that concrete individuals obtain an ever-increasing actualization of their form. To the Eleatic philosophers, linking change to being was tantamount to proclaiming the existence of nonbeing: by showing that change alone allows each being to actualize its essence, Aristotle granted change a precise ontological function.

But while this interpretation endowed change and movement with meaning, it did not do so without further qualification. Defining movement as the functioning of what is potential as potential was tanta-

---

[16]Ibid., I, 8, 191 b 10 ff.
[17]Although it is founded in being, change can be produced only by an external agent; ibid., VIII, 4, 254 b 33 ff., and 259 b 3 ff.
[18]Ibid., I, 7, 191 a 12–14.
[19]Ibid., I, 8, 191 b 10 ff.
[20]Ibid., III, 1, 201 a 10–11.

mount to asserting that movement is primarily a way of proceeding from nondetermination (the starting point) to determination (the end point), so that what sense it has is mainly derived from its termini. Thus we have change or movement in the full sense whenever an uninstructed man becomes educated, whenever a heavy body approaches the center of the earth, and, in general, whenever a subject x, qualified as non-a, acquires the quality a.[21] Now, the fact that movement would be unthinkable without the help of two precise termini, the *terminus a quo* and the *terminus ad quem*, has extremely important implications.

How should we treat change as a physical phenomenon? What precise status must we grant it? In attempting to answer these questions we must be careful to avoid two pitfalls. The first is to confuse change with the actual existence of what is being changed. Thus the bronze intended for the statue is already fully actualized as bronze, and it is not in its substance that it is being changed.[22] The second error is to identify change with its end result. Thus, if we call "buildable" whatever is capable of becoming a house, its actualization will be the building itself; once the latter is finished, it will have ceased to be buildable, and the change will have come to an end.[23] Movement is a potential action, but precisely because it is potential, it cannot be part of the final state for which it has prepared the way. It must be considered as an intermediate process, bound up with primary being (the essence) but in no way identical with it. Movement is supposed to be a sort of imperfect actuality for the reason that the potentiality, whose actuality it is, is incomplete. And therefore it is hard to grasp what movement is, for it must be classified under "privation" or under "potentiality" or under "pure actuality," and none of these appears possible.[24] Affecting as it does potential being, initiated by privation and tending toward action, movement is thus by definition a transitory process, destined to vanish once the action has been accomplished.

All this can be justified by a dialectic analysis. Let us suppose that movement were not a finite process involving concrete subjects, but that it had an existence of its own. In that case, nothing can prevent us from endowing movement with all sorts of attributes and indeed

---

[21]See D. Ross, *Aristotle's Physics*, p. 22.
[22]*Physics* III, 1, 201 a 27 ff.
[23]Ibid., I, 201 b 5–15. Cf. ibid., V, I, 224 b 15–16: whiteness is not a movement; the movement that ends in whiteness is whitening.
[24]Ibid., III, 2, 201 b 33–202 a 1.

from speaking of the movement of a movement or of the change of a change. Now, it is easy to show that this supposition makes nonsense of all the foregoing arguments and merely serves to turn change and movement into unintelligible phenomena once again.[25] To make this clearer, let us assume that change, now subject to change itself, could be said to "become hot, or cold, to change its place, or to increase or decrease."[26] The actualization corresponding to this process of heating, cooling, and so on, would have to be referred to the change itself, which would therefore become its own end, lose its ontological function, and hence become deprived of sense. A movement whose initial and final termini would themselves be movements is another example of the same process,[27] one that according to Aristotle is no less absurd than the first, for if a movement could directly generate another movement, then the very idea of movement—which is not an essence, and hence does not strive toward an end—becomes quite meaningless.[28] And even if we admit that some kind of order could govern the pure succession of movements or changes, the fact that every movement or change would then tend toward another movement or change, and no longer toward a fixed terminus, would inevitably destroy the elementary distinction between movement or change and what it generates, for example, between the construction of a building and the completed edifice, or between "the movement of learning and learning itself."[29] Being familiar with the idea of acceleration, the modern reader may well find this kind of argument somewhat confusing. Aristotle's reasoning was, however, a perfectly natural consequence of his own definition of change. To him the acceptance of the possibility of the movement of a movement was tantamount to endowing motion with a degree of independence, which meant severing its links with primary being (essence) and hence surrendering the concept of becoming to the irrational status it enjoyed among the Eleatic, and to a lesser extent among the Platonic, philosophers. By the same token, however, his rejection of the movement of a movement demonstrates to what extent Aristotle's own analysis of

---

[25]Ibid., V, 2. The interpretation of this chapter is difficult. Aristotle joins his critique of the idea of the movement of a movement to a demonstration that there can be no change or movement (in the sense of κίνησισ, see below) in respect of the substance. His critique is thus part of a wider argument, which makes his train of thought extremely hard to follow.

[26]Ibid., 225 b 18–19.

[27]Ibid., 225 b 21–23.

[28]Ibid., 225 b 28–29.

[29]Ibid., 226 a 15.

local motion forced him, from the very outset, to steer a course completely at variance with that of classical mechanics. In particular, his refusal to grant movement any degree of autonomy prevented him from treating movement as a state, and as such as equivalent to rest.

The fact that change must always take place between two termini does more than impose a transitory and finite character on each movement. It also determines the classification of the various modes of change and, in particular, helps to elucidate the role of local motion in Aristotle's system. However, it is not enough to say that change always takes place between two termini; a complete description of change must also include a precise definition of these termini and of their interrelations. Now, according to Aristotle the termini are of two kinds, and of two kinds only: they may be contraries or contradictories. In the first case, that is, in the case of the kind of movement to which Aristotle refers as *κίνησισ*, change represents the acquisition by a substance *x*, qualified as non-*a*, of the property *a*: in the second case, however, something more radical occurs: something non-*x* becomes the substance *x*, or the substance *x* becomes something non-*x*. This type of change no longer affects the transformation of the state of substances (change of quality, quantity, or place) but involves their generation or destruction.[30] Thus under the general heading of change, Aristotle combines two types of processes, both equally real but vastly different in their ontological functions.[31]

The following scheme sums up his classification of changes (based on his list of categories) and also shows in what way the science of change is bound up with ontology, the science of being qua being:

Change (*μεταβολή*)
1. In respect of substance (or as between contradictories): generation (*γένεσισ*) and corruption (*φθορὰ*)
2. Affecting the state of existing substances (or as between contraries):
   κίνησισ
   - in respect of quality (*ἀλλωιώσισ*)
   - in respect of quantity (*αὔξησισ* and *φθίσισ*)
   - in respect of place (*φορά*)

From this scheme, Aristotle's view of what problems the study of local motion must set out to solve becomes perfectly clear. First and foremost, change of place (or *φορά*), that is, what we call motion, has been

---

[30]Ibid., V, 1, 225 a 12 ff.

[31]Ibid., V, 1, and 4.

put on a par with alteration, growth, or diminution.[32] Not only is local motion (or any other change) not a state, but it cannot legitimately be studied in isolation from alteration or change in respect of quality. Hence the science of change[33] is first and foremost the science of *κίνησισ*, and to ignore this would be to risk losing sight of its true nature. Did Aristotle therefore consider the separate study of local motion a pointless exercise? We believe that several reasons militate against this view. For though he did not grant local motion de jure precedence over quantitative or qualitative change, he nevertheless granted it de facto priority. Thus he considered it the chief movement of "living beings in the degree to which they have attained to their most fully developed nature," and as such "prior to all other forms of movement in the order of complete being."[34] Moreover, all other types of change or movement spring, directly or indirectly, from local motion: thus the origins of alteration itself and of growth or diminution (which involve prior alteration) must be sought in the approach or withdrawal of a moving principle.[35] Admittedly, this interdependence is limited, since neither alteration nor growth and diminution can be explained as such in terms of local motion. For all that, "if there must always be movement, then there must always be, first among movements, local motion."[36] This explains why Aristotle's theory of local motion should have acquired the status of a special doctrine, worthy of being examined and studied for its own sake. Thus when Galileo referred to it in his early works, no less than in his later years, he invariably treated it as a clearly defined and sound body of propositions, and one that he never felt obliged to divorce from its original ontological perspective. To reconstruct Aristotle's treatment of local motion while disregarding its links with the other forms of *κίνησισ* is admittedly a labor of abstraction and simplification, but it is nevertheless perfectly legitimate, provided only that it is aimed not so much at resurrecting Aristotle's approach as such as to analyze the conceptual universe from which (and also in opposition to which) classical mechanics was founded.

However, though we can make a separate study of Aristotle's con-

[32]Cf. Aristotle's analysis of condensation and rarefaction. Ibid., IV, 9.

[33]Let alone of change in respect of substance, which does not concern us here.

[34]Ibid., VIII, 7, 261 a 13 ff. It should be noted that celestial bodies, whose substance cannot be generated or corrupted, experience no changes other than local motion.

[35]Ibid., 260 b 1–5.

[36]Ibid., 260 b 5–6.

ception of local motion, we cannot ignore the general frame in which local motion takes place or the precise nature of the bodies concerned in it. In any treatment of motion as an independent state this oversight would have no other effect than once more to ignore the actualizing function from which it is inseparable and hence to fly in the face of Aristotle's doctrine. Before we can even begin to analyze the special phenomenality of movements or changes, before we can classify and describe them, however sketchily, we must first determine the precise physical context in which Aristotle placed them; this calls for an examination of his cosmology and theory of the elements.

Local Motion and Cosmological Order

Cosmology first of all provided mechanics with the idea of an organized world. The idea of the cosmos, which emerged in Miletus in the sixth century B.C. and was perhaps formulated by Anaximander, had undergone important modifications by the time it came down to Aristotle. But though the dominant idea had always been the general order of the universe, the stress had originally been laid on the interdependence of all its vital members, quite especially of men and nature, and on the remarkable seasonal regularity of celestial and meteorological processes.[37] The concept gradually became "naturalized" and lost its moral content; at the end of a long process, now difficult to reconstruct although some of its stages can be recognized in the writings of Philolaus and Melissus,[38] the spatial order of the universe gradually gained ascendancy over the temporal order, while the belief in the necessary connection between all living beings began to fade away. As a result, the idea of the cosmos came to represent the certainty that "nothing natural is unordered; on the contrary, nature is uniformly principled by order."[39] Soon afterward another conception was added to the first: not only was the universe an organized whole and therefore complete; it was also spherical and symmetrical. Though we cannot attribute this observation to anyone in particular,[40] we do at least know that Anaximander combined it with the principle of sufficient reason to deduce the immobility of the earth at the center of the universe. Xenophanes, Empedocles, and Parmenides reasserted the spherical nature and the symmetry of the

---

[37]Charles Kahn, *Anaximander and the Origins of Greek Cosmology*, pp. 190–191.
[38]Ibid., pp. 228–229.
[39]*Physics* VIII, 1, 252 a 11–12.
[40]Kahn, *Anaximander*, p. 78.

universe; and it was by explicit reference to Parmenides that Aristotle justified his belief that the universe was a limited whole, "being equal in every direction from the middle."[41] Moreover, Aristotle came very close to Anaximander when, in his *On the Heavens*, he said of the hemisphere above our heads that it is "the same all round, and that the center is equally related to every part."[42]

Nevertheless, there are a number of indications that Aristotle's cosmos differed from that of his precursors. To the finitude and sphericity that previously characterized the whole, he added six directions: up and down, right and left, and forward and backward. The objection that this was a naïve form of anthropocentrism and that his distinctions were too dependent on subjective experience carried little weight with him; nor did another—that the spherical nature of the world must make it "uniform in all directions."[43] Aristotle dismissed these arguments, one based on psychological and the other on deductive reasons, on the grounds of common experience. Even if the center of the world were below us in a purely figurative sense, it would still differ in essence from its extremity; moving away from the center to the extremity and moving away from the extremity to the center are completely opposite actions, reflecting the fundamental opposition of the two directions concerned. "We, however, apply 'up' to the extremity of the world, which is both uppermost in position and primary in nature; and since the world has both an extremity and a center there clearly must be an up and a down."[44] This sentiment was echoed in the *Physics:* "In nature each of these directions is distinct independently of our own position. Not any chance direction is 'up' . . . not any chance direction is 'down,' " and one of the best demonstrations of this fact is that "these distinctions of place do not only differ conventionally but also depend on the way in which bodies act."[45] The four remaining directions, though less important for the analysis of local motion, are defined no less objectively: we give the name of "right-hand" to the side from which the stars rise, and the name of "left-hand" to the side where they set;[46] similarly, "forward" is the direction of the motion of the stars (from left to right), and "backward" the opposite direction.[47] If

[41] *Physics* III, 6, 207 a 16–17.
[42] *On the Heavens* IV, 1, 308 a 27–28.
[43] Ibid., 308 a 19.
[44] Ibid., 308, a 21–24.
[45] *Physics* IV, 1, 208 b 15 ff.; cf. III, 205 b 31–35. The idea of "up" and "down" can also be found in Ptolemy's *Almagest*, 1, 7.
[46] *On the Heavens* II, 2, 285 b 16–17.
[47] Ibid., II, 5, 287 b 22–288 a 12.

we add these six directions to the definitions introduced by Aristotle's precursors, we obtain an *ordered structure* which we may legitimately describe as the most fundamental element of Aristotelian cosmology. Aristotle clearly showed that he regarded this structure as taking logical, if not chronological, precedence over material bodies when he said that "we must suppose the world to resemble a being of the class whose right differs from the left in shape as well as in function, yet which has been enclosed in a sphere."[48]

However, his addition of directions to the traditional cosmos did not exhaust Aristotle's personal contribution, for he also based the ordered structure of the universe on another factor, equally irreducible and fundamental. Thus sense experience shows that "all natural bodies and magnitudes are capable of moving of themselves in space,"[49] and combining this observation with the proposition that straight and circular lines are the only simple magnitudes,[50] he concluded that all motion in space must be either naturally straight or circular, or else a combination of the two.[51] The fact that he introduced this idea so abruptly must not blind us to its novelty or importance, for it means quite plainly that the orderly structure of the universe is based not only on the elements we have just defined but also on a no less basic but quite distinct element, namely, natural motion in a straight or circular line. The most cursory comparison of Aristotle's position with that of the Atomists and particularly that of Democritus will show how radically the cosmological picture had been changed. Thus, according to Democritus, the world consists exclusively of an infinite number of indivisible elements in simple motion; as a result of this motion the atoms, though at first mixed together, gradually disaggregate and then congregate by virtue of their affinities to form organized bodies—the world and all the separate bodies it comprises. Now, Aristotle presents these philosophers with the following dilemma: Are the motions of the atoms subject to some order or are they completely disorderly? If they are subject to some order, that is, if they are capable of generating the cosmos, then the latter must have had an orderly structure *ab initio*, for how else could bodies have moved without constraint to "find rest in their proper places, and therefore form the same arrangement as they do now, the heavy traveling toward the center, and the light away

[48]Ibid., II, 2, 285 b 1–3.
[49]Ibid., I, 2, 268 b 14–16.
[50]Ibid., 268 b 19–20.
[51]Ibid., 268 b 17–18.

from the center?"[52] If the answer is that the motions of the atoms are disorderly, it must follow that "with this disorderly motion some of the elements might have united in those combinations which constitute natural bodies like bones or flesh."[53] But this is tantamount to saying that order can spring from disorder, and proportion from what is lacking in proportion. If we reject the idea that nature can be generated from its contrary, then we cannot possibly attribute the construction of the cosmos to disorderly local motions. Hence the only solution—since the existence of natural motions is an immediate and irreducible fact—would be to fit them directly into the a priori organization of the cosmos, taking care to avoid any suggestion of anteriority. Now, in an organized world, simple natural motions can assume three forms: (1) circular motion around the center of the world, and rectilinear motion (2) away from the center and (3) toward the center.[54]

The foregoing remarks may have helped the reader appreciate the central idea on which Aristotle based his cosmology: the fusion, based partly on logic and partly on experience, of the orderly structure of the cosmos and the three simple natural motions into a single, indivisible whole. This point is worth stressing. Aristotle was at one with a great many of his precursors in postulating an absolute cosmological order, both intelligible and also the guarantor of stability. However, there was this essential difference between his cosmological order and theirs: while his precursors, and especially Philolaus and Melissus, treated this order primarily as a frame that ensured the coordination of the bodies contained in it, Aristotle completed this frame with his three simple motions, thus elevating the latter to the rank of *cosmological premises* responsible for the order of the world, in much the same way as the six directions he had postulated earlier. The general cosmological view on which his natural philosophy is based was therefore anything but static: it combined the idea of a clearly defined and orderly structure with the idea of three natural motions, thanks to which the cosmos was able to preserve its organization and harmony.

Aristotle's cosmological theory helps us, first of all, to define the most

[52]Ibid., III, 2, 300 b 18–25. The same criticism could also be leveled against Plato's *Timaeus*.

[53]Ibid., 300 b 21–28.

[54]Because it allows of the specification of motions, the formal organization of the cosmos enjoys logical priority over them, but this in no way affects their status as original and irreducible entities; ibid., I, 4, 271 a 25–27.

general condition of local motion. Change, as a simple process of actualization, can only be conceived as a factor ensuring the transition from one state to another. When applied to local motion, this conception implies the existence of predetermined places between which local motion can take place. That only the orderly structure, which Aristotle has made the central principle of his cosmology, can fulfill this role is amply reflected in his critique of the idea of an infinite world as well as of the void.

Let us first take the hypothesis of an infinite world. Its constituent elements would necessarily have to be infinite in magnitude, and so would its natural places.[55] Hence, whatever motions led the elements to their natural places would have to be infinite as well; but this would be absurd, for how could a body "be moving to a place where nothing can ever in its movement actually arrive?"[56] Nor was this the only difficulty. If the world were infinite, an infinite distance would separate such of its bodies as are set infinitely apart. But observation shows us that all celestial bodies describe a circular motion around the earth in 24 hours: hence, were the world indeed infinite, the furthermost of the celestial bodies would cover an infinite distance in finite time, which is a manifest contradiction.[57] Finally, in an infinite world, all the directions characteristic of cosmic motion—up and down, forward and backward, and right and left—would cease to exist: "Where there is neither center nor circumferences, and one cannot point to one direction as up and another as down, bodies have no place to serve them as the goal of their motion."[58] Lacking these natural termini, the idea of local motion would become meaningless. And since any movement lacking a starting and an end point must needs be unintelligible, we have to assume that in an infinite world every body would persist indefinitely in its original state or, in Aristotle's own words, that "it will remain in itself."[59] Much the same argument applies to the idea of the void, that is, to space lacking any qualities. As far as the celestial bodies are concerned, the void would once again be a place without directions and particularly without up or down, because by definition it lacks such

---

[55]Ibid., I, 7, 274 b 8–10. What Aristotle had in mind were such primary elements as water and fire.
[56]Ibid., 274 b 17–18.
[57]Cf. ibid., I, 5, 271 b 29 ff.
[58]Ibid., I, 7, 276 a 8–10.
[59]*Physics* III, 5, 205 a 12–205 b 24. These are by no means the only arguments Aristotle marshals against the idea of an infinite world; thus Chapters 5, 6, and 7 of Book I of *On the Heavens* are devoted entirely to this subject.

distinctions; all places to which local motion could translate these bodies having disappeared, there would be no reason for any bodies to move to one place rather than to another and why it should be subject to any motion whatsoever.[60] As if this were not enough, the same cause —the absence of a natural terminus—which prevents the motion of a body in a void, also makes it impossible that, once put into motion, it should ever stop. "Hence, a body either would continue in its state of rest or would necessarily continue in its motion indefinitely, unless interfered with by a stronger force."[61] It would, of course, be quite wrong to consider this remark a brief glimpse of the principle of inertia, forgotten almost as soon as it was taken; to Aristotle the idea of a motion capable of infinite duration was bound to appear just as untenable as the idea of the movement of a movement. For both cases, Aristotle set out some of the most important consequences of the treatment of movement as an autonomous phenomenon rather than as a simple effect on bodies; the reason why he rejected these consequences and held firmly to the idea of motion as a process was his conviction that the latter alone was compatible with the ontological situation, no less than with the most fundamental demands of natural philosophy.

However, cosmology does not simply endow the concrete (real) motions with a general basis for existence but also gives rise to two theories, on which the entire analysis of these motions must depend: (a) the necessary presence in bodies of irreducible motor qualities on which all downward or upward motions depend, namely weight and lightness, and (b) the opposition between natural and violent movements and consequently the opposition between movement and rest in general.

What is the meaning of "light" and "heavy"? "Light bodies," Aristotle tells us, "naturally tend upward and heavy bodies downward."[62] Now, it is the word "naturally" that forms the crux of the whole argument. What does Aristotle mean when he makes lightness and weight the principles of natural motion? His critique of the solutions offered by Plato and the Atomists provides us with a first clue. To Plato, lightness and heaviness (weight) were simple quantitative concepts: "The heavier that which is made up of a greater number of identical parts, the lighter that which is made up of fewer."[63] This explains why two

---

[60]*Physics* IV, 8, 214 b 31–215 a 1.
[61]Ibid., 215 a 20–22.
[62]Ibid., IV, 4, 212 a 24–26; cf. III, 5, 205 b 26–27; *On the Heavens* IV, 1, 308 a 29–31.
[63]*On the Heavens* IV, 2, 308 b 5–7.

heterogeneous bodies of equal volume differ in weight, and also why the more voluminous of two homogeneous bodies is heavier than the other.[64] As for the Atomists, while they also regarded bodies as being composed of elementary particles, they made use of a special hypothesis to account for the fact that heaviness is not necessarily a simple function of volume: "Bodies are made light by the void contained within them, and this accounts for the larger bodies sometimes being the lighter."[65] Despite this difference, Plato's and the Atomists' theories were clearly inspired by the same idea: that the heaviness and lightness of bodies are relative qualities, depending on their respective densities and on nothing else. No wonder that Aristotle raised almost the same objections to both interpretations: by treating lightness and heaviness as relative terms, his precursors had failed to base upward and downward motion on natural foundations. Let us assume that the lightness of a body derives exclusively from its consisting of fewer material particles than do neighboring bodies; a large fire, because it contains a greater number of particles than a small one, will then have to move upward more slowly or even drop to the ground;[66] similarly, a large quantity of air can be heavier than a small quantity of water and take its place beneath the latter, which is contrary to observation.[67] Nor does the idea of the "void" improve matters. If the lightness of a body were entirely dependent on the ratio of its empty to its full portions, then there would have to be "a quantity of fire in which the solid and full portion exceeds the solid parts contained in a given small quantity of earth,"[68] so that under certain conditions the earth would move upward and fire downward. To interpret lightness and heaviness in a relative sense, as Plato and the Atomists do, is therefore tantamount to robbing such irreducible properties of the cosmos as upward and downward movements in a straight line of their natural foundation.

We now have a better idea of what Aristotle meant when he described the light as that which tends naturally upward and the heavy as that which tends naturally down. He was clearly referring to principles or tendencies, such that bodies in which they are contained will in all cases (and no matter what the environmental changes) move either upward or downward.[69] In other words, lightness and weight

[64]We shall pass over the special difficulties Aristotle saw in the geometrical nature of the Platonic representation.

[65]*On the Heavens* IV, 2, 308 b 30–309 a 11.

[66]Ibid., 308 b 16–21.

[67]Ibid., 308 b 21–26.

[68]Ibid., 309 a 30–32.

[69]Aristotle also called these tendencies "potentials." Ibid., IV, 1, 307 b 29.

must be conceived as *absolute qualities*, theoretically independent of the matter with which they are associated. Now there is only one way of defining lightness and heaviness in this absolute manner, namely by relating them directly to two a priori directions of the world—up and down. A light body, that is, a body which naturally tends to move upward, would then be one with an inherent tendency, impressed upon it[70] by the top of the world, to return to the latter. Similarly, a heavy body, that is, a body that naturally tends toward the center is such thanks above all to a certain quality, or *gravity*, impressed upon it by the bottom of the world. "For light and heavy bodies tend to their respective places,"[71] that is, either to the circumference of the world or else to its center.[72] Finally, some bodies may move upward on some occasions and downward on others; this is because, being compound bodies, they contain lightness and heaviness in variable proportions: thus wood falls through the air because it is heavier but rises in water because it is lighter.[73] Aristotle's entire theory was therefore a direct consequence of his basic cosmological premises, and the eighth book of the *Physics* says so explicitly. If there are things that by nature invariably move away from the center, and others that move toward it, "the reason is that they naturally tend in distinct directions. This is precisely what it means for them to be light or heavy—namely that they tend upward or downward, respectively."[74] His solution was certainly consistent, but by defining lightness and heaviness as irreducible qualities and refusing to consider them as the direct consequences of the relative density of bodies, Aristotle nevertheless sowed the seeds of a confusion, the traces of which could still be found in the writings of Galileo.[75]

---

[70]Or more precisely on its form, for Aristotle knew nothing of forces acting at a distance.
[71]*Physics* VIII, 4, 255 b 11–12.
[72]Aristotle also rejected Empedocles' explanation that bodies are impelled toward the center by a vortex; cf. *On the Heavens* II, 13, 295 a 29 ff.
[73]Ibid., IV, 4, 311 b 1–13.
[74]*Physics* VIII, 4, 255 b 15–17; cf. *On the Heavens* IV, 4, 312 a 6–7.
[75]Aristotle undoubtedly knew that there was a connection between lightness, heaviness, density, and rarity, and he realized that "heavy = dense and light = rare (dense differing from rare in that the former contains a greater quantity in an equal bulk)" (*On the Heavens* III, 1, 299 b 7–11). Density thus played some part in downward motion and rarity in upward motion, but to Aristotle density and rarity were no more than secondary oppositions between bodies, as witness the theory of condensation and rarefaction expounded in the *Physics* IV, 9. A body is dense or rare primarily by virtue of its form; in other words, some forms involve the principle of lightness, and the matter in which these forms find their expression, far from impeding that principle, allows it to manifest itself. Lightness and heavi-

Finally, it is unquestionably on the orderly structure of the cosmos and on simple upward and downward motions that Aristotle based his distinction between natural and violent motion, and that between motion and rest in general. Cosmology provides an a priori location of *natural places* between which and toward which bodies can move; it also implies that every body has an intrinsic principle of upward and downward motion. From this it follows that a body having the quality of lightness and being impelled upward or a body having the quality of heaviness and being impelled downward will execute motions in harmony with the order of things, that is, its respective motions will be natural; conversely, a body having the quality of lightness yet impelled downward or a heavy body moving upward will be describing unnatural and violent motions. Hence there is no justification for attributing, as is sometimes done, the natural or violent nature of their motions exclusively to the nature of bodies; lightness and heaviness being qualitatively distinct principles and not merely the consequences of the quantity of matter involved, the origin of the opposition between natural and violent motions has to be sought among the a priori principles governing the organization of the cosmos. Now, in treating it as such, Aristotle was in fact introducing an essential distinction between motion and rest. The transition from the idea of natural motion to that of natural rest is a simple one: if light bodies tend to move upward and heavy bodies tend to move downward, then a light body having reached the end of its ascent, or a heavy body having reached the end of its descent, will have attained a state of natural rest. Now, what do we mean when we say that a body has attained a state of natural rest? Just this: that it will put up an active resistance to all attempts to displace it. It is of course difficult to imagine the nature of Aristotelian resistance because, unlike the inertia of classical mechanics, it cannot be reduced to the simple tendency of a body to persist in its present state; according to Aristotle, the resistance shown by a body has its true origins in the fact that its form can be actualized only in its natural place, and it is thanks to this actualization that it will always show the same resistance to motion as nature shows to anything tending to disturb its order. Hence, motion and rest can never have the same ontological signifi-

---

ness cannot therefore be reduced to rarity and density; it is for this reason that the concept of specific weight, of which Aristotle had at least an inkling, could play no part in his evaluation of the magnitude of motions.

cance for material bodies; inasmuch as "motion toward its proper place is for each thing motion toward its proper form,"[76] motion remains the normal means by which the essence of things can attain its full expression. Natural rest, on the other hand, because it is an actual rather than a potential state, must necessarily enjoy priority over motion.[77] It is perhaps this point more than any other which reveals the extent to which Aristotle's theory of local motion depends on his cosmological premises, and also the extent to which the construction of classical mechanics was dependent on the demolition of that cosmology.

Local Motion and the Theory of the Elements

Though it helped to assign the natural termini of local motion, Aristotle's a priori definition of the cosmological order called for some amplification. In particular, once motion is defined as the functioning of "what is potential as potential," it cannot be conceived in the absence of an active moving principle which, having initiated it, will maintain it from one moment to the next. This idea, expressed earlier by Plato,[78] implies that all motions are partly dependent on the action of a mover. But inasmuch as it has been defined as a process of actualization, there can be no local motion in the absence of concrete mobiles,[79] which it helps to distribute in accordance with their essence. It follows that bodies provided with predetermined qualities that are as irreducible to genetic explanations as the orderly structure of the cosmos remain indispensable. Defined as the functioning of what is potential from the ontological point of view, local motion considered from the "mechanical" point of view thus appears as the *combined functioning of mover and moved*. It is this interpretation, one of the most characteristic of Aristotle's science of local motion, which we must now examine more closely.

To begin with, is it not a contradiction in terms to speak of the "combined functioning of mover and moved"? If motion is the functioning of what is potential, while the mover functions in actuality, what is the precise relationship between the action by which the mover

[76] *On the Heavens* IV, 3, 310 a 34–b 1.
[77] At least in the sublunary world. As far as the heavenly bodies are concerned, we shall see that local motion corresponds to the actual rather than the potential state. However, since this results from a material distinction, the opposition between motion and rest is in no way attenuated.
[78] *Timaeus*, 57 e.
[79] *Physics* III, 1, 200 b 32–33.

actualizes what is potential and that other action which is none other than the motion itself? Are we not dealing with two irreducible actions, that by which the mover creates an effect and that by which the moving body expresses that effect? But if we distinguish the action of the moving body from that of the mover, are we not conferring a measure of autonomy on motion, which would be incompatible with Aristotle's previous conclusions? This problem is therefore crucial, so much so that Aristotle's entire theory of motion stands or falls with its solution.[80]

Let us suppose, first of all, that the actions of mover and moved are in fact distinct. Then in what subjects do they reside? Can we assign the activity to the agent (mover) and the passivity to the recipient (the moved)? If so, the recipient, while possessed of motion would not itself be moved (since the action of the recipient is an "activity" in quite a different sense). Or should we assign both actions to the recipient? But this would be no less absurd, since in that case the functioning of a thing would not be inherent in it; moreover, the moving body would undergo two motions at once (that by which the mover actualizes the moving body, and that by which the moving body is actualized). Having thus shown why the actions of mover and moved cannot be different, Aristotle was ready to offer his solution of the problem. To appreciate it, we must return to his theory of actual and potential existence. Now, that theory in no way prevents the action of one thing from residing in the other, which it actualizes: thus the action by which a teacher instructs his pupil is unquestionably the teacher's own, but it is actualized in the pupil, not in the master. Hence, it is perfectly conceivable that the "same action appertains to two things," provided only that we remember that what unites them is not an identical essence but the link between a potential and an actual thing. In other words, the functioning of a potential being is simply the actualization of the manner in which an actualized being acts upon it. Now this analysis applies equally to the relation between mover and moved, for the action of the moving body is simply the actualization of the action of the mover or, as it were, its reflection. What is involved is one and the same action, as seen through the eyes of mover and moved in turn, and these two viewpoints are identical, much as "the road from Thebes to Athens and the road from Athens to Thebes are the same." Thus,

---

[80]For this entire discussion, see ibid., III, 3.

while they must never be confused, the actions of mover and moved are a single reality.

This justification of motion as the common act of mover and moved has a number of important consequences. To begin with, it shows once more why motion cannot be a *state* capable of maintaining itself; by affirming that the action of the mover is inseparable from the action of the moved, Aristotle was asserting that the presence of a mover is indispensable not only to the initiation of a motion but also to its conservation. To explain a motion, we must therefore first discover the mover on which its conservation depends and whose variations alone can account for possible variations in the motion. Thus, while the fundamental problem of classical mechanics would be to explain why a motion, once begun, should not continue indefinitely and uniformly in a straight line, Aristotle's fundamental problem was to ensure its continuation, since, in his view, once the mover ceased to act, the motion must come to an end.[81] This approach led to a radical depreciation of the study of motion in its spatiotemporal development: because he referred the effects of motion primarily to the mover, Aristotle was unable to elaborate such dynamic concepts as impulse and kinetic energy. Moreover, the primary role which he attached to the mover in the production of motions diverted attention from a dimension that is the very basis of a true knowledge of motion, namely speed. Admittedly, Aristotle did speak of speed and even considered it the only factor by which motions could be compared to one another. Thus, he claimed it was "evident that every change can be attributed to speed or slowness."[82] But he did not recognize speed, so to speak, as anything but the direct result of the action of the mover on the moved, and he failed to appreciate that it was a magnitude sui generis and hence to distinguish clearly between uniform and nonuniform motions.[83] Stripped of any direct attributes, motion could be characterized only from the outside and not in its own phenomenality.

Moreover, the preceding analysis was confined to the part played by the mover in that combined action of a mover and moved which is local motion. Hence, it may be called an abstraction, since it is only

---

[81]Thus, as we shall see, he attributed retardation to the gradual exhaustion of the mover.

[82]*Physics* IV, 14, 222 b 31–32; cf. ibid., IV, 10, 218 b 13–14; and VI, 2, 232 b 21–22.

[83]In Chapter 2 we shall show by what remarkable steps fourteenth-century scientists succeeded in filling this gap.

through the physical nature of the moved, which provides the material basis of the mover's action, that the magnitude and effects of the latter can be evaluated. It is to Aristotle's theory of the elements that we must now look for a definition of the physical nature of moving bodies.

Just as Aristotle did not invent the idea of the cosmos, so he was not the first to explain the material nature of bodies in terms of qualitatively distinct elements. Plato had done so previously, and so had Empedocles, whose theory of the four elements may be said to have foreshadowed that of the Peripatetics.[84] But even if the content of that theory was not completely new, Aristotle parted company with his predecessors both by endowing the elements with new properties and also by his a priori definition of their existence. Rejecting the mathematical approach of Plato, who in his *Timaeus* had recourse to the theory of geometrical means, Aristotle relied on the orderly structure of the cosmos and even more so on the simple motions associated with it. With the help of the latter he was able to demonstrate that the elements could neither be reduced to a single one nor be infinite in number. For if we postulate the existence of only one element, as the early Ionian philosophers did, we are immediately faced with a major stumbling block, namely the existence of more than one simple motion. Thus, if "all bodies were one substance," they would not be moving in distinct ways[85] or in distinct directions. The Atomists' doctrine of the existence of an infinite number of elements[86] (as attributed to them by Aristotle) is no less untenable. Aside from the logical objections to this kind of solution,[87] the mere fact that "there is not an infinite number of simple motions, because the directions of movement are limited to two and the places also are limited," provides clear proof that "there cannot be an infinite number of elements."[88] This critique, taken in conjunction with the assertion that "a simple body must have a simple motion"[89] indicates by what method alone Aristotle felt it was possible to determine both the number of elements and their properties.

From the assumption that "the motions of simple bodies are limited

[84]For the origin of this theory, see Kahn, *Anaximander*, pp. 134–159.
[85]*On the Heavens* III, 5, 304 b 14 ff.
[86]Ibid., III, 4, 303 a 12.
[87]Ibid., 303 a 20 ff. In particular, this solution struck Aristotle as being incompatible with the arguments of mathematics.
[88]Ibid., 303 b 3 ff.
[89]See ibid., I, 3, 270 b 26–31; also I, 7, 274 b 1–5.

in number and that each of the elements has a particular motion assigned to it,"[90] and from the fact that rectilinear motion in the sublunary world takes two simple forms, it necessarily follows that there must be two simple elements moving naturally from and toward the center of the world, one upward and the other downward; one will have the absolute quality of lightness and the other the absolute quality of heaviness. The existence and identification of the second pose no problem, since everyone readily admits that bodies move downward by virtue of their own weight, and the earth, as experience patently shows, has just this characteristic.[91] However, not everyone accepts the existence of elements that are light in themselves. Nevertheless, the same reasons that persuade us to conceive of lightness as an active tendency to upward motion must also convince us of the existence of an element whose natural motion is upward—a case in point is fire, since "any chance portion of fire moves upward."[92] Any lingering doubt must disappear with the realization that if fire lacked this absolute quality of lightness there could be nothing to stop some other body from rising to the top of everything, and there is no evidence of the existence of such a body.[93] Earth and fire, then, are the two primary elements that are imposed on the sublunary world by the orderly structure of the cosmos and by simple motions. All the same, Aristotle realized that his argument was incomplete, because observation also shows that "the same bodies do not appear to be heavy (or light) in every place."[94] To save the appearances and to explain why certain bodies must, as a result of their relative lightness or relative weight, hold a natural position between fire and earth, Aristotle was therefore forced to introduce the existence of intermediate elements. These elements, two in number, are water and air; since their nature is mixed and yet both have weight (fire being the only absolutely light element), their lightness and weight must perforce be relative. "Since there is one body only which rises to the top of everything, and another which sinks beneath everything, there must exist two others which both sink beneath something and rise to the top of something else. The kinds of matter, therefore, must be the same in number as the bodies, namely four."[95]

So far we have been considering only simple rectilinear motions.

---

[90]Ibid., I, 8, 276 b 8–10.
[91]Ibid., IV, 4, 311 a 21.
[92]Ibid., 311 a 19–20.
[93]Ibid., 311 b 21–27.
[94]Ibid., 311 b 1–3.
[95]Ibid., IV, 5, 312 a 28–31.

However, in addition to these, simple circular motion also formed a basic part of Aristotle's cosmology. Now, since each of the four Aristotelian elements has a natural motion, and since no body can have more than one motion, it follows that circular motion cannot be natural for any of them.[96] Hence, unless we deny the obvious existence of circular motion, we are left with only one alternative, which is to describe circular motion as the natural motion of an additional element. "If circular motion is natural to anything," Aristotle contended, "it will clearly be one of the simple and primary bodies of such a nature as to move naturally in a circle, as fire moves upward."[97] Thus the same principle that allowed him to deduce the existence of the four sublunary elements—earth, fire, water, and air—now forced Aristotle to accept the idea that "there exists some physical substance besides the four in our sublunary world, and moreover that it is more divine than, and prior to, all these."[98] But what were its precise properties?

Heavy bodies, as we saw, are these that tend naturally toward the center, and light bodies those that, attracted by the extremity, move away from it. But a body that is naturally endowed with circular motion will neither approach the center nor recede from it; consequently it can have neither weight nor lightness.[99]

Let us take another look at rectilinear motions; not only are they contrary to each other, but they also take place between contrary termini, leading that which is potentially heavy to the center, where it best actualizes its form, and leading that which is potentially light to the extremity. Hence, rectilinear motion is appropriate only to bodies incapable of actual existence in their present state, and hence subject to generation and corruption. The characteristics of circular motion, on the other hand, are entirely different. Since every point on the trajectory it describes has the same value (it is in effect both the end and the beginning of the motion), it cannot possibly lead a body from contrary to contrary.[100] Now, since all rectilinear motions represent contraries and since nothing can have more than one contrary, it follows that natural motion about a center must be a motion without a contrary.[101] For all these reasons, it is quite impossible to apply the concept

---

[96]Ibid., I, 2, 269 b 1–2.
[97]Ibid., 269 b 2–6.
[98]Ibid., I, 3, 269 b 26–32.
[99]Ibid., 269 b 26–32.
[100]*Physics* VIII, 8, 264 b 9 ff.
[101]In this analysis we are following *On the Heavens* I, 3, 270 a 12 ff.

of contraries to bodies in natural circular motion: their motion takes place, by definition, between identical states, so that a body whose motion is without a contrary can rightly be said to be without a contrary itself. But in that case what is the precise function of circular motion, and what is the precise physical composition of bodies subject to it? Aristotle's answer was as bold as it was logical. Since the notion of contraries makes no sense in the case of bodies in circular motion around a center, it follows that such bodies, lacking as they do potential existence, must necessarily enjoy an actual and immediate existence, that is, they must exist in the fullness of their essence; unlike sensible, alterable, and corruptible bodies, they are ungenerable and unengendered, unalterable, and impassive. Their matter can therefore be subject to no influences other than local motion. In other words, circular motion is the typical way in which this matter enjoys actual existence (thus it is called "topical matter," ὕλη τοπίκη), so that even in this extreme case local motion preserves its actualizing function and remains inseparable from the realization of essential or complete being.

Whatever may be thought of this argument, its importance must not be underestimated. As the only bodies subject to circular motion are celestial bodies,[102] Aristotle's theory of the elements introduces an absolute split between heaven and earth into the very heart of natural philosophy. This split was to have far-reaching effects, for once rectilinear and circular motions became associated with bodies differing in essence, these motions themselves were bound to appear as being fundamentally distinct, and hence as not lending themselves to an analysis by a single set of principles and concepts. This made the construction of a unified science of motion an a priori impossibility. Moreover, by equating circular motion with the actual existence of celestial bodies, Aristotle underlined the transitory nature of rectilinear motion and thus confirmed the superiority of rest, the only state compatible with the actual existence of terrestrial bodies. The split between heaven and earth thus became one of the fundamental cornerstones of Aristotelean

[102]Chapter I, 3 of *On the Heavens*, in which the properties of the element corresponding to circular motion are set out, does in fact identify that element with the celestial bodies, but Aristotle fails to offer any justification for this identification and presents it as being self-evident. Is this then another factor which by its very nature does not lend itself to verification by experience? Or is it so obvious an idea that Aristotle supposed it would be accepted without question by all his readers? In either case, the argument is equivocal, and the conclusion of the chapter goes far beyond what its contents permit.

cosmology, so much so that Galileo made its refutation a sine qua non of the success of his new science, devoting to it the entire First Day of the discussions constituting his Dialogue.[103]

Aristotle's theory of the elements had still further repercussions on mechanical thought. The reader will recall that Aristotle was led to the conclusion that natural downward motion was the characteristic motion of the element earth and, more generally, of all predominantly heavy bodies, and that natural upward motion was the absolute characteristic of the element fire and the relative characteristic of all bodies combining lightness with weight. The functions of these motions were defined accordingly: they served to return to the center those parts of the element earth and also those parts of the element water that had been removed from it and to return to the extremities those parts of the element fire and also of the element air that had been forced downward. Thus the principal function of rectilinear motion was to restore at all times the natural arrangement of the elements demanded by the order of the world but constantly impeded by the forces of generation and corruption.[104] Counteracting the disorder introduced by the latter, rectilinear motions thus tend to restore the cosmological order, even though, in the very nature of things, they cannot do so completely. By their action, order is reaffirmed dynamically, in the face of disorder, and the world is enabled to preserve both its unity and its coherence.

We have shown that lightness and weight, as potentials, were thought to depend directly on natural downward and upward motion. Hence they had to be logically anterior to the primary elements, earth and fire. Nevertheless, the very manner in which Aristotle introduced these elements also made them appear as the concrete expressions of weight and lightness, that is, as the *movers* to which upward and downward motions are due in the real world. Hence, and despite the fact that the mover was said to be radically distinct from matter, there was no reason why matter should not serve to evaluate the mover or to act as its concrete equivalent. This compelled Aristotle to postulate a direct link between the motion of a body and its physical constitution; a motion must always be a motion of a particular body, constructed of a partic-

---

[103] Galileo's refutation will be discussed in Chapter 4.

[104] Aristotle attributed generation and corruption to the sun, which by activating contraries produces a differentiation within the elements; without that action the elements would remain immobile in their concentric orbs around the center of the world; see *Generation and Corruption* II, 10.

ular element: the idea of the abstract study of motion, disregarding the nature of the bodies involved, struck him as being no less absurd than that of a motion capable of persisting by itself, that is, without the constant intervention of a mover. The material composition of a body is therefore a decisive factor in determining the direction in which it moves, no less than the modalities of its motion (such as its acceleration). Nor is that all. If the elements are, in fact, the concrete expressions of such natural movers as weight and lightness, it is obvious that of two bodies constructed of the same element the more voluminous should also contain more "weight" or "lightness," and hence would tend to return to its natural place more rapidly. In other words, the idea that the natural motion of a body—its free fall, for example—is the faster, the greater its weight was not simply, as is so often claimed, a mere empirical prejudice on Aristotle's part but also a direct consequence of the most general principles of his cosmology.

However, in order to grasp the full impact of the theory of the elements on the science of motion we must also examine, however briefly, Aristotle's views of the immobility of the earth and the mobility of the heavenly bodies. The immobility of the earth was deduced from two propositions we have already met: the existence of a center of the universe and the existence of the natural motion that restores all fragments of the element earth to that center, simply because "it is natural for the whole to be in the place toward which the part has a natural motion."[105] Moreover, every body can have only one natural motion (for otherwise a thing would at one and the same time also be its opposite); from this it follows that any motion other than downward would, as far as the earth is concerned, have to be violent, thus running counter to the eternal order of the universe, and for that very reason incapable of persisting. Situated as it is in its natural place, the earth must of necessity stay where it is. Its center coincides with that of the universe; hence, whenever a heavy body falls toward the center of the earth, it is, in fact, moving toward the center of the universe and only incidentally toward that of the earth.

Once the immobility of the earth had been established, the mobility of celestial bodies raised no special theoretical problem. Hence, Aristotle

---

[105]For the entire analysis that follows, see *On the Heavens* II, 14, 296 b 25 ff.

felt it sufficient to show to what extent circular—as opposed to rectilinear—motion was in full accordance with the continuous and eternal nature of the celestial motions. Thus rectilinear motion could not be continuous (if only because of the finite nature of the world) unless it turned back every time it has reached its end; and "turning back requires a pause," since otherwise one and the same point would at one and the same time be the end of one motion and the beginning of another.[106] This difficulty disappears automatically in the case of circular motion, where moving away from *A*, "a body is forthwith in position for motion toward *A*, since it is then in motion toward the point at which it will arrive. It does not simultaneously execute contrary or opposite motions."[107] In addition to being uninterrupted, circular motion also has a potential for indefinite duration; the sole reason for the finite nature of rectilinear motion is the opposition of the termini between which it occurs; in the case of circular motion, on the other hand, any point of the trajectory is simultaneously the beginning, the middle, and the end of a motion, "with the consequence that a body in circular motion is in some sense both always and never at a beginning and an end."[108] To these considerations, based on the properties of the circle, Aristotle added a number of optical arguments. Some he drew from direct experience—for instance, that heavy objects thrown forcibly upward in a straight line come back to their starting point,[109] whereas if the earth were itself in motion they would not. Other examples he borrowed from astronomy: the stars always rise and set over the same regions of the earth, and they would certainly not do so if the earth were subject to any motion whatsoever.[110] Ptolemy took up this argument and refined it in an attempt to show that the earth was not only immobile but also at the center of the universe: if it were not, he argued, one side of the heavens would seem nearer to us than the other and the stars would look larger there; if it were on the celestial axis but nearer to one pole, the horizon would bisect not the equator but one of its parallel circles; if the earth were outside the celestial axis, the ecliptic would be divided unequally by the horizon.[111] We are not detracting from these arguments, which lend observational support to philosophical deductions, when we point out that, as far as Aristotle

[106] *Physics* VIII, 8, 262 a 17 ff.
[107] Ibid., 264 b 10–13.
[108] Ibid., VIII, 9, 265 a 27 ff. This also explains why nothing resembles rest so much as a sphere turning on its own axis.
[109] *On the Heavens* II, 40, 296 b 23–24.
[110] Ibid., 296 b 2–6.
[111] Ptolemy, *Armagest* I.

was concerned, they remained secondary considerations; to him the philosophical arguments were the only conclusive ones.[112] This was the case to such an extent that the optical arguments in favor of the immobility of the earth were questioned long before Copernicus and Galileo, without anyone ever doubting the geocentric nature of the universe.

The most interesting aspect of Aristotle's solution was the manner in which he arrived at it. Ever since the beginning of the fourth century B.C., celestial motions had been the subject of precise observations. As a result, astronomers were put in possession of a set of reliable data, and not simply of a jumble of empirical knowledge, with which to approach the problem of celestial motion in general and to establish the order of the principal celestial bodies in particular. In these circumstances, would it not have been reasonable to expect a purely rational inquiry into the possible motions of the earth and the heavens, that is, an inquiry based above all on geometric analyses and on such basic mathematical arguments as the principle of simplicity? By affirming, on the contrary, that only a "physical" explanation—one based directly on the intrinsic properties of substances—could account for physical phenomena, Aristotle turned his back on a natural philosophy in which rational reconstructions of the observed phenomena are the only valid explanations. To be sure, he was not lacking in arguments: physical bodies subject to motion were essentially different from mathematical objects deprived of motion; no physicist basing himself on mathematical premises would ever be able to grasp the real world; thus motion must remain an inexplicable mystery to him. How, for example, would he be able to justify the fact that different bodies move in distinct ways or that certain bodies always tend upward and others always downward? Or how could he possibly hope to decide between the mobility or immobility of the earth? How would he be able to show that things are as he says they are and not otherwise?[113] On the other hand, the physicist who bases his conclusions on essences, who assigns principles

---

[112]According to Duhem, two other arguments in favor of the immobility of the earth were implicit in the Aristotelian view: the fact that, in order to turn on its own axis, a sphere needs an immobile center (*On the Heavens* II, 3, 286 a 13–14); and the fact that the theory of places calls for the presence of a fixed body in the center of the universe so as to localize the eighth sphere; cf. P. Duhem, *Le mouvement absolu et le mouvement relatif*, pp. 8 ff.

[113]This was certainly the gist of Aristotle's criticism of Anaximander in *On the Heavens* II, 13, 295 b 10 ff. In the same way, he accused Plato of attempting, in the *Timaeus*, the impossible task of reconstructing reality out of elements that by definition were not part of reality; ibid., III, 8.

of change to changeable things, eternal principles to eternal things, principles of corruption to corruptible things—the physicist, in short, who in a general way chooses his principles so that they "are of the same genus as what falls under them"—would have no difficulty in solving these problems.[114] His findings would be superior to all others, and this could mean only that the mathematical (or astronomical) analysis of natural phenomena could, at best, play a subordinate role: it might be descriptive but could never be explanatory. A text by Geminus, quoted by Simplicius in his commentary to the *Physics* of Aristotle, provides a remarkable illustration of this point. "To physical science," Geminus held, "belongs the examination of the nature, power, quality, birth, and decay of the heavens and the stars. . . . Astronomy, on the other hand, is incapable of dealing with such primary matters; it makes known the arrangement of the heavenly bodies, having first declared that the order of the heavens is regular." What then was the precise task of the astronomer? According to Geminus, he must find an answer to the question "Why do sun, moon, and planets appear to move unequally?" And that answer is "Because when we assume their circles to be eccentric or the stars to move on an epicycle, the appearing anomaly can be accounted for, and it is necessary to investigate in how many ways the phenomena can be represented, so that the theory of the wandering stars may be made to agree with the etiology in a possible manner. Therefore also a certain Heracleides of Pontus stood up and said that when the earth moved in some way and the sun stood still in some way, the irregularity observed relatively to the sun could be accounted for. In general, it is not the astronomer's business to see what by its nature is immovable and of what kind the moved things are, but, framing hypotheses as to some things being in motion and others being fixed, he considers which hypotheses are in conformity with the phenomena in the heavens. He must accept as his principles from the physicist that the motions of the stars are simple, uniform, and regular."[115] Physical analysis, helped by mathematics, could certainly try to "save the appearances" by providing the necessary geometrical models from them, but it was up to physical analysis alone to

[114]Ibid., III, 7, 306 a 9–11.

[115]Simplicius in *Aristotelis physicorum Libros quatuor prioris commentaria*, ed. Diels (Berlin, 1882), pp. 291–292. We have been quoting from J. L. E. Dreyer, *A History of Astronomy from Thales to Kepler*, pp. 131–132.

fix the guiding principles of a true explanation of the physical world, because it alone was concerned with the essences.

Finally, the "physical" justification of the immobility of the earth throws light on another characteristic feature of traditional mechanics: its rejection of the idea of relativity, and hence of any of the questions this idea entails. Of course, we do not claim that Aristotle was unaware of the optical relativity of motion or of the possibility of explaining all motions by the displacement of the object under observation or of the observer. However, the way in which he linked motion with the essence of bodies made this possibility quite irrelevant to physical science. To Aristotle, accepting the idea of relativity would have been tantamount to denying the actualizing function and hence the intelligibility of local motion. The idea of motion could have no meaning except in relation to a body, and being moved could never be the same thing to a body as standing still. This point cannot be emphasized strongly enough, because some of Aristotle's remarks would seem to suggest that he accepted the idea of relativity—for instance, in his assertion that "every mover that is a physical agent is moved."[116] In fact, this assertion has nothing in common with the idea of relativity. Thus weight, which enables a heavy body to move toward the center can be "stimulated" by the removal of external obstacles: the resulting downward motion of a body is a fully determined "affection" of that body, and part of its own nature. Similarly, though the mover, ensuring the circular motion of all celestial bodies, partly reflects the action of a Prime Mover, the motion of celestial bodies remains real enough and in no way relative to the possible motion of the earth. Now, it is beyond doubt that by excluding relativity from his physics, Aristotle robbed mechanics of a host of fruitful problems, as the subsequent history of that science shows only too clearly. Thus it was by the methodical development of the implications of the idea of relative motion that Galileo was able to place cosmological thought on completely novel foundations and thus to propound some of the most basic principles and concepts of classical science.[117] But before that happened, the absolutist interpretation of motion continued to cripple mechanics and to impede its progress for two thousand years.

---

[116] *Physics* III, 1, 201 a 19 ff.

[117] These problems will be examined in Chapter 5.

The Continuous Nature of Local Motion

Being a transitory process dependent on both a mover and a moving body, motion has yet another characteristic aspect, namely, continuity. This subject, to which Aristotle devotes a whole book of his *Physics*, is important for several reasons. One of the grounds on which the Eleatics, speaking through Zeno, declared that motion was irrational and contrary to all reason was precisely its continuous nature. It was in order to meet this challenge squarely that Aristotle set out to demonstrate beyond all doubt that the continuity of motion was perfectly intelligible. Moreover, it was in connection with this analysis that Aristotle came closest to a description of motion as such, and that he realized that in treating of continuity, it was no longer enough to consider the cosmological order or the actions of mover and moved. Hence, if we wish to set precise limits to Aristotle's theory of local motion, we must pay careful heed to this treatment of the problem of continuity. The subtle use he made, inter alia, of the conclusions of the great Greek geometers in the fifth and fourth centuries B.C. incidentally sheds a great deal of light on an aspect of traditional physics that is far too often ignored.

Analysis of continuous magnitudes sprang from two distinct problems. One of them is bound up with the discovery of irrational magnitudes. The authors of the first great mathematical synthesis, the Pythagoreans, had postulated the absolute correspondence (commensurability) of spatial magnitudes with numbers. Guided by their conception of the latter —not as an abstraction in the modern sense but as a set of units lending themselves to spatial expression—they believed that the relationship between any two magnitudes could be invariably reduced to their numerical ratio. The first theory of proportions to be built on this basis (it forms the essential part of Book VII of Euclid's *Elements*) expressed the conviction that geometrical properties simply confer upon space relations that have an essentially numerical character.[118] However, the discovery of irrationals served to shake this conception to its very foundations, for, if the side of a square and its diagonal were in the ratio of one number to another, it would follow that the same number must be both odd and even.[119] If we add that other examples of irrational magnitudes were soon to follow (thus Theodorus showed that the square root of

---

[118] G. Milhaud, *Les philosophes géomètres de la Grèce*, p. 102.

[119] T. L. Heath, *A History of Greek Mathematics*, Vol. I, pp. 90–91.

every nonsquare number from 3 to 17 is incommensurable with 1),[120] the reader will readily appreciate that in the years from 430 to 420 B.C. continuous magnitudes came to be treated as being generally inexpressible by numbers, and that a veritable split in mathematical thought ensued.

The speculations on infinitesimals indulged in by some mathematicians at about the same time were bound to lead to much the same results. While we cannot tell precisely who introduced these speculations and why, their underlying principle was clearly the hypothesis of indivisibles, that is, the theory that every continuous magnitude, if divided and subdivided far enough, could be reduced to indivisible and homogeneous particles. More concretely, a line can be considered as constructed of points, an area as constructed of lines, and a volume as constructed of areas. This hypothesis was, of course, purely speculative, and the ideal division to which it referred had nothing to do with the material divisions that can be performed in the real world. Indivisibles seem to have been used in two distinct ways. First, they were applied directly, that is, in strict accordance with the picture of continuous magnitudes they suggested; it is quite possible, for example, that Democritus arrived at his computation of the volume of a solid (such as a pyramid or a cylinder) by treating it as the sum of an infinite number of parallel planes or infinitely thin laminae.[121] Second, and much more subtly, the doctrine of indivisibles, by giving concrete expression to the idea of proceeding to a limit, appears to have provided the necessary principle to the earliest attempts to solve the problems of the quadrature of areas or the cubature of volumes. Aristotle himself tells us that Antiphon made an attempt to square the circle.[122] He inscribed a square in a circle, on each side of which he then described an isosceles triangle with its vertex on the arc of the smaller segment of the circle subtended by an arc. Next, he repeated the same operation on each side, on the assumption that in the end the area of the entire circle would be used up, and that he would be left with a polygon the sides of which would, owing to their smallness, coincide with the circumference of the circle. The only rational basis for this was the idea that continuous lines are constituted of homogeneous points, and this was the central idea underlying the theory of indivisibles. In the same way, Democritus may have

---

[120] *Theaetetus*, 146 e to 148 b.

[121] Heath, *History of Greek Mathematics*, Vol. I, p. 180.

[122] *Physics* I, 2, 185 a 14–17.

employed this theory to find the volume of a cone by increasing indefinitely the number of sides in a regular polygon forming the base of a pyramid.[123]

Nevertheless (and however fruitful they may sometimes have been),[124] these speculations also raised difficulties that could not be ignored. In principle, they amounted to the supposition that it was perfectly legitimate, at a given moment, to put a stop to the infinite divisibility of continuous magnitudes. This supposition, as Zeno of Elea showed, was utterly absurd. It is, in fact, impossible to conceive of a continuous magnitude, however small, that cannot be divided further. Hence, it follows that the indivisibles that are supposed to constitute the continuous magnitudes can have no size themselves (nil-elements). Now, how can the addition of these nil-elements to anything, no matter how many times, increase its size? If, on the other hand, infinite division were to lead to parts having some size, then, because there is an infinite number of such parts, the original magnitude that was divided would have to be infinite in size as well.[125] The conclusion is therefore self-evident: as it serves to deprive continuous magnitudes of all dimensions, or conversely to make them infinitely large, the idea of indivisibles must be rejected out of hand. However, if we do so, the problem of continuous magnitudes remains unsolved: after all, the theory of indivisibles was designed to restore a measure of harmony to mathematical thought by reconciling number and continuous magnitudes with the help of infinitesimals. In showing the absurdities to which this idea could lead, Zeno merely increased the break introduced by the discovery of irrationals. In particular, his critique of the concept of indivisibles clearly proved the impossibility of defining the nature of continuous magnitudes by means of discontinuous processes. But in that case the idea of continuity becomes unintelligible, and so does every phenomenon based on continuous processes. This would seem to be the meaning of Zeno's famous paradoxes. In particular, if we assume that length and time are infinitely divisible, we cannot possibly hope to explain that motions can have a beginning or an end.[126] Conversely, if we assume that they consist of infinitely small but distinct parts, we cannot explain the con-

[123]Heath, *History of Greek Mathematics*, Vol. I, p. 180.

[124]Archimedes himself sometimes relied on indivisibles; ibid., Vol. II, pp. 27–30.

[125]Ibid., Vol. I, pp. 274–275.

[126]The *Dichotomy* and the *Achilles*.

tinuation of a motion.[127] Considered in itself, motion thus proves to be completely impervious to rational analysis.

Aristotle was quite undaunted by all this confusion. Not only did he dispose of Zeno's paradoxes, but he conceived of continuity as a mode of being perfectly conceivable in itself. What had happened in the interim? Thanks to the work of Tannery, Zeuthen, and Heath, we can retrace the fundamental steps by which fifth- and fourth-century Greek geometers, notably Eudoxus, managed to place the science of continuous magnitudes on solid foundations. The first two propositions of Book X of the *Elements* provide an excellent summary of their ideas. Zeno's arguments had shown that any attempt to analyze the continuous in terms of the discontinuous must come to grief as a result of infinity; noting this, Eudoxus, by a stroke of genius, turned what had been a source of difficulty and contradiction into a definition, or rather into an axiom of his science. The first proposition of Book X of Euclid's *Elements* has preserved that axiom for posterity: two unequal magnitudes being set out, if from the greater there is subtracted a magnitude greater than its half, and from that which is left a magnitude greater than its half, and if this process is repeated continually, there will be left some magnitude that will be smaller than the lesser magnitude set out. This axiom recalls another by Archimedes, which is, in fact, no more than its complement: "Given two lengths, there is always a multiple of the smaller such that it exceeds the length of the greater."[128] Since there is nothing in either case to prevent our choosing as the smaller a magnitude as small as we wish, Eudoxus' proposition clearly amounts to a rejection of the infinitesimal; it bases the science of continuous magnitudes on the axiomatic assertion that they cannot be reduced to the discontinuous. The specific nature of the continuous thus having been established, all Euclid had to do when constructing a general theory of the continuum was to provide a rigorous definition of irrational or incommensurable magnitudes. This is precisely what he did in the second proposition of Book X, where he formulated a general criterion of incommensurability, undoubtedly basing himself on the work of Theodorus and Theaetetus:[129] "If, when the lesser of two un-

[127] The *Arrow*.
[128] P. Boutroux, *Les Principes de l'analyse mathématique*, Vol. I, p. 74.
[129] Heath, *History of Greek Mathematics*, Vol. I, pp. 202 ff.

equal magnitudes is continually subtracted in turn from the greater, that which is left never measures the one before it, the magnitudes will be incommensurable." The proof of this proposition is remarkable. It is based on the usual operation for finding the greater common measure and amounts to demonstrating that, in the case of two incommensurable magnitudes, this operation never comes to an end, while the successive remainders become smaller and smaller until they are less than any assigned magnitude. Clearly, Euclid was using much the same approach as Eudoxus: the infinite, which had constituted an insuperable barrier for the Pythagoreans no less than for Zeno, was transformed from a sign of unintelligibility into a criterion by which continuous processes could be defined. This radical change in viewpoint made possible the construction of a unified science of the continuum: rational and irrational magnitudes now came to be treated as species belonging to a common genus to which the same method could be applied, namely, the general theory of proportions (commonly attributed to Eudoxus) as set out in Book V of Euclid's *Elements*. Hence, the problem of continuous magnitudes had been solved in Aristotle's day, but with the help of a split between magnitude and number which could not fail to have severe repercussions on Greek science. Moreover, Eudoxus had admittedly provided geometry with the tools it needed to describe everything appertaining to magnitude in the most rigorous terms, even enabling it to pave the way for a mathematical science of nature—as witness the statics and hydrostatics of Archimedes—since, under certain conditions[130] it allowed the construction of proportions between heterogeneous magnitudes. It nevertheless remains a fact that the renunciation of a systematic correspondence between continuous magnitudes and numbers (which is expressed inter alia by the proposition that "if magnitudes have not to one another the ratio which a number has to a number, the magnitudes will be incommensurable")[131] stood plainly in the way of the mathematical definition of such essential physical magnitudes as velocity or acceleration and thus proved a grave obstacle to the construction, if not of physical science as such, at least of quantitative physical science.

---

[130]Euclid, *Elements*, Book V. Definition 5: "Magnitudes are said to be in the same ratio, the first to the second and the third to the fourth, when, if any equimultiples whatever be taken of the first and third, and any equimultiples whatever of the second and fourth, the former equimultiples alike exceed, are alike equal to, or alike fall short of, the latter equimultiples respectively taken in corresponding order."

[131]Ibid., Book X, proposition 7.

Now it is this very conception that lay at the root of Aristotle's approach to continuous magnitudes. His own definition, namely, that "anything continuous is divisible into parts infinitely divisible,"[132] is a more philosophical echo of Eudoxus' axiom. Aristotle, like the mathematicians, thus barred the way, from the very outset and axiomatically, to the introduction of infinitely small, indivisible quantities. However, he deserves credit for the way in which, by his definition of the relation between infinity and continuous magnitudes, he brought the theory of Eudoxus to a remarkable conclusion. In particular, his ontological analysis conferred much wider scope on Eudoxus' contribution. Thus he argued that since the division of a given magnitude, even though continued to infinity, always leaves us with finite parts (which when added together often enough will again exceed any given magnitude), infinity must have a potential rather than an actual existence.[133] However, as he went on to explain, there are two distinct ways of "potential being": first of all, that by which the bronze is potentially a statue or the uneducated man is potentially a scholar. This type of potentiality is characteristic of the relation between living creatures or things and the qualities they are capable of assuming. But there is also a second type of potentiality—namely, that illustrated by the relationship between games, contests, etc., and existence. As simple attributes of beings or substances already actualized, their kind of existence is nothing but a continual coming and going away,[134] so that they remain in a perpetual state of potentiality. Now, it is in just this way that the infinite must exist in continuous magnitudes.[135] Its true nature is therefore clear: just as a game has no reality independent of the players' moves, so the infinite cannot exist without the process of division to which continuous magnitudes lend themselves by their very essence. It is, in fact, no more than an attribute of these processes and merely reflects their property of being infinitely renewable: "The infinity of numbering may be understood in terms of the continual bisection of a line. . . . But there is no separate infinite number, nor is the infinity of numbering a permanent actuality; instead, it is a process."[136] Hence, the fact that continuous

---

[132]*Physics* VI, 1, 231 b 15–16.
[133]Ibid., III, 6, 206 a 18.
[134]Ibid., 206 a 22.
[135]Ibid., 206 a 18–21.
[136]Ibid., III, 7, 207 b 10–14. All divisions of continuous magnitudes do not, moreover, make the infinite appear; to bring it out, we must add "a part determined by a constant ratio" to "a determinate part of a finite magnitude" (ibid., III, 6, 206 b 7–9). This is tantamount to constructing a decreasing geometrical series such as $1 - (1/2 + 1/4 + \ldots + 1/2 + \ldots)$. Apart from by division (*κατά διαίρεσιν*), infinity can also be constructed by addition (*κατά*

magnitudes are compatible with endlessly repeated divisions in no way implies that they are composed of an infinite number of elementary parts, actualized by such divisions. Lending themselves to an endless series of divisions, continuous magnitudes thus contain the infinite as a potential; in actuality, by contrast, they never contain anything but a finite number of finite parts.[137] Having thus removed all traces of ambiguity, Aristotle gave to the idea of indivisibility a new lease on life, and furthermore a precise status in its relation with continuous magnitudes. In particular, since the division of continuous magnitudes must result in finite parts, a point on a line can be no more than the limit of these parts; it serves to separate the parts of the line but is not itself contained in it. This excludes any possibility of reconstructing continuous magnitudes with the help of infinitely small but determinate parts.

Aristotle found it quite simple to deduce the continuity of motions from this perfectly coherent interpretation. Since no motion can be divorced from a linear trajectory, their respective parts must be in the same ratio to one another. Hence, in order to demonstrate the continuity of motion, we need merely prove the absurdity of all attempts to represent the trajectory and the motion as being simultaneously composed of indivisible parts. Let a line *ABC* be composed of the indivisible parts *A*, *B*, and *C*, and let *DEF* represent the motion of a body *X* traveling over this line.[138] Now suppose that the indivisible part of the motion *D* corresponds to the indivisible part of line *A*, and the indivisible parts of the motion *E* and *F* to the indivisible parts of lines *B* and *C*. Then for our subsequent argument we need make use of just one principle, namely, that anything moving from somewhere to somewhere cannot, while in motion, be moving and at the same time have completed its motion to its destination. Hence, in the particular case of the relation between any one indivisible part of the motion, say *D*, and any one indivisible part of the line, say *A*, the moving body *X* must either have completed its passage through *A* after passing through it (which is what normally happens), and its motion must be divisible, which is contrary to the hypothesis; or else, in order to save the in-

---

πρόσθεσιν), that is, by the summation of parts decreasing in a given ratio. Infinity by addition also applies to the construction of the set of positive integers.

[137] Ibid., III, 6, 207 a 1–3.

[138] Ibid., VI, 1, 231 b 18–232 a 18.

divisibility of the motion *D*, we must postulate that *X* was passing through *A* at the same time as it had completed its motion through it, which is plainly absurd. Let us now consider the distance *ABC* as a whole; if *D*, *E*, and *F* remain indivisible parts of the motion, whatever applies to *D* must equally apply to *E* and *F*. In that case, we shall finish up with a motion without motion or, as Aristotle puts it, with a motion consisting of discrete atomic movements. In other words, the body *X* would have "completed a movement without being in motion." Denying the continuity of motion is therefore tantamount to dropping the distinction between motion and rest.

However, this analysis does not yet bring out the full consequences of the continuous nature of motion. To do that, Aristotle had to take a further step. Though motion necessarily involves spatial magnitudes, it is on its relations with time that its concrete evaluation must be based. This calls for a brief digression. Aristotle's own definition of time, namely, "the number of precessions and successions in process,"[139] suggests that its continuity can be deduced directly from that of motion; this is precisely what he succeeded in doing with the help of an elegant demonstration.[140] To follow it, we need merely remember that, of two bodies moving with unequal speed, the faster one will cover a greater distance in the same interval of time, or an equal distance in a shorter time. Let *A* and *B* be two moving bodies, with *A* traveling faster than *B*. Then if *B* covers the distance *CD* during the time *fg*, the faster body *B* will have covered the same distance in the shorter time *fj*. Now during the time it takes *A* to traverse *CD*, the slower body *B* will only have traversed *CE*. Moreover, if it takes *B* the time *fj* to traverse *CE*, body *A*

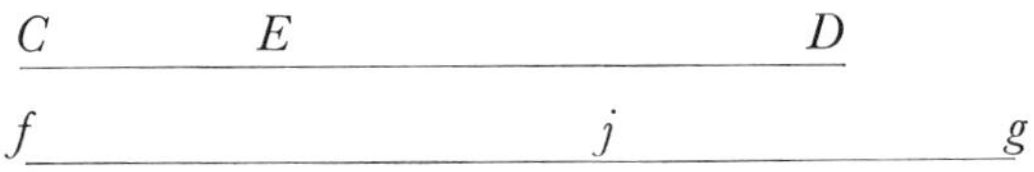

C *E* *D*

f *j* *g*

will have covered the same distance in a time less than *fj*, and again, during the interval *fj* the body *B* will have covered a distance less than *CE*, and so on. In other words, to every new division of the time taken by the slower body, there corresponds a new division of the time taken by the faster one, and this ad infinitum. It follows that time, like length and motion, must be absolutely continuous, and that it can contain

[139] Ibid., IV, 11, 219 b 1.

[140] Ibid., VI, 2, 232 b 20–233 a 10; see also ibid., 233 b 15–31.

infinity (by division or addition) only as a potential, not as an actual function.[141] Moreover, just as a line cannot be generated by the infinite repetition of points, so time cannot be generated by an infinite succession of distinct moments. Like the point, a moment is simply a limit introduced from the outside; its sole purpose is to help us determine and compare durations. Or as Aristotle himself put it, "the [momentary] now is somehow an extreme end of the past, since it has nothing of the future, and somehow an extreme end of the future since it has nothing of the past; it is, we maintain, a limit of both."[142] In other words, a moment, though occurring in time, cannot be part of it. Like the point, it is an indivisible entity, in the new sense given to that term by the concept of "indivisible."

This analysis has a direct bearing on motion; first of all, it can be used to demonstrate the general impossibility of instantaneous motion. Thus if a body could move during the moment *n*,[143] its motion might be quicker or slower. Were it quicker, the body would cover, say, the distance *AB*; were it slower, it would cover only a part of that distance, say, the distance *AC*. But obviously, if quicker, it would take less time to cover *AC* than *AB*; a shorter moment would suffice, which means that the moment would have to be divided. But such division is impossible since, by definition, a moment has no duration. Instantaneous motion would therefore be motion, not in time but in the limit of time, and such motion is impossible.[144]

This first conclusion leads directly to two others concerning the beginning and end of a motion. Let us first consider the latter, that is, the moment in which the change produced by the motion has been completed: this moment can be shown to be without duration.[145] For let us call it *AC*, and let us suppose that it did have a duration. Then, by dividing *AC* at *B*, we should obtain two partial moments, *AB* and *BC*, and at once be faced with a double impossibility. Either the motion is completed in the partial moment *AB* (or the partial moment *BC*) and the last moment of the motion would no longer be *AC* as we have assumed it to be, or else the motion is completed simultaneously in *AB*

---

[141]Ibid., 233 a 17–21.

[142]Ibid., VI, 3, 233 b 35–234 a 1.

[143]Ibid., 234 a 24–31.

[144]This is why every doctrine implying the idea of instantaneous motion must be rejected, as must the doctrine of the void (*Physics* IV, 8, 215 b 12 ff.) and of the infinity of the world (*On the Heavens* I, 6, 273 b 27 ff.).

[145]For this analysis, see *Physics* VI, 5, 235 b 13–236 a 6.

and in *BC*. Once again we should no longer be able to say that it is completed in the moment *AC*. It follows that the last moment of a motion must be timeless, that it can be no more than an indivisible limit separating motion from rest. Nor is it possible to attribute a duration to the initial moment of a motion. For supposing *AD* were the duration during which a body was first set in motion, then that duration could always be divided into two parts, and the same difficulty would reappear: either the motion is not yet completed in either of the two parts and *AD* cannot be the first moment, or else it is completed in both parts and the second must necessarily have followed the first, or finally it is completed in one part only, so that *AD* is again no longer the first moment.[146] The same indivisible limit that separates motion from final rest therefore separates it from initial rest as well.[147]

These reflections enabled Aristotle to dispose of Zeno's contention that the continuity of motion detracts from its intelligibility.[148] The attack was three-pronged. A first blow was aimed at the very formulation of the problem. Zeno had contrasted the infinite number of parts into which space can be divided with the finite time it takes to traverse them. Formulated in this way, the problem made no sense. But Zeno forgot that, thanks to its continuity, time, like length, stretches into infinity, and above all that this infinity, far from being "infinite in quantity," is merely "infinite in divisibility." A length infinite in quantity (supposing it existed) could not possibly be spanned in finite time, in contrast to an infinitely divisible length, "for time itself is also infinite in this sense."[149] Nevertheless, this argument had, above all, an *ad hominem* value,[150] and though it established the possibility and intelligibility of motion *in hypothesi continui*, it did not imply that motion must necessarily take place in a finite period of time. Hence, Zeno's arguments called for a further refutation, and, to adduce it, Aristotle had

---

[146]Ibid., 236 a 13–27.

[147]Similarly, it can be shown that there is no first moment during which a moving body comes to rest; ibid., VI, 8, 238 b 31 ff.

[148]Let us recall that, to Aristotle, Zeno's paradoxes were the direct result of a failure to define continuous magnitudes correctly. Aristotle was concerned mainly with Zeno's first two paradoxes (which involved the continuity of space) and particularly with the *Dichotomy*, of which the *Achilles* struck him as being no more than a more dramatic form. It was Gregory de St. Vincent who, in the seventeenth century, seems to have been the first to consider that both the *Dichotomy* and the *Achilles* posed a special mathematical problem, the determination of the limit of a convergent series.

[149]*Physics* VI, 2, 233 a 21–31.

[150]Ibid., VIII, 8, 263 a 15–17.

recourse to the distinction between actual and potential infinity. Zeno had argued that because length was infinitely divisible, it must embrace infinity as an actual principle. However, before the divided parts, assumed to be infinite in number, can exist in actuality, a mental operation must be performed; in the case of motion this operation involves the stopping of the moving body.[151] The resulting dilemma is obvious: either the body comes to rest, in which case the motion ceases to be continuous,[152] or else the body does not stop, in which case it cannot have traversed an *actually* infinite number of points; at most it can have traversed a *potentially* infinite number of points, which it can do in finite time.[153] Aristotle's third refutation of Zeno was based on the nature of continuous magnitudes, as defined by the geometers, and it was doubtless this refutation that he considered the most important of all. Motion, as we showed, cannot take place instantaneously, since in order to traverse a distance, however small, a body needs a finite portion of time. Suppose then that a body traversed the finite distance *AB* in the infinite time *cd*.[154] Let *AE* be a part of *AB;* then the body will cover the distance *AE* during a part of infinite time, that is, in a finite time interval. Now, if we select another part of *AB* equal to *AE*, this part of the motion would also, for the same reason, take a finite period of time, regardless of whether the motion is uniform or not.[155] Now if the elementary part *AE*, however small, is added a sufficient number of times to itself, it will cover or exceed the distance *AB*, or, for that matter, any given magnitude; "consequently [the body] *X* will have traveled the distance *AB* during a finite time."[156] Aristotle's argument was clearly based on Eudoxus' axiom, in the particular form used by Archimedes, and not on the Peripatetic theory of continuous magnitudes. Spurred on by this demonstration, Aristotle went on to deal Zeno a final blow: in the same way that he had proved that a finite distance cannot be traversed in infinite time, he now proved that an infinite distance cannot be covered in finite time.[157] In both cases the refutation was based on the application of the general definition of continuous magnitudes to the case of motion: however small the partial distances traversed by the moving body in however long a time, any one of these

[151]Ibid., 262 a 22–26.
[152]Ibid., 263 2 23–29.
[153]Ibid., 263 b 3–9.
[154]Ibid., VI, 7, 237 b 23–238 a 18; cf. VI, 2, 233 a 31-b 14.
[155]Ibid., VI, 7, 237 b 34.
[156]Ibid., 238 a 17–18.
[157]Ibid., 238 a 20–30.

partial distances, if repeated a sufficient number of times, will always cover the total distance.

Apart from being of interest in itself, this critique was in perfect agreement with Aristotle's earlier arguments. In the first place, by demonstrating that the continuous nature of motion in no way detracts from its intelligibility, he brought to a successful conclusion what he had begun with his analysis of the constituent principles of being; his was the first attempt to construct a science of motion and to present an organized body of propositions on this subject. However, his analysis of continuous processes did more than round off his general arguments: it also lent additional weight to his fundamental thesis that motion, inasmuch as it has an ontological function, must of necessity be a transitory process, differing in essence from the state of rest. His demonstration that motion cannot have an initial or final moment in time was particularly telling in this respect: by treating these two moments as the timeless limits of the preceding or subsequent states of rest, his theory of continuous processes formally excluded the possibility of a gradual transition from rest to motion or from motion to rest: far from weakening the opposition between motion and rest, his theory therefore served to reinforce it. Moreover, the particular way in which he established the continuity of motion confirmed what strikes us as being a direct consequence of his definition of motion as the "combined action of mover and moved." Aristotle, in fact, proceeded by a reduction to absurdity: if we postulate that motion is composed of elementary parts,[158] we are at one and the same time denying the self-evident fact that if a moving object is displaced from one point to another, it must necessarily have covered the distance separating these points. In other words, he presented the continuity of motion, not as a property of its spatiotemporal function, but as a consequence of its links with a spatial trajectory, whose continuity he had already demonstrated. This step was in full accord with such other of his contributions as the rejection of the concept of the movement of movement, or the demonstration that a mover is needed to maintain all motions. Movement as such is a potential action, and anything we can legitimately say about it must be based on such objectively determined realities as movers or spatial

[158]That is to say, as Aristotle shows elsewhere, of parts whose limits touch and are contained in each other; *Physics* VI, 3, 227 a 11–12.

magnitudes—hence, the rather paradoxical appearance of Aristotle's reflections on the subject of continuity. For while these reflections enabled him to introduce a whole series of propositions on motion as a process of actualization, and hence on its subsequent course, they did not lead him to isolate or analyze that course as such. Far from it—they stood in the way of any such examination, as witness his rejection of the idea of instantaneous motion. Not that Aristotle was being inconsistent even in this respect: to him an instant was just what it is—the ideal limit of a time interval, but without duration, and hence incapable of sustaining motion. All the same, it certainly is not true that the idea of instantaneous motion is deprived of all sense. Thus when classical mechanics established the relation between an infinitely small space and infinitely short time, that is, the ratio ds/dt, it did in fact employ that very idea. Hence, the modern reader may gain the impression that Aristotle's analysis of the continuity of motion did not so much bear on motion as such as on space and time. Galileo, in particular, was to draw quite different conclusions from the continuity of motion, but the idea of velocity would then have undergone a radical revision that made it possible to consider motion without explicit reference to movers, moving bodies, or trajectories.[159]

In any case, Aristotle's failure to study the continuity of motion as such finds confirmation in his endeavor to place the idea of continuous motion on firm physical foundations. Being no more than an "affection" of the moving body, motion as such appeared incapable of ensuring its continuation, and so did space and time, which, though inseparable from motion, were not part of the original order of things. Hence, the inherent continuity of motion could be explained only by reference to a material factor, capable of communicating its action to what it moves. Aristotle arrived at this factor in a highly original way: he asserted that all movements resulted from the combined action of a resisting and a motor force. In natural motion the resistance was due to the *elementa media* (air and water); in violent or forced motions it was due to the weight of the moving body. In the first case the air that a heavy body must traverse on its downward path is at rest in its natural place. Like any body at rest, it has an active and continuous aversion to all forms of displacement and opposes its own inertia to the

---

[159]For Galileo's own conclusions, see Chapter 6.

motion of the body. Since, therefore, the relation between the moving body and the medium is identical to the relationship between two bodies, one of which tries to impress a violent motion on the second while the second is in a state of rest, the resistance offered by the medium to the moving object may be likened to the resistance that one force puts up to another. Because the moving object is continuously subjected to the action of this force, it must constantly surmount it or else come to a stop. To the dynamic power of moving bodies, the medium thus opposes not a physical resistance, but its own dynamic. It is the latter that gives rise to the continuity of motion: the moving body cannot leap from one position to the next without first passing through a host of intermediate positions; it must successively traverse all the points constituting its trajectory. The medium, though opposed to the motion, is therefore a sine qua non of its very existence. Now, the necessary presence of a resisting force has important consequences. First of all, it compels the physicist to consider resistance in all his computations of the magnitude of motions. But it is above all the part the resistance plays in the actual production of motions which lends it its special significance. Modern mechanics knows the decelerating action of the medium in the form of friction, but this could not have been Aristotle's own view. Far from merely slowing down the motion, the resistant force, according to him, preserved its physical reality from one moment to the next, and this precisely by forcing the moving body to overcome the opposing force. In other words, the effect of the resistance on the motion is one of division and not one of mere subtraction. As far as Aristotle was concerned, therefore, velocity was not so much a dimension of the motion as a measure of the superiority of the mover over the resistance or, if you like, of the degree of penetration of the moving object into the medium.

Joined to the idea of the incompatibility of motion and rest, this interpretation of the *successio motus* once again proved the primacy of the cosmological order and of the theory of the elements. Because it was predetermined in its modalities, motion, even before its commencement, could never acquire the rank of a state or enjoy that essential prerogative of all states which is the ability to persist in the absence of external obstacles. Not that Aristotle divorced motion completely from the idea of conservation; he took the view that, thanks to its actualizing function and its restoration of order, motion helped the universe to preserve its proper constitution. But this type of conservation also implied the transient character of motion, except in the case of celestial bodies, which are themselves part of the cosmological order. Classical mechanics may therefore be called the exact reversal of the Aristotelian position, for instead of basing the cosmological order on an a priori structure, it founded that order on motion itself and then went on to confer upon the latter the conservation Aristotle had attributed to the former.

Attempt at a Descriptive Analysis

Although rooted chiefly in ontology and cosmology, Aristotle's analysis was nevertheless a first attempt to produce a concrete definition of both the chief aspects of local motion and also of the most important factors determining its production. A large number of these aspects and factors are already familiar to us, for they constitute an essential part of Aristotle's general treatment of "mechanics." In other words, no matter what their subsequent influence may have been, the propositions we are about to examine, far from being the results of inductive research, were merely the application and, as it were, the particularization of a set of general philosophical principles.

Aristotle's first attempts to apply these principles to the case of motion were classificatory. Altogether he made three such attempts, which, by and large, remained uncoordinated. His most fundamental distinction was unquestionably that between natural and violent or forced motions. Springing directly from his cosmological view, this distinction was of necessity bound up with his conception of the ontological function of motions. Aristotle himself summed up the general idea in the following words: "A body moves naturally to that place where it rests without constraint, and rests without constraint in that place to which it naturally moves. It moves by constraint to that place in which it rests by constraint, and rests by constraint in that place to which it moves by constraint."[160] How does this definition lead to the determination of natural and violent motions? By making it possible to decide whether or not the mover generating the motion is natural to the body under consideration. Natural motions pose no special difficulties in this respect, for they comprise the displacements of things "moved by themselves," that is, of living beings together with the downward motion

[160] *On the Heavens* I, 8, 276 a 23–26.

of heavy bodies and the upward motion of light bodies, since "this is precisely what it means for them to be light or heavy, namely, that they tend upward or downward, respectively."[161] Violent motions, which are mostly due to the action of living beings on inanimate objects, are more difficult to define. Nevertheless, two major categories can be distinguished according to whether or not the mover remains in direct contact with the moving body.[162] The first category can be subdivided into four types: namely *pulling* (in the sense that, for instance, a lever shifts a weight);[163] *pushing continuously*, as when a mover impels something away from itself and follows its motion, *carrying*, and *rotating*. The last two types are more complex than the first two, which enter into them as basic elements. It is also to pulling and pushing that we must attribute such movements as expansion, combination, separation, and so forth. The second category of violent motions is much smaller but poses a number of extremely difficult problems. It includes only the case of *pushing away* (that is, "when the mover does not follow the course of the thing it has moved") and of *throwing* (that is, "when the mover impels a thing to move away from it with a motion more violent than any which the thing thrown would naturally have").

In addition to this classification, based on the contact of, or the lack of contact between, mover and moved, Aristotle introduced another one, this one based on the nature of the trajectory. Just as violent and natural motions constitute two distinct classes, so *rectilinear* motions differ fundamentally from *circular* ones. This difference is rooted in the organization of the cosmos: simple rectilinear and circular motions appertain to distinct regions of the world and play distinct parts in them. More generally, motion in a straight line and motion in a circle are based on irreducible geometrical distinctions; for "local motion is divided into kinds if its path is divided into kinds."[164] This explains why Aristotle expressly rejected the idea of comparing rectilinear and circular motions by their respective velocities, explaining that this approach might lead to the conclusion that "there might be a straight line which would be equal to a circle, but a circle and a straight line are not comparable."[165] Thus, while motion in a straight line can by its

---

[161] *Physics* VIII, 4, 255 b 15–16.
[162] For this entire analysis see ibid., VII, 2, 243 a 15–244 a 14.
[163] Ibid., VIII, 4, 255 a 21–22.
[164] Ibid., VII, 4, 249 a 14–16; cf. ibid., V, 4, 227 b 14–21.
[165] Ibid., VII, 4, 248 b 5–6.

very nature take place only between contrary termini, circular motion can be continuous; moreover, while rectilinear motion cannot persist without interruption, circular motion can last indefinitely. Now the finite nature of rectilinear motion has a remarkable consequence. Imagine a heavy body impelled upward; since its (violent) motion will come to an end with the exhaustion of the agent, it will quite naturally make way for a downward motion. Now, Aristotle contends that at the precise point where the motion is reversed the moving body must be momentarily at rest, and this for a purely logical reason: one and the same point would have to be the *terminus ad quem* of one motion and the *terminus a quo* of another, both at the same time, which is contradictory.[166] Nor was this his only objection. In reversing its path, the moving object must actualize a point as the end of its trajectory, and only a rest can, as we know, actualize a point on a straight line. To assume that a point can be actualized as the terminus of a motion without coming to a halt would be tantamount to postulating that motion in general has the power of actualizing all the points along its trajectory, and we should be back with Zeno's paradoxes. However, though in perfect agreement with Aristotle's general analysis of continuous magnitudes, the thesis of an intermediate rest introduced a difficulty that was to encumber mechanics until the seventeenth century; we still find traces of it in the earlier writings of Galileo, and its elimination may be considered a measure of progress toward the construction of a unified science of motion. In any case, by refusing to compare rectilinear with circular motions, Aristotle conferred a paradoxical autonomy on the trajectory: far from treating it as the result of a motion, he held that the trajectory was no less predetermined than the world in which the motion took place.

Aristotle also classified motions by their regularity or lack thereof. This classification was by far the most confused of all three, and it is impossible to decide by what criterion he determined whether a particular motion was uniform (ὁμαλήσ) or nonuniform (ἀνώμαλοσ). He took at least two factors into account. The first and most important seems to have been the shape of the trajectory. A motion was said to be nonuniform if any part of its path taken at random could not be fitted upon other parts taken at random; in this connection Aristotle used the

---

[166]Ibid., VIII, 8, 264 a 16–21.

unfortunate example of a spiral. But his intentions were clear: uniformity and nonuniformity were concepts that owed their meaning to the path described by the moving body.[167] True, to this entirely external factor he added another, one that was more significant, namely, speed or rate (*το τάχοσ*) of a motion: "Movement at a uniform rate is uniform; at a nonuniform rate nonuniform."[168] But he immediately limited the scope of this observation by adding that the "differences of rate which accompany all the different species of movement do not constitute species or differentiae of movement."[169] In other words, though he considered it part of the distinction between uniform and nonuniform motions, Aristotle did not think that speed constituted a sufficient criterion. A passage in *On the Heavens* in which the successive phases of nonuniform motions are set out may help the reader in following his argument. All nonuniform motion was said to pass through three phases, namely, intensification (*ἐπίτασισ* or *intensio*), climax (*ἀκμή* or *summa velocitatis*), and retardation (*ἄνεσισ* or *remissio*).[170] Natural nonuniform motion (the fall of heavy bodies) involves only the last two phases, since the climax lies at the end point of the motion; in violent movement by pushing away (*ἄπωσισ*), as when a cart is impelled along an even plane, the climax will be at the source of the motion, and the motion will experience a gradual retardation; finally in the case of throwing, the climax will be at the end of the ascendent phase, itself subject to constant intensification.

We thus obtain six classificatory principles of local motion, which enable us to construct eight combinations, of which six correspond to real motions. Table 1 sums up Aristotle's general views on this subject.

From the foregoing remarks it follows that Aristotle believed that motion, as a simple process of actualization, could be conserved only in the presence of a mover. Its physical reality, moreover, required the presence of a resisting force thanks to which the action of the mover could be actualized from one moment to the next.

There is at least one case in which Aristotle thought that the definition of the moving force presented few difficulties—the case of living beings. Here, the mover was the soul, whose force not only was bound up with the moving body but resided within it. The only problem was the precise manner in which the soul acts. Had it the power to initiate

---

[167] Ibid., V, 4, 228 b 21–25.
[168] Ibid., 228 b 27–28.
[169] Ibid., 228 b 28–30.
[170] *On the Heavens* II, 6, 288 a 16 ff.

Table 1. Aristotle's Principles of Motion

1. Natural Rectilinear Uniform	No motion	3. Natural Curved Uniform	Motion of the celestial spheres
2. Natural Rectilinear Nonuniform	Spontaneous motions of heavy bodies	4. Natural Curved Nonuniform	No motion
5. Violent Rectilinear Uniform	Motions due to pulling, pushing, etc., under the action of a constant force	7. Violent Curved Uniform	Same as 5
6. Violent Rectilinear Nonuniform	Same as 5, but under the action of a variable force	8. Violent Curved Nonuniform	Same as 6

a motion by itself, or did it require the assistance of an agent? Aristotle's answer was rather surprising: observation showed that the movement of living beings need not always be attributed to their own actions: "Many processes are possibly, or perhaps even necessarily, set up in an organism by the environment, and some of these arouse thought or desire which, in turn, affect the organism as a whole."[171] Though it was difficult to explain precisely how the soul could be moved, Aristotle believed that the process could be likened to that of the "desirable which moves without being moved itself."[172]

The problem changes completely when we look at the motion of inanimate objects. Because of the special links between mover and moved, and also as a direct consequence of his construction of the elements according to their weight and lightness, it was only to be expected that Aristotle should have tried to base his definition of the mover, at least in the case of natural movements, on the physical composition of the

[171] *Physics* VIII, 2, 253 a 7–20.
[172] *Of Generation and Corruption* I, 6, 323 a 25 ff.

bodies concerned. In fact, his definition was based on two axioms: the axiom of an external agent: whatever moves is moved by some agency outside itself (or as the Schoolmen were to put it: *quidquid movetur ab alio movetur*); and the axiom of "togetherness"; the mover must always be in contact with what has moved (*oppportet movens esse cum moto*).[173] As we shall see, these two axioms were to introduce grave complications into the description of moving forces.

In the case of natural movements, the identification of the mover poses two distinct problems, which have to be solved in turn: to what general moving force must natural motion be attributed, and what is the cause responsible for the characteristic increase in its speed (one hardly dares speak of acceleration so early in the history of science)? Let us begin with the first problem. The moving force cannot possibly be identified with matter or form, because this would contradict the axiom of the external agent; moreover, as Aristotle himself was quick to note, the movements of inanimate objects could then no longer be distinguished from those of living beings, for we should have to grant that inanimate objects had the power of setting themselves in motion or bringing themselves to a stop as animals can, which is absurd.[174] Should the moving force therefore be identified with the properties of heaviness or lightness which determine the nature of the elements? But since these properties are merely the result of two preexisting points, namely, the top and bottom of the world, this explanation would involve recourse to actions at a distance, and this would contradict the axiom of togetherness. Posed in this way, the problem could not be solved, nor did Aristotle succeed in doing so. He merely gave us two brief indications as to the efficient cause of natural movements. In one passage the first cause of such movements is expressly said to reside in the heavens;[175] in the other the movement of an object is attributed to the incidental removal of an obstacle in its path.[176] However, these factors, which at best define the cause of natural movements, tell us nothing about the forces or magnitudes involved. In fact, when it comes to the quantitative evaluation of these forces, Aristotle, as might be expected, falls back on the weight

---

[173]For the first axiom see *Physics* VII, 1, 241 b 24; for the second, see ibid., VIII, 6, 259 b 3–20. The axioms apply equally to living beings.

[174]Ibid., VIII, 4, 255 a 5 ff.

[175]*Meteorology* I, 2, 339 a 30–33. This became the *causa generans* of the Schoolmen.

[176]*Physics* VIII, 4, 255 b 24–29. This became the *removens prohibens* of the Schoolmen.

of bodies as expressed by their physical constitution. Now, while this is admittedly a practical solution, it has the double disadvantage of being contrary to one of the axioms of the system and of reducing the mover to the moved.

As for the second problem (the regular increase in the speed of natural motions), it proved to be more intractable still, so much so that we can only speculate with Simplicius about a plausible solution. Aristotle himself, in a passage of *On the Heavens*, seems to link the increase in the rate of the motion of a heavy body to the fact that its weight becomes the more actualized the closer its path takes it to the center of the universe.[177] Does this mean that a heavy body grows heavier in motion, that the increase in its rate of motion is the direct result of its increase in weight? In that case the idea that speed is proportional to the moving force would be saved, but not the axiom of the external agent.

The situation is slightly less confused in the case of violent motions. Whenever throwing is not involved and the mover remains at one with the moving object, the motor is generally a muscular force, acting either directly or indirectly through a simple engine. The case is not nearly so clear-cut in the case of projection, for here a special effort must be made to save the axiom of togetherness of mover and moved. Plato had earlier offered an explanation, namely, that a moving body expels the air from its place, and that what is thus expelled keeps on displacing its neighbors by antiperistalsis, so that the path of the moving body is unimpeded.[178] Though he rejected this idea, Aristotle agreed that only the medium could ensure the continued motion of a projectile. But precisely how? In his rather vague solution, he seems to suggest that the mover stirs up the air or some other medium in its immediate vicinity, and that this medium is capable of being moved and of communicating its movement to something else.[179] In other words, the medium, while ceasing to be moved as soon as the mover stops moving it, nevertheless continues to function as a mover, though with ever-decreasing power, until it finally becomes exhausted. At this point the violent motion comes to a stop and makes way for the natural motion that will restore the moving body to its natural place. In this way, Aristotle was undoubtedly able to save

[177] *On the Heavens* I, 8, 277 a 27–28; cf. Simplicius, *In Aristotelis de Caelo commentaria*, Heiberg (ed.), Chap. VIII, p. 284. The same interpretation was also offered by Themistius and Joannes Philoponus; cf. Duhem, *Système du monde*, Vol. I, pp. 265 f.

[178] *Timaeus*, 79 a.

[179] *Physics* VIII, 10, 267 a 2–5, and more generally 226 b 27–267 a 21.

both of his axioms, but he failed to specify the source from which the medium derives its ability to receive and transmit the moving force. Did he take the view that the inherent lightness of the air was stimulated under the initial impulse of the throw and, in turn, caused it to stimulate the upward motion?[180] This solution, though in general accord with his basic assumptions, gave rise to keen controversies from the very start, so much so that attempts to improve upon it resulted in some of the most interesting studies of medieval physics.

We saw that in order to base the reality and notably the continuity of motion on firm physical foundations, Aristotle felt it necessary to consider the action of resisting forces in addition to the forces we have just been discussing. To that end, he distinguished between external resistances—for example, the medium in the case of natural motion—and internal resistances, capable of manifesting themselves by their tendency to restore the body to its natural place or by impressing upon it an active reluctance to leave that place. As A. L. Maier has so rightly pointed out, the first tendency is particularly marked in the case of vertical motion, and the second in the case of horizontal motion.[181] In heavy bodies, the internal resistance resides in the weight, so that the same cause can be a source of motion or a source of rest depending on the case. The roles that medium and rest play in resisting forces are thus opposite to the roles they play in moving forces.

We must now examine Aristotle's attempts, pursued so conscientiously by his successors, to evaluate the magnitude of motions. Since these attempts brought traditional mechanics closest to the idea of a quantitative analysis, it is important to define their limits. Now, Aristotle was convinced that all local motions, at least those falling into one and the same category, could be compared to one another by means of their respective speeds. "Every change," he wrote, "is evidently fast or slow."[182] Not that he tried to evaluate speeds directly; as we showed, he never succeeded in defining speed as a characteristic dimension of motion. Instead, he treated it as a general feature of the magnitude of motions, and hence as a simple means of comparing one motion to another. What he was mainly concerned to establish, therefore, was to what factors and to what physically determined concrete realities it is possible to ascribe

[180] *On the Heavens* III, 2, 301 b 17 ff.

[181] A. Maier, *Die Vorläufer Galileis im XIV. Jahrhundert*, p. 68.

[182] *Physics* IV, 14, 222 b 31–32.

the differences in the magnitude of motions which our senses reveal to us. This—and the point is worth stressing—was the only way in which Aristotle himself viewed the problem. However, he realized that the magnitude of a motion can be determined either by its effects, that is, by the distance traversed, or else by its causes. Still, it would be quite mistaken to think that Aristotle, by this distinction, was anticipating, to however slight a degree, the modern distinction between kinematics and dynamics. Motion being nothing in itself, he felt that it could not be made the subject of an independent investigation; moreover, as we said earlier, he made no distinction between motion as seen from the respective viewpoints of mover and moved. Thus when he estimated speeds by the effects of motions, he did so for one purpose only: to consolidate and if possible improve upon what estimates could be called from the causes.

Now the effects, that is, the distances traversed, are much easier to measure than are the causes, that is, the forces. And this was precisely what Aristotle bore in mind when he tried to determine the respective speeds of two bodies in uniform motion (he invariably reasoned about bodies, never about points). The greater speed, he believed, could be recognized by the fact that the faster of two moving bodies travels a greater distance in equal time, or an equal distance in shorter time.[183] Now, however summary it may be, this approach does in fact enable one to isolate among the effects an objective factor with the help of which it is possible to compare the speeds of different motions, or rather of different uniform motions, for nonuniform motions had never been properly defined by Aristotle, as witness his failure to specify their effects in any detail. Thus in the only passage with any possible relevance, he merely contended that a body moving at variable speeds might take a very long time to traverse a very short distance, for as its speed decreased, it might cover increasingly smaller distances in increasingly longer periods of time.[184] Moreover, this passage was merely intended to prepare a refutation of Zeno's arguments, so that the analysis of nonuniform motions must be considered one of the most obvious cracks in the conceptual edifice Aristotle handed down to his successors.

It was a natural consequence of his conception of speed that Aristotle should have tried to define the magnitude of motions primarily by refer-

---

[183]Ibid., VI, 2, 232 a 23–b 20.

[184]Ibid., VI, 7, 237 b 28–238 a 11.

ence to their causes. To that end he simply tried to isolate, from case to case, the moving and resisting forces and then to define the relationship between their respective speeds. Now, in the case of heavy bodies, it is possible to identify the moving force with the bodies' weight, for it is thanks to its weight that every heavy body has a tendency (ῥοπή) to move in a downward direction; its rate of progress is in direct proportion to its weight.[185] Aristotle's insistence on this point leaves us in no doubt about his views on the matter: of two homogeneous bodies, the larger will always be the faster;[186] of two bodies with the same volume but made of different substances, the lighter will also be the slower.[187] As for the resisting force, which, as we saw, is represented by the medium, it could apparently be evaluated by the density of the latter, to which the velocity must be in inverse ratio.[188] These conclusions can be summed up by means of a very simple formula: in natural motion, $v = W/M$, where W is the absolute weight of the body, and M the resistance of the medium.[189]

The case of violent motions does not differ basically, even though different factors must be chosen to determine the order of magnitude of the speeds. The moving force is no longer the weight of the body but the force responsible for the pull, push, or throw; naturally the speed must be directly proportional to that force, though Aristotle himself was quite unable to measure it. The resisting force too no longer resides in the medium but in the weight of the moving body. Once again, Aristotle's conception can be expressed with the help of a simple formula: in a violent motion $v = F/W$, where F is the initial force and W the weight of the body. It should, however, be added that this formula applies only under certain conditions. Thus the ratio F/W must always be greater than unity, or else there would be no motion; it must also be constant, or else the speed would not remain uniform. Moreover, the force must exceed a certain minimum,[190] for, as Aristotle explains, if the weight of a

---

[185] Ibid., IV, 8, 216 a 11–16; *On the Heavens* I, 6, 273 b 30–274 a 2.
[186] *On the Heavens* IV, 2, 309 b 12–15.
[187] Ibid., IV, 1, 308 a 29–33.
[188] *Physics* IV, 8, 215 a 24–b 10. The shape of the body must also be taken into account; cf. *On the Heavens* IV, 6.
[189] This is, of course, an anachronistic way of putting it. Aristotle himself could only have compared the velocities themselves, and hence could only have arrived at such formulas as $V/V_1 = W/W_1$ and $V/V_1 = M_1/M$. It follows that motion in the void is impossible: if $W = 0$, the ratio W/M becomes indeterminate.
[190] Cf. the case of the drop of water; *Physics* VIII, 3, 253 b 13–19. This mistaken view was shared by all classical mathematicians, especially by Pappus.

body were halved, it would cover the same distance in half the time, or twice the distance in the same time; if the force and the weight were both halved, the times and distances would remain the same, but if one decreased the force and not the weight, it would be impossible to say that the force would impel the motion in any particular amount or in any particular period of time.[191]

In short, it would be a grave mistake to describe these formulas as "laws"; not only do they fail to show us how to proceed from the study of the effects to that of the causes and to express both in the same analytical language, but the Aristotelian analysis of the causes did not even purport to discover how the moving force engenders motion or how the resisting force slows it down. Nevertheless, the relations we have set out are of some importance, for they can be combined into a single formula, $v = F/R$, where F is the moving force associated with the weight in the one case, and with muscular exertion in the other, and R the resistance of the medium or the weight itself. We are in no way trying to gloss over the difficulties of this attempt to correlate natural with violent motions when we maintain that this formula provides a correct enough expression of what Duhem has called the fundamental law of Peripatetic dynamics, namely, that a constant force impresses upon a body a uniform speed directly proportional to the force itself and inversely proportional to the particular resistance that body encounters. As Duhem has put it: "If a certain force (ἰσχύσ) or power (δύναμισ) moves a certain body with a certain speed, twice that force or power would be needed to move the same body with twice that speed."[192]

In the preceding pages we have tried to produce a rational reconstruction of Aristotle's mechanics by showing on what foundations it was built and from what sources its most important conclusions were drawn. If we adopt the viewpoint of all those who, for so many centuries, accepted it at face value, we shall see that Aristotle's contribution can be divided into three distinct groups of principles and closely connected propositions: (1) a set of propositions deriving directly from Aristotle's general conception of motion and constituting the *intrinsic analysis of local motion;* (2) a second set of propositions deriving from cosmology,

[191]For this entire analysis, see ibid., VII, 5, 249 b 30–250 a 28.

[192]P. Duhem, *Etudes sur Léonard de Vinci*, Vol. III, p. 58; see also his "De l'accélération produite par une force constante," p. 861.

Table 2. Scheme of Aristotle's Mechanics

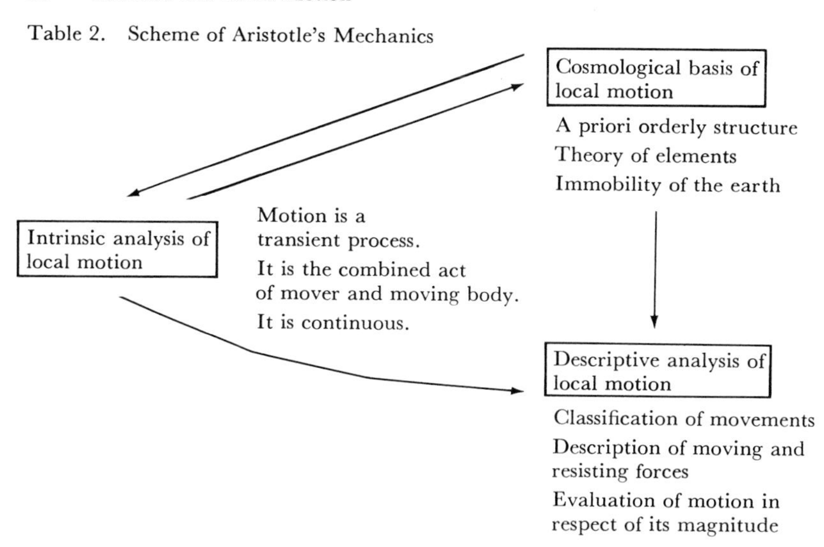

determining both the framework and also the bodies without which motion is inconceivable, and constituting what we may call the *cosmological basis of local motion;* and (3) a third group of propositions whose aim it was to describe local motion in its effective modalities and constituting, as it were, the *descriptive analysis of local motion.* In Table 2 we have tried to schematize the structure of Aristotle's mechanical ideas and to show how their elements relate to one another.

There is little doubt that this system was highly coherent. Its most remarkable aspect, however, was the manner in which it fitted the analysis of motion into a cosmological frame, or more precisely into the only such frame as at the time was compatible with both the demands of reason and the evidence of the senses. Much more than the vague agreement of Aristotle's conclusions with the facts (for who has ever seen fire rising into the lunary sphere or bodies dropping at a rate directly proportional to their absolute weight?), it was the solidity of his cosmological substructure that ensured the longevity of Peripatetic mechanics. It was this aspect, for instance, that was responsible for the persistence of so unfruitful a dogma as the theory of weight and lightness, a dogma that for centuries stood in the way of the general adoption of Archimedes' theory of specific gravity. It alone explains why certain conclusions of Aristotle's mechanics could be questioned by some of the greatest minds (for instance, those concerning moving forces or motion in the void) without bringing the whole edifice down. It finally explains why the Aristotelian doctrine of motion was able to survive the demise of his ontology, so much so that even Galileo still felt impelled to refute it step by step, at a time when the substantial forms had long since lost their significance. The rise of classical mechanics was therefore dependent not simply on the emergence of new concepts and a reinterpretation of the experimental evidence but also on the demolition of an edifice whose various sections were cemented so firmly together.

2 The Fourteenth-Century Tradition

If, as so many references tend to show, Galileo's approach to mechanics is difficult to grasp without some acquaintance with Aristotle's natural philosophy, it cannot, a fortiori, be appreciated in isolation from the Schoolmen's contribution. That this fact has been generally overlooked is largely Galileo's own fault. To begin with, he held up the "superhuman genius" of Archimedes against the authority of Aristotle in order to mark his determination that all his own discoveries should be based on mathematics; moreover, by failing to mention any Peripatetic contributions other than those of Aristotle himself and only very occasionally of some of his own contemporaries, he apparently tried to disown any debt he might have owed to the Middle Ages. Nevertheless, a number of references in his earlier writings and his use, in the treatment of both motion and equilibrium, of propositions in flagrant contradiction to the letter, though not the spirit, of Aristotle's doctrine, show clearly how greatly he was influenced by other factors. Thus the publication of his *Juvenilia* in the late nineteenth century[1] proved beyond any doubt that the contribution of Archimedes was by no means Galileo's sole inspiration.

All these remarks would have remained so many generalizations without the remarkable attempts by Pierre Duhem and Anneliese Maier[2] to resurrect the work of fourteenth-century natural philosophers.[3] Today we know that within two generations, at Oxford and Paris, kinematics, or the science of pure motion, was transformed more radically than at any time since Aristotle or before Galileo. At Oxford, the revival of mathematical studies led, in 1320, to the emergence of an entirely new language especially adapted to the analysis of motion, thanks largely to the writings of Thomas Bradwardine, William Heytesbury, Richard Swineshead (the famous "Calculator"), and John Dumbleton. Their contribution, which was widely acclaimed by their contemporaries, was soon afterward developed at Paris, where Jean Buridan and his pupils

[1]*Opere di Galileo Galilei*, Vol. I. *Juvenilia* was the title Favaro gave to Galileo's student notebooks, written in the years 1584–1586 at Pisa. Their contents will be examined at the end of this chapter.

[2]Special mention must also be made of the work of Professor Marshall Clagett of the University of Wisconsin. In his *Science of Mechanics in the Middle Ages*, Professor Clagett has presented numerous previously unpublished fourteenth-century texts, from which we shall be quoting at some length.

[3]Thirteenth-century contributions to statics will be examined in Chapter 3.

Albert of Saxony, Marsilius of Inghen, and Nicole Oresme formed the nucleus of a school whose influence spread swiftly to all parts of Europe and persisted until Galileo's day.[4]

Now, while few modern writers question the role of fourteenth-century concepts in the genesis of Galileo's contribution, they are far from unanimous in their assessment of its importance. The wide scope of the innovations made at Oxford and Paris, the broad attack on Aristotelian ideas, and also the fact that many seventeenth-century authors themselves used the same language as their fourteenth-century precursors, make it very difficult to decide, without closer scrutiny, whether these contributions were mere modifications of Peripatetic mechanics or the first faltering steps of modern science. But, difficult though it may be, that decision can be made, provided only that we do not look upon the Mertonians and Parisians as mere precursors of Galileo and instead try to fit their most important writings on kinematics into the philosophical framework that supported them.

Analysis of Motion by Its Effects

It was chiefly their study of motion with respect to magnitude that made the contribution of fourteenth-century physicists so highly original. To begin with, they replaced Aristotle's vague formulations[5] with a pre-

---

[4]Thomas Bradwardine, who flourished in 1321–1349, was the author of the *Tractatus proportionum seu de proportionibus velocitatum in motibus* (1328) and possibly of the *De Continuo*. William Heytesbury, who flourished from about 1330, wrote the *Regulae solvendi sophismata* and the *Tria predicamenta de motu*. Richard Swineshead was the author of the *Regulae de motu locali* and of the famous *Liber calculationum* (written between 1328 and 1350). John Dumbleton, the least-known member of this group, wrote the *Summa de logicis et naturalibus*. All these men taught at Merton College; thus they are known as the Mertonians. For more complete details of their contributions, see M. Clagett, *Science of Mechanics in the Middle Ages*, pp. 199 ff.

Jean Buridan, Rector of the University of Paris in 1328 and again in 1340, wrote the *Summulae logicae*, the *Questiones super libros quattuor de caelo et mundo*, and the *Questiones super octo physicorum libros Aristotelis*. Commentaries on Aristotle's *Physics* and *On the Heavens* were also written by Albert of Saxony (Rector of Paris University in 1357 and of Vienna University in 1365), and by Marsilius of Inghen (Rector of Paris University in 1361 and 1375 and later of Heidelberg University (cf. E. Gilson, *History of Christian Philosophy in the Middle Ages*). Nicole Oresme, Grand Master of the College of Navarre in 1356, Master of Theology in 1362, died as Bishop of Lisieux in 1382. In addition to the *Tractatus de configurationibus qualitatum*, he was the author of a book on money, the *De mutationibus monetarum*, of the *Questiones in libros Euclidis Elementorum*, of the *De proportionibus velocitatum in motibus*, and of two French works, the *Livre de divinations*, which was an attack on astrologers, and a *Traité du Ciel et du Monde*.

[5]See Chapter 1.

cise language. Thus Bradwardine distinguished expressly between the study of motion *sub specie causae motivae* (that is, dynamics) and that of motion *sub specie effectus* (that is, kinematics) and noted that the proportions obtaining between speeds may be determined by reference to either causes or effects.[6] Hence, there appeared two distinct fields of analysis: *Penes quid attenditur velocitas tanquam penes causas? Penes quid attenditur velocitas tanquam penes effectus?* (How can speed be evaluated in terms of the causes? How can speed be evaluated in terms of the effects?) However, the originality of these fourteenth-century writers was not confined to a simple improvement of Aristotle's terminology. Aristotle, as we saw, had already remarked that it was possible to explain the differences in the speed of several moving bodies in terms of the effects, but because he never considered motion in its specific dimension, he had to content himself with elementary remarks on the subject. Fourteenth-century writers, on the other hand, devoted all their efforts to filling this gap, and the success of their endeavors was such that the analysis *quoad effectus* was quickly to surpass the analysis *quoad causas* in both importance and significance. This success was due largely to their treatment of speed as an intensive magnitude, which enabled them to base the analysis of motion on its spatio-temporal effects.

Posing the problem of intensive magnitudes is in many ways tantamount to posing a riddle. Not that it is particularly difficult to identify them: aside from such extensive magnitudes as length or weight, which can be measured directly, Aristotle had already distinguished magnitudes that, though they can increase or decrease, do not lend themselves to direct measurement and serve as qualitative attributes of substantial forms. Of these, the best-known examples were the prime qualities set out in *On the Heavens*—hotness and coldness, dryness and wetness—to which we might add color and brightness.[7] However, great difficulties arose with attempts to analyze increases and decreases in these magnitudes—or their *intension* and *remission* as such changes were then called. Take the intension of heat. It must have a fixed intensity or degree at any one instant, yet no sooner has one degree been reached than it is superseded by the next, and heat, which can never have more than one

[6]Bradwardine, *Tractatus*, p. 64. Richard Swineshead expressed himself in similar terms; see Clagett, *Science of Mechanics*, pp. 208 and 245.

[7]Medieval writers also treated the theological virtues as intensive magnitudes.

fixed degree, always assumes the form of an indivisible reality. Nor does it help to treat heat as an extensive magnitude, for such magnitudes can increase or decrease only by the addition or subtraction of parts. When heat becomes more intense, however, the fact that at any given moment it can only have one fixed degree and no other makes it almost impossible to express the increase in quantitative terms: what possible significance can be attached to the addition of "parts of heat" when the process of increase or intension involves the destruction of successive degrees by definition and when the present degree can never be equal to the arithmetic sum of the preceding ones? Similarly, it is impossible to explain the decrease, or remission, of heat by removal of distinct parts. For what is one to make of increases or decreases in which the degrees expressing the intensity of the quality are individualized and destroyed in turn, so that they cannot be treated as additions and subtractions of the ordinary kind?[8]

It seems to have been Duns Scotus who, by completing the Aristotelian analysis of local motion with his theory of the *ubi*, was the first to suggest a comparison capable of elucidating the problem of the intension and remission of intensive magnitudes. By introducing his *ubi*, he aimed first of all at the elimination of an anomaly in the science of change, namely, that it based the determination of changes in quality and quantity on factors residing within the body, but the determination of all cases of φορά on factors residing without the body. To remove this distinction and also to strengthen the unity of the concept of κίνησισ, it was posited that, as a body moves to a new place, an *ubi* becomes actualized in it, and also that local motion bestows upon the moving body a series of constantly renewed *ubi*.[9] Now there was an obvious resemblance between the succession of *ubi* in local motion and the succession of degrees in intension and remission, and Miss Maier has shown most convincingly how the idea of applying the intension and remission of intensive magnitudes to the case of local motion arose quite naturally in the first half of the fourteenth century.[10] Much as one *ubi* continually succeeds another in local motion, so one degree was thought to succeed another in intension and remission; again, much as one *ubi* is incompatible with

---

[8]The Schoolmen's general solution of this problem will be examined in due course; for the moment we shall confine our remarks to their treatment of local motion.

[9]P. Duhem, *Le Système du monde*, Vol. VII, pp. 304 ff.

[10]A. Maier, *Zwei Grundprobleme der scholastischen Naturphilosophie*, pp. 59–61.

its successor (*termini motus sunt incompossibiles*), so the degree characterizing an intensive magnitude at any one instant is continually destroyed to make way for the next.

Conversely, it follows that local motion might itself be due to the action of a quality in the same way that, say, heating or cooling involves the intension or remission of the quality of hotness or coldness. The identification of that quality posed no problem at all: it was bound to be that by which all philosophers since Aristotle[11] had compared different types of local motion, namely, by their speed or velocity. Moreover, hardly had it been conceived than the idea of treating velocity as an intensive magnitude had to take root, for a number of reasons. First of all, it crowned the Scotist efforts to unify the science of change: the theory of the *ubi* had placed the determination of local motion into the moving body, and the treatment of velocity as an intensive magnitude made it possible to attribute this motion, like all other forms of change, to the action of a quality. Moreover, this interpretation seemed justified by experience and observation. Oresme, for example, noted that the idea of differences in the intensity of motions was suggested directly by experience, and he even stipulated that "*intensio*" [in motions] is more clear and palpable as it is commonly used [than *extensio*] and so is prior in our cognition to *extensio* and perhaps with respect to nature."[12]

But no matter what the precise sequence of ideas, one fact is quite certain: by 1320 Oxford physicists were treating velocity as an intensive magnitude, subject to intension and remission. For them, it became the *qualitas motus* or *intensio motus*, and they went on to distinguish the quality of a motion (that is, its velocity) from its quantity (that is, the distance traversed). Thus Bradwardine asserted that the motions of bodies traveling through a given medium might "bear the same proportion" qualitatively and only differ quantitatively.[13] Marsilius of Inghen also contended that in motion, as in all phenomena due to the action of a quality, a sharp distinction must be made between extension and intensity. More subtly still, Oresme even went on to distinguish the quality of a given

---

[11]See Chapter 1. In what follows, the term "velocity" is of course used without vectorial connotations.

[12]Oresme, *Tractatus de configurationibus qualitatum* I, 3, as quoted in Clagett, *Science of Mechanics*, p. 370.

[13]Bradwardine, *Tractatus*, p. 118. The language of fourteenth-century writers was often ambiguous. Thus *motus* was variously used to refer to both motion and velocity; and *intensio* to increase or intensity. *Intensio motus* was also used to refer to the magnitude underlying motion, that is, to speed.

velocity from its quantity. These ideas were to prove extremely fruitful; above all, they made possible the direct study of local motion in its spatio-temporal effects through one of its specific dimensions. To Aristotle, speed had been of physical significance only with respect to motor and resistant forces; in other words, having defined velocity by the ratio F/R, he was bound to treat it as something external to motion. Now, once velocity was treated as an intensive magnitude, it became *coextensive* with motion, which could therefore be analyzed in its very process. Admittedly, the resulting conceptualization was far from mathematical: constructed as an independent entity, speed also received the status of a quasi-physical object. Its definition as *intensio motus* must therefore not be confused with its definition as the ratio of space to time but should be considered a translation into philosophical language of the intuitively grasped fact that local motion is akin to the process of remission and intension. For all that, the approach of fourteenth-century physicists constituted an immense advance over all that had gone before. In defining velocity as an intensive magnitude, the Mertonians and Parisians conferred a precise meaning on the idea of *changes in velocity*, and hence felt free to treat the latter as the cause and explanation of changes in motion. This made it possible, for the first time, to proceed to the study of nonuniform motions, a study whose repercussions could still be felt in the work of Galileo three centuries later.

Before they could tackle that study properly, however, our fourteenth-century writers had to take one further step: although their treatment of velocity as an intensive magnitude clearly introduced the idea of changes in speed, it did not yet enable them to describe or conceive of these changes as such. Lacking the appropriate mathematical techniques, the Mertonians and Parisians accordingly turned to philosophy, from which they borrowed the concept of latitude (*latitudo*).

This concept had been introduced by thirteenth-century philosophers, St. Thomas Aquinas chief among them, to mark the distinction between substantial forms and qualities; thus when they defined the latter by their "latitude," they wished above all to contrast the variability of qualities with the immediate perfection of forms. The term "latitude" was therefore more or less synonymous with capacity for change, and the problems it posed were purely ontological.[14] Fourteenth-century physi-

---

[14]Duhem, *Système du monde*, Vol. VII, pp. 484–486.

cists, in their turn, applied "latitude" to the description of intensive magnitudes, but they gave it two entirely new connotations. Intensions and remissions always result in differences between an initial and a final intensity, and it was in order to express these differences that the Mertonians first used the term "latitude," thus applying it not to the *capacity for change* but to the *process of change* itself. They accordingly spoke of velocities subjected to increasing or decreasing latitudes or variations,[15] and many of them used quantitative expressions from the outset.[16] However, the analysis of nonuniform motions demanded a further step for, indispensable though it was, the distinction between velocity and changes in velocity was still no more than a preliminary: what was needed in addition was an explanation of the precise nature of these changes. The originality of the Mertonians was that they succeeded in providing this explanation once again, though in a different sense, by recourse to the concept of latitude. Suppose a given velocity has been changing positively or negatively during a fixed period of time. We can then divide that period into equal parts and also divide the total change of velocity into corresponding fractions. If now we apply the term "latitude" to these elementary, increasing or decreasing, variations, we shall obtain an extremely accurate classification and description of motions. Leaving aside uniform motions, that is, motions without variation, let us now consider nonuniform motions, or difform motions, as they were then called. To begin with, let us imagine a difform motion involving a steady increase in velocity. In the simplest case, equal changes will be produced during each successive time interval; as these changes are uniform, the resulting motion can be described as being "uniformly difform."[17] Again, if the changes in the velocity are not uniform, that is, if the velocity increases by unequal latitudes (*latitudines*) in equal intervals of time, the resulting motion will be "difformly difform."[18] In that case it was possible to distinguish further between motions in which the

[15]The increase in latitude was referred to by the verb *adquirere* (*adquirere latitudinem motus*), and the decrease by the verb *deperdere* (*deperdere latitudinem motus*); see excerpts from the writings of William Heytesbury and John of Holland, in Clagett, *Science of Mechanics*, pp. 241, 249, etc.

[16]For instance, Heytesbury (Clagett, *Science of Mechanics*, pp. 277, 279, and 282–283), Swineshead (ibid., p. 298), and Dumbleton (p. 317).

[17]Heytesbury put it as follows: "Uniformiter enim intenditur motus quicumque, cum in quacumque equali parte temporis, equalem acquerit latitudinem velocitatis"; see *Regulae solvendi Sophismata*, as quoted in Clagett, *Science of Mechanics*, p. 241.

[18]"Difformiter vero intenditur aliquis motus, vel remittitur, cum majorem latitudinem velocitatis acquirit vel deperdit in una parte temporis quam in alia sibi equali"; ibid., p. 242.

changes in velocity were themselves subject to uniform variations (uniformly difformly difform motions) and those in which no law could be assigned to the changes (nonuniformly difformly difform motions). It is easy to see that the same remarks apply to decreases or remissions in velocity.[19] This classification was regarded so highly that it remained practically unchanged until the seventeenth century.[20]

Though it was an original contribution of the Mertonians, the theory of nonuniform motions found its completion in the work of Nicole Oresme. In defining and classifying nonuniform motions in accordance with the modes of increase or decrease in velocity during successive intervals of time,[21] the Oxford physicists had shown that they had some inkling, at least implicitly, of the idea of acceleration. However, such concepts as *latitudo*, *motus*, and *velocitas* (the first two of which, moreover, were used in several senses) often failed them when it came to explaining basic principles. Oresme surmounted many of these difficulties by introducing a new term, namely, *velocitatio*, which, he explained, had the same rela-

[19]Heytesbury gave the following definition of a uniformly difformly decreasing motion: "Et uniformiter etiam remittitur motus talis, cum in quacumque equali parte temporis, equalem deperdit latitudinem velocitatis"; ibid., p. 241.

[20]Galileo was familiar with this classification, and the term "difform motion" recurs several times in his writings. In his *Juvenilia* (*Opere*, Vol. I. p. 110) the definition was presented in slightly different, but basically similar, form. Thus a quality was said to be uniformly difform if the differences between the intensities of its successive parts were equal, which happens if the first part attains two degrees, the second four, the third six, and so on. A quality was said to be uniformly difformly difform if the differences between the intensities of its successive parts, though unequal, were nevertheless subject to a law of proportions, such that, say, the first part was like four, the second like six, the third like nine, and so on. Nonuniformly difformly difform motions were such that the differences between intensities of their successive parts obeyed no law of proportionality. The same approach had been used much earlier by Jacobus de Sancto Martino (*Tractatus de latitudinibus formarum*), whose geometrical representation was directly inspired by Oresme. See Clagett, *Science of Mechanics*, pp. 399–401.

[21]Apart from studying changes in velocity *quoad tempus*, to which we shall confine ourselves here, fourteenth-century physicists also studied changes in velocity *quoad subjectum* (or *quoad partes mobilis*). Considering a line (radius) in uniform rotation about a fixed center, and marking on it points equal distances apart, they showed that, because the differences in rates of rotation of these points were equal, the motion can be said to be uniformly difform (see Albert of Saxony's analysis in Duhem, *Système du monde*, Vol. VII, pp. 478 ff.). Albert proposed to reserve the terms "uniform" and "difform" for the study of motion *quoad subjectum*, and the terms "regular" and "irregular" for the analysis of motion *quoad tempus*. In fact, all these authors used the terms "uniform" and "difform" in a broad sense. Despite the concern of fourteenth-century writers (and Oresme in particular) with the study of motion *quoad subjectum*, we shall not consider it here, because its influence was very slight.

tion to *velocitas* as increases in velocity had to velocity itself.[22] His approach was similar to that of the Mertonians, but much simpler and more rational. Thus his assumption that the character of nonuniform motions depends on their *velocitatio* made it possible to distinguish between uniform *velocitatio*, which produces uniformly difform motion; uniformly difform *velocitatio*, which produces uniformly difformly difform motion; and difformly difform *velocitatio*, which produces nonuniformly difformly difform motion. Oresme defined the crucial role of *velocitatio* in his classification of nonuniform motions as follows: "Every velocity is intensible or remissible. Continuous increase (of velocity) is called *velocitatio*. And indeed this acceleration or augmentation of velocity can take place either more quickly or more slowly. Thus sometimes it happens that velocity is increasing while acceleration is decreasing; sometimes both are increasing at the same time. And similarly acceleration of this sort can sometimes take place uniformly or difformly in various ways."[23] Needless to say, Oresme's concept of *velocitatio* was a long way from the modern idea of acceleration; in particular, he failed to consider instantaneous changes in speed; by *velocitatio* he simply referred to the manner in which increases in speed occur throughout the duration of a particular motion. For all that, it seems legitimate to translate *velocitatio* by "acceleration"; while it did not yet describe the *rate of change in velocity*, in the sense that classical mechanics later defined it, Oresme's *velocitatio* nevertheless referred to variations in speed as such, and hence transformed the latter into a distinct object of thought.

Moreover, the treatment of speed as an intensive magnitude also made it possible to describe motion, still *ab intrinseco*, in quite a different way. Thus, in the case of a speed subject to intension, not only is it possible to give an overall description of the changes involved, but it also follows that the speed must have a fixed intensity at any one moment. Now, all fourteenth-century writers assumed that this was something essential for the study of variations in motion. Their language, however, was far from clear; thus while the Mertonians applied the term "degree" (*gradus*) to these instantaneous intensities, Oresme referred to them either as *intensio gradualis* or, giving that term yet another shift in

[22]Oresme, *Tractatus de configurationibus qualitatum*, II, 5, as quoted in Clagett, *Science of Mechanics*, p. 376. The verb *velocitare* was also used by Heytesbury, but on one occasion only and without further comment (*Regulae solvendi sophismata*, ibid., p. 241).

[23]Oresme, *Tractatus*, as quoted in Clagett, *Science of Mechanics*, p. 356.

meaning, as *latitudines*.[24] But no matter what the formula employed, it clearly served to introduce a new concept, that of instantaneous velocity,[25] which must therefore be counted among the major contributions of medieval science. Its spokesmen had come a very long way since Aristotle, who had never treated speed as a concept sui generis but had simply reduced the continuity of motion to the continuity of space. Moreover, since he considered moments as limits of time without duration, he had perforce to dismiss the very idea of instantaneous speed as a meaningless concept. The new conception of speed as an intensive magnitude completely altered this picture, for it enabled physicists to make a direct study of speed, and hence drove it home to them that the effects of a motion depend on the value which its intensity assumes from one moment to the next. However, here again we must avoid interpretations based on hindsight. In fact, the modern idea of instantaneous speed seems to express the magnitude of a given speed at a fixed moment. For the rest, however, the case is the same as it is for speed in general: as a pure conceptualization, the instantaneous speed of fourteenth-century writers was nothing but the translation of an intuitively grasped physical fact into a language borrowed from philosophy. And how, in fact, could it have been otherwise? Following more or less blindly in the footsteps of the Greek geometers, on whose fundamental ideas they wrote long commentaries,[26] the Mertonians and Parisians had no means of proceeding to a mathematical study of changes in speed; the essential instrument was not provided until three and a half centuries later by the introduction of the infinitesimal calculus. The only path open to them was therefore that of philosophical analysis, and it redounds greatly to their credit that they were able to profit from such rough and ready concepts as *intensio* and *latitudo*. But, important though it

---

[24]This particular usage of the term *latitudo* was rarely, if ever, employed at Oxford; at the most we can say that Heytesbury used certain expressions in which *latitudo* might have referred to the magnitude of speed as well as to its variations; cf. Clagett, *Science of Mechanics*, p. 242 (line 101). Oresme, by contrast, used *latitudo* to refer to the instantaneous magnitude of a given speed, and objected to all those who used it to describe overall changes in speed (cf. *Tractatus* I, 3, in Clagett, p. 371). In fact, the introduction of the concept of *velocitatio* brought the word *latitudo* into closer correspondence with its etymological origins. The word *latitudo* thus had at least three meanings in the fourteenth century, and their confusion makes it impossible to grasp the theories of both the Parisians and the Mertonians.

[25]The expression *velocitas instantanea*, though not unknown, was used very infrequently.

[26]Bradwardine and Oresme each wrote a *Tractatus de proportionibus*.

undoubtedly was, their contribution was bound to be severely limited. Between the medieval concept of instantaneous speed and the modern view there is the same difference that obtains between Oresme's *velocitatio* and the definition of acceleration as the rate of change in speed. In modern mechanics, the speed of a point at any given moment is the differential coefficient of space with respect to time, and the quotient expressing it, namely ds/dt is simply the limit of the ratio of the finite space Δs to the finite time Δt, as Δt approaches zero;[27] in other words, what is involved is a mathematical ratio defined by a certain operation, and nothing else. In the absence of this language, the medieval concept remained bound up with a qualitative and intuitive substrate—in this case the idea of intensive magnitudes—which gave it its meaning but also restricted its scope.

These qualifications must not, however, be allowed to detract from the fruitful manner in which fourteenth-century writers handled their new conception. It was through his attempts to describe speeds by their instantaneous magnitudes that Oresme arrived at his ingenious graphical representations of the main types of change. But above all, it was thanks to the same idea that the quantitative study of motion *quoad effectus* first took shape. Thus all the Mertonians no less than the Parisians devoted much effort to discovering the correct means of measuring speeds at a fixed moment and, though Swineshead still failed to bring out the uniform character of the motion by which this measurement could be effected,[28] Heytesbury was quite explicit on this point. "It clearly follows," he wrote, "that such a nonuniform or instantaneous velocity is not measured by the distance traversed but by the distance which would be traversed by such a point if it were moved uniformly over such or such a period of time at that degree of velocity with which it is moved in that assigned instant."[29] Oresme took much the same line: "I say that this degree of velocity is more intense or greater in an absolute sense, by which, during the same interval of time, motion acquires or loses a

---

[27]More precisely, if v is the instantaneous velocity, we have

$$v = \lim_{\Delta t \to 0} \frac{\Delta s}{\Delta t} .$$

In the case of rectilinear motion, where the velocity at a given moment is simply the derivative of space with respect to time, we could also write

$$v = \lim_{\Delta t \to 0} \frac{f(t + \Delta t) - f(t)}{\Delta t} .$$

[28]Cf. Clagett, *Science of Mechanics*, p. 245.

[29]Ibid., p. 241.

greater part of its own perfection."[30] Though these views were purely theoretical, they nevertheless showed to what great extent medieval attempts to arrive at a quantitative interpretation of nonuniform motions were a direct result of the new approach to speed.

Nor were such unprecedentedly clear descriptions and classifications of local motion the only consequences of the new method, as witness the following comment by Swineshead: "It ought to be known that the intensification of speed (*intensio motus*) is to motion in the same ratio as it is to the space traversed; for just as space is traversed thanks to motion, so motion is increased by the intensification of its speed."[31] All this suggests that if the changes on which the *intensio motus* depends could be expressed in quantitative terms, then it ought also to be possible to determine the order of magnitude of the displacement of a body in nonuniform motion. The very logic of their investigations thus forced fourteenth-century physicists to make the first attempts at establishing a clear link between speed, or changes in speed, and linear magnitudes. Because the methods they used were no less interesting than the results, we shall first take a brief look at the former.

Here there were important differences between Oxford and Paris. The Mertonians, the older of the two schools, were the first to express physical magnitudes systematically by means of literal and numerical symbols, for instance referring to fixed amounts of heat or speed as *a* or *b*. At the same time, they also assigned a numerical order of magnitude to changes in speed (which were said, for example, to vary from 0 to 8 degrees or to double their intensity in a given interval of time). This use of symbols, which, as they claimed, made it possible to transpose motions *in terminis*, led to a simplification and generalization of the arguments and at the same time formed the basis of a specialized language for dealing with physical problems.[32]

Once these conventions had been introduced, precise methods for evaluating the effects of the main types of change could be developed. Since they are the most regular, and hence the simplest, it is natural to begin with uniformly difform motions. The very manner in which fourteenth-century writers posed this problem introduced the concept of instantaneous velocity directly: they asked themselves whether a body

---

[30]Ibid., p. 355 (*Tractatus*. II, 3).

[31]Swineshead, *De Motu*, as quoted in Clagett, *Science of Mechanics*, p. 244.

[32]Possibly this language was introduced by Bradwardine; cf. A. Maier, *Die Vorläufer Galileis*, pp. 113–116.

whose motion is subject to a uniform change assumes one instantaneous value among many, such that, had it been moving at a uniform rate, it would have traversed the same total distance in the same time. Let us see how Swineshead dealt with this problem.[33] Let two bodies *A* and *B* be animated by identical uniformly difform motions, but one increasing and the other decreasing. *A* will experience a uniform increase in speed up to a terminal degree *c*, while *B*, starting from the same degree *c*, will experience a uniform decrease in speed until it attains the initial degree with which *A* started its motion. Let *e* be the mean degree, that is, the mean value of either variation. Swineshead then went on to quantify the solution on the assumption that the proportions obtaining between two successive symmetrical parts of two quantities must also obtain between these quantities themselves,[34] in which case the two bodies in uniformly difform motion must, for symmetrical reasons, cover the same distance in the same time as they would have covered in the course of a uniform motion with a speed equal to twice their mean speed, that is, $2e$. Swineshead's approach, though indirect, was altogether remarkable. Unable to express the rate of change of a uniformly difform motion in mathematical terms, he first substituted a uniform motion and then, by considering the hypothetical case of two identical uniformly difform motions, one of which increased while the other decreased, reduced a complex mechanical problem to one of simple arithmetic.

Progress was even more marked with uniformly difformly difform motions, that is, with motions in which the speed is subject to a uniformly difform variation. The solution offered by the Mertonians must be considered one of the earliest attempts to express intensive magnitudes directly in mathematical language. Take any interval of time during which a change in speed takes place; divide that interval into "proportional parts," that is, divide it in such a way that successive parts form a geometric progression. Now, since the ratio commonly employed was $\frac{1}{2}$ (*proportio subdupla*) the first proportional part was put equal to half the total time, the second to a quarter, the third to an eighth, and so on, the whole forming the series

[33]See Clagett, *Science of Mechanics*, pp. 298 f.

[34]Swineshead himself formulated this principle as follows (ibid., p. 300); "If two sets of quantities be compared term for term so that from each comparison arises the same proportion, then, if the sum of the first set is compared with the sum of the terms of the second set, the sums will have the same proportion as that of the individual term for term comparisons."

$$\frac{1}{2} + \frac{1}{4} + \frac{1}{8} + \cdots + \frac{1}{2^n} + \cdots .^{35}$$

Next, imagine a motion which, during the first proportional part of time, has the intensity a, and let this intensity be changed at an (arbitrarily) fixed rate during each of the successive proportional parts. The convergent series formed by the proportional parts of time (the limit of which was known to the Mertonians) can then be used to determine the order of magnitude of the total change in speed and hence of the space traversed. An example from Swineshead may serve to illustrate the procedure.[36] He considered the case of a fixed interval of time and divided it into proportional parts in the ratio 1:2; if the motion has the constant intensity a during the first of these parts (or in half the total time), its "quantity of perfection" will therefore be $\frac{1}{2}a$. In the second proportional part, the intensity will be doubled, in the third part it will be tripled, in the fourth part it will be quadrupled, and so on. Hence, the total variation can be represented by the series

$$\left(\frac{1}{2} \cdot a\right) + \left(\frac{1}{4} \cdot 2a\right) + \left(\frac{1}{8} \cdot 3a\right) + \left(\frac{1}{16} \cdot 4a\right) + \cdots + \left(\frac{1}{2^n} \cdot na\right) + \cdots = 2a,$$

or

$$a\left(\frac{1}{2} + \frac{2}{4} + \frac{3}{8} + \frac{4}{16} + \cdots + \frac{na}{2^n}\right) = 2a.$$

This time the method used was direct: by combining the regular increase in intensity with the summation of a convergent geometric series, Swineshead was able to quantify the variations characteristic of uniformly difformly difform motions, and hence to establish a clear link between the distance traversed and changes in speed. The crucial step from a purely conceptual and qualitative analysis to a quantitative analysis had been taken.

[35]In these *calculationes*, the time was no longer divided into equal parts, as it was in the treatment of uniformly difform motions. Hence, there was a clear discrepancy in the conceptual analysis of these motions and in the mathematical treatment of the changes involved. The method was nevertheless of great interest.

[36]Cf. Carl B. Boyer, *The Concepts of the Calculus*, p. 76.

It would of course be wrong to exaggerate the scope and accuracy of his method. To begin with, by expressing Swineshead's purely verbal account in symbolic form[37] we have made it appear far more incisive than it was; moreover, it should be remembered that, like Archimedes, fourteenth-century scientists had no clear idea of the concept of a limit. Unfamiliar with the modern idea of proceeding to a limit (and merely assuming that the summation of a series would be continued indefinitely), the Mertonians, no less than the Parisians, had to use indirect methods to obtain the final result of their convergent series—for example by showing that the latter resembled series with known limits, and treating them as such. Thus Swineshead established that the limit of the series

$$a\left(\frac{1}{2} + \frac{2}{4} + \frac{3}{8} + \frac{4}{16} + \cdots\right)$$

was $2a$ by demonstrating that the intension so expressed was equivalent to the easily computed intension represented by the formula

$$2a\left(\frac{1}{2} + \frac{1}{4} + \frac{1}{8} + \cdots\right) = 2a.$$ [38]

Based as it was on such trials and errors, this approach despite its inherent interest did not, therefore, lend itself to generalization but remained confined to particular cases. Nor could it lead to measurements: it expressed neither the real value of the speed of a motion nor the absolute value of its increase, but simply the order of magnitude of its variation. This was admittedly a great step ahead: since the distance is directly proportional to that order of magnitude, the method made it possible to determine the additional path covered as a result of changes in speed. For all that, the new method failed to transcend the traditional one (which was concerned with determining the order of a magnitude rather than with direct measurement), and hence cannot be said to have fallen into the province of mathematical physics. It is nevertheless true to say that the Mertonians opened up a new path in natural philosophy, and one that proved so successful that their *calculationes* spread rapidly to the rest of Europe, where they became so entrenched that at the end of the

---

[37]For typical excerpts from the *Liber calculationem*, see ibid., pp. 76 ff.

[38]This, at least, is our interpretation of Swineshead's argument as quoted in ibid., p. 78.

seventeenth century Leibniz was still able to assert that Swineshead was the first to introduce mathematics into natural philosophy.[39]

The Parisians, for their part, looked to geometry rather than to arithmetic for a direct link between distance and changes in velocity. Thus, at the start of his *Tractatus de configurationibus qualitatum*, Oresme asserted that every measurable thing except discrete quantities could be conceived "in the manner of continuous quantity," and hence be subjected to measure and proportion.[40] This, he believed, was particularly true of time and of the instantaneous intension of speed, the two magnitudes on which the representation of changes in speed depend. Both, as Oresme explained, could be represented by straight lines. "Although time and a linear magnitude are not comparable in quality," he wrote, "yet there is no ratio found between one time and another that is not also found between lines, and vice versa. . . ."[41] And what applies to time applies equally to intension: "Whatever kind of a proportion is found between one intension and another—among intensions of the same kind—a similar proportion is found between some one line and another and conversely."[42] With the help of geometrical representations, two incommensurable magnitudes could thus be compared, introduced into the same figure, and hence subjected to the same treatment.

Starting from these principles, Oresme went on to construct a system of great originality. Suppose we wish to represent changes in the intension of velocity during a fixed period of time. Since speed is a *qualitas successiva*, we may consider the time as the "support" of its changes, and as such represent it by a horizontal line or "longitude."[43] On each point of this line, we can construct a perpendicular, representing the "latitude" or intension of the velocity at that point—a line twice as high as another would then represent twice as great an intension. Now, Oresme believed that a large enough number of latitudes would allow one to correlate every type of variation to a geometrical construction, serving to *transpose its particular process of intension or remission in the order of extension*. As a result, every variation could be associated with a figure

---

[39]Letter to Theophilus Spizelius, April 7, 1670, *Opera Omnia* (ed. Dutens) V, p. 346; letter to Thomas Smith, 1696, ibid., p. 567.

[40]Cf. Duhem, *Système du monde*, VII, p. 535. Latin text in Clagett, *Science of Mechanics*, p. 368.

[41]Clagett, *Science of Mechanics*, p. 377 (*Tractatus* II, 8).

[42]Regardless of whether or not these intensions are measurable. Ibid., p. 368 (*Tractatus* I, 1).

[43]Ibid., p. 370 (*Tractatus* I, 3).

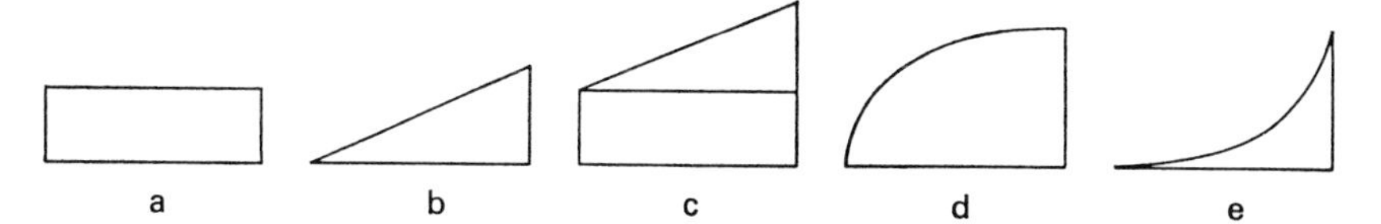

Variations of motions (left to right). (*a*) Uniform motion: (*b*) and (*c*) uniformly difform motions; (*d*) uniformly difformly decreasing *velocitatio;* (*e*) uniformly difformly increasing *velocitatio.*

on which its most characteristic properties were immediately apparent.

Thus uniform motion, that is, motion in which the speed does not vary, would be represented (see figure) by a rectangle (*a*); and uniformly difform motion, that is, motion subject to uniform acceleration, by a right-angled triangle whenever the minimum speed is zero (*b*) and by a trapezium whenever the minimum speed is greater than zero (*c*). Finally, curvilinear figures with a convex or concave "summit line" would serve to represent difform variations.[44]

Oresme's *configurationes* thus provided a clear visual image of the possible modes of variation in speed. However, his aim was not purely descriptive; two further steps enabled him to use his *configurationes* to correlate the distance traversed with the variation in speed. We saw that each point of the line representing the time (or longitude) could be associated with a segment or "latitude" expressing the intensity of the speed at a given moment; since the "longitude" was made up of an infinite number of points, the number of segments, too, had to be infinite. Now, by considering the total number of instantaneous velocities of the motion during its intension or remission, Oresme arrived at a new concept, namely, that of the amount of speed *(quantitas velocitatis)*[45] —in other words, the sum, or at least the synthesis, of all the intensities the speed has run through. In that case, the *configurationes* representing the total change in speed will also represent the (total) amount of speed resulting from that change. The new concept was admittedly far from clear, and Oresme's method of summing the instantaneous velocities of

[44]An extensive account of Oresme's system can be found in Duhem: *Système du monde*, Vol. VII, pp. 534–550. Miss Maier has published long extracts from Oresme's Latin text in *Zwei Grundprobleme der scholastichen Naturphilosophie*, pp. 88–109, as has Clagett (*Science of Mechanics*, pp. 367 ff.). In Appendix 1 the reader will find additional notes on Oresme's contribution and a discussion of Duhem's interpretation of his work.

[45]Clagett, *Science of Mechanics*, p. 375 (*Tractatus* II, 3).

a motion was, strictly speaking, devoid of both mathematical and physical sense. For all that, it stood him in good stead. Lacking the analytical techniques that would one day emerge with the creation of the infinitesimal calculus, he had no means of proceeding to a direct evaluation of the effects of changes in speed on the distance traversed, but he knew perfectly well that though these effects were dependent on the mode of variation, they were also dependent on the absolute value of the speed at any particular moment. In these circumstances, the best way of grasping the action of the process of variation was surely to try to take in *simultaneously* all the successive values the speed had traversed. By conserving and "summing" these values, would one not obtain a first overall indication of the magnitude and hence of the process of variation? Oresme's subsequent interpretation shows that he used the concept of *quantitas velocitatis* in just that way. In a passage, the importance of which has rightly been stressed by A. L. Maier, he did in fact implicitly liken the quantity of speed to the total speed, that is, to the maximum value assumed by the speed in the course of the motion.[46] Now, in the Middle Ages, the total speed was thought to be proportional to the distance traversed during a given period of time:[47] hence likening the quantity of speed to the total speed was tantamount to considering the space traversed under the action of a process of variation as being directly proportional to the area of the configuration expressing this process, or rather to evaluating the relative distance traversed during a given period of time by means of the relative magnitude of the part of the figure corresponding to that period. In short, Oresme

[46]Maier, *Die Vorläufer Galileis*, p. 129. Oresme's argument (*Tractatus* III, 8) took the following form: "Eodem modo si aliquod mobile moveretur in prima parte proportionali alicuius temporis taliter divisi aliquali velocitate, et in secunda moveretur in duplo velocius, et in tertio triplo, et in quarta quadruplo, et sic consequenter in infinitum semper intentendo, *velocitas totalis esset precise quadrupla ad velocitatem prime partis*, ita quod illud mobile in tota ora pertransiret precise quadruplum ad illud quod pertransivit in prima medietate illius hore." (Similarly, if any moving body were moved with some velocity in the first proportional part of a period of time so divided [into proportional parts] and were moved with a double velocity in the second [proportional part] and with a quadruple velocity in the fourth [proportional part] and continually increasing velocity in this way to infinity, the "total velocity" would be precisely quadruple the velocity of the first part, so that a body in the whole hour would traverse four times as much distance as it traversed in the first half of that hour.) See Clagett, *Science of Mechanics*, p. 381.

[47]It was said, for example, that two bodies had the same total speed when, at the end of the same period of time, they had traversed the same distance. See Maier, *Die Vorläufer Galileis*, p. 118.

had hit upon a way of expressing elementary mechanical problems in simple geometrical terms.

His method quite naturally gave rise to a host of commentaries, and many writers have considered it a form of integration, albeit of a purely graphic and intuitive kind.[48] Thus, in modern analysis, the area between the curve of a motion, the abscissa, and the (initial and terminal) ordinates represents the integral of the speed during the period of time interval under consideration; and the distance traversed is treated as an area, just as Oresme did. This is, however, where the resemblance between them and Oresme ends, and in any case there is no need to step outside the traditional methods to explain Oresme's procedure. In the method of *indivisibles*, to which, in fact, he confined himself, an area was thought to be made up of an infinite number of lines, just as a *configuratio* was thought to be made up of an infinite number of "latitudes," or instantaneous speeds erected on an infinite number of points along the "longitude"; hence, when he posited that an area was equivalent to the infinite number of lines of which it was composed, and also that the successive intensities of the speed were proportional to the distance traversed, Oresme had all the elements needed to identify that distance with an area. In so doing he kept well within the traditional frame; he was, in fact, using an argument with which even the Greeks had been familiar and one that would still be used in the seventeenth century by such men as Kepler and Galileo.

In any case, thanks to his concept of *quantitas velocitatis*, Oresme, like the Mertonians, was able to express the distance traversed by a moving body in terms of its speed or at the very least to evaluate the order of magnitude of that distance in terms of the magnitude of the total change in the speed. And because geometry provided him with a more striking, and above all with a handier instrument than did infinite series, it was only natural that his method should have been greeted with such enthusiasm and that it, rather than the *calculationes* of the Mertonians, should have presided over the birth of modern mechanics.

Calculationes and *configurationes* thus enabled medieval physicists to determine the ratio of the distance traversed to changes in speed. This determination may be said to represent the highlight of fourteenth-

---

[48]Ibid, pp. 130 f.

century mechanics and to provide clear proof of its great fruitfulness.

The most striking results undoubtedly bore on the case of uniformly difform motion. Fourteenth-century authors were concerned primarily to establish the "effect" of uniformly difform motions by reducing them to uniform motions of a fixed intensity. This idea originated at Oxford about 1330; Heytesbury expressed it as follows: "Whether it commences from zero degree or from some [finite] degree, every latitude [that is, change in velocity] as long as it is terminated by some finite degree and as long as it is acquired or lost uniformly, will correspond to its mean degree [of speed]. Thus the moving body, acquiring or losing this latitude uniformly during some assigned period of time, will traverse a distance exactly equal to what it would traverse in an equal period of time if it were moved uniformly at its mean degree [of speed]."[49] This proposition, which might be called the "law of the mean degree," was equivalent to Galileo's first theorem on uniformly accelerated motion: If Vo is the initial velocity, Vt the final velocity, and s the distance traversed, then $s = \frac{1}{2}(Vo + Vt) \cdot t$. Attempts to prove this proposition were as numerous as they were varied. We shall ignore the purely syllogistic proofs[50] so favored by the Mertonians and restrict our remarks to those based expressly on the quantification of the process of change, beginning with Swineshead's proof. The reader may recall that this proof was based on the correlation of two identical uniformly difform motions, one increasing and the other decreasing at the same rate. Swineshead contended that by virtue of the symmetry of the two motions the sum of their velocities at any given moment must always have the same value, that is, twice the mean value of either, or $2e$. By establishing further that the ratio of two successive symmetrical parts of two quantities also applies to these two quantities taken as a whole, he was able to show that the two *uniformly difform* velocities under consideration were equivalent to a uniform motion with a speed equal to $2e$. Now, since the two motions were symmetrical, it followed that either of them must "correspond" to a uniform motion of the same duration, and with a speed equal to e, that is, to the mean intensity of the variation.[51]

Interesting though it was, this proof nevertheless remained indirect, as did all the Mertonian attempts to prove the mean-speed theorem.[52]

---

[49]Clagett, *Science of Mechanics*, p. 277.

[50]Among them Heytesbury's proof; ibid., pp. 287–288.

[51]Ibid., pp. 263 ff.

[52]Cf. Dumbleton's proof, as discussed by Clagett (ibid., pp. 313 ff.).

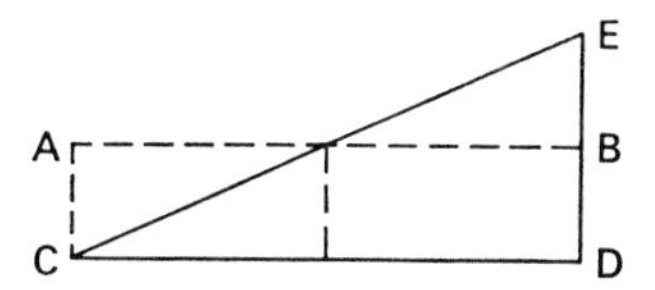

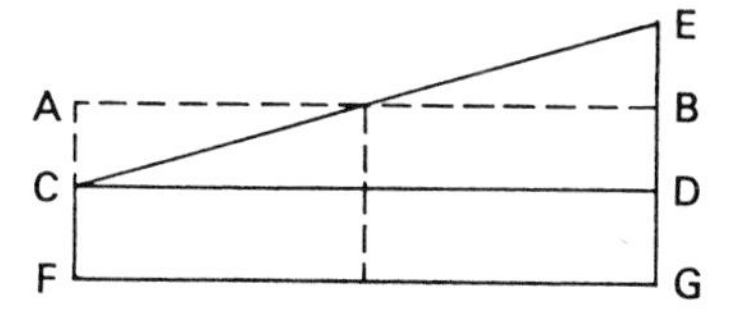

Configurationes representing uniformly difform motions.

Oresme, on the other hand, was able to provide a direct proof of that theorem. His formulation was very broad: "Every uniformly difform quality [the subject remaining identical] is equivalent to a uniform quality whose degree of intensity is the degree of the point in the middle of the same subject."[53] Thus, in the case of the *configurationes* representing uniformly difform motions (see figure), if the original velocity is zero, the triangle *CDE* by which Oresme represented the distance traversed by a body in uniformly difform motion must obviously be equal in area to the rectangle *ABDC*, representing the distance that would have been traversed in the same time interval by a body in uniform motion, that is, by a body whose velocity has the same intensity throughout its motion as the changing speed has at the middle instant in time ($t/2$).[54] The proof is just as direct in the case of a change in speed starting from an intensity greater than zero: the trapezium *CEFG* is quite obviously equal in area to the rectangle *ABGF*.

To this proposition most of our authors added a second one, concerning the ratio of the respective distances traversed in the first and second parts of the total time. Let us once again quote Heytesbury: "When a motion increases uniformly from zero degree [of speed] to some degree [of speed], the distance it will traverse in the first half of the time will be exactly one third of that which it will traverse in the second half of the time."[55] This "law of distances" could be derived from the law of the "mean degree," and this was precisely the procedure Heytesbury adopted when he contended that a uniformly difform motion lasting for an hour will during the first half of that hour correspond to a uniform motion with a velocity half that of the mean (velocity) of the total motion, and hence a third that of the uniform

[53]Oresme, *Tractatus* III, 7, as quoted in Clagett (ibid., pp. 379 f.). If the quality is speed, the subject must obviously be time. For a more general discussion, see Appendix 1.

[54]Ibid., pp. 379 f.

[55]Ibid., p. 288.

motion corresponding to the second half hour of the uniformly difform motion; it follows that the distance traversed during the first half of the motion must be a third of the distance traversed during the second half. Needless to say, the converse applies to the case of a uniformly decelerated motion.[56] As for Oresme, although his *Tractatus* did not say so explicitly, there is good reason to suppose that he deduced the second law from the simple analysis of the configuration representing a uniformly difform motion: the part of his figure corresponding to the first half of the time must obviously be equal to a third of the part corresponding to the second half of the time. This interpretation is borne out by the remarkable way in which he handled the "law of distances" in his *Questiones in libros Euclidis Elementorum*, where, considering the case of a uniformly difform quality, he asserted that if the line of extension (representing the subject or support of the quality) is divided into equal parts, the successive parts of the quality thus obtained (that is, the distances in the case of velocities) are as the series of odd numbers *(sicut series imparium numerorum)*. In generalizing the formulation of the Mertonians in this way, Oresme anticipated Galileo's second corollary to the law of uniformly accelerated motion.[57]

As might be expected, the analysis of difformly difform motions did not produce any propositions with anything like the same degree of accuracy—the infinite number of possible rates of change and the lack of general methods constituted too many obstacles in its path. Moreover, fourteenth-century physicists quite generally confined their efforts to determining the order of magnitude of the "effects" of uniformly difform changes (uniformly difformly difform motions).

Their basic procedure was, as we saw, to divide a fixed period of time into proportional parts, most often in the ratio 1:2. The simplest mode of change was clearly that in which the speed was doubled in each proportional part of the time. The result could be expressed by the series

$$\left(\frac{1}{2}a\right) + \left(\frac{1}{4}2a\right) + \left(\frac{1}{8}3a\right) + \cdots + \frac{1}{2^n}na + \cdots = 2a.$$

In the second half of the time, a body in uniformly difformly difform

[56]Ibid., pp. 280–281 and 288–289; cf. Swineshead's proof, ibid., pp. 300–301.

[57]See Clagett, *Science of Mechanics*, pp. 344 f. Galileo probably had no direct knowledge of Oresme's writings; in any case, he never mentioned Oresme by name.

motion would accordingly cover three times the distance it had covered in the first half of the time, and, as Swineshead remarked, the mean speed of the total motion would be that engendered by the change during the second proportional part of the time.[58] Fourteenth-century physicists also knew this type of change in another form: they considered the case of a body moving with uniform speed a during a given period of time, but so accelerated during the next, equal period of time that its speed was doubled in the first proportional part of the period, tripled in the second, quadrupled in the third, etc. The result was the series

$$(1a) + \left(\frac{1}{2}2a\right) + \left(\frac{1}{4}3a\right) + \left(\frac{1}{8}4a\right) + \left(\frac{1}{2^{n-1}}\,na\right) + \cdots = 4a.$$

The "effects" of this motion are the same as those we have just described, and the distance traversed during the second interval of time is equal to three times that traversed during the first interval.[59] It is also possible to imagine other types of change, either by assuming that the intensity varies in a more complex manner than by successive doublings during proportional parts of the time divided in the ratio 1:2 or else by considering successive doublings of the velocity during parts of the time divided in ratios other than 1:2.[60] Oresme examined a number of such cases, and this type of exercise quickly became an essential aspect of the Schoolmen's activity, though the increasing number and subtlety of the examples treated did not prove of any particular benefit to the science of motion.[61] The Italian universities which assumed the mantle of Oxford and Paris ensured the faithful transmission of all these doctrines, and a page in Galileo's *De motu*[62] bears witness to the longevity of the language introduced by fourteenth-century physicists.

---

[58]As quoted in Boyer, *Concepts of the Calculus*, p. 76. Note that this type of change produces the same result as a uniform change.

[59]Oresme (*Tractatus* III, 8, in Clagett, *Science of Mechanics*, pp. 381 f.) produced an interesting illustration of this series in the language of *configurationes*, showing that a given speed can be increased to infinity while the quantity of speed (and hence the total distance traversed) remains finite. This case, he pointed out, was identical with that of a finite surface, which "can be made as long and as high as you wish" without an overall increase in the area, "so long as the second dimension is decreased proportionally."

[60]For examples and references, see Boyer, *Concepts of Calculus*, pp. 86 f.

[61]As their complexity grew, these problems, moreover, ceased to be the exclusive province of mechanics and became increasingly of purely mathematical interest, giving rise to the first systematic treatment of infinite series.

[62]*De motu* (*Opere*, Vol. I), pp. 331 f.

Having tried to reconstruct the medieval contribution to the study of motion *quoad effectus* as objectively as possible, we must now try to assess its precise role in the history of mechanics and to define its limitations, the more so as the novelty of their ideas and the importance of some of their results has made it tempting to treat the Mertonians and Parisians as pioneers of modern mechanics. And, indeed, can it not be said that they were the first to express in clearly formulated language a great many of the ideas and principles that Galileo was to employ three centuries later and that, thanks to such concepts as *latitudo*, *velocitatio*, and *velocitas instantanea* and to such laws as that of the mean degree and that of distances, they laid the foundations of a science of motion which seventeenth-century scientists did no more than refine?[63] It is to this problem that we shall now turn our attention. Let it be said from the outset that our own interpretation is not free from the taint of equivocation. Thus an adequate account of the contributions made by the men of Oxford and Paris called for a systematic analysis, and this meant introducing a unifying thread that cannot, in all truth, be said to have run through the work of these fourteenth-century authors; more seriously still, it also meant abstracting their contribution from the general philosophical context of which it was an integral part. Now, it is our firm conviction that only this systematization and artificial abstraction could have led so many commentators to go beyond the facts and to maintain that fourteenth-century physics is of a piece with seventeenth-century mechanics.

To begin with, the ideas we have just examined constitute anything but a coherent whole. Thus the study of uniformly difformly difform motions was thought to be quite distinct from that of uniformly difform motions. Moreover, there was a similar lack of unity in the study of uniformly difform motions themselves. Certain texts may suggest that our authors deduced the "law of distances" from the "law of the mean degree," much as Galileo did, and as was probably required for a systematic study of these motions. In fact, the Mertonians as often as not deduced the second law from the first; thus, while Heytesbury followed the first procedure,[64] Swineshead took the opposite course;[65] Dumbleton on one occasion produced a direct proof of the law of the mean degree and on another employed the law of distances without bothering to

[63]This is the view of Duhem, Clagett, and many other authorities.

[64]See Clagett, *Science of Mechanics*, pp. 280 f.

[65]Ibid., pp. 300–302.

prove it.[66] As for Oresme, though he did offer a justification of the law of the mean degree, he failed to deduce the law of distances from it, as hindsight tells us he ought to have done.

This lack of coherence is not at all surprising when we consider the reason why fourteenth-century physicists were so interested in the study of motion *quoad effectus*. On this point, their writings are quite unambiguous: their preoccupation with qualities subject to change (for instance, with speed) was due primarily to their concern to determine the order of magnitude of the effects by reference to a uniform intensity. Thus they would say that a uniformly difform motion "corresponds" to the mean degree of its speed; and in the case of uniformly difformly difform motions they would establish the effect of the changes by reference to the uniform motion that they believed to take place during the first proportional part of the time. Now, this approach was in no way comparable to Galileo's attempt to base mechanics on geometry; far from leading to the direct study of "difform" motions, it merely tried to explain them by a *reduction* to uniform motions. Moreover, fourteenth-century writers never even questioned the characteristic attitude of Peripatetic science, whose sole concern, as we have stated repeatedly, was the determination of the order of magnitude of phenomena, and never the construction of mathematic models. It is because they failed to see the fundamental difference between this and Galileo's procedure that some commentators, seizing upon the resemblance that Oresme's treatment of the law of the mean degree bore to Galileo's first theorem of uniformly accelerated motion, felt entitled to conclude that the latter was a direct consequence of the former, and that seventeenth-century mechanics was an offshoot of fourteenth-century kinematics.

However, besides being limited by the traditional approach, the description of motion was also not an autonomous discipline. The reader should bear in mind that this entire chapter is based on what to a fourteenth-century man would have been an incomprehensible abstraction, namely, the study of changes of qualities exclusively *quoad tempus*. Now, in the fourteenth century this study, which is the only approach of interest to modern science, was, as we saw,[67] only one sector of a much wider field in which the analysis of changes *quoad subjectum*, or of the distribution of qualities in the subject, played at least

[66]Ibid. For the first case see pp. 319–321, for the second case see pp. 322–324.

[67]See footnote 21, above.

as important a role[68] and in fact preceded it. The law of the mean degree, for instance, was in a sense no more than the translation into the language of *quoad tempus* analysis of a much earlier *quoad subjectum* principle. This becomes particularly clear from an early thirteenth-century treatise by Gérard of Brussels.[69] Examining the case of a rotating radius, every segment of which must have a distinct linear velocity, he wondered whether there might not be one whose linear velocity was such that if the radius were allowed to move parallel to itself with the same speed the point had when rotating, it would traverse the same area in the same time it normally did in one revolution. This problem clearly involved the reduction of a uniformly difform motion to a uniform motion of translation. The required point, as Gérard showed, was the median point of the radius.[70]

Much as it was bound up with the *quoad subjectum* analysis, so the study of changes in speed *quoad tempus* was also thought to be inseparable from the study of changes in other qualities. Thus the concept of latitude was applied to hotness or coldness as well as to speed, and so was the whole classification of possible modes of change: a uniformly difform change in hotness was believed to merit the same attention as a uniformly difform change in speed. The same terms and the same methods and propositions were applied to both (for instance, the law of the mean degree or the law of the quantity of perfection.)[71] In other words, the idea of local motion was never divorced, either conceptually or methodologically, from that of qualitative change: fourteenth-century writers still fully accepted the Aristotelian concept of *κίνησισ*. Nor can it even be claimed that their treatment of speed as an intensive magnitude did more than consolidate the traditional point of view. Thus while it unquestionably allowed for the study of nonuniform motions, the fact that it led to the treatment of speed as a quality of motion also reinforced the prevailing belief that local motion was part and parcel of the science of change in general.

---

[68]Book I of Oresme's *Tractatus* was devoted to the study of changes in quality *quoad subjectum* (or *secundum partes mobilis*).

[69]*Liber de motu*. Gérard of Brussels lived about 1187–1260.

[70]In his analysis of other problems, Gérard tried to reduce curvilinear to rectilinear motions by trying to determine which point of the moving body was such that its speed, when applied uniformly to the entire body, would enable the latter to traverse the same distance in the same time. Cf. Clagett, *Science of Mechanics*, pp. 185 ff.

[71]To which we have been referring as the "law of distances."

There are thus many good reasons why the Mertonian and Parisian approach to the study of motion *quoad tempus* cannot be considered a direct anticipation of classical mechanics. There remains yet another—and much subtler—source of confusion. It concerns the intrinsic approach of fourteenth-century physicists to increases and decreases in speed. No doubt, they formulated and developed the idea of a total over-all quantitative evaluation of the effects of these processes; however, evaluating the order of magnitude of an effect is something quite different from upholding a quantitative interpretation of the processes of remission and intension as such. The problem we have to answer is therefore whether or not fourteenth-century quantifications of changes in speed were genuinely based on a quantitative view of the underlying processes.

Before we can do so we must first revert to the quandary in which the intension and remission of intensive magnitudes placed fourteenth-century writers, puzzled as they were by both the apparently insurmountable gap between intensive and extensive increases or decreases and by the successive destruction of the degrees through which a quality had to pass in the course of intension and remission. What were they to make of processes that could not be treated as so many additions or subtractions of parts and in which the degrees of intensity were continually particularized and destroyed? The Schoolmen came up with two solutions: while some of them adopted the Aristotelian view that there was strict separation between qualities and quantities, others tried to effect a reconciliation between these two "genera." The first group included a number of thirteenth-century writers, St. Thomas Aquinas, Henry of Ghent, and Durand of St. Pourcain chief among them. While St. Thomas attributed intension and remission to the increasing or decreasing participation of subjects in the particular qualities,[72] Durand, for his part, attributed intension and remission to the qualities themselves. The greatest and smallest aspects of an accidental form, or quality, "are merely the essence of the form at a certain essential degree of perfection or at another degree."[73] Thus Durand believed that heating was due, not to the acquisition of increasingly higher degrees of the quality of hotness, but to that of an infinite number of distinct qualities. "At

---

[72]For the doctrine of St. Thomas Aquinas, see Duhem, *Système du monde*, Vol VII, pp. 482–486; for Durand of St. Pourcain, see ibid., pp. 493 f.

[73]Ibid., p. 495.

every moment," Duhem explained in his commentary, "one type of heat is destroyed and another, more perfect, type is generated; nothing of the first subsists in the second."[74] The process of intension was therefore thought to involve a continuous series of generations and destructions. Or to use a comparison that was greatly favored in the thirteenth century, the growth of an accidental form was said to resemble the lengthening of the days in the summer, not by the addition during one and the same day of a shorter to a longer duration, but by the succession of increasingly longer days.

Duns Scotus and his disciples John of Basoles, Franciscus de Mayronis, and Franciscus de Marchia put forward a different solution. Following in the footsteps of William of Ware,[75] all these Scotists assumed that it was possible to divide every quality into a formal part and material parts. The formal part was that without which no form could be considered as such, and it persists as long as the form itself is preserved. The material parts referred to the successive values the form assumed in a given body—for example, its successive degrees of hotness.[76] Consequently, it was in respect of its material parts and of these parts only that the intension and remission of a quality was thought to take place. This distinction was sharpened further when the Scotists likened the relationship between form and its concrete degrees to that obtaining between a species and its individual members, and defined a degree of any form (for example, heat) as a "limited individual of that form."[77]

With the help of these concepts, the Scotists proceeded to treat the intension and remission of qualities as purely additive processes. *Intensio*, they contended, takes place *per additionem* and hence in a manner comparable to the increase of extensive magnitudes. "Charity," John of Basoles explained, "like all forms susceptible to *intensio* and *remissio*, increases by the apposition of a new degree of the same nature as the preceding degree."[78] A quality was accordingly thought to become more intense in the same way that a mass grows larger, that is, by the quantitative addition of parts.[79] This interpretation raised a number of special difficulties, for the idea of increases (or decreases) by the apposition of distinct parts had to be reconciled with the continuous existence of the

[74]Ibid.
[75]Maier, *Zwei Grundprobleme*, pp. 48 ff.
[76]Cf. Duhem, *Système du monde*, Vol. VII, p. 507 (quoting John of Basoles).
[77]Ibid.
[78]Ibid. This was also the view of Franciscus de Marchia; ibid., p. 519.
[79]Maier, *Zwei Grundprobleme*, pp. 79 ff.; cf. Duhem, *Système du monde*, Vol. VII, p. 506.

quality considered as an indivisible magnitude. The Scotists simply noticed this fact. Thus John of Basoles explained that "the new degree is added to the preceding degree in one and the same subject; accordingly it constitutes a unique individual of the same form, but one that is more perfect than that which went before."[80] At best, the Scotists asserted that the addition of one degree of heat to another is like the addition of one mass of water to another and not like the accumulation of separate stones.

In any case, this Scotist conception, which was incidentally shared by William of Ockham,[81] and universally known in the fourteenth century, highlights the problem with which we are here concerned. More precisely, when the Mertonians tried to express changes of speed in mathematical terms, and when Oresme introduced the concept of quantity of speed, may it not be argued that they adopted the same view of the intrinsic course of the processes of intension and remission as did the Scotists? Consider for instance, their analysis of uniformly difformly increasing motions. If they really thought that positive and equal changes in speed during successive and equal intervals of time were due to the regular addition of new degrees of speed to the total speed attained so that each of these degrees would correspond to what we call acceleration, then it would indeed be correct to say not only that fourteenth-century physicists had succeeded in conceptualizing the variation of speed as such (as witness, for example, their use of *velocitatio*), but also that they were able to reconstruct this variation from an elementary quantity of speed and hence considered the latter essential factor in the interpretation of motion. In that way, what is considered one of Galileo's greatest contributions would have to be antedated by almost three centuries.[82]

Our own view about this matter is quite straightforward: not only is there no single text that entitles one to assert that the Mertonian and Parisian treatment of changes in speed was in any way influenced by the Scotists, but there is also no shred of evidence that their general

---

[80]Duhem, *Système du monde*, Vol. VII, p. 507.
[81]Cf. ibid., pp. 103 f. and 511.
[82]This is implicit in Duhem's assertion that the Mertonian and Parisian contributions followed directly from the Scotist solution; ibid., pp. 530–533. Clagett takes much the same line (*Science of Mechanics*, p. 206) and moreover contends that Galileo's definitions of uniform motion and uniform acceleration had their "almost exact Merton counterparts" (p. 252).

methods and concepts were based on an implicit adoption of Scotist arguments. Duhem's view is based on hindsight: because a synthesis between Scotist speculations and the positive study of changes in speed ought to have proved fruitful, he and many other modern writers took it for granted that such a synthesis was in fact produced at the time. However, let us take another look at the *calculationes* of the Mertonians. Their aim, as we saw, was to correlate the characteristic change involved in a motion with the quantity of perfection acquired by virtue of that motion; now, although the summation of proportional parts to which they were thus led implied a fixed increase or decrease in speed, they paid no attention to the precise physical (or ontological) manner in which this increase or decrease was effected. Thus they assumed without further ado that the speed might be doubled in the first proportional part of the time, tripled in the next, and so on, but they never bothered to explain precisely how these successive intensifications come about. Moreover, their arbitrary choice of modes of variation proves better than anything else their total lack of interest in this crucial problem.

And what about Oresme? Can we assume that his concept of *quantitas velocitatis*, which played so important a role in connecting the distance traversed with the process of change, bore any resemblance to the summation of determined, elementary quantities of speed? This interpretation, too, is completely unjustified: Oresme never set out to sum elementary speeds which, added together would give the total speed, but intended only to sum the successive intensities (no doubt, infinite in number) through which the speed had to pass. What he was solely concerned with was to establish an overall picture of the variation in speed. That Oresme lacked even an implicit quantitative conception of the process of intension is borne out by one of the rare passages in which he dealt with the philosophical aspects of the problem: "There is neither plurality nor a real accumulation of degrees, as some have asserted. The quality persisting throughout the process of alteration is a successive quality, differing from one moment to the next."[83] Thus, if Oresme did have a definite attitude to the ontological problem of intension, it was in no way that of the Scotists but rather that of Durand of St. Pourcain,

[83] *Tractatus* II, 13, as quoted in Maier, *Zwei Grundprobleme*, p. 86: "Nec est ibi realis multitudo sive superpositio graduum, prout aliqui opinantur. Talis igitur qualitas toto illo tempore in quo est alteratio est una qualitas successiva et in qualibet ejus parte est alia et alia." Our own conclusion agrees with that of A. Maier, who did not, however, attach the same importance to the problem as we have.

who held that a quality subject to intension is totally renewed during each successive part of the time.

In the fourteenth century there was thus no quantification of the concept of acceleration in its proper sense; indeed we should be doing violence to that term if we claimed that the Mertonians and Parisians had any real understanding of the very idea of acceleration. They admittedly realized that a given speed can be subjected to various changes, and they described the latter; but they invariably treated the changes themselves as total processes, distributed over the entire duration of the motion—they never conceived of instantaneous variations produced by the addition of a fixed elementary quantity of speed. And this very failure must suffice to refute the rash claims that their contributions and Galileo's constitute a continuous whole. If we remember that both the Mertonians and Parisians were primarily concerned with evaluating the order of magnitude of the effects of motion and that they continued to treat local motion as a mode of change (κίνησισ), we cannot but conclude that it needed several further and decisive steps before the threshold of classical science could be crossed.

The Problem of Motive Forces

While Aristotle's contribution to the analysis of the effects of motion had been extremely meager, his attempts to distinguish between different types of motive force succeeded only in causing confusion among his heirs and successors. In particular, his assertion that a constant speed calls for the application of a constant force and his two basic axioms (of the external agent and of substantial contact) vitiated all attempts to arrive at a logical or even at a probable solution of the problem of motive forces. Concerned as they were to keep to the facts, fourteenth-century physicists were thus compelled, willy-nilly, to reexamine Aristotle's procedure.

Paradoxically, it was the subject of violent motions that first persuaded them to take this step. Aristotle's own explanation of projection was in fact so defective that serious objections to it had been raised even in antiquity.[84] It was not, however, until the fourteenth-century that a

[84]For the objections of John Philiponus, see Duhem, *Système du monde*. Vol. I, pp. 380 ff. It is almost certain that Buridan was unfamiliar with Philiponus's arguments. At the beginning of the fourteenth century, however, the Scotist Franciscus de Marchia put forward a thesis that while not completely opposed to Aristotle's, partly foreshadowed Buridan's; cf. Maier, *Zwei Grundprobleme*, pp. 161–200. Buridan himself never referred to that thesis.

reasoned critique of the traditional view was put forward, quite especially by Buridan, who set out logical as well as physical arguments against it. On logical grounds, he pointed out first that there was no reason to endow air with the capacity to move (every part communicating its motion to the contiguous one) and to deny that capacity to the projectile. Moreover, he thought it exceedingly difficult to reconcile the obvious resistance of the air with its alleged motive power.[85] On physical or experiential grounds, he objected that a smith's wheel or a top does not stop moving when the surrounding air is cut off by a cloth; that a flexible body bends backward when violently moved; that a stone, though more difficult to carry, travels farther than a feather; and that the air, moved by our hands never touches us as roughly as does a projectile launched with the same force. For all these reasons and for a host of others as well,[86] Aristotle's solution had to be rejected and another put in its place, namely, the famous "impetus" theory with which Buridan's name is associated to this day. Because it was based on a very simple principle and also because its history has been described at length by many other authors, we need do no more than outline its central idea. Basically, Buridan discarded the Aristotelian axiom of the external agent and invested the projectile itself with the powers Aristotle had attributed to the air, namely, the conservation of the force initiated by the mover. "Therefore, it seems to me that it ought to be said that the motor in moving a moving body impresses in it a certain *impetus* or a certain motive force of the moving body [which impetus acts] in the direction toward which the mover was moving the moving body, either up or down, or laterally, or circularly. And by the amount the motor moves that moving body more swiftly, by the same amount it will impress in it a stronger impetus. It is by that impetus that the stone is moved after the projector ceases to move. But that impetus is continually decreased by the resisting air and by the gravity of the stone, which inclines it in a direction contrary to that in which the impetus was naturally predisposed to move it. Thus the movement of the stone continually becomes slower, and finally that impetus is so diminished that the gravity of the

[85]Buridan's objections were put forth in his *Questiones super octo physicorum libros Aristotelis* (Question 12 in Book VIII). We are following the Latin text as published in Maier, *Zwei Grundprobleme*, pp. 207–214; Clagett's translation, *Science of Mechanics*, pp. 532–538, and Duhem's translation, *Système du monde*, Vol. VIII, pp. 203–205.

[86]See particularly Duhem, *Système du monde*, Vol. VIII, pp. 203–205.

stone wins out over it and moves the stone down to its natural place."[87] Buridan's "impetus" was therefore no more than a certain quality impressed in the moving body by the motor, directly proportional to the latter, and hence capable of prolonging the violent motion for as long as its effects are great enough to overcome the combined resistance of the air and of gravity, which invariably exhausts these effects after a finite interval.[88] Defined in this way, the impetus concept not only served to explain the motion of projectiles but a whole range of other motions that had posed so many intractable problems in traditional physics. Thus it helped to account for the continued motion of the smith's wheel after the smith had withdrawn his hand, the spinning of a top, the bouncing of a ball, the vibrations of strings, etc. So great was the persuasive force of the new concept that it was adopted by all of Buridan's disciples and successors, whose opinions differed in only two, but essential, points: whether or not the impetus was associated with an initial period of acceleration, and whether or not its exhaustion was due solely to the resistance of the air and the gravity of the moving body. Buridan himself rejected the idea of initial acceleration and, as we saw, attributed the exhaustion of the impetus exclusively to the resistance of the air and the gravity of the body. This was also the opinion of Albert of Saxony. Oresme, by contrast, accepted the idea of the initial acceleration and held that the impetus spent itself of its own accord.[89] It was in these two related but not identical forms that the impetus theory came down to the seventeenth century.

The whole theory could easily have been treated as a mere redefinition of motive force as applied to the case of violent motions, had not Duhem suggested a possible filiation between Buridan's contribution and certain major concepts of classical mechanics. Now, as long as impetus is simply treated as the motive force impressed in the body, where

---

[87]Clagett, *Science of Mechanics*, pp. 534 f.

[88]The impetus concept raised two philosophical problems that Buridan brushed aside in turn: first, by attributing the impetus to an intrinsic principle, was he not transforming the (forced) motion of projectiles into a natural motion? And second, was the impetus identical with motion and, if not, was it something successive or something permanent? Cf. Duhem, *Système du monde*, Vol. VIII, pp. 207–209, and Maier, *Zwei Grundprobleme*, pp. 225–227.

[89]For the views of Albert of Saxony, see Duhem, *Système du monde*, Vol. VIII, pp. 216–220. For Oresme, see Maier, *Zwei Grundprobleme*, pp. 236–290, and more briefly her *Die Vorläufer Galileis*, pp. 137–141. No mention of Buridan's theory was made at Oxford, except for a violent critique by Swineshead.

it becomes increasingly weaker, there is indeed no problem. But two of Buridan's own remarks suggest the possibility of quite a different interpretation. On the one hand he believed that the impetus impressed in the moving body and ensuring its continued motion was proportional to the speed of the mover; on the other, he also believed that it was proportional to the quantity of matter contained in the projectile. "In a dense and heavy body," Buridan wrote, "other things being equal, there is more of prime matter than in a rare and light one. Hence a dense and heavy body receives more of that impetus."[90] This argument is particularly important, since several of Buridan's other texts suggest that he may have had a precise idea of the concept of quantity of matter, so much so that Duhem felt justified in concluding that he must have equated the latter with the product of the density and volume, thus directly heralding the modern concept of mass.[91] The problem is therefore clear: Once the impetus, which was originally no more than the power of the motor, had passed into the moving body, did it not change its meaning when Buridan announced its double proportionality to the initial speed and to the quantity of matter? Did it not change the status of motive cause for that of a symptom or attribute of motion? Under the cover of the concept of impetus, an entirely new idea would thus see the light of day: that of a *dynamic dimension proper to motion*. True, the concept of impetus, interpreted in this way, confused two distinct notions—impulsion and kinetic energy, but Buridan's text was clearly explicit enough for Duhem to consider it a first expression of what Galileo later called the *impeto* of a motion, and Descartes the quantity of motion, concepts which despite their ambiguity must be treated as falling in the province of classical mechanics. "Galileo's mechanics," Duhem asserted, "was the adult form of a living science of which Buridan's mechanics was the larva."[92]

This view is, however, quite untenable, for a very simple reason: it is impossible to invest motion with a dynamic effect so long as matter has not been defined in terms of inertia. In order to explain why a body will continue to move over a certain distance after the motor has ceased to act upon it—in other words, before such concepts as kinetic energy and impulsion can have any meaning—it must first be shown that matter has a tendency to conserve its particular state of motion or rest;

[90]Clagett, *Science of Mechanics*, p. 535.

[91]*Système du monde*, Vol. VIII, pp. 211–215.

[92]Ibid., p. 200.

once a body has been set into motion, the problem is not why its motion should continue but why it should come to a stop. Traditional mechanics, whose approach to this subject Buridan never rejected, expressly adopted the opposite view. Far from being indifferent to rest and motion, matter (at least in the sublunary world) was thought to have a spontaneous tendency to rest and consequently an active and persistent resistance to motion.[93] The role of impetus was precisely to overcome that resistance, as expressed concretely in the gravity of the body; impetus alone, or rather the motive force whose permanence it ensures, enabled the projectile to continue on its path, and once the impetus had disappeared, the violent motion was directly transformed into the natural motion of free fall. Motion thus had no greater significance for Buridan than it had for Aristotle. That being the case, how could Buridan possibly have invested it with a dynamic dimension of its own? And more generally how could the impetus concept possibly have transcended the approach of traditional mechanics when its very raison d'être was rooted in a view of matter that was the precise antithesis of the classical one? Nor did Buridan himself ever think otherwise. "The impetus," he explained, was "a certain thing of permanent nature, distinct from the local motion in which the projectile is moved";[94] far from being a dimension coextensive with motion, it was simply a "quality naturally present and predisposed for moving a body in which it is impressed."[95]

However, Buridan did not simply apply his impetus concept to the case of violent motions; generalizing his first results, he went on to use it in reformulating the problem of the motive force in natural motions. Ever since Aristotle, the free fall of heavy bodies had posed two distinct problems: to what conjoined motor must that motion be attributed, and what factors were responsible for its characteristic acceleration? The first problem was the easier of the two to resolve. By attributing violent motions to an impetus, Buridan had clearly rejected the axiom of the ex-

[93]We cannot even agree with the view of Maier (*Die Vorläufer Galileis*, p. 68) that Aristotelian and Scholastic mechanicians were familiar with the idea of inertia in the case of bodies at rest, for the tendency to remain in that state was never attributed to their indifference to the state of motion. Maier was, however, quite right in rejecting Duhem's interpretation (ibid., pp. 142–147).

[94]Duhem, *Système du monde*, Vol. VIII, p. 208.

[95]Ibid.

ternal agent; hence, he had no difficulty in treating gravity, that is, the force residing within the moving body, as the motor responsible for the natural fall of bodies.[96]

That left the problem of acceleration. According to the traditional view, which Buridan fully shared, acceleration could be explained only in terms of a simultaneous increase in the intensity of the motive force. To account for that increase, Buridan asserted that the force of gravity, by its very action, engenders in the moving body a complementary force or impetus, which by adding its effect to gravity keeps reinforcing the latter. "One must imagine," he wrote, "that a heavy body not only acquires motion unto itself from its principal mover, i.e., its gravity, but that it also acquires unto itself a certain impetus with that motion. This impetus has the power of moving the heavy body in conjunction with the permanent natural gravity. . . . So, therefore, at the beginning, the heavy body is moved by its natural gravity only; hence it is moved slowly. Afterwards it is moved by the same gravity and by the impetus acquired at the same time; consequently it is moved more swiftly. And because the movement becomes swifter, therefore the impetus also becomes greater and stronger, and thus the heavy body is moved by its natural gravity and by that greater impetus simultaneously, and so it will again be moved faster; and thus it will always and continually be accelerated to the end."[97] This interpretation had a number of positive consequences. Thus, despite the links it preserved with the fundamental law of Peripatetic dynamics, it led to the rejection of the idea that the acceleration of freely falling bodies involves an effective increase in their gravity. Buridan himself was careful to explain that the acquired impetus represents a purely accidental form of gravity and hence a provisional addition that in no way alters the natural gravity of bodies.[98] With one stroke he had thus overthrown all the theories on which the acceleration of heavy bodies had previously been based, that is, either on the increase of their own gravity or on the action of the ambient air.[99] But the most

---

[96]Ibid., p. 285. "Gravity is the principal mover."

[97]Clagett, *Science of Mechanics*, pp. 560 f. Cf. Maier, *Zwei Grundprobleme*, pp. 219 f.

[98]Duhem, *Système du monde*, Vol. VIII, p. 285; and Maier, *Zwei Grundprobleme*, p. 220.

[99]Typical upholders of the first tendency included not only Aristotle and Themistius but also St. Thomas Aquinas, Albertus Magnus (Duhem, *Système du monde*, Vol. VIII, pp. 238–245), and Roger Bacon (ibid., pp. 68–73, 240–242, and 245–248). The second tendency, which went back to Averroës –(ibid., pp. 232–238), was upheld by Giles of Rome (p. 262), John of Jandun (pp. 263 f.), and Walter Burley (pp. 267–269).

important aspect of Buridan's contribution was unquestionably his explanation that the acceleration of heavy bodies was due, not to their greater proximity to the center of the earth, as had previously been thought, but to their increasing distance from their point of departure. Thus, in contrast to all his predecessors, he remarked pointedly that the reason why a falling body moves ever more swiftly is that "as it continues to descend, it moves further and further away from the starting point of its motion."[100] His attempt to describe the increase in velocity in purely physical terms made Buridan the first to formulate the free fall of heavy bodies in a meaningful way, and in that respect too his impetus theory played an important role in the history of mechanics.

However, even here it is far from certain that his theory really bore the germs of classical mechanics within it. Though it was undoubtedly a step forward, Buridan's approach to the problem of acceleration in natural downward motion was anything but clear-cut. In particular, it would seem that he constantly attributed a double role to the impetus, treating it not only as a result of the free fall but also as a cause of acceleration: during the first part of the motion, the body's weight engenders an impetus which, acting in conjunction with that weight, increases the speed of the moving body; as a result of that acceleration, the impetus is increased further, and so on. But, in using this argument, was Buridan not investing motion with a capacity for accelerating itself and thereby using a circular argument? Maier, who has examined this problem, refuses to admit that Buridan could have been guilty of so grave a logical lapse.[101] The error, she believes, is due to the mistaken view that Buridan saw a direct link between gravity and impetus or between impetus and motion, when all he was trying to establish was that gravity initiates a motion, thanks to which it engenders the original impetus in conjunction with which it leads to an increase in velocity. During the next period of the motion the increased motive force produces a faster motion and an impetus greater than the first, which, added to the product of gravity and the first impetus, helps to engender a further increase in speed and hence enables the body, during this second period, to traverse a greater distance than it did during the first. However, even this interpretation does not remove two major problems. First of all, Buridan did not explain how gravity, which resides within the body, could produce

---

[100] Ibid, p. 281.

[101] *Zwei Grundprobleme*, pp. 230–231. This difficulty was first mentioned by Buonamico in the sixteenth century; cf. A. Koyré, *Études galiléennes* I, pp. 37 f.

an impetus from within, as it were. But, above all, Buridan's theory, though apparently coherent when viewed very broadly, becomes almost unintelligible when we look to it in order to discover what happens during the first moment of the fall. Since a constant force can engender only a constant velocity, it follows that, as the heavy body first begins to move, its velocity must remain uniform for no matter how short a time. Whence the dilemma: either that consequence must be rejected and we must suppose that the motion is difform from the outset[102]—in other words, that it occurs in the absence of any impetus—and Buridan's entire explanation falls by the wayside; or else we must suppose that the original motion is uniform for however short a period of time, in which case again Buridan's theory cannot be said to explain the fall of heavy bodies.[103]

Albert of Saxony's attempts to develop Buridan's analyses fully corroborate this impression. Though he tried to determine the cause of the increase in speed, Buridan never asked himself how this increase came about. Albert, for his part, clearly realized that two problems remained to be solved. The first concerned the precise manner in which the increase in speed was effected: Was it increased arithmetically, so that the speed was doubled, tripled, quadrupled, etc., in successive equal parts of the motion? Or was it an increase by the addition of proportional parts, for example, in the ratio 1:2? He had no difficulty in rejecting the second alternative, for it led to the absurd consequence that a downward motion, even when extended to infinity, could never exceed a fixed limit.[104] The second problem involved the independent variable to

[102]Though he evidently did not know into what category to place it, Buridan had perforce to treat the free fall of bodies as a difform type of motion.

[103]Buridan's arguments were adopted almost word for word by Piccolomini, Scaliger, and Benedetti in the sixteenth century. Thus Benedetti wrote: "Every body moved naturally or violently receives in itself an impression and impetus of movement, so that separated from the motive power it would be moved of itself through a space of some time. For if a body is moved naturally, it will always increase its velocity, since the impression and impetus in it would be always increased. This is so because it has contact with the motive power perpetually" (quoted in Clagett, *Science of Mechanics*, p. 663). Duhem considered this passage an anticipation of the modern principle that "a motor of constant force does not determine a constant velocity, as Aristotle and the Peripatetics asserted it did, but a velocity subject to constant acceleration" (Duhem, "De l'accélération produite par une force constante," p. 885). In fact, Benedetti, no less than Buridan, merely used the idea of impetus to explain the intensification of the motive force.

[104]Duhem, *Système du monde*, Vol. VIII, pp. 294 f.

which the increase in speed must be attributed, and Albert plumped for space rather than for time.[105] Perhaps this unfortunate choice, which Galileo too was to adopt in his early years, was due partly to the persistence of the traditional opinion that the center of the universe was the natural place for all heavy bodies, so that their motion was dependent on their position in space. This explains the failure to appreciate that the fall of heavy bodies must be placed in the category of uniformly difform motions. This failure was not, moreover, confined to the fourteenth century; it persisted until Galileo's day, and the only one who might perhaps be said to have appreciated the correct nature of the free fall of heavy bodies was Domingo de Soto. We say "perhaps" because this sixteenth-century physicist merely alluded to the matter and moreover in a passage in which he applied the term "uniformly difform" to changes in uniformly difform motions[106] themselves.

Fourteenth-century discussions of motive forces were therefore of far less interest than the corresponding studies of changes in velocity. Although the impetus concept led to a reexamination of most of the problems to which we have alluded, it failed to solve them or even to pose them in the correct way. This is borne out once again by the fact that all fourteenth-century writers distinguished between a rectilinear and a circular impetus,[107] thus continuing to harp on one of the most characteristic theses of traditional mechanics. Intended as it was to describe appearances more faithfully, the impetus concept was therefore at best no more than a temporary solution.

Fourteenth-Century Mechanics and Cosmology

The reader would, however, have but an incomplete picture of fourteenth-century mechanical thought if he had gained the impression that it was purely confined to the study of motion *quoad effectus* and *quoad causas*[108] or to the examination of motive forces in the sublunary world. Another, and most original contribution of fourteenth-century writers was that they were the first to discuss problems relating to the celestial world in mechanical terms, and this reexamination of cosmological problems led

---

[105]Ibid., p. 295.

[106]De Soto's reference is mentioned in P. Duhem's *Études sur Léonard de Vinci* III, pp. 267 and 555–558. For the actual quotation see Clagett, *Science of Mechanics*, p. 555.

[107]Duhem, *Système du monde*, Vol. VIII, pp. 205–222.

[108]For the study of motion *quoad causas*, that is, for Bradwardine's law, see Appendix 2.

them to a number of conclusions that differed markedly from those of the traditional doctrine.

For many centuries the motion of the celestial spheres had been no more than a routine problem. Clearly adapted as it was to the principles of natural philosophy, Aristotle's solution did not seem to demand any real modifications. Since each sphere was in uniform and eternal motion, its motor, too, had to be eternal and immutable and since eternity excludes potential existence, it also had to be nonmaterial. This theory was never seriously put in doubt before about 1330, and philosophers as different in their other views as Franciscus de Marchia and William of Ockham accepted it as the only plausible explanation.[109]

The problem, however, assumed a new form as soon as Buridan, continuing his revision of dynamic ideas, tried to express it in terms of his impetus concept. At first sight, his endeavors seem hard to justify. If the impetus impressed upon a moving body depends on the body's mass, it follows that—to warrant this extension of the theory—celestial matter must at least be possessed of weight. Yet Buridan, who considered celestial matter a simple substance, formally rejected the idea of its identity with elementary matter.[110] However, he observed that if by matter we simply refer to what is "capable of being here at one moment and there at the next," then celestial matter is indeed matter and hence capable of receiving an impetus. This difficulty having been put out of the way, let us suppose that God, during the creation of the world, had impressed on each of the celestial spheres a rotational impetus corresponding to the motion he wished to communicate to it. We know that, according to Buridan, an impetus becomes exhausted solely by virtue of the resistance it meets from the gravity of the moving body and the friction of the air. Now, since no such resistances are present in the heavens, it follows that there is nothing to "corrupt or exhaust the impetus,"[111] so that once it has been impressed on the celestial spheres the latter must needs continue to move uniformly. Aristotle's Motor Intelligences thus lost their raison d'être: the motion of the universe preserves itself without outside assistance.

This explanation was unquestionably an improvement on the older conceptions, and we can see why Duhem considered it the first precise

[109]Duhem, *Système du monde*, Vol. VIII, pp. 325–328.

[110]Ibid., p. 335.

[111]Ibid.

affirmation about the conservation of motion. For when Buridan claimed that the impetus was immutable, was he not asserting that motion was immutable as well, seeing that "in a given moving body the changes in the magnitude of the impetus precisely follow the changes in the motion"?[112] In his approach to the motion of the celestial spheres, that is, to bodies in uniform rotation, Buridan would then have anticipated the modern concept of angular momentum[113] and the principle of its conservation by several centuries. Whereas we had to search the texts carefully in order to identify the impetus of terrestrial bodies with impulsion or kinetic energy, it seems rather easy to credit Buridan with the first inkling of the inertia of matter. Moreover, his conviction that the motion of the celestial spheres was governed by the principle of conservation seems to confer a new significance upon various remarks devoted more especially to rectilinear and circular motions on earth. Take, for example, the ninth question in his Commentary on Book XII of Aristotle's *Metaphysics:* "A projectile is set in motion whenever the impetus is greater than the resistance; that impetus would be conserved indefinitely were it not diminished and destroyed by some contrary thing that resists it, or inclines the moving body to adopt a contrary motion.[114] Elsewhere he noted that a smith's wheel having received an impetus might "well" continue to move indefinitely if the impetus were not subject to corruption or alteration.[115] These passages are so remarkable that we are bound to wonder to what extent Buridan realized that a motion can persist indefinitely under certain circumstances.

However, if we place Buridan's contribution in the general context of his physical system, it becomes quite impossible to accept this conclusion, for we should once again have to accept the fact that he was identifying motion with impetus. Now what we refused to grant in the case of projectiles, we cannot, a fortiori, grant in the present case, for if we did we should once again be ignoring the fact that Buridan invariably made a clear distinction, when treating the motion of celestial spheres no less than of terrestrial bodies, between motion and impetus, refusing, in particular, to ascribe to the former the permanent nature

[112]Ibid.

[113]Angular momentum is the product of the moment of inertia and the angular velocity of the body.

[114]Duhem, *Système du monde*, Vol. VIII, p. 335.

[115]Ibid., p. 337. This passage occurs in Buridan's *Questions on Aristotle's Four Books of the Heavens and the World.*

he attributed to the latter. For Buridan, the impetus was the cause of motion; it was the cause, not the effect, which according to him was alone conserved. To his mind, a motion could continue only thanks to the constant motive force which the impetus impressed upon the moving body. Let us repeat: Buridan's idea of impetus was in full accord with the axiom that a uniform velocity demands the presence of a constant motive force. Thus all those who see in his view that the impetus is conserved in the motion of the celestial spheres a rudimentary enunciation of the principle of the conservation of momentum are, in fact, trying to trace back the origins of modern mechanics to its very antithesis.[116] But, at the same time, it is even less justified to maintain that Buridan accepted the view that, in the absence of all obstacles, terrestrial bodies would conserve their motions indefinitely. To do so, he would have had to repudiate the most fundamental beliefs of traditional mechanics, namely, that there could be no motion at all in the absence of a resisting force, and a fortiori no conservation of motion. For only the presence of a resistance ensured the successive, and hence temporal, character of motion: without that resistance, if it did take place at all, the motion would have to be instantaneous, so that a moving body would find itself in a place without having had to go there. But how

[116]Maier has drawn attention to the causes of Duhem's mistake. The impetus concept appeared in the writings of most sixteenth- and early seventeenth-century writers, but in two distinct forms. Some of them, while fully accepting the explanation which this concept provided of the motion of projectiles, tried to reconcile the new concept with Aristotle's doctrine—among them Agostino Nifo, Alessandro Piccolomini, Julius Caesar Scaliger, Domingo de Soto (who claimed he had discovered the idea in the writings of St. Thomas Aquinas), and the Spanish and Portuguese Jesuits, particularly at the University of Coïmbra. Others, by contrast, considered the idea of impetus incompatible with Aristotle's doctrine and used it as a refutation of traditional mechanics, but only after subjecting Buridan's contribution to a radical transformation. This group included Telesio, Bruno, Benedetti, and above all Galileo. Duhem, in his analysis, considered only the second of these two currents; as a result he was misled into thinking that Galileo and the other founders of the new mechanics had stepped straight into the shoes of their fourteenth-century precursors. But as we have just pointed out, all sixteenth-century Schoolmen accepted the idea of impetus, with or without distinguishing its finer shades of meaning, and in no way deemed it incompatible with Aristotle's teachings. Having failed to appreciate this fact, Duhem also failed to realize to what great extent philosophers of the second current had transformed the impetus concept (cf. Maier, *Zwei Grundprobleme*, pp. 294–305). The impetus theory in the form Buridan had given it was incidentally so much of a piece with traditional mechanics that the seventeenth-century Jesuit Honoratius Fabri used it, in the name of the School, to refute Descartes' physics (ibid., pp. 312–313).

could a motion possibly be thought to conserve itself when the very factors on which its possibility depends are precisely those that must lead to its destruction? Hence, when Buridan remarked that a smith's wheel might "well" continue to move indefinitely once every possible resistance had been removed, he was in no way discovering the inertia of matter; he was simply stating an impossibility, much as Aristotle had remarked that if one granted the existence of the void, there was no reason why a body once set in motion should ever come to rest. This of course does not mean that Buridan was mistaken. A rotating wheel will continue to go around, and a ball traveling along a plane surface will continue to move in a straight line for quite some time; these are correct observations which the theory of impetus helps to explain despite the inherent tendency of matter to come to rest. However, between the vague realization that something may "well" be conserved and the a priori affirmation that motion *must* be conserved in specific cases, there is the precise difference that distinguishes an observation from a scientific principle.[117]

As Duhem, in particular, has shown, fourteenth-century writers were also familiar with many previously unknown cosmological problems. In 1277, Stephen Tempier, Bishop of Paris, condemned 219 articles drawn from the doctrine of Aristotle and Averroës, two of which bore directly on cosmological questions. Thus he rejected as being contrary to the faith the proposition that God could not have created a plurality of worlds *(Quod prima causa non posset plures mundos facere)*, together with the proposition that God could not have caused the world to move in a straight line because, in so doing, it would leave a void *(Quod Deus non possit movere Caelum motu recto. Et ratio est quia tunc relinqueret vacuum)*.[118] The first proposition clashed too sharply with all the convictions inherited from the past for its condemnation to lead to important dis-

[117]Nor does the application of the impetus concept to the motion of the celestial spheres mean, as Duhem has suggested (*Système du monde*, p. 340), that Buridan rejected the distinction between motion in the sublunary and the superlunary worlds. Thus he explicitly refused to identify celestial with elementary matter and indeed drew a sharp line between them (p. 333) when he conceded the conservation of impetus in the celestial spheres but not in terrestrial bodies.

[118]Duhem, *Système du monde*, Vol. VII, p. 205. Duhem has rightly stressed the weakness of the second argument—Aristotle never contended that the world occupied a void.

cussions.[119] The second proposition, by contrast, raised two new problems: How was it possible to conceive of a motion without either a *terminus a quo* or a *terminus ad quem;* moreover, if the rectilinear motion of the world did indeed create a void, ought one to grant the possibility of motion in the latter? Although he was not in the least concerned with science,[120] Bishop Tempier's decree—because it cited "divine cases" in direct opposition to several axioms of physics—nevertheless suggested the possible revision of various established ideas and so opened a breach through which new concepts could be insinuated.

Now, it was a fact that the rectilinear motion of the world was contrary to the Aristotelian view according to which all motions had to have an initial and a final terminus. Tempier's hypothesis—which dispensed with an initial no less than a final terminus—thus involved the paradoxical existence of a motion deprived of any function. In these circumstances there was only one way of endowing motion with any reality, namely, to define it, not in terms of what it actualizes, but directly as such. The Scotists were the first to come up with such a definition. Thus Duns Scotus himself affirmed that the motion of the eighth sphere was of a certain form that did not have to be considered "in relation to any other body, contained in it or containing it."[121] Developing this idea, Franciscus de Mayronis defined local motion as a process of *successio*, quite distinct from the acquired reality (or *successivum*).[122] While it was never treated as a *res permanens*, motion thus tended to become a more clearly defined entity. Thus to illustrate their conception, the Scotists made it a point to replace the usual definition of motion as a *forma fluens* (fluent form) by that of a *fluxus formae* (flux of form).[123]

However, it was in the form that the Parisians were to give it that the new conception first offered a real challenge to the traditional conception. Once again, the essential contributions came from Buridan and Oresme. Proceeding to a simultaneous examination of the motion of the eighth sphere and the rectilinear motion of the world, Buridan

[119] Duhem has described the essential aspects of the matter in Vol. IX of his *Système du monde*, pp. 363–430. The problem was examined almost as a matter of ritual by all the Schoolmen and was still discussed when Galileo was a student at Pisa; cf. *Juvenilia*, in *Opere*, Vol. 1, pp. 28 ff.

[120] E. Gilson, *History of Christian Philosophy in the Middle Ages.*

[121] Duhem, *Système du monde*, Vol. VII, p. 307.

[122] Ibid., p. 314.

[123] Maier has traced this interpretation back to Albertus Magnus, who examined it, only to reject it again (see *Die Vorläufer Galileis*, pp. 12–15).

asked himself whether a motion in which the moving body did not behave "sometimes in one way and sometimes in another toward some immobile thing or external object," that is, in which it did not necessarily shift its position with respect to another body,[124] might nevertheless be possible. His reply was formal: A body can move even in the complete absence of immobile bodies. In other words, motion was an intrinsic change of behavior (*se habere aliter ac aliter intrinsece*)[125] or, as Albert of Saxony put it, "to every mobile body in local motion we must attribute a flux or inherent motion which that body acquires successively."[126] Oresme seems to have gone even further. He began by rejecting several traditional definitions of motion. Thus he questioned the dictum that moving "is to become differently disposed in comparison with some place or some other body at rest." If, as Aristotle contended, rest is the privation of motion, it could play no part in the definition of the latter. Does moving then mean "to become differently disposed in comparison with another body, be it moving or not"? This cannot be the right answer either, for the motion of the world demands an infinite and immobile space in which it can move and in which it must be alone. Even if there were two worlds, nothing in the above argument would be changed, for, as Oresme explained, had God created only two bodies, *a* and *b*, and had He moved both in the same fashion, they would still be disposed in the same way with respect to each other, from which it follows that moving does not mean "being differently disposed in comparison to bodies."[127] What, then, is local motion? Oresme's answer was remarkable: "It is to become differently disposed in oneself relative to the *space conceived as immobile*; for it is in respect of such space that the *ysnelleté* (velocity) of the motion of the [moving] bodies must be measured." And further on, he went on to say: "And again it would appear that what is called local motion is something quite other than the body thus moved, for it is a fact that the body is always differently disposed in respect of space conceived as immobile."[128] Although he based his own on Buridan's definition, Oresme thus went a great deal further than the latter. For while Buridan had merely asserted that motion is a certain flux, intrinsic to the moving body and differing from it,

---

[124]Duhem, *Système du monde*, Vol. VII, pp. 353 f.
[125]Ibid.
[126]Maier, *Zwei Grundprobleme*, p. 22.
[127]For this entire discussion see Duhem, *Système du monde*, Vol. VII, p. 300.
[128]Ibid.

Oresme examined the relative motions of two bodies, without particular physical properties, through homogeneous and infinite space. This suggested to Duhem that Oresme had discovered Newton's concept of absolute space.[129] But might we not also claim with equal justification that, because he characterized motion exclusively by its relations to "space conceived as immobile," Oresme had departed so far from the Aristotelian tradition as to advance to the threshold of the modern view that motion is a state? For what other meaning can be attached to the motion of two bodies deprived of gravity and moving through infinite and neutral space than that it is of the same nature as rest?

No one can deny the importance and novelty of Oresme's contribution, which involved a capacity for abstraction and idealization as great as that embodied in Galileo's earlier arguments. Moreover, the conception of motion on which it was based was highly compatible with the spirit of modern mechanics. There is, however, one irrefutable reason why Oresme's approach cannot be considered a rudimentary expression of the modern idea of motion being a state: in discussing changes in speed, he never once made use of that definition. Though his *Tractatus* was more abstract than Aristotle's *Physics*, it nevertheless remains a fact that Oresme, like the Mertonians, confined all his arguments to finite motions or to changes limited in time; the law of the mean degree or the law of distances, as he himself used them, applied exclusively to finite motions whose termini could be accurately established. Hence, we are entitled to claim that the analysis of motion *quoad causas*, just like the analysis *quoad effectus*, in no way reflected the least appreciation that motion is a state and that Oresme never even conceived of the possibility of a motion that could perpetuate itself without the presence of one motive force or another.[130] This is borne out once again by his concern to demonstrate that motion was an accidental attribute of the moving body, albeit an *accidens intrinsecum*: "And such motion [= local motion] is an accident, not something that can be separated from everything else and exist in itself."[131] In other words, as far as Oresme was concerned, the translational motion of the world remained a purely theological problem and as such fell outside the province of physics. Hence it is probably mistaken to liken Oresme's "immobile space" to Newton's absolute space: when Oresme referred

---

[129]Ibid., p. 301.

[130]Oresme accepted Bradwardine's law without qualification; see Appendix 2.

[131]Duhem, *Système du monde*, Vol. VII, p. 301.

to space as "the immensity of God Himself,"[132] he was simply echoing Bradwardine's view that space is the infinite residence of the divine power.[133] We may therefore take it that Oresme's infinite, *immobile* space, benefiting as it did from God's immutability, simply provided the world's motion with fixed and successive termini, whose presence Oresme still considered indispensable to any real understanding of motion. Thus while Newton used absolute space and absolute time to lend objective foundations to the distinction between relative and absolute motions,[134] it seems most likely that Oresme introduced the idea of immobile space primarily as a substitute for the external references he had been forced to abandon.

As we have remarked, not only did Stephen Tempier's decree lead to a reexamination of the usual definition of motion; by asserting the possibility of the rectilinear motion of the world it also called for a new look at the problem of the void and of motion within it. Was this not an opportunity, at least in one specific case, for revising the traditional ideas on the dynamic conditions of motion?

And, indeed, most late thirteenth- and early fourteenth-century natural philosophers except for the Averroists,[135] took a fresh interest in the problem of the void. There were apparently two distinct attitudes. In the more prevalent, the void was treated as a simple consequence of the rectilinear motion of the world. This view was shared, for instance, by Henry of Ghent and Richard of Middleton,[136] and was expounded particularly well by Duns Scotus. Considering the problem posed by an ultramundane void, Duns Scotus tried to define the latter by reference to the space occupied by concrete bodies. He contended that treating the void as a space with positive dimensions would be tantamount to turning it into a body and, since the world had to be placed in it, to fitting two bodies into one and the same place. Since this is impossible, the void could be only a possible means for receiving positive dimensions,[137] in other words, potential space or, as Buridan was to put it, "something that does not have to yield its place in order to receive

[132]Ibid., p. 298.

[133]Cf. A. Koyré: "Le Vide et l'espace infini au XIVe siècle," in *Archives d'histoire doctrinale et littéraire du Moyen Age* (1949), pp. 45–91. We shall return to this question.

[134]By which he referred to such motions as are capable of revealing new forces.

[135]Including particularly John of Jandun; cf. Duhem, *Système du monde*, Vol. VIII, pp. 95 ff.

[136]For Henry of Ghent, see ibid., pp. 38 f.; for Richard of Middleton, pp. 41 f.

[137]For Duns Scotus, see ibid., pp. 45–47.

natural bodies."[138] Two writers, but only two, did not share this prudent approach: Bradwardine and Oresme. The former was the first to admit that an infinite void might surround the world. As a disciple of St. Anselm, he seems to have based this view on the very idea of God. Being perfect and omnipotent, God must be omnipresent throughout eternity, and since He is also infinite, it follows that an infinite space containing an infinite number of places must have existed even before the creation of the world. "Consequently," he went on to explain, "God is necessarily, eternally, and infinitely present in the infinite and imaginary space."[139] The same view was also expressed by Oresme, though in a more secular form and with Pythagorean and Stoic overtones. "And therefore, outside the Heavens, there exists an empty and incorporeal space, quite distinct from any filled and corporeal space; much as the duration we call eternity is of a different kind from the one we call temporal, even if the latter were perpetual."[140]

The acceptance of the void was, however, of physical interest only inasmuch as it went hand in hand with a reexamination of the idea of motion. In that respect, Stephen Tempier's decree reintroduced a situation that Aristotle had formally rejected, namely, that motion could take place in the absence of resisting forces. Now these forces, as we know, were needed to guarantee the successive character, and at the same time the reality, of motion. The very logic of the problem thus called for a redefinition of the *successio motus* and of its physical basis, and hence for a reassessment of the effective role of the medium; otherwise there seemed to be no way of avoiding the conclusion that the translational motion of the world must be instantaneous.

The solutions presented by the only two writers who genuinely envisaged the possibility of motion in a space devoid of resisting forces were far from radical. Thus when Duns Scotus had conceded the possibility of motion in the void, he contrived to reintroduce a resisting force or, more generally, a physical element on which the successive character of motion could be based once again. This element he discovered rather paradoxically in the body itself. "Every body, be it heavy or light," he explained, "has an active power which conveys it

---

[138] Ibid., p. 54. Buridan added elsewhere that he found the idea of the void unintelligible.

[139] Cf. Koyré, "Le Vide et l'espace infini," pp. 81–90.

[140] Duhem, *Système du monde*, Vol. VIII, p. 58. Extract from Oresme's commentary on Aristotle's *Heavens and the World*.

to the place it is destined to occupy by nature."[141] He therefore likened the rest of the body to a dead mass that had to be pulled along by the mover: in that way he was able to resurrect a resisting force which, though present within the body, nevertheless sufficed to ensure the succession of its motion.[142] William of Ockham, for his part, tried to locate the resistance in the space traversed or, more precisely, in its divisibility and extension. "Clearly, therefore," he explained, "the successive character of local motion is derived from the very fact that space is possessed of magnitude."[143] Not only did this explanation fail to introduce greater clarity into the problem, but Ockham, like Duns Scotus, was quite unable to consider the void outside the traditional framework, and hence to proceed to a methodical exploitation of the fruitful consequences of the new hypothesis.

The reader might object that, even in this restricted form, the new ideas served to uphold, against Aristotle, the possibility of motion in the void, until such time as this "divine case" would become a genuine physical problem. In this respect too the originality of the fourteenth-century contribution has been greatly exaggerated. The possibility of motion in the void had, in fact, been accepted, even in antiquity, particularly by John Philoponus, and fourteenth-century writers were familiar with his views, if not directly, then at least through Averroës' critique of Avempace's (Ibn Bajja's) commentary.[144] For Philoponus, gravity was a "self-subsisting quality of bodies," "the active cause of downward motion"; and, being the sole cause of that motion, it was capable of sustaining the successive nature of motion even without the help of a resisting force. Moreover, Philoponus believed that the medium had a purely decelerating effect, and "if there were such a thing as a void, distinct from all bodies, there would be nothing to prevent these bodies from moving in a certain time, or even from moving more slowly or more swiftly."[145] Relegated as it was to a marginal existence while Peripatetic physics continued to hold full sway, this theory never-

[141]Ibid., p. 82.

[142]Ibid., pp. 82–85. Duhem considered this solution an anticipation of the modern distinction between weight and mass. However, to Duns Scotus the "mass," far from ensuring the conservation of motion, would be precisely what impedes it.

[143]Ibid., p. 87. This solution was adopted by most Franciscans.

[144]Philoponus' commentary was apparently unknown until its publication in the sixteenth century. For Averroës' critique of Avempace, see ibid., pp. 10 ff.

[145]For John Philoponus, see Duhem, *Système du monde*, Vol. I, pp. 352 ff.

theless provided an alternative solution for those philosophers who were anxious to dispose of the objections to the void hypothesis. Moreover, it seems highly doubtful that the problem of motion in the void weighed heavily on the minds of fourteenth-century physicists, seeing that even the two staunchest defenders of the idea of the void, namely, Bradwardine and Oresme, failed to raise it in their writings. Hence, we have good reason to claim that Buridan's approach was the most typical of all: having analyzed the solutions offered by Aristotle and Avempace and having demonstrated their incompatibility, he declared that he was quite unable to choose between two solutions equally incapable of proof.[146]

The idea of motion in the void reintroduced yet another old question: If motion in the void was a possibility, would bodies fall in it with the same or with different speeds? John Philoponus and Avempace provided a general solution: If weight were the real mover and if the medium played no more than a retarding role, bodies would fall through the void with a velocity directly proportional to their respective motive powers, that is, to their weight. "Since bodies possess a greater or lesser downward tendency in and of themselves, clearly they will preserve this difference in themselves even if they move through a void. The same space will consequently be traversed by the heavier body in shorter time and by the lighter body in longer time, even though the space be void."[147] Was this a reference to the individual weight of each body or to its specific weight? Neither Philoponus nor Avempace gave an unequivocal answer to this question, but the context, particularly in the case of the former, suggests that what they had in mind was the individual weight.[148] Fourteenth-century writers also treated this problem in a highly confused way; Buridan, for example, refused to make any pronouncements,[149] and only Albert of Saxony stated clearly that "mixed bodies of the same composition" move equally fast through the void; this would make him the first to formulate the idea that bodies of the same specific weight but of different individual weights would fall in the void with the same velocity. However, since elsewhere he rejected Avempace's hypothesis of the void,[150] the importance of his remark should not be exaggerated.

---

[146]Duhem, *Système du monde*, Vol. VIII, pp. 100–102.

[147]Clagett, *Science of Mechanics*, p. 433, and Duhem, *Système du monde*, Vol. I, p. 364.

[148]Duhem, *Système du monde*, Vol. I, pp. 368 f.

[149]Duhem, *Système du monde*, Vol. VIII, p. 109.

[150]Ibid., p. 103.

Though the "divine cases" may therefore have had the positive effect of reviving certain anti-Aristotelian tendencies of classical thought and of persuading fourteenth-century writers to incorporate these tendencies into their arguments, the idealization of mechanical problems to which they gave rise was in no way comparable to that which Galileo introduced under the influence of Archimedes. Not only did his idealization fail to introduce any truly novel concepts, but, because it was limited to the hypothesis of the translational motion of the world, it in no way altered the fundamental structure and constitution of the latter, that is, it left the real basis of traditional mechanical thought intact. This probably explains why the speculations about the essence of motion or the void to which it gave rise did not impinge upon the sphere of genuine mechanical problems.

* * *

The existence, or rather the persistence, of a strict division between the different fields of research—this, we believe, is the chief impression to emerge from our analysis of fourteenth-century mechanical thought. Skillful though they proved to be in putting forward new solutions, fourteenth-century physicists seemed incapable of applying these solutions outside the narrow field with which they were concerned. Here some of our readers might object that fourteenth-century treatment of velocity as an intensive magnitude or the application of the concept of impetus to natural motions disproves our claim. However, in neither case was there a genuine extension of the old boundaries. The traditional concept of change (κίνησισ) in no way precluded the treatment of velocity as a quality: moreover, by the introduction of the idea of an accidental type of gravity whose own force combined with that of natural gravity, it was possible to apply the impetus theory to the free fall of heavy bodies without making the slightest modification to the fundamental law of Peripatetic dynamics. However, things are quite different when the fields are truly distinct. Had there been any real attempts to link the new interpretation of changes in velocity to the Scotist conception of *intensio* and *remissio*, Galileo's analysis of uniformly accelerated motion would indeed have been antedated by three centuries. Neither the Mertonians nor the Parisians ever gave up the idea that the natural motion of heavy bodies could be interpreted only by the analysis of motion *quoad causas*; consequently, none of them saw fit to apply the new theory of difform motion to the clarification of the possible changes in, and effects of, such motion. In the same way, they failed to forge any links between the analysis *quoad causas* as modified by Bradwardine[151] and the analysis of motion *quoad effectus*; and not a single one of them so much as suspected that the two could be expressed in a common language. Neither their attempts to redefine motion nor their hypothesis of the void, as a divine case, proved any exception to this rule: motion remained a finite process within set limits, and the need to introduce resisting forces among the dynamic conditions of motion was never seriously questioned. In this particular sphere, Oresme's approach was probably the most typical. Being of a remarkably liberal mind and far less encumbered with the letter of the Aristotelian dogma than any of his contemporaries—on several occasions he substituted Platonic and Stoic concepts—Oresme too never succeeded in fusing his various solutions and critiques into a coherent body of propositions; his opposition to Aristotelian mechanics thus remained fragmentary and, when all is said and done, rather ineffective. We shall illustrate this point by a final example: When discussing the problem of the plurality of worlds, Oresme expressed his clear preference for the Platonic conception of gravity as expressed in the *Timaeus* and hence apparently abandoned the idea of natural places;[152] for all that, however, he never felt compelled to reject the Aristotelian theory of lightness and weight or the distinction between natural and violent motions.[153] While they were admittedly far less concerned with ontological questions than their thirteenth-century predecessors, fourteenth-century writers thus kept just as closely to the traditional separation and isolation of related problems; the very fact that their physical ideas were invariably expressed in commentaries on the works of Aristotle proves once again that this school and the almost totally rigid frame in which they worked were bound to inhibit the renewing powers of their own contribution from the very outset.

Doubtless, this fact also explains why none of their ideas was definitely accepted. Indeed, with the exception of the description of the variations (*latitudines*) of speed and of their effects, none of the theories developed at Oxford and Paris was adopted unreservedly by even their contemporaries. Thus the old theories continued to receive the same

---

[151]See Appendix 2. Bradwardine replaced the equation $v = F/R$ with an expression of the type $v = \log(F/R)$.

[152]Duhem, *Système du monde*, Vol. IX, p. 405 (*Timaeus* 61 d–64 a).

[153]See the extract from his *Commentary on Aristotle's Heavens and the World* in Duhem, *Système du monde*, Vol. VIII, pp. 300 ff.

attention and reappeared regularly in all the commentaries; even the concept of impetus, widely accepted though it was, never succeeded in ousting its Aristotelian counterpart, and this applied equally well to Bradwardine's law and to the theory of the void and its possible consequences for the fall of heavy bodies. In short, there is every justification for the view that fourteenth-century physics (like its elaborations in the fifteenth and sixteenth centuries) had a noncumulative character: the introduction of a new concept into one particular sphere never led to the transformation of the rest or the reformulation of the entire problem. Fourteenth-century innovations were ranged round the Aristotelian and Averroist core like so many fresh layers round an old cake.

To assess the precise nature of the medieval contribution to mechanics, we can therefore do no better than to offer a term-for-term comparison of the respective solutions proposed by Aristotle and by fourteenth-century physicists. If we do so, we find that the most important innovations of the latter undoubtedly fell into the province of what we have called the descriptive analysis of motion. Thanks to the renewal of the study of motion *quoad effectus*, that analysis was modified on two distinct levels. First of all, it was enriched with a new definition and classification of nonuniform motions, the treatment of speed as an intensive magnitude, and the formulation of such new concepts as *latitudo*, *velocitatio*, *velocitas instantanea*, and *quantitas velocitatis*. The fruitfulness of these new concepts was such that instead of Aristotle's sparse remarks on the subject, it became possible to formulate the two "laws" of the mean degree and of the distance traversed for uniformly difform motions and also to evaluate the effects of some of the more complex variations associated with uniformly difformly difform motions. Moreover, although it had a less rewarding future, the new impetus concept contributed to the clarification of the problem of motive forces by suppressing the role of the medium in the case of violent motions and by suggesting a less arbitrary explanation of acceleration in the case of natural motions. But while it is correct to stress their positive aspects, we cannot ignore the piecemeal character of these fourteenth-century contributions. Even in the case of the descriptive analysis of motion many of the old ideas remained unchanged; thus Bradwardine's law served merely to refine Aristotle's own law; the idea of resisting forces survived, and so did the distinctions between natural and violent mo-

Table 3. Medieval Reactions to Aristotelian Concepts

	Aristotle	14th Century
1. Classification of motions	a. Natural and violent b. Rectilinear and circular c. Uniform and nonuniform	a. No change b. No change c. Completely changed description
2. The problem of dynamic forces	a. Motor forces –in natural motions –in violent motions b. Resisting forces	a. Analysis changed by the introduction of "impetus" b. No change
3. Laws and propositions	a. Scanty comments as to the "effects" of motion b. Formula of the type $v = F/R$ to characterize the motion *quoad causas*.	a. Law of the mean degree and law of distance applicable to uniformly difform motions b. Bradwardine's new formula of the type $v = \log(F/R)$

tions, and between rectilinear and curvilinear motions. Table 3 compares the respective contributions of Aristotle and his fourteenth-century successors.

However, it is chiefly by its approach to the *intrinsic analysis* of motion and to its *cosmological foundations* that the pre-Galilean character of medieval mechanics can be recognized. In particular, the intrinsic approach of Aristotle's *Physics* remained unchanged. Could it have been otherwise? Aristotle, as the reader will recall, refused to entertain the idea that motion was a state on the grounds that there could be no such thing as the movement of a movement; now the new distinctions between speed and motion and between speed and *velocitatio* contained an implicit refutation of that argument and could therefore have led Aristotle's successors to the realization that motion was indeed a state and not a process. However, not only did they fail to make this deduction, but they linked the only case in which the idea of motion as a state might have emerged with considerations alien to the description of changes in speed. Nor did any of them question the important idea of the unity of motion (that is, of the identity of motion when viewed from the respective sides of mover and moved, and hence the impossibility of making a clear distinction between the cause of motion and its dynamic dimensions), as Duhem wrongly contended they did when discussing the impetus concept. There was the same lack of progress in the treatment of the continuity of motion; thus while some fourteenth-century writers considered velocity a quality capable of change and thereby put themselves in possession of the means for analyzing the continuity of motion by reference to one of its specific dimensions, and not merely to space or time, none of them profited from this advantage to abolish the opposition between motion and rest, that is, to define rest as "motion of infinite slowness."

More significant still was the persistence of the traditional cosmological picture. Neither the idea that the world was based on an a priori order nor the fact that motion served to preserve that order was ever seriously challenged. Now it was in these ideas that the opposition between motion and rest found its ultimate justification, as did the finite nature of all motions, and the impossibility of dissociating natural motions from the bodies they affected. In this sphere also Oresme was far ahead of any of his contemporaries. Thus in his discussion of the diurnal motion of the earth he clearly discerned the relativity of motion and concluded that it was impossible to determine by any experiment whether it was the heavens that are moved or the earth.[154] He even anticipated some of the views of Copernicus and Galileo when he tried to meet some of the usual objections to the motion of the earth based either on observation[155] or on Holy Writ.[156] He also noted that, once it has been assumed that the earth rotates, the rest attributed to the eighth sphere is more in keeping with the perfection of the heavens,[157] and that it is more rational to replace a "diverse and outrageously great" number of operations with a single one.[158] Nevertheless, he had hardly raised the problem and thus suggested the first truly significant transformation of the traditional cosmological picture when he went on to point out that he had broached the subject only by way of a "diversion."[159] Now, this strange twist would not have worried Oresme's con-

[154]Duhem, *Système du monde*, Vol. IX, pp. 330–332. Oresme's arguments were presented in his *Commentary on Aristotle's Book of the Heavens and the World*. Buridan had previously advanced the same argument (ibid., pp. 345–346).

[155]Ibid., pp. 332 f. Oresme considered the case of an arrow projected upward and explained that it returned to its starting point because if the earth were moved with a diurnal motion, the air through which the arrow passes would revolve as well and pull the arrow along with it. This explanation, which assigns a motor role to the medium, ran counter to the impetus theory and thus involved Oresme in a clear contradiction. Buridan was more consistent when he rejected the whole argument (ibid., pp. 350 f.).

[156]Ibid., pp. 336 f.

[157]Ibid., pp. 337 f.

[158]Ibid., pp. 339 f.

[159]Ibid., p. 341.

temporaries unduly, for Oresme at no time rejected the theory of natural motions and the theory of the elements from which the Peripatetic school had deduced the immobility of the earth. Possible though the diurnal motion of the earth might therefore have appeared to be, it was ruled out by cosmological premises to which Oresme himself continued to subscribe. In these circumstances, any discussion of the motion of the earth was bound to be no more than a formal exercise, and as such without influence on mechanics. This explains why fourteenth-century excursions into cosmology left the theory of motion completely untouched; it needed several further revolutions of thought for the new ideas to bear any fruit.

That Galileo was familiar with them is beyond doubt in any event. Faithfully transmitted by fifteenth- and sixteenth-century writers,[160] they represented a fairly considerable part of the teaching he received at Pisa between 1583 and 1586. Though his masters included a number of orthodox Aristotelians (whose method he criticized sharply in his *De Motu*),[161] Buonamico and others gave him a full account of the theses of both "ancients" and "moderns" on the essential aspects of physics.[162] The *Juvenilia*,[163] a book of the utmost importance to anyone interested in Galileo's intellectual development, clearly reflect this cross-current. Two significant passages show that Galileo's early framework was still wholly Aristotelian. Thus, after a detailed discussion of all the contrary opinions, he came down squarely in favor of the simplicity and incorruptibility of the heavens and the need for motor intelligences.[164] Moreover, he described the diverse motions of the celestial spheres (whose number was thought to be eleven) in accordance with the writings of traditional astronomers and justified the immobility of the earth at the

---

[160]The Mertonian and Parisian ideas were brought to Italy by Blasius of Parma; they were taught and discussed throughout the fifteenth century at Padua, then the intellectual center of Europe, by Paul of Venice, Giovanni da Fontana, Johannes de Marchanova and Gaetano de Thienis, who initiated a debate on Swineshead's *Liber calculationum*, and reviewed all the arguments in favor of a quantitative physics; cf. Clagett, *Science of Mechanics*, pp. 645 ff; and J. H. Randall, "Scientific Method in the School of Padua," in *Journal of the History of Ideas* I, 1940. For the sixteenth century, see Duhem, *Études sur Léonard de Vinci* III, pp. 182 ff., and Maier, *Zwei Grundprobleme*, pp. 293–304.

[161]*De Motu*, in *Opere*, Vol. I, p. 285.

[162]For Buonamico's discussion of the projection and acceleration of heavy bodies, see Koyré: *Études galiléennes*, Vol. I, pp. 18–41.

[163]This, as we mentioned earlier, was the title under which Favaro published Galileo's student notes of ca. 1584. These notes were probably based on Buonamico's lectures.

[164]*Opere*, Vol. I, pp. 55–110.

center of the universe by predominantly optical arguments, most of which could be traced back to Ptolemy.[165] Purely Aristotelian, too, was the *Tractatus de elementis*, which is included in the *Juvenilia*. It deals with the number and nature of the elements and in it Galileo describes the primary qualities and their modes of action with exceptional erudition.[166] However, the *Treatise on the Latitude of Forms*, which appears between these two fragments takes us straight back to the fourteenth century, and though the "questions" posed in it bore on qualities in general and not on velocity in particular, we may take it from the context that even at that early period of his career Galileo was familiar with medieval speculations about changes in speed and their effects. Thus he made a careful distinction between various types of *latitudines* (or changes)[167] and described the main features of intension and remission.[168] It is, moreover, remarkable that in his discussion of the ontological problem of the intension or remission of forms, Galileo (or his teacher) should have adopted the Scotist solution; the intension of a quality—of hotness, for instance—was said to take place by the addition of degrees, in the same way that in an extensive magnitude "a quantity combines with a preexisting quantity,"[169] so we must also conclude that "intension is due to the production of a new quality, in such manner that the part previously acquired coexists with the new part."[170] A paragraph devoted to the possible plurality of worlds[171] and a great many allusions to the Mertonians and Parisians bear further witness to the important role their contributions played in the philosophical and scientific education of the young Galileo.

[165]Ibid., pp. 38–55.
[166]Ibid., pp. 122–177.
[167]Ibid., p. 120. See also footnote 20, above.
[168]Ibid., pp. 117 f.
[169]Ibid., p. 116.
[170]Ibid., p. 117.
[171]Ibid., pp. 28–31.

3 The Preparatory Years (1589–1602)

Neither the Aristotelian nor the medieval heritage, which Galileo's doctrine was largely to overthrow, can, of course, tell us how and from what problems he was led to the idea of a mathematicized science of local motion; influences from quite a different source are alone capable of explaining the emergence of a new conceptual framework. Now these influences are not difficult to pinpoint. The most obvious and decisive was that of the Greek geometers. In 1584, while still a student at Pisa, Galileo began to read Euclid, and in 1585, on his return to Florence, he added the study of Archimedes, beginning with the mathematical treatises and a few years later going on to the statics and hydrostatics. Far more than even Euclid, it was Archimedes who, by his skill in applying mathematics to the analysis of natural phenomena, decided Galileo's true vocation. Thus in 1586 the young Galileo constructed a hydrostatic balance and carefully reproduced the experiments by which Archimedes had established the density of various substances;[1] in 1588 he wrote a brief essay on the centers of gravity of several solids.[2] While these contributions failed to introduce novel concepts, they nevertheless make it clear that as early as 1587 or 1588 Galileo was already familiar with the essential contributions of the Greek mathematicians. At the same time, other writers also aroused his interest, among them Tartaglia, who in the 1540s had written a new account of the motion of projectiles[3] and a few years later had republished one of the most important medieval texts of statics,[4] thus ensuring the transmission of an intellectual current whose importance Galileo was fully to appreciate. But, above all, ever since the publication of the works of Archimedes in 1543, general interest in mechanical problems had grown apace, and though Guido Ubaldo del Monte still confined his attention to statics and to simple machines,[5] Giovanbattista Benedetti, many of whose theses Galileo was to adopt, tried to construct a general science of dynamics, in which the basic

---

[1] *La Bilancetta*, in *Opere di Galileo Galilei*, Vol.I, pp. 215 ff.

[2] *Theoremata circum centrum gravitatis solidorum*, in ibid., pp. 187 ff.

[3] *Nova scientia inventa da Nicolo Tartalea* (Venice, 1537); and *Quesiti et inventioni diversi di Nicolo Tartalea* (Venice, 1546). For Tartaglia's contribution see A. Koyré, *La dynamique de Nicolo Tartaglia* republished in the *Études d'histoire de la pensée scientifique* (Paris, 1966), pp. 101 ff.

[4] *Jordani opusculum de ponderositate* (Venice, 1565).

[5] In the *Mecaniche dell' Illustrissimo Sig. Guido Ubaldo de' Marchesi del Monte* (Venice, 1581); and *In duos Archimedis aequeponderantium libros paraphrasis* (Pesaro, 1588).

principles of Archimedean hydrostatics were wedded to the medieval concept of impetus.[6] Greek mathematics, contemporary works on statics, and Benedetti's synthesis—besides traditional physics—these were the chief sources of Galileo's inspiration during his preparatory years and provided him with the essential instruments for the transformation of the old approach and for the gradual elaboration of a new and original method of analysis.

Luckily, two of Galileo's youthful contributions have been preserved, and it is thanks to them that we can follow the development of his thought from close quarters. One was written in about 1590; the other in 1595. In 1589, after several abortive applications, Galileo was invited to take the chair of mathematics at the University of Pisa. From 1589 to 1592 he delivered a series of lectures, presumably on the works of Euclid, on geocentric cosmology, on Ptolemy's planetary theory, and finally, at the request of his students, on astrology.[7] By then he was taking an increasing interest in the problem of motion; his *De Motu*, which summarized the results of these early investigations, also included his first attempt to apply Archimedean methods to a problem in philosophical physics. In 1592, for somewhat obscure reasons, Galileo left Pisa for Padua, where he remained until 1610. Although we now know that many of the discoveries set out in his major works of 1632 and 1638 were made during his stay in Padua,[8] we have no reason to think that when he left the conformist University of Pisa for the much more liberal atmosphere of Padua, his ideas underwent a sudden change; his writings during these early years suggest quite the contrary. True, his field of endeavor expanded because, in addition to his university duties, he now devoted himself to practical studies and even tried his hand at engineering.[9] However, the only theoretical work he produced during that period was mostly synthetic and preparatory: *Le Mecaniche*, which dates from about 1595 and was revised about 1600, was intended as a clear and coherent account of statics, and it was only in its last few pages that

[6] *Demonstratio proportionum motuum localium contra Aristotelem* (Venice, 1554); *Diversarum speculationum mathematicarum et physicarum Liber* (Turin, 1585).

[7] Emil Wohlwill, *Galileo Galilei und sein Kampf für die Copernicanische Lehre*, Vol. I, pp. 83–84. In 1591 Galileo wrote several (no longer extant) commentaries on Ptolemy's *Almagest* (cf. Vol. I, p. 304).

[8] See A. Favaro, *Galileo Galilei e lo studio di Padova*, pp. 315 ff.

[9] In 1593 he wrote a Treatise on Fortifications; subsequently he invented a machine for raising water and was granted a patent (*Opere*, Vol. 19, pp. 126 and 202); in 1597 he invented a "military and geometrical" compass.

Galileo applied the results of his analyses to the revision of some crucial concepts of dynamics.[10] Like the *De Motu*, *Le Mecaniche* is therefore of particular interest to anyone anxious to retrace the path that led Galileo to the transformation of the traditional view. As to the date on which he concluded this preparatory phase, we unfortunately cannot assign it with any great accuracy. In 1602, writing to Guido Ubaldo del Monte, Galileo mentioned two discoveries: the isochrony of pendular swings and the proposition that, in the *Discourses* (1638), was to become the law of chords: "If from the highest or lowest point in a vertical circle there be drawn any inclined planes meeting the circumference, the times of descent along these chords are each equal to the other."[11] Although nothing entitles us to state categorically that these were recent results, we nevertheless feel justified in treating the year 1602, in which Galileo felt sure enough of himself to speak quite openly of his personal contribution, as the end of this preparatory phase.[12]

The *De Motu* as a Watershed

The *De Motu*,[13] which sums up Galileo's reflections during his stay in Pisa, sets out explicitly to examine and correct Aristotle's doctrine on natural motion and the motion of projectiles. The project was therefore far from

[10]The work, which was not published until 1649 (under the title *Della scienza mecanica e della utilità che si traggono di quella*), had been circulating much earlier in manuscript form. An excellent English translation (from which we shall be quoting at length) by Stillman Drake has been published by the University of Wisconsin Press, Madison, 1960. Our page references are to Volume II of the Edizione nazionale.

[11]*Discorsi e dimostrazione matematiche interne a due nuove scienze* (*Discourses*), p. 221. The letter to Guido Ubaldo del Monte appears in *Opere*, Vol. X, p. 99.

[12]It is interesting to note, moreover, that in 1597, in a letter to Jacopo Mazzoni, Galileo defended the doctrine of Copernicus and the Pythagoreans (Vol. II, p. 198); on August 4 of the same year he also wrote a letter to Kepler in which he wholeheartedly endorsed the heliocentric solution and deplored the fact that there were "so few who study the truth instead of engaging in idle philosophizing" (Vol. X, p. 67). However, nothing entitles us to think that at this early stage he had already looked for a mechanical proof of the Copernican doctrine, and that is why, though we do not underestimate the importance of Galileo's semipublic acceptance of Copernicanism, we have chosen 1602, the year in which veritable discoveries were first made public.

[13]This is the title under which Favaro (*Opere*, Vol. I) has combined a series of studies which, in the form of questions, served to review the main problems posed by natural and violent motion; since the order in which these questions are presented is not necessarily that adopted by Galileo himself, we shall not try to follow it. Most of the questions contained in the *De Motu* were subsequently arranged in dialogue form. The reader is referred to I. E. Drabkin and Stillman Drake's excellent English translation of *Galileo Galilei, On Motion and on Mechanics* (University of Wisconsin Press, 1960.)

original, the less so as Galileo confined himself in the main to comparing Aristotle's solution with that of other authorities selected indifferently from antiquity and the Middle Ages. For all that, the *De Motu* represents a milestone in Galileo's intellectual development, for its simultaneous treatment of natural and projectile motions may be considered a first step in the construction of a general science of dynamics. Moreover, Galileo's systematic recourse to Archimedean principles shows his determination to replace the historical and dialectical method of the philosophers with a direct and quantitative method. In short, it is a faithful mirror of Galilean thought in rudimentary form.

For Aristotle, the speed of a body falling freely through a given medium was determined by two main factors: its total weight and the resistance of the medium;[14] even without discussing the merits of this choice of principle, it is easy to see that neither of the two relations to which it gave rise (that is, that the speed is directly proportionate to the weight of the falling body and inversely proportional to the density of the medium) could be substantiated. Thus, if we imagine that the water on which a large piece of wood and a small piece of the same wood are normally afloat is gradually made lighter, so that in the end it becomes lighter than the wood, the two bodies will obviously sink at the same rate, for "since the volume of the water to be raised by the large piece of wood is equal to that of the wood itself, and similarly with the small piece, those two quantities of water which are raised by the respective pieces of wood have the same ratio to each other in their weights as do their volumes."[15] This conclusion was clearly incompatible with Aristotle's view. Moreover, if two bodies of the same material but of different volumes are joined together, and if the smaller (and hence lighter) did indeed have a smaller velocity than the larger (and heavier), then the combination also would have to travel more slowly than the larger alone, and this would be self-contradictory.[16] Again, if we observe two bodies of different materials, we shall find that, say, a very large and suitably weighted bladder filled with air will fall very slowly, while a very small leaden sphere will fall very quickly.[17]

Aristotle's second principle might seem to have been based on more

[14]We are deliberately using this vague formulation because Aristotle failed to distinguish between the decelerating effects of the medium resulting from its density and those resulting from friction.

[15]*De Motu*, p. 264.

[16]Ibid., pp. 265–266.

[17]Ibid., pp. 266–267.

solid foundations than his first, since, superficially at least, it looks as if a denser medium must necessarily tend to decrease the speed of a falling body, while a rarer medium must tend to increase it. However, as Galileo went on to show, this assumption is no more justified than the first. Thus, let there be a moving body *O* and two media, *A* and *B*, say, water and air, and let the rareness of *B* be 8 and that of *A* (in which *the moving body does not sink but floats*) be 2. Now, suppose the body *O* is such that its speed in air, that is, in *B*, is 4; then, if Aristotle is right and the ratio of the speeds is indeed equal to that of the densities of the media,

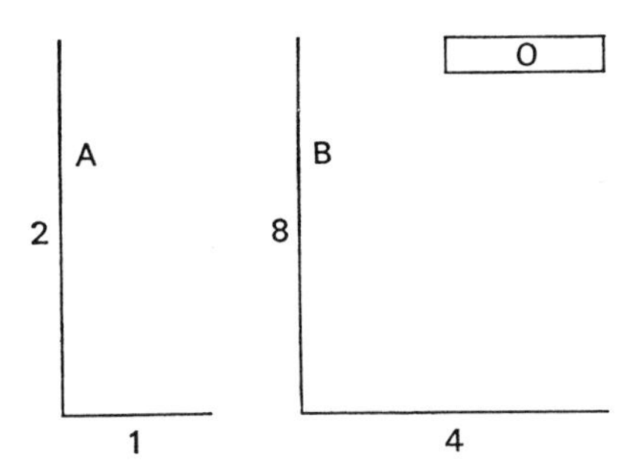

the body will have to move through medium *A* with a speed of 1 when by definition it floats on the surface. Hence, the assumption that the speed of a falling body is inversely proportional to the density of the medium leads to the absurd conclusion that a body will sink in a medium that has the same rareness as the body itself.[18] Nor is it true to say that increases in speed invariably reflect decreases in the density of the medium, for some bodies can travel more quickly in a dense medium than they can in a rare one. Thus an inflated bladder will move more swiftly through water than it will through air.[19] Last but by no means least, the traditional view is incompatible with the obvious fact that a body does not merely move more slowly but also reverses its direction whenever the density of the medium exceeds its own.[20]

Under these conditions, would it not have been best to discard the two factors by which Aristotle and his disciples had tried to adduce a dynamic explanation of natural motions? Galileo did not think so, for he too believed that it was the joint effect of the weight of moving bodies and the resistance of the medium which alone could provide a true

[18]Ibid., pp. 268–269; cf. *Dialogue* version, pp. 399–400.

[19]*De Motu*, p. 261

[20]Ibid., pp. 398–399.

account of the natural motions of heavy bodies and lead to the correct evaluation of their speeds.[21] Although he did not describe it in these words, it is clear from his general remarks that he felt Aristotle's error lay in considering the actions of the moving body and of the medium *successively* rather than *conjointly*. This alone explains why he attributed a greater velocity to the larger of two homogeneous bodies falling through one and the same medium, instead of realizing that since their volumes vary as their weights, they have to overcome resistances varying in the same proportions. Had he used the correct approach, he would also have noted that of two heterogeneous bodies of equal weight (such as a small piece of lead and an inflated bladder) the larger would be more strongly affected by the action of the medium and hence would descend more slowly in it than the smaller. In that case, he might also have realized that what determines the direction and speed of a body traveling through a given medium is not its weight or the resistance of the medium considered in isolation, but its weight or lightness relative to the medium.[22] Once the problem has been stated in these terms, two questions have to be answered: (1) On what basis must the weight of the moving body and the resistance of the medium be evaluated? (2) How must their combined action be understood?

A very simple argument, proceeding imperceptibly from statics to dynamics, provided the answer to the first question. Let a balance be represented by the line *ab* pivoting about the point *c*, the center of *ab*, and let two weights *e* and *o* be suspended from the points *a* and *b*.[23] Now, in the case of the weight *e* there are three possibilities: it may be at rest, move upward, or move downward. If it is heavier than *o*, it will move downward; if it is less heavy it will move upward, "not because

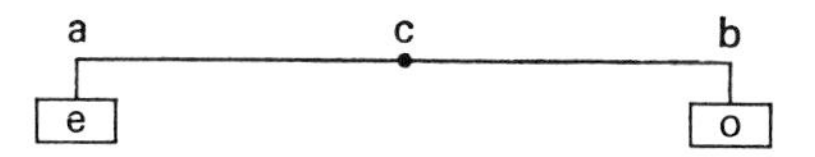

it does not have weight, but because the weight of *o* is greater." Finally, if the weight of *e* is equal to that of *o*, then *e* will move neither upward nor downward. Let us now suppose that *e* measures the weight of a moving body and *o* that of an identical volume of the medium in which *e* is placed. If *e* is heavier than *o*, the body will again move downward,

[21] Ibid., pp. 253–254.
[22] Ibid., p. 262.
[23] Ibid., pp. 257–258.

and if it is lighter than o the body will again move upward; if $e = o$, there will be equilibrium, and the body will remain motionless in the medium.[24] Now this is precisely what happens whenever *any* body moves through or in *any* medium, that is, whenever it impinges on a volume of the medium equal to its own. To move downward, its density or specific weight must therefore be greater than that of the medium; otherwise, there will be no motion at all or the body will be thrust upward.[25] Because they provide a measure of their own forces, the *specific weights* of the moving body and of the medium are thus the two fundamental elements needed for the dynamic study of the motion of freely falling bodies.

But how can we determine the speed of a body if it is due to the joint action of its specific weight and that of the medium? While statics had guided him in the preceding argument, it was to hydrostatics that Galileo looked for an answer to the second question. Let us follow Archimedes and examine the behavior of a solid immersed in water. If the two have the same density, the body will float beneath the surface of the liquid, neither sinking nor rising; if the body is rarer than the medium it will rise partially above the surface, and from Proposition 5 of Archimedes' *On Floating Bodies* (Book I) we know to what precise extent: any solid lighter than a fluid will, if placed in the fluid, be so far immersed that the weight of the solid will be equal to the weight of the fluid displaced.[26] In the third case, that is, when the body is of greater density than the fluid, it will sink to the bottom. Let us now look more particularly at what happens when a solid is forcibly immersed into a liquid of greater density than itself: the liquid, which exerts an upward pressure on the solid, will immediately thrust it to the surface. Archimedes had shown that this upthrust is equal to the difference between the weight of the solid and that of the fluid displaced,[27] and Galileo

[24]This argument is of course not strictly correct unless the balance is placed in a vacuum.

[25]"We define as equally heavy two substances which, when they are equal in size [that is, in volume], are also equal in weight. Thus, if we take two pieces of lead which are equal in volume and equal also in weight, we shall have to say that they are equally heavy. And clearly, therefore, we must not say that wood and lead are equally heavy. For a piece of wood which weighs the same as a piece of lead will far exceed the piece of lead in volume. Again, one substance should be called heavier than a second substance if a piece of the first, equal in volume to a piece of the second, is found to weigh more than the second." Ibid., p. 251.

[26]Ibid., pp. 254–257. Galileo proved the first two cases by *reductio ad absurdum*.

[27]*On Floating Bodies* I, Proposition 6.

translated that proposition into the language of dynamics: he treated the pressure as a dynamic force, or more precisely as a measure of the force with which the medium tends to thrust the body upward. Now, since the pressure exerted by the medium on the solid is a direct consequence of the difference between their specific weights, Galileo found it easy to evaluate the magnitude of the upthrust and of the velocity it imparts to the body.[28] Now this conclusion could easily be extended to the case of bodies whose specific weight was greater than that of the medium: to that end, the pressure exerted by the medium need merely be treated as a property of the moving body: the downward pressure of the latter exceeds the upward pressure of the former in precisely the same ratio as the weight of the moving body exceeds the weight of the fluid displaced. "Hence it is clear that if we find in what media a given body is heavier, we shall have found media in which it will fall more swiftly. And if, furthermore, we can show how much heavier that same body is in this medium than in that, we shall have shown how much more swiftly it will move downward in this medium than in that."[29]

Galileo's earliest contribution to the study of natural motion can therefore be summed up as follows: While bodies lighter than the medium in which they are immersed are thrust upward by a force (and hence with a speed) equal to the excess of the specific weight of the medium over their own, bodies heavier than the medium in which they are immersed will be impelled downward with a force equal to the excess of their own specific weight over that of the medium.[30] Thus if a solid has a weight of 8 and the weight of an equal volume of water is 4, the solid will move downward in the water "with a speed and facility that may be represented by 4." Again, if the weight of the medium is merely 3, the "facility" of the motion will increase to 5, etc.[31] For Aristotle, the speed of a moving body was expressed by the geometric ratio of its weight to the resistance; Galileo, for his part, in subtracting the specific weight of the medium from that of the moving body, was able to replace Aristotle's geometric expression with a simple arithmetical one. In other words, while Aristotle's law governing the behavior of

---

[28]In this, he was greatly aided by the traditional assumption that the force is proportional to the speed; in other words, Galileo still accepted the fundamental law of Peripatetic dynamics.

[29]*De Motu*, p. 262. In this argument Galileo failed to consider the effect of friction, and it was not until 1638 that he tried to make good this omission.

[30]Ibid., pp. 274–275.

[31]Ibid., p. 287.

freely falling bodies might be expressed by some such formula as $V = kW/M$ (where W is the absolute weight of the body, M the resistance of the medium, and k a factor of proportionality), Galileo's dynamics as set forth in the *De Motu* can be expressed by the formula $V = k(W - M)$ (where W is the specific weight of the body and M the specific weight of the medium);[32] if V is positive, the body will sink, if it is negative, the body will rise.[33]

While Galileo's theory of natural motion was clearly inspired by Archimedes, his analysis of projectile motion was rooted in fourteenth-century ideas. Rediscovering the arguments of Buridan and Oresme for himself, Galileo set out, first of all, to prove that no motor role can be assigned to the medium. If Aristotle was right, could an arrow really be set in motion by a bowstring in the face of a strong wind? Nor do even those parts of the medium which are in direct contact with the projectile move forward in the same direction, for when a ship sails against the current, the water touching the hull does not follow it any more than

---

[32]Cf. E. A. Moody, *Galileo and Avempace: Dynamics of the Leaning Tower Experiment.* Republished in *Roots of Scientific Thought* (Basic Books, New York, 1957). This formula had to remain of purely theoretical importance until such time as the absolute specific weights of the moving body and the medium could be determined. Now, when he wrote the *De Motu*, Galileo had not yet made any attempt to proceed to that determination. He did, however, examine the problem at some length in his *Discourses.*

[33]Historians have shown that Galileo's arguments are more or less identical with those Benedetti had put forward several years earlier. (See G. Vailati, "Les speculations de Benedetti sur le mouvement des graves," in *Scritti*, pp. 161 ff; cf. A. Koyré, *Études galiléennes*, Vol. I; and "Benedetti, critique d'Aristotle" in *Mélanges offerts à Étienne Gilson*, Paris, 1959.) Thus Benedetti had asserted that only the specific weight of a heavy body determined its speed of descent in a given medium and had shown that two heavy bodies of the same specific weight but of different volumes will travel with the same speed through the same medium (Koyré, *Études galiléennes* I, pp. 46–51 and pp. 351–355; Benedetti's argument was based on Archimedes' proof of the principle of the lever. A similar argument had previously been used by Cardano (*Opus novum de proportionibus* V, par. 110). Benedetti made further use of Archimedean hydrostatics when he tried to arrive at a general determination of the speed of a solid immersed in different media (see Vailati, "Speculations de Benedetti," pp. 165 ff.) Galileo may also have been influenced by a much older tradition going back to John Philiponus and Avempace (cf. Moody, *Galileo and Avempace*, pp. 196 ff.). According to Vailati, moreover, Benedetto Varchi had produced yet another refutation of Aristotle's dynamics about the middle of the sixteenth century (*Scritti*, pp. 168 f.). For all that, there is little doubt that Galileo was the first to try systematically to base dynamics on statics, and that he alone produced a justification of this approach.

the rest of the sea.[34] Moreover, if the medium supported the motion, why would an iron ball fired with the same force as a ball of tow be flung over a very much larger distance than the latter? And why should a sharper and more elongated arrow travel further than a thicker arrow when it is less exposed to the motor action of the medium?[35] Or consider a marble sphere, perfectly round and smooth, which can rotate on an axis the ends of which rest on two supports. Now, suppose someone twists both ends of the axis with his fingertips. Not only will the rotation continue for some time, but the air around it will remain quite motionless, for if a flame is brought near the rotating sphere, it will be neither extinguished nor disturbed.[36] Aristotle's explanation not only flew in the face of the facts; it could not even be reconciled with some of his own remarks on forced or violent motion. Thus he himself had asserted that forced motion is swifter in the middle of the motion than at the beginning.[37] Hence, if the parts of the air that move the body are *A*, *B*, *C*, *D*, and *E*, then the part *C* of the air, under the impulse of *B*, would be moved more swiftly than *B* and would in turn communicate to *D* a motion swifter than its own, and so on. Therefore, forced motion would be constantly accelerated, there being no reason why this increase in speed should stop. In the same book, Aristotle also affirmed that if what is moved is neither heavy nor light, its motion will be only by force and that this motion, offering no resistance, would go on forever.[38] But the air itself is by nature possessed of neither heaviness nor lightness; hence, if it is responsible for the motion of projectiles, the latter, participating in the motion of their medium, would perforce continue to move ad infinitum, and most probably at a uniform speed.

But if the traditional solution had to be discarded, what was the correct explanation of the motion of projectiles? Galileo's answer was in no way original, but based on a principle John Philoponus had formulated in the sixth century: Since it is absurd to suppose that the projector communicates its motive power to the air, why not assume that it communicates it to the projectile itself? And, following in the footsteps of Buridan and the Parisian Nominalists, Galileo went on to

---

[34]*De Motu*, pp. 308–309.
[35]Ibid.
[36]Ibid.
[37]*On the Heavens* II, 6, 288 a 22; *De Motu* p. 308.
[38]*On the Heavens* III, 2, 301 b 1–4; *De Motu*, p. 309.

argue that the motion of a projectile was due precisely to the impetus that the projector impresses on the projectile and that tends to conserve the motive force of the former in the latter.[39] However, Galileo did not simply copy the ideas of his precursors but also tried to improve upon their definition of the impressed force. Rather than treat it as a quality (which might have served to reintroduce the occult qualities of the Schoolmen), he preferred to define it by its effects on the projectile. In other words, he equated *impetus* with the *privatio gravitatis* by which the projectile is deprived of its weight and rendered light;[40] even though he described the resulting lightness as accidental or preternatural, he nevertheless claimed that it resembled "natural" lightness in that it was capable of engendering upward motion.[41] A stone that is set in motion by a force, Galileo explained, loses its "natural and intrinsic weight in the same way as when it is placed in media heavier than itself."[42] But how precisely does an impetus become impressed upon a moving body? To answer this difficult question, Galileo was forced to resort to analogies. An impetus bestows a provisional lightness upon the moving body in the same way that "the iron is moved, in an alternative motion, toward heat so long as the iron is in the fire and is deprived by it of its coldness."[43] Though more direct, the example of the striking clock was not much more illuminating. "The bell is struck by the striking object; the stone is moved by the mover. The bell is deprived of its silence, the stone of its state of rest. A sonorous quality is imparted to the bell contrary to its natural silence; a motive quality is imparted to the stone contrary to its state of rest. The sound is preserved in the bell when the striking object is no longer in contact; motion is preserved in the stone when the mover is no longer in contact."[44] Perhaps that is as far as Galileo could go along this rather unsatisfactory road; nevertheless, his

---

[39]It is not certain that Galileo would have acknowledged Buridan as the father of this theory; in any case, he at no time mentioned any of the Parisian authors by name. He must, however, have heard about the theory from Buonamico (*De Motu*, Book X) and from Borrius (*De Motu gravium et levium*), two of his teachers in Pisa from 1583 to 1586. We know that Buonamico attributed the impetus theory variously to John Philiponus (this might have been justified) and to Albertus Magnus and St. Thomas Aquinas (this was mistaken); cf. Koyré, *Études galiléennes* I, pp. 24 ff.

[40]*De Motu*, p. 310.

[41]The terms "lightness" and "weight" must be understood in a relative sense.

[42]Ibid., p. 312. The similarity between Galileo's treatment of this case and that of a solid forcibly immersed in a denser medium is clear proof of his determination to construct a unified science of dynamics.

[43]Ibid., p. 310.

[44]Ibid.

faltering steps led him to appreciate the twofold nature of the impetus: it not only concentrates the energy of the motor in the mover but also bestows a new quality upon the latter.

However, he thought that this quality was purely provisional; that is a point on which it is well worth dwelling at some length. Galileo, in fact believed he could prove that there cannot exist two points of time in the course of projectile motion such that the impetus is the same in both. Let *ab* be the upward trajectory of the projectile, and let us assume that it were possible to mark off on it two points *c* and *d* at which the impressed force has the same value. "Since, then, the motive force is the same at *c* as at *d*, the medium the same, the body the same, and the line in which the motion takes place also the same, it follows that the body

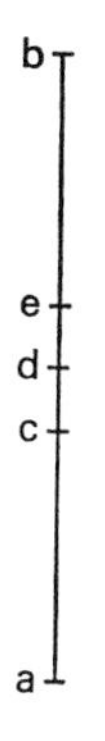

will move from point *d* at the same speed as it moved from point *c*."[45] But it was assumed to have moved from *c* to *d* with a constant velocity so that it will also be moved from *d* on a line *de*, equal to *cd*, with the same speed; there is no reason for the impressed force to remain the same from *c* to *d* but not from *d* to *e* since, as we saw, the body, the medium, the force, and the direction of the motion remained unchanged.[46] It follows that, unless the impetus is continuously diminished, a projectile must continue to move indefinitely with a uniform velocity, which, Galileo contended, was absurd.

Galileo's argument, though perfectly clear, nevertheless failed to explain the decrease in impetus in a satisfactory manner. Thus, unlike Buridan's impetus, Galileo's was not dissipated in overcoming the resist-

---

[45]Ibid., p. 315. [46]Ibid.

ance of the medium; rather it spent itself in maintaining the motion. "The impressed force gradually diminishes in the projectile when it is no longer in contact with the projector; the heat diminishes in the iron when the fire is not present."[47] Was Galileo misled by the analogy with qualitative changes? All we can say with certainty is that he abandoned the fruitful idea that an impetus can conserve itself in the absence of a resistant medium. In other words, he adopted Oresme's view in preference to that of Buridan and Albert of Saxony, and this unfortunate choice proved clearly how ambiguous this impetus concept really was, and hence how unlikely it was to pave the way for the concept of inertia.[48]

By contrast, it was to Buridan's ideas that Galileo appealed when he affirmed that the impetus is proportional to the quantity of matter,[49] and quite particularly when he rejected the Aristotelian theory of the *quies media*. According to Aristotle, as the reader may recall, two contrary motions cannot be continuous; an interval of rest (*quies media*) must necessarily occur at the turning point. Against this view, still defended by some of his contemporaries,[50] Galileo mustered a whole series of arguments. Some of these he took from fourteenth-century philosophy;[51] others were based more directly on mechanical considerations, for instance, the following, which Galileo borrowed from Copernicus: Imagine two circles, the center of each of which is carried on the circumference of the other. Now, whenever one of the two circles revolves more rapidly than the other, a point on the circumference of the first circle will move back and forth continuously in a straight line over the same path.[52] The most interesting arguments, however, were developed by Galileo himself. He demonstrated, first of all, that the assumption of the *quies media* leads to a number of absurd consequences. Suppose a stone is thrown up from *a* to *b*, and then falls back naturally from *b* to *a*; if this stone is

[47]Ibid., p. 310.
[48]Cf. Koyré *Études galiléennes* I, p. 58.
[49]*De Motu*, p. 320. Galileo explained, inter alia, that a bullet would travel the further the higher the line of fire, because when fired upward it presses down more on the powder so that a greater part of the motor force of the powder becomes incorporated in it (ibid., p. 338).
[50]Buonamico among them; see Koyré, *Études galiléennes* I, p. 26.
[51]When a small pebble is thrown up from below against a large stone falling from a tower, the stone will not be sufficiently blocked by the pebble to allow the latter to be at rest for any interval of time (*De Motu*, p. 326). The same example was previously used by Buridan; cf. P. Duhem, *Système du monde*, Vol. VIII, p. 286.
[52]Copernicus, *De revolutionibus orbium coelestium* III, ch. 4. Benedetti used a similar example (Koyré, *Études galiléennes* I, p. 51) based on technical considerations, thus showing what role technical observations must have played in the critique of traditional mechanics.

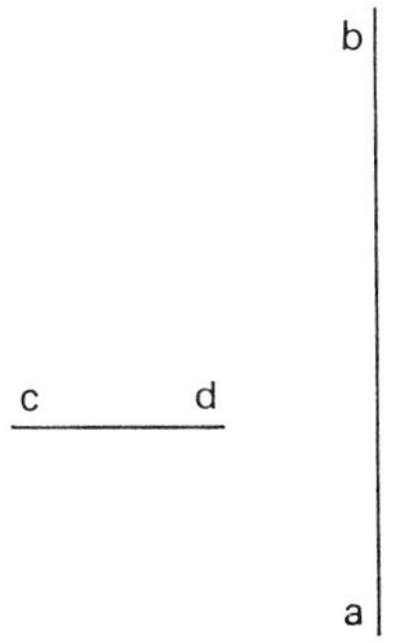

b

c d

a

at rest at *b* for an interval represented by the line *cd*, the external projecting force (that is, the impetus) must, throughout the interval *cd*, be equal to the weight of the body. However, the latter does not change, so that the impetus at the moment *c* must necessarily be equal to the impetus at the moment *d*. But the force at moment *c* sustains the body throughout the time *cd*, and there is no reason why it should not be able, at moment *d*, to sustain the same stone through another interval of time *cd*, and so on. Accepting the *quies media* thus amounts to bestowing a perpetual rest on the stone in the point *b*.[53]

Galileo's second argument was even subtler. Since the upward motion of a heavy body is due to the excess of the impetus over its natural weight, there must be, at any one time, a fixed ratio between its weight and the impetus; and since the impetus decreases gradually, the ratio must necessarily pass through an infinite number of decreasing values. Now it is obvious that the projectile does not stop at any of these values (their infinite number would cause the motion to continue for an infinite time). Hence, there is no reason why it should stop at the value 1—the upward motion is followed directly by the downward motion.[54]

The rejection of the *quies media*, as we have said, marked a return to

---

[53] *De Motu*, p. 327.

[54] Ibid., pp. 327–328. That the ratio of impetus to weight passes through an infinite number of decreasing values before reaching equality was not the subject of any special analysis, and Galileo did not seem deterred by the possibility that a body might cover this infinite number in finite time; the reason was that he fully shared Aristotle's view on Zeno's Dichotomy. In his *Dialogue Concerning the Two Chief World Systems* (1632), Galileo would deal with this problem explicitly and would again adopt Aristotle's solution; in the dialogue version of the *De Motu* (Vol. I, p. 392) he moreover thought it self-evident that a body traversing a surface must cross an infinite number of lines without coming to rest on any one of them.

Buridan's theory, but the particular way in which Galileo applied the impetus concept to the acceleration of freely falling bodies differed radically from the use to which this concept was put in the fourteenth century.[55] Thus, in considering the case of an impetus impressed upon a projectile, Galileo argued that if it is greater than the weight of the projectile, the latter will move in an upward direction until such time as the diminishing impetus is precisely equal to the downward pull of the weight, at which time the motion is reversed, and the projectile begins to fall. Now, according to Galileo, the impetus itself will not have vanished at this point—only that part of it which exceeds the pull of the weight; the remaining fraction will continue to be impressed in the projectile, too weak to oppose the downward motion, but still too strong to allow the downward motion to attain its full velocity. In other words, the natural motion succeeding the forced motion is not completely "natural" at first. Before it can become so, the remaining fraction of the impetus must disappear and with it the "accidental lightness it entails": the acceleration corresponding to this disappearance thus expresses the gradual recovery by a moving body of its natural speed.[56] This unusual theory, first advanced by Hipparchus in antiquity, enabled Galileo to explain the increase in speed[57] associated with natural motions that follow upon forced motions, no less than with natural motions starting from rest. In the second case, the chief difficulty is to determine to what precise cause the provisional loss in weight, as the only possible explanation of the acceleration, must be attributed. Since the medium cannot be involved here, Galileo was left with only one possible answer: The provisional loss in weight must be due to the obstacle which prevents the body from falling. Suppose we released a stone from a tower: Before it begins to fall, our hand must in some way have held it up with a force equal to its weight. That force will play precisely the same role as does the impetus in projectile motion; deprived by our hand of its natural gravity, the stone will start with a motion of less than its proper speed, which it must gradually recover. In that case again acceleration would simply express the transition from an artificially low speed to the body's natural one.[58]

---

[55]Cf. Vailati, *Scritti*, pp. 173–174.
[56]*De Motu*, pp. 120–121.
[57]Galileo used the medieval expression *causa intensionis*.
[58]This explains why heavier bodies fall less swiftly at the beginning of their descent than do lighter bodies. They have to destroy more of the accidental force that impels them in the opposite direction; but once the resistance has been overcome, they will fall more rapidly. *De Motu*, pp. 334–335.

This interpretation is interesting for two reasons. First of all, it turns acceleration into a transitory phase of the motion, uncharacteristic of the natural fall of heavy bodies. Had these bodies but time enough to eliminate all the accidental lightness opposing their weight, they would undoubtedly fall at a uniform rate directly proportional to their specific weight.[59] Did Galileo really accept this conclusion? The question is extremely difficult to answer, because no text written at the time entitles us to assume that he had already begun to treat naturally accelerated motions as uniformly accelerated motions. On the other hand, this explanation clearly demonstrates Galileo's determination to construct a unified science of dynamics. The traditional distinction between natural and violent motions was considerably reduced in that a single cause, residing within the moving body, was now held responsible for the slowness of forced motions at the end and of natural motions at the beginning. "But the cause of the slowness of forced motions at the end is the smallness of the amount by which the projecting force exceeds the resistant weight. . . . Similarly, we must hold that the cause of the slowness of natural motion at its beginning is the smallness of the amount by which the cause of natural motion exceeds the cause of forced motion, that is, by which the weight that presses the body down, exceeds the lightness, that is to say, the impressed force which impels the body up."[60] In other words, one and the same concept now presided over the dynamic explanation of natural and of forced motions; despite the ambiguous nature of such expressions as "accidental lightness" or "privation of weight," it was weight and changes in weight that, according to Galileo, were alone responsible for the directions of motions and also for changes in their velocity.

Critical and synthetic in its scope, the *De Motu* was thus a work of considerable importance to Galileo's future development, as witness the the fact that several of its themes reappeared forty-five years later in the *Discourses* (First Day): for example, that like bodies will fall through like media with the same velocity; and the (still partial) role of the medium in producing the differences in speed displayed by heterogeneous bodies.

The approach, too, was new, for as Galileo himself put it, "the method

[59] *De Motu*, Dialogue version, pp. 406–407. Galileo even tried to show that acceleration is the result of an optical illusion.

[60] *De Motu*, p. 322.

we shall follow in this treatise will be always to make what is said depend on what was said before and, if possible, never to assume as true that which requires proof."[61] Thus the use of the principle of sufficient reason (for instance, in demonstrating the continuous diminution of the impetus or in rejecting the *quies media*) bares witness to the fact that Galileo, unlike the Peripatetics, was determined to apply mathematical arguments to physical problems. Let us therefore try to determine which parts of Galileo's youthful work pointed to the future and which were simply rooted in the past.

On the credit side, the *De Motu* produced a convincing refutation of the Aristotelian idea that bodies are absolutely heavy or light. In Book IV of *On the Heavens* Aristotle had argued that if all bodies were formed of the same substance so that their weight and lightness would be purely relative, it would follow that a large enough quantity of fire must be heavier than a small quantity of air, and a large enough quantity of water must be heavier than a small quantity of earth. In that case, since weight is the principle of downward motion, there would be nothing to prevent a large quantity of fire from falling through the air or a large quantity of water from sinking below certain parts of the element earth. Yet experience shows that any portion whatever of earth sinks in water, and that any portion of fire rises in air.[62] Now, according to Galileo, this argument was mistaken in two respects. The first was that Aristotle, in treating of the natural and reciprocal motions of two bodies, say, a large bag of tow and a small piece of lead, completely ignored their volumes, although it is only by reference to these that valid comparisons can be made.[63] Aristotle's second error lay in believing that it was possible to determine the weight of an element "in its own region": to assume that water has less weight than lead because a piece of lead invariably sinks in it, is to assume that the weight of water can be determined in water. It is only when we weigh an element in a medium rarer than itself (for example, water in air, or air in a vacuum) that we can arrive at valid conclusions, namely, that a large enough mass of air can weigh more than a small mass of water or earth.[64] It follows that there is no such thing as absolute lightness but that "there is a single matter in all

---

[61]Ibid., p. 285.

[62]*On the Heavens* IV, 2, 310 a 11-13; IV, 5, 312 b 19 ff.

[63]*De Motu*, pp. 291–292.

[64]Ibid., pp. 291 f. In his *Discourses* (1638) Galileo mentioned a method of weighing air in air (cf. Vol. VIII, pp. 125 f.).

bodies, which is heavy in all of them" and has a natural inclination to move toward the center;[65] in short, weight and lightness are relative qualities; "no body is devoid of weight, but all bodies are heavy, some more, some less, according as their matter is more crowded together and compressed, or diffused and spread out."[66] Thus the distinction between lightness and weight which was at the very heart of Aristotle's theory of natural motions was transformed by Galileo into a distinction between density and rarity as reflected in the specific weights of the substances under consideration.

This approach, which in a sense marked a return to Platonic ideas,[67] had the immediate advantage of putting an end to the artificial division between natural upward and downward motions. If by "natural" we refer to such motions as follow from the very nature of the moving bodies, then the term must be reserved for downward motions; upward motions, resulting as they do from the higher density of the medium, that is, from an external cause, must perforce be violent, the bodies being "extruded" by the pressure of the medium. Now this conclusion had immediate repercussions on cosmology. For if all bodies are heavy, all that is needed is a center around which they will dispose themselves in decreasing order of weight, so that the idea of a natural "up" without which the intrinsic lightness of certain bodies cannot be explained, loses its raison d'être. Moreover, Aristotle's a priori cosmological order was shown to be the result of interactions on the part of bodies and directly explicable in terms of their distinct specific weights. Thus, the construction of a unified science of local motion went hand in hand with the rejection of the idea that the cosmos had an orderly structure determined a priori. To what extent did the *De Motu* reflect this consequence? After the elimination of natural upward motions, the most obvious step would have been to proceed to the elimination of the contrast between natural and forced motions, and there is good reason to believe that Galileo was thinking of doing just that: he compared the privation of weight due to the impetus to the lightening effect that a body receives from the fluid into which it is plunged, and he explained the increase in speed associ-

---

[65] *Ibid.*, p. 362. On pages 252–253 Galileo even made a veiled allusion to the Atomists, "those same philosophers, who were perhaps wrongly criticized by Aristotle."

[66] Ibid., p. 360.

[67] *Timaeus*, 61 d–64 a. Although the concept of specific weight is not mentioned explicitly in the *Timaeus*, Galileo liked to invoke Plato's views on this point; cf. *Discourse on Bodies in Water*, Vol. IV, p. 85.

ated with natural motions by arguing that "speed and slowness are a consequence of weight and lightness."[68] However, these were meant to be no more than analogies, and specific weight, as a cause of motion, remained radically distinct from impetus. The most fruitful parallels were drawn elsewhere—for example, when he assigned to one and the same cause the retardation of violent motions and the acceleration of natural motions; when in his rejection of the *quies media* he argued that a forced motion can follow directly upon a natural motion; or finally when he viewed the rotation of a sphere about a horizontal axis as a result of a kind of juxtaposition of the natural motion of the descending parts and the forced motion of the rising parts.[69]

The second great contribution of the *De Motu* was the assertion that motion in a vacuum was a distinct possibility. On this point too Galileo remained adamant throughout his career. In particular, he contended that it was Aristotle's mistaken view of the role of the medium which alone had caused him to deny the possibility of motion in the void. Thus the erroneous assumption that the velocity of a moving body is inversely proportional to the resistance of the medium led Aristotle to the conclusion that in a medium of zero resistance, such as the vacuum, $V_1/V_2 = M_2/0$, so that the velocity of a body would become infinite, which is contrary to experience.[70] However, if the resistance of the medium merely decreases (but does not divide) the motive force by robbing it of part of its power, motion in the vacuum ceases to be an impossibility and can then be represented by a formula of the type $V_1/V_2 = (W_1 - M_1)/(W_2 - 0)$.[71] Moreover, Aristotle had been mistaken in thinking that motion in a vacuum would have to be instantaneous—this because of another misconception. Aristotle's argument, Galileo asserted, could be summed up as follows: "A void is a medium infinitely lighter than every plenum; therefore motion in it will be infinitely swifter than in a plenum; therefore such motion will be instantaneous."[72] While it is perfectly true that a void is infinitely lighter than a plenum, it is just as wrong to conclude from this fact that motion in it would be infinitely swift as it would be

[68] *De Motu*, p. 318.
[69] *Dialogue* version, p. 373; cf. *De Motu*, pp. 304–307.
[70] *Physics* IV, 8, 215 b 12–216 a 8; *De Motu*, pp. 277–278. The symbols V_1 and V_2 refer to the respective velocities of the moving bodies; M_2 refers to the density of the plenum.
[71] *De Motu*, p. 279. Here W_1 and W_2 represent the specific weights of the two moving bodies and M_1 the specific weight of the plenum.
[72] Ibid., p. 281.

to conclude that a plenum must have infinite weight.[73] In fact, infinity resides neither in the void nor in the plenum but in the number of intermediate densities; similarly, there is an infinite number of intermediate speeds between that of a body in a given medium and that in the void. "Hence it follows, not that motion in a void is instantaneous, but that it takes place in less time than the time of motion in any plenum." In this respect, speed is comparable to a geometrical continuum, for here too, while it is possible to interpose an infinite number of intermediate lines between two lines *a* and *b*, of which *a* is the greater, line *a* cannot be said to be infinitely greater than line *b*.[74]

The acceptance of the idea of motion in the void had other consequences than the rejection of a prejudice of traditional physics; in particular, it made possible a better evaluation of speed differences between falling bodies. Now in non-vacuous media that evaluation is exceedingly difficult to make, since not only do the speeds of bodies traveling in different media change, but the greater the difference in the density of two media, the greater the distortion of the velocity ratio. Suppose there are two solids *a* and *b* and their respective weights are in air 8 and 6. If the same volume of air has a weight of 1, the ratio of the velocities would be $V_a/V_b = 7/5$; in water, on the other hand, whose weight is, say 4, the ratio of the velocities V_a/V_b would be 4/2.[75] In these conditions, only a medium in which the body suffers no decrease in weight, that is, in which its specific weight can be determined directly, can serve as an accurate frame of reference. Now, the void alone supplies such a frame because it is only in the void that the dynamic force inherent in the specific weight of bodies can express itself without impediment. Admittedly, this early reference to the void was still purely theoretical and a far cry from the dynamics of the *Discourses*. However, its importance must not be underestimated, since without the void, there can be no abstract study of motion, and above all no study of motion compatible with the mathematical resources available to Galileo at the time. Moreover, by accepting the possibility of motion in the void, Galileo rejected the role the medium was thought to play in the continuity and successive nature of motion; while he did not yet affirm the principle of the conservation of motion, he nevertheless removed yet another obstacle in its path and thus broke the hold of cosmology over physics

---

[73] Ibid.

[74] Ibid., p. 282.

[75] Ibid., p. 295.

even further. By contending that motion could take place in the absence of resisting forces and that it could not be truly depicted in any other way, the *De Motu* cleared the way for the eventual treatment of motion in the void, not as a particular case of natural motion but as the ideal geometrical case, of which concrete motions were but particular instances.

Hoewver, a great distance still separated the *De Motu* from the *Discourses.* For while Galileo realized full well that the true speed of bodies could be determined only in the vacuum, he also still believed that it depends exclusively on the specific weight of the moving body and hence rejected the idea that all bodies would fall through the void with the same speed.[76] Although for different reasons Galileo, no less than Aristotle, thereby rejected the idea that under certain conditions all bodies might fall with one and the same speed.[77] In fact, Galileo had not yet passed beyond the stage of a concrete dynamics, which treated motion as closely bound up with the nature of bodies. Admittedly, he had defined this nature in terms of the bodies' specific weight, but the latter still retained the power of determining the speeds of bodies directly. Close to Aristotle in this respect, Galileo never renounced the latter's fundamental axiom that a constant force engenders a uniform speed. The theory of acceleration illustrates this point almost by way of a caricature. For when the subsisting fraction of the impetus, which initially retards the return motion of a projectile, has vanished, the only active force to persist is that inherent in the weight of the body. The resulting motion in the plenum no less than in the void must be therefore uniform. Now, neither the statics nor the hydrostatics on which the *De Motu* relied so heavily could help Galileo to sever the traditional link between speed and weight; statics could help him toward a better evaluation of the factors responsible for the behavior of the moving body and of the medium; hydrostatics could suggest that the direction and speed of the motion depend on the relationship between the specific weight of the

[76]Ibid., p. 283. Moreover, in the single reference to the pendulum contained in the *De Motu* he remarked that if two weights, one of wood and the other of lead, were suspended by a cord, the lead weight "will certainly move back and forth for a longer interval of time;" ibid., p. 335.

[77]This suffices to disprove the legend that as early as 1590, Galileo had dropped different bodies from the top of the Tower of Pisa to demonstrate that they traveled down with the same speed; A. Koyré, *Galilée et l'expérience de Pise* (reprinted in *Études d'histoire de la pensée scientifique*, pp. 192 ff.), and R. Giaccomelli, *G. Galilei giovane e il suo "De Motu,"* pp. 9 ff.

moving body and that of the medium; in either case his dynamic analysis of natural motion remained based on considerations of weight alone. Moreover, it may very well be that hydrostatics was partly responsible for Galileo's failure, in the *De Motu*, to distinguish between the two forms in which the medium can affect the motion of a body, that is, by friction and lightening, for the former plays but little part in the motion of a solid through water. This very failure not only led Galileo to discard Buridan's ideas on the conservation of impetus but also made it even more unthinkable that all bodies should fall through the void with the same velocity.[78]

Moreover, though Galileo attributed the relative arrangement of different bodies in the universe to their specific weights, clearly defining that "such places have been determined for them by nature,"[79] and though he unified the conception of matter by ridding it of distinctions based on Aristotelian qualities,[80] he nevertheless preserved the traditional cosmological picture, albeit he chose to reconstruct rather than adopt it as such. Thus he still contended that the four elements were ranged concentrically around the center of the world, according to a natural order and in so doing perpetuated the alleged precedence of rest over motion. And while he banished the idea of the "supreme above" from the universe, he nevertheless retained the idea of the center of the world as the natural terminus of the motion of all heavy bodies and hence as an essential element in the dynamic study of motion. This may explain why Galileo, like Albert of Saxony before him, would encounter such difficulties when, many years later, he tried to determine the independent variable that must be assigned to the motion of bodies in free fall. It is not even certain that the cosmology implicit in the *De Motu* was completely devoid of finalist overtones; thus, certain passages suggest that the order of the world, far from being no more than the direct result of the mechanical interactions of bodies, is also due to causes of quite a different kind. It is remarkable, Galileo noted, that the rarer elements, that is, those enclosing matter in an ample space, are precisely those which occupy parts at the greatest distance from the center of the

---

[78]Among the other weaknesses of the *De Motu*, we might also mention Galileo's assent to a theory first propounded by Albert of Saxony and endorsed by Cardano and Tartaglia in the sixteenth century, namely, that the trajectory of a projectile consists of three distinct parts; *De Motu*, p. 337; cf. A. R. Hall, *Ballistics in the Seventeenth Century*, pp. 82–84.

[79]*De Motu*, p. 252.

[80]Ibid.

world, much as in a sphere the spaces become narrower as we approach the center and larger as we recede from it. "It was therefore with both prudence and justice that nature determined as the place of earth, one that was too narrow for the other elements, that is, near the center; and for the other elements places more spacious in proportion as the matter of each of these elements was rarer."[81] Though it would be quite wrong to take this semimythical language at face value, it is nevertheless beyond doubt that at the time he wrote the *De Motu* Galileo still believed that the arrangement of the elements reflected not only a mechanical necessity but also a certain "harmony."[82]

All this explains why the *De Motu*, despite its professed concern with mathematics, did not approach the study of natural motions in a manner that may be said to be radically different from that adopted by fourteenth-century writers. It is a fact that Galileo confined himself to the study of motion *quoad causas* and that, like Buridan or Oresme, he failed to see any connection between that study and the study of motion *quoad effectus*. And though there is good reason to believe that at the time he was already familiar with the medieval theory of *latitudines* (or variations in velocity), he made no attempt to apply that theory to *concrete* motions. Galileo, like his fourteenth-century precursors, never asked himself how precisely the force expressed in the fundamental dynamic relation (in his case, $W - M$) engenders speed in the moving body; or how the speed of the body increases under the influence of that force—in brief, how the course of the *intensio motus* in space and time can be described in the language of dynamics. Nor did the *De Motu* propose to go beyond evaluating the order of magnitude of the speed of moving bodies; it made no attempt to explain the action of the motive force in terms of increases in space with time. Moreover, mathematics served Galileo mainly as a source of simple examples with which to illustrate conclusions obtained by nonmathematical arguments.

The precise status of the *De Motu* is thus somewhat ambiguous. Though it appealed to Archimedes' geometrization of statics and hydrostatics, it failed to go beyond the concrete dynamics of the traditional type. Before dynamics could achieve the status Archimedes had bestowed upon statics, it had first to be provided with concepts of its own and not sim-

[81]Ibid., p. 253. In another version (ibid., p. 344), Galileo expressed the same idea in a form that recalled the myth of the demiurge in the *Timaeus*.
[82]Ibid., p. 253.

ply with concepts borrowed from other branches of science. It was in his *Mecaniche* that Galileo took his first steps in this direction; this was by no means its least important aspect.

Le Mecaniche I: The Tradition Mastered

In the *Mecaniche*, Galileo combined three distinct trends in statics into an orderly and rational theory of simple machines: the Peripatetic current, whose origins can be traced back to the *Mechanical Problems*, a work attributed to Aristotle but actually a product of his school; the Archimedean current; and a thirteenth-century current originated by Jordanus Nemorarius, who produced the first proof in statics based expressly on the idea of virtual displacements, and a correct explanation of motion on an inclined plane.

The *Mechanical Problems* of the Peripatetic school contained the earliest known analysis of why small forces can move great weights by means of the lever.[83] Treating this problem implicitly as one in statics, the author showed that the ratio of the weight moved to the weight moving it is in the inverse ratio of the distances from the center.[84] However, this was merely an observation, not a principle on which an explanation could be based; the explanation, in fact, went back to the problem of the balance, which, in turn, was based on a reflection about the circle.[85] Thus, in a rotating circle, the greater the distance of a point on its diameter from the center, the greater the circle that point will describe, and the greater also its speed.[86] The author then went on to produce a somewhat confused analysis which, however, went to the very heart of his doctrine: a body suspended from the end of a rotating radius is subject to two simultaneous displacements: a "natural," tangential motion, and a "forced" motion towards the center of the system. Now, the greater the radius, the greater the tangential motion[87] and hence the velocity of the suspended body; it is the increase in velocity associated with the greater distance from the center of rotation that explains why

---

[83]*Mechanical Problems*, 850 a 30–32 (Aristotle's *Minor Works*, Loeb Classical Library, London, 1955).
[84]Ibid., 850 b 1–2.
[85]"The facts about the balance depend upon the circle, and those about the lever upon the balance." Ibid., 848 a 11–13.
[86]Ibid., 848 a 15 ff.
[87]The same idea was also propounded by Blasius of Parma (P. Duhem, *Les origines de la statique*, Vol. 1, p. 150), and by Guido Ubaldo del Monte (ibid., p. 218).

a small force can balance a much larger one.[88] Passing on to the lever, the author then shows that equilibrium will be established whenever the lengths of the arms are such that the bodies suspended from their ends can move with speeds inversely proportional to their weights, that is, whenever we have $W_1/W_2 = V_2/V_1$. This approach was clearly a dynamic one: the author's intention was to describe the equilibrium of a system in terms of the motions of its extremities; hence one may be tempted to agree with Duhem and Vailati that the *Mechanical Problems* contained the germs of the principle of virtual velocities.[89] Nevertheless (and contrary to Duhem's view), we must insist that the *Mechanical Problems* failed to treat the products of the weights and their respective velocities as independent entities and, a fortiori, to base the state of equilibrium on the equality of these products. Thus, when they attributed the latter to the respective motions of the force and the resistance, they did so on the basis of experimental findings rather than of axiomatic principles.

The general treatment of equilibrium was however not the only contribution of the *Mechanical Problems*, for it also described the behavior of devices more complex than the lever, among them the wedge, the pully, the tackle, the forceps, and the nutcracker.[90] Despite its obscurities and omissions, the analysis was completely consistent in that it reduces the behavior of all these implements to that of the lever. Thus, as we saw, the facts about the balance depend upon the circle and those about the lever upon the balance; moreover, nearly all the other problems of mechanical movement could be shown to depend upon the lever as well.[91] Hence, though it failed to adduce a systematic theory of simple machines, the *Mechanical Problems* nevertheless blazed a trail that Galileo was able to explore to the fullest. In any case, there is little doubt that Peripatetic statics continued to hold sway until the classical period, as witness, inter alia, the references to it in the medieval *Liber karastonis*,[92]

---

[88]"Now the greater the distance from the fulcrum, the more easily it [the weight] will move." *Mechanical Problems*, 850 b 2–3.

[89]Duhem, *Les origines*, Vol. I, p. 8, and Vailati, *Scritti*, p. 96. It should, however, be stressed that the *Mechanical Problems* never treated virtual displacements but only virtual velocities.

[90]For the wedge, see 853 a f.; for the pulley, 853 a ff.; for the forceps, 854 a 33 ff.

[91]Ibid., 848 a 11–15.

[92]The *Liber karastonis* was a Latin translation from the Arabic of a ninth-century treatise by Thābit ibn Qurrā, which in turn professed to be a revision of a Greek work entitled *Causae Karastonis;* cf. Ernest A. Moody and Marshall Clagett, *The Medieval Science of Weights*, pp. 79 ff.

in the works of later writers (from Blasius of Parma to Guido Ubaldo del Monte), and by Galileo himself.

The Archimedean tradition was of quite a different kind. Because Archimedes set out to transform statics into a demonstrative science, he refused, in his theory of the lever, to treat equilibrium in terms of the virtual motions of the motive force and the resistance. Adopting the same procedure that had proved so fruitful in his geometric treatises, he began with seven postulates, the first three of which were based directly on experience and described the equilibrium and disequilibrium of the lever in terms of symmetry.[93]

1. Equal weights at equal distances are in equilibrium; and equal weights at unequal distances are not in equilibrium but inclined toward the weight that is at the greater distance.
2. If, when weights at certain distances are in equilibrium, something be added to one of the weights, they are not in equilibrium but incline toward the weight to which the addition was made.
3. Similarly, if anything be taken away from one of the weights, they are not in equilibrium, but incline toward the weight from which nothing was taken.

Of the last four postulates, these three involved the idea of a center of gravity, which played a crucial role in the actual proof:

4. When equal and similar figures coincide if applied to one another, their centers of gravity will similarly coincide.
5. In figures that are unequal but similar, the centers of gravity will be similarly situated.
7. In any figure whose perimeter is concave in [one and] the same direction, the center of gravity must be within the figure.

Though the exact importance of the sixth postulate, namely, "If magnitudes at certain distances be in equilibrium, [other] magnitudes equal to them will also be in equilibrium at the same distances," seems less obvious at first sight, it was, as we shall see, by no means less great than that of the other postulates.

On the basis of these postulates and five propositions,[94] Archimedes

---

[93] "On the Equilibrium of Planes" in *The Works of Archimedes*, ed. T. L. Heath, pp. 189–190.

[94] Of the first five propositions, note particularly the fourth, in which Archimedes established that the center of gravity of a magnitude made up of two other equal magnitudes is at the middle point of the line joining their centers of gravity; and the fifth proposition, which states that if three equal magnitudes have their centers of gravity on a straight line at equal distances, that of the entire system will coincide with that of the middle magnitude. In Corollary 1 of Proposition 5 this idea was extended to any odd number of magnitudes; in Corollary 2,

was able to enunciate his law of the lever (Proposition 6): "Two commensurable magnitudes balance at distances reciprocally proportional to their weights." From the preceding postulates and demonstrations the guiding thread of the argument becomes perfectly clear: equilibrium will be maintained whenever, if two unequal weights have been suspended from unequal arms, the fulcrum coincides with the common center of gravity of the two weights; in other words, whenever the centers of gravity of the two bodies are placed at distances from the fulcrum inversely proportional to their respective weights. As for proof, it consisted in the reduction of this general case to the simplest case, namely, to that in which two equal weights are suspended from equal arms and hence are in equilibrium for reasons of symmetry.

Suppose the magnitudes *A* and *B* to be commensurable and the points *A* and *B* to be their centers of gravity. Let *DE* be a straight line so divided at *C* that $A/B = DC/CE$.[95] It must then be shown that if *A* be suspended from *E*, and *B* from *D*, *C* will be the common center of gravity of the two taken together.

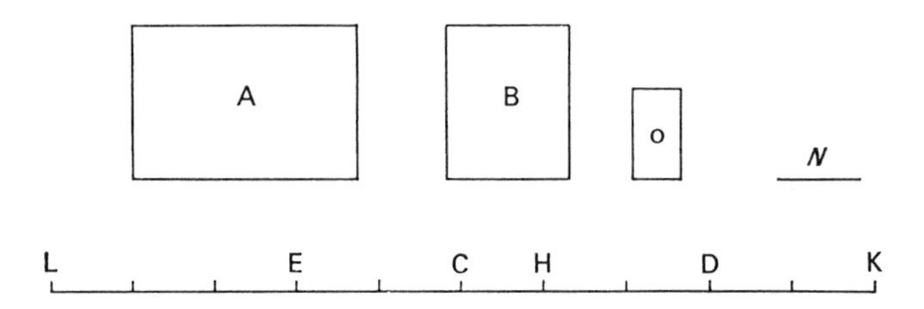

Since *A* and *B* are commensurable, so are *DC* and *CE*. Let *N* be a common measure of *DC* and *CE*; and make *DH* and *DK* each equal to *CE*, and *EL* (on *CE* produced) equal to *CD*. Then $EH = CD$, since $DH = CE$. Therefore *LH* is bisected at *E*, as *HK* is bisected at *D*. It follows that *LH* and *HK* must each contain *N* an even number of times.

Take a magnitude *o* such that it is contained as many times in *A* as *N* is contained in *LH* and as many times in *B* as *N* is contained in *HK*. Now divide *LH* and *HK* into parts each equal to *N*, and *A*, *B* into parts each equal to *o*: the parts of *A* will be equal in number to the parts of *LH* and the parts of *B* to those of *HK*. From the center of each of the

---

Archimedes asserted that if the central magnitude is neutralized or removed, in the system described, the center of gravity of the entire system will remain unchanged.

[95] Ibid., pp. 192–193.

segments N thus obtained suspend a magnitude o; then (Prop. 5, Cor. 2) the center of gravity of all the magnitudes o on LH will lie in E, and that of all the magnitudes o on HK will lie in D; hence, we may take it that A itself is applied at E, and B itself is applied at D. But the system formed by the parts o of A and B together is a system of equal magnitudes, even in number and placed at equal distances along LK. Moreover, since $LE = CD$ and $EC = DK$, it follows that $LC = CK$, so that C is the middle point of LK and consequently the center of gravity of the system placed on LK. Thus A acting at E and B acting at D will balance about the point C. By extending this proof to the case of two incommensurable magnitudes, Archimedes was able to give it full generality.

This proof, with its appeal to geometrical symmetry, as the reader may know, has been severely critized by Ernst Mach.[96] Archimedes had based his entire proof on the assumption that, once the two systems of weights o, corresponding to the distances LH and HK, are suspended, their respective actions can be determined from the position of their centers of gravity, E and D, so that the former can be likened to the magnitude A and the latter to the magnitude B. Now this argument presupposes that the action of a body on a lever depends exclusively on its weight and on the distance of its center of gravity from the fulcrum. In other words, Archimedes not only made implicit use of the principle of static moment (the equality of the moments A and B being the cornerstone of the proof) but also assumed that the latter can invariably be expressed as the product of W and L,[97] or rather as Mach himself put it, that the function $Wf(L)$ must necessarily assume the form WL. Archimedes' geometrical argument was therefore based on an extra-geometrical or mechanical principle which assumed that which had to be proved.

There are two ways of looking at Mach's criticism. We may first conclude that Archimedes was not aware that his argument was founded on a nongeometrical premise. If that was indeed what Mach meant,[98] it is hardly compatible with the facts. For, as we saw, Archimedes stated

[96]Ernst Mach, *Die Mechanik in ihrer Entwicklung historisch-kritisch dargestellt* (revised edition, 1908); English translation by T. J. McCormack, *The Science of Mechanics*, pp. 13 ff.

[97]Where W is the weight of the body, and L the distance of its point of suspension from the fulcrum.

[98]This is the view of Mach's interpretation given by E. J. Dijksterhuis in *Archimedes*, p. 293.

quite clearly that, by virtue of the position of the magnitudes *o* on the segments *N*, the center of gravity of all the parts of *A* will be at *E*, and the center of gravity of all the parts of *B* at *D*, so that *A* itself will act at *E*, and *B* itself at *D*, even though neither *A* nor *B* are given in their original form. Now it is mostly unlike that so rigorous a thinker as Archimedes would have argued in this way without realizing that his proof was based on the implicit assumption that the action of a magnitude on a lever depends solely on its weight and on the position of its center of gravity. And, in fact, let us take another look at Postulate 6: "If magnitudes at certain distances be in equilibrium, [other] magnitudes equal to them will also be in equilibrium at the same distances." This postulate may appear tautological; it is nevertheless true that it appears among a set of others containing explicit references to the center of gravity. Hence, are we really doing violence to Archimedes' thought if we claim that by the expression "other magnitudes equal to the first," he really meant *magnitudes of different form but of equal weight, whose center of gravity is at the same distance from the fulcrum?* There is little doubt that if we complete the sixth postulate in the spirit of its immediate context, we can absolve Archimedes from the taint of logical error.[99] However, this addition does not dispose of Mach's criticism in full. It is a fact that, in assuming that the action of a body on a lever depends solely on its weight and the position of its center of gravity, Archimedes took the principle of the lever for granted. In fact, his mistake was not so much that he smuggled mechanical considerations in through the back door as that he failed to conceptualize them and more particularly that he failed to make explicit the principle of static moment. Not only would his proof have been greatly simplified, but it could easily have been extended to all levers (as Galileo succeeded in doing) instead of remaining confined to levers of the first class. Reflecting as it does Archimedes' overriding concern with geometrization, this lack of generality also explains why Archimedean statics never succeeded in ousting the Peripatetic variety, despite the obvious shortcomings and confusion of the *Mechanical Problems*. It is, in fact, quite impossible to construct a general theory of statics covering the case of the straight lever, the bent lever, the inclined plane, and the equilibrium of floating bodies from the postulates contained in the *Equilibrium of Planes*; new hypotheses and principles

---

[99]Cf. ibid., p. 295.

had to be introduced for every new case.[100] In that respect, unsatisfactory though its method of exposition undoubtedly was, the Peripatetic analysis followed a more direct and far broader path.

For all that, Archimedes was the first to raise statics to the level of an autonomous science;[101] he was also the first to grasp the full importance of the center of gravity. Hence, we need not be astonished to find that the revival of Archimedean statics in the second half of the seventeenth century, as reflected particularly in the writings of Maurolicus, Commandinus, Guido Ubaldo del Monte, and Luca Valerio, should first have taken the form of a systematic inquiry into centers of gravity. Thus Maurolicus determined those of the pyramid, the cone, and the paraboloid of revolution; and Commandinus those of the latter, together with those of the truncated pyramid, the sphere, and the cylinder. All these writers applied the method of exhaustion in the spirit of Archimedes. Galileo himself was to discuss centers of gravity in 1587, and though his method, like that of his predecessors, was based on a double reduction to absurdity and hence brought no innovation, his exposition brought an original heuristic approach.[102] What these various contributions did above all was to restore mathematical statics to a clear place of honor. Finally, it should also be remembered that Archimedes, with his theory of the lever, had succeeded for the first time in expressing a physical problem in geometrical terms; by treating the arm of a lever as a weightless beam and the weights as plane figures, and by neglecting the convergence of the verticals toward the center of the world, he not only demonstrated but justified the idealization demanded by the con-

---

[100]As the example of Stevin shows only too clearly: a strict Archimedean in his treatment of the lever, Stevin based his theory of the inclined plane on a completely different principle, namely, the impossibility of perpetual motion. Now, that principle and its underlying dynamic intuition were much closer to the Peripatetic and medieval traditions than to the Archimedean.

[101]Duhem, *Les origines*, Vol. I, p. 11.

[102]Costabel (*Revue d'histoire des sciences*, Vol. VIII, 1955, pp. 116 ff.) has shown that Galileo's heuristic procedure was much broader than that of Archimedes and Commandinus; thus he reduced the determination of the center of gravity of a paraboloid of revolution to that of a straight line on which weights in arithmetic progression had been placed at equal distances—a problem that is easy to resolve: the center of gravity is found two-thirds of the way from the lighter end of the straight line. By then, however, Stevin was already using more direct procedures; he dispensed with the double reductio ad absurdum proof and boldly concluded that inasmuch as the difference between two points could, by continued subdivision, be shown to be less than any given quantity, there could be no difference; cf. Carl Boyer, *The Concepts of the Calculus*, pp. 100–102.

struction of mathematical physics. It is, in any case, certain that Archimedes provided by his example at least the model for, though not the sole inspiration of, Galileo's *Mecaniche.*

A fresh analysis of the lever and the first correct theory of the inclined plane constituted the essential contribution of medieval statics and thus of the third current informing the *Mecaniche.* Though little is known about the life of its chief exponent, Jordanus Nemorarius, his influence was wide and considerable.[103] As his theory of the lever shows quite plainly, Jordanus drew his main inspiration from dynamic considerations. Thus, like the Peripatetic author of the *Mechanical Problems*, he based the determination of equilibrium on the virtual motions of the ends of the balance beam. However, his arguments were far more rigorous and led to an important transformation, as witness the following proof of the principle of the lever:[104]

> Let the balance beam be *ACB* and the suspended weights *A* and *B*; and let the ratio of *B* to *A* be as the ratio of *AC* to *BC*. I say that the balance will not move in either direction. For let it be supposed that it descends on the side of *B*, and let the line *DCE* be drawn obliquely to the position of *ACB*. If then the weight *D* equal to *A* and the weight *E* equal to *B* are suspended, and if the line *DG* is drawn vertically downward and the line *EH* vertically upward, it is evident that the triangles *DCG* and *ECH* are similar, so that the proportion of *DC* to *CE* is the same as that of *DG* to *EH*. But *DC* is to *CE* as *B* is to *A*; therefore *DG*

---

[103]Medieval statics is embodied mainly in two works, the *Elementa Jordani super demonstrationem ponderum*, containing seven suppositions and nine propositions and the much more highly elaborated *Liber Jordani de ratione ponderis*, containing seven suppositions and fifty-five propositions, divided into four parts of which the first is the most interesting (Part 4, which dealt with dynamic problems will be discussed later; see footnote 148). According to the most recent conjectures, Jordanus, who seems to have taught mathematics at the University of Toulouse, flourished in 1230–1240; cf. Moody and Clagett, *The Medieval Science of Weights*, pp. 121–123. Duhem does not think that Jordanus could have been the author of *De ratione ponderis*, on the grounds that the book contradicts theorems of the *Elementa Jordani;* he contends that the real author was some unknown precursor of Leonardo. However, his argument is not convincing: Jordanus himself could easily have improved on his earlier text during the years that must have elapsed between the writing of the two; cf. Moody and Clagett, pp. 171–172. A third work, the *Liber Jordani de ponderibus*, contains seven propositions in statics and a Peripatetic commentary. According to Duhem, it is no more than a paraphrase of the *Elementa Jordani* (cf. Duhem, *Les origines*, Vol. I, p. 128), but Moody and Clagett (pp. 145–148) feel this could have been an independent work, and possibly even prior to the *Elementa*.

[104]*Liber Jordani de ratione ponderis*, Proposition 6; as quoted in Moody and Clagett, *The Medieval Science of Weights*, p. 183.

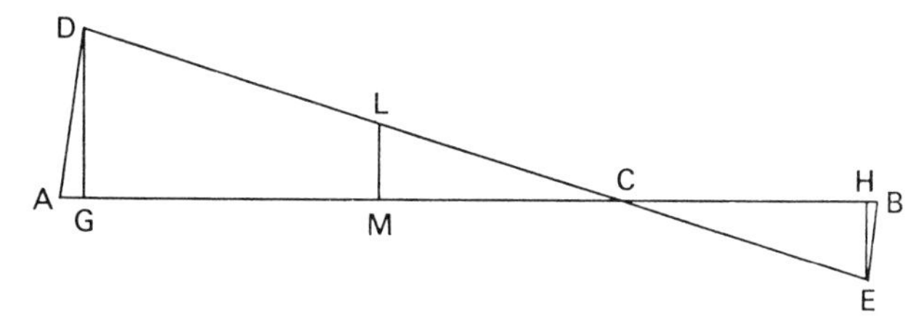

is to *EH* as *B* is to *A*. Then suppose *CL* to be equal to *CB* and to *CE*, and let *L* be equal in weight to *B*; and draw the perpendicular *LM*. Since *LM* and *EH* are shown to be equal, *DG* will be to *LM* as *B* is to *A* and as *L* is to *A*. But, as has been shown, *A* and *L* are inversely proportional to their contrary (upward) motions. Therefore what suffices to lift *A* to *D* will suffice to lift *L* through the distance *LM*. Since therefore *L* and *B* are equal, and *LC* is equal to *CB*, *L* is not lifted by *B*; and consequently *A* will not be lifted by *B*, which is what is to be proved.[105]

Jordanus, like his Peripatetic precursor, therefore based his proof of equilibrium on the respective displacements of the two weights. But whereas the Greek author of the *Mechanical Problems* had considered the speeds, Jordanus considered the distances: according to him two weights are in equilibrium whenever the ratio of the vertical distances traversed by them during the movement of the system is inversely proportional to the ratio of the weights themselves. Now by substituting distances for velocities, Jordanus not only made his analysis far more lucid but took the first step toward enunciating the general principle of virtual work, which, as the reader knows, is the true foundation of modern statics.[106] Can we then consider Jordanus as the real author of the principle, though he never stated it explicitly? To agree with Vailati that we can is certainly not justified—to appreciate this fact we need only note how, in his treatment of the bent lever, Jordanus continues to argue in terms of *finite* displacements.[107] Consider a bent lever *BCA*, with two arms,

[105]In other words, if the two weights *A* and *B* were not in equilibrium, but if one of them (for example, *B*) lifted the other, we would also have to assume that it was capable of moving through a distance equal to that through which the weight *L*, suspended at the same distance from the fulcrum, descends; but two equal weights suspended at equal distances from the fulcrum are necessarily in equilibrium.

[106]Duhem has stressed this point in *Les origines*, Vol. I, pp. 121–123 and 147; and Vailati has pointed out that the author of the *De ratione ponderis* based his determination of the equilibrium conditions of any mechanism to which weights are applied, on the product of these weights and their changes in height once the system has begun to move; cf. Vailati, *Scritti*, pp. 864–865.

[107]*De ratione ponderis*, Proposition 8; Moody and Clagett, *The Medieval Science of Weights*, p. 186.

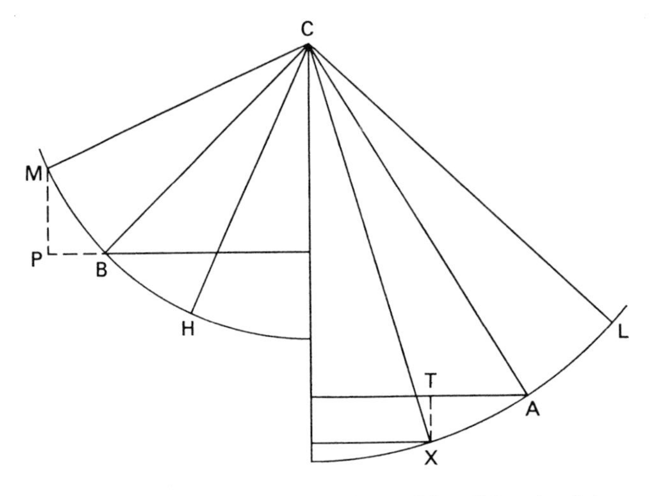

BC and *CA*, of unequal length, and let *C* be the fulcrum. Jordanus claims that if *B* and *A* are equidistant from a vertical drawn through the fulcrum, equal weights suspended from *B* and *A* will be in equilibrium. Let *CX* and *CL* be two radii making equal angles with the longer arm and *CM* and *CH* two radii making equal angles with the shorter arm *CB* and also with the angles *XCA* and *ACL*. Now assume that the weight *A* moves the equal weight *B*, and that the arms *CA* and *CB* will move so as to coincide with *CX* and *CM*, respectively. When the weight *A* has reached *X*, it will have descended through the vertical distance *TX*, while the weight *B*, on reaching *M*, will have been lifted through the vertical distance *PM*, which is clearly greater than *TX*. Hence, the descending weight would have lifted an equal weight over a larger distance than it has traversed itself, *which is impossible.* Now, though this proof again is based on the idea of virtual displacemets, the situation is quite different from that which we met in the case of the straight lever: instead of neutral equilibrium, we now have stable equilibrium, and the ratio of the moving force to the resistance is no longer equal to that of the space traversed by the resistance to the space traversed by the force.[108] In other words, the work done by the moving force is no longer equal to the work of the resistance, at least when we consider *finite* and not *infinitely small* displacements. Hence, it was only by virtue of the *indirect* form his reasoning took that Jordanus arrived

[108]Duhem, *Les origines*, Vol. I, p. 142.

at his conclusions; this very fact shows clearly that he did not employ or conceive of anything resembling the modern definition of work. Important though his contribution undoubtedly was, he failed to construct statics on a general principle.

Having made this point, we believe it is reasonable to say that the great originality of medieval statics lay in its treatment of the theory of the inclined plane. A truly novel idea presided over the analysis: that of *gravitas secundum situm*, or positional gravity. The new concept was clearly of experimental origin, for as even the simplest observation shows, a body on an inclined plane does not merely act by virtue of its own weight but also by virtue of the "quality" of its position. Thus Tartaglia, paraphrasing Jordanus,[109] asserted that "a body is heavier or lighter than another according to its place or position, because the quality of that place renders it more [or less] heavy than the other, even though both have the same weight." However, the *De ratione ponderis* did not confine itself to this rather vague statement; it began with several "suppositions" defining the concept of *gravitas secundum situm* and formulating the dynamic approach in which the theory would be developed. We can do no better than quote them in full:[110]

1. The movement of every weight is toward the center (of the world) and its force is a power of tending downward and of resisting movement in the contrary direction.
2. That which is heavier descends more quickly.
3. It is heavier in descending to the degree that its movement toward the center (of the world) is more direct.
4. It is heavier in position when in that position its path of descent is less oblique.
5. A more oblique descent is one which, in the same distance, partakes less of the vertical.
6. One weight is less heavy in position than another if it is caused to ascend by the descent of the other.
7. The position of equality is that of equality of angles to the vertical, or such that these are right angles, or such that the beam is parallel to the plane of the horizon.

These suppositions tell us clearly what we are to make of the concept of *gravitas secundum situm:* translated into modern terms (and if we liken weight to force) it refers precisely to the component of that force di-

---

[109]Nicolas Tartaglia, *Quesiti et inventioni diversi*, as quoted in Vailati, *Scritti*, p. 100.

[110]As quoted in Moody and Clagett, *The Medieval Science of Weights*, p. 174.

rected along its path of possible movement. More simply, for Jordanus himself, positional gravity represented a fraction of the natural weight of the body, namely that fraction which had to be counterbalanced on an inclined plane. Defined in that way, the concept of *gravitas secundum situm* had the double advantage of posing the problem of the inclined plane in clear terms and of offering a direct solution.[111] We need only recall Pappus's approach to the same subject, to appreciate this fact more fully.[112] Pappus too had based his argument on the changes in gravity which a body experiences on an inclined plane and had tried to determine these changes in terms of the force which the body exerts on a horizontal plane, that is, of the resistance it was thought to put up to motion on such a plane. The concept of *gravitas secundum situm*, on the contrary, provided the correct solution directly: it based the determination of the changes in power experienced by a heavy body on the total power of its gravity *along the vertical* and not on the action of a fictitious resistance acting along the horizontal. In fact, once it is realized that *gravitas secundum situm* is proportional to the obliquity of the plane, that is, to the ratio of its height to its overall length, the problem of the inclined plane can be solved without difficulty.

And this was precisely the procedure Jordanus adopted and generalized in his *De ratione ponderis*. Having first shown that the *gravitas secundum situm* of a given body remains unchanged throughout any one inclined plane,[113] he advanced the following proposition:[114] "If two weights descend along diversely inclined planes, then, if the inclinations are directly proportional to the weights, they will be of equal force in descending." Let *ABC* be a horizontal line, let *BD* be erected vertically on it and

[111]In fact, Jordanus first used the concept of *gravitas secundum situm* in his treatment of levers, asserting that the positional gravity of a body varies according to the amount of vertical descent which it can accomplish through a movement compatible with the constraint system. In the *Elementa*, this particular approach went hand in hand with confused or false propositions (Propositions 6 and 7; cf. Duhem, *Les origines*, Vol. I, pp. 118–121); in the *De ratione ponderis* the errors were corrected, but since the application of the concept of positional gravity to the lever was abandoned soon afterward, we have thought it unnecessary to complicate our exposition by delving further into the matter. It remains a fact, however, as Vailati has noted (*Scritti*, pp. 102–103 and 864–865), that the great importance Jordanus attached to virtual displacements in all his proofs was bound up with the new concept of positional gravity.

[112]For Pappus, see Duhem, *Les origines*, Vol. I, pp. 184–186.

[113]Proposition 9: "Equality of the declination conserves the identity of weight." See Moody and Clagett, *The Medieval Science of Weights*, p. 189.

[114]Ibid., p. 191.

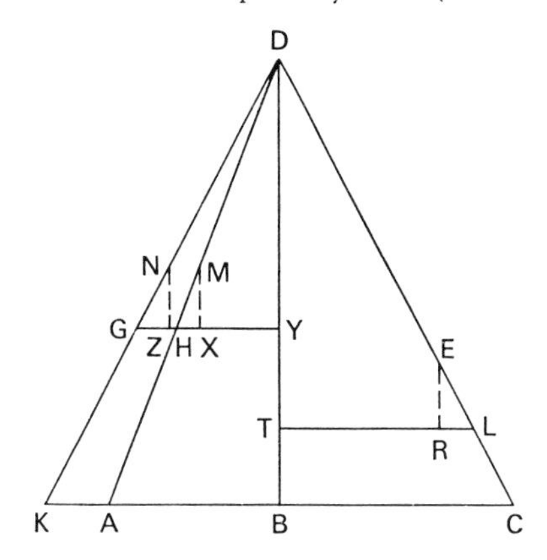

from *D* draw two lines *DA* and *DC*, with *DC* of greater obliquity. "I then mean by proportion of inclinations, not the ratio of the angles, but that of the lines taken to where a horizontal line cuts off an equal segment of the vertical." Now place the weight *E* on *DC* and the weight *H* on *DA*, and let *E* be to *H* as *DC* is to *DA*. "I say that those weights are of the same force in this position."

The proof was conducted along the same lines as the analysis of the straight lever. "For let *DK* be a line of the same obliquity as *DC* and let there be on it a weight *G* equal to *E*. If then it is possible, suppose that *E* descends to *L* and draws *H* up to *M*. And let *GN* be equal to *HM*, which in turn is equal to *EL*. Then let a perpendicular on *DB* be drawn from *G* to *H*, which will be *GHY;* and another from *L* which will be *TL*. Then on *GHY* erect the perpendiculars *NZ* and *MX;* and on *LT* erect the perpendicular *ER*. Since then the proportion of *NZ* to *NG* is as that of *DY* to *DG* and hence as that of *DB* to *DK*, and since likewise *MX* is to *MH* as *DB* is to *DA*, *MX* will be to *NZ* as *DK* is to *DA*—that is, as the weight of *G* is to the weight of H." Therefore, to assume that *E* can move *H* is tantamount to assuming that it can move a weight equal to itself situated on a plane of the same inclination and possessed of the same *gravitas secundum situm;* since this is impossible, it follows that *E* will not be able to lift *H* to *M*, and that the two weights will remain in equilibrium.

Because of the use it made of the concept of virtual displacements and also because of its solution of the problem of the inclined plane, the *De ratione ponderis* may be said to constitute not only a highly original but also a highly important contribution to statics. Small wonder, therefore, that it continued to make its influence felt until the end of the sixteenth century. In the fourteenth century in particular, its arguments were reflected in the writings of both the Mertonians and the Parisians; a little later Blasius of Parma drew from it the inspiration for his *Tractatus de ponderibus*,[115] and, as Duhem has shown so convincingly, Leonardo da Vinci also derived his basic ideas on statics from Jordanus' work. A new edition of *De ratione ponderis*, revised by Tartaglia,[116] appeared in 1565, and there is little doubt that it was with the help of this version that Galileo was initiated into medieval statics.

Galileo opened his study of simple machines in the *Mecaniche* with a careful definition of all the concepts he proposed to use. Thus he defined weight or gravity as "that tendency to move naturally downward which, in solid bodies, is found to be caused by the greater or lesser abundance of matter of which they are constituted."[117] However, the effect of a solid body on a mechanical system does not depend on its weight alone; it also depends on its distance from the center of rotation or, in the case of the inclined plane, on the obliquity of the latter. To that end, Galileo introduced the concept of moment (*momento*), which he defined as "the tendency to move downward caused not so much by the heaviness of the movable body as by the arrangement which different heavy bodies have among themselves." Thus, he contended, it was "through such moment that a less heavy body will often be seen to counterbalance some other one of greater heaviness, as in the steelyard a little counterweight is seen to raise a very heavy weight, not by excess of heaviness but rather by its distance from the suspension of the steelyard. This, combined with the heaviness of the lesser weight, increases its moment and its tendency to go downward, with which it may exceed the moment of the other, heavier weight. That moment is defined as that impetus to go downward that is composed of heaviness, position, and of anything else by which this tendency may be caused."[118] This last definition is remarkable in that it enabled Galileo

[115]Cf. Moody and Clagett, *The Medieval Science of Weights*, p. 233; and Duhem, *Les origines*, Vol. I, pp. 147 ff.

[116]*Jordani opusculum de ponderositate, Nicolai Tartaleae studio correctum* (Venice, 1565). Tartaglia himself died in 1557, and the work was completed by Curtius Trojanus.

[117]*Le Mecaniche*, Vol. II, p. 159.

[118]Ibid., p. 159; English text from I. E. Drabkin and Stillman Drake: Galileo Galilei, *On Motion and on Mechanics*, p. 151. Since Galileo had introduced the idea of forces other than weight (for example, muscular force, pp. 169 and 186), he obviously did not confine the

to apply the concept of moment in three distinct ways. First (and this application followed most obviously from the definition), moment can be used to refer to the force exerted by a heavy body as determined by its distance from the axis of the system; in that case it could be expressed by some such formula as $W \times L$ (where W represents the weight, and L the distance from the center of rotation) and would therefore be equivalent to the concept of static moment.[119] Second, moment was used in the *Mecaniche* and in several of Galileo's other works, to express a product such as $W \times V$, where W represents the weight of the body and V the velocity of its virtual motion; in this second sense the concept clearly reflects the dynamic approach of the *Mechanical Problems* and the works of Jordanus, and thus bears witness to Galileo's interest in all the currents that combined in the development of the science of equilibrium.[120] In his analysis of the inclined plane, Galileo used moment in yet a third sense, namely, to express increases or decreases in the intensity of the tendency with which a heavy body tends to move downward on an inclined plane.[121] Finally, Galileo added a definition of the center of gravity as "that point in every heavy body around which parts of equal moments are arranged," thus using *momento* in its first sense. It follows that if a heavy body is suspended from that point, it will not tilt in any direction; "the parts to the right will balance those to the left, the parts to the fore those to the rear, and those above will balance those below."[122]

---

idea of moment to motion of heavy bodies but also used it to express the magnitude of any physical force acting in any direction; however, since he believed that every force can be replaced by a heavy body suspended vertically (p. 169), we may take it that he applied the concept of moment chiefly to the tendency of heavy bodies to move naturally downward, as expressed in their weight.

[119]Galileo was thus the first to make explicit use of the concept of static moment, which Jordanus had used implicitly in his study of the bent lever (cf. Duhem, *Les origines*, Vol. I, pp. 142 f.). This particular usage of moment presided over Galileo's treatment of all simple machines.

[120]Galileo used the concept of moment in the second sense in his *Discorso intorno alle cose che stanno in su l'acqua* (*Discourse on Bodies in Water*), Vol. IV, pp. 68–69; in his *Dialogue on the Two Great Systems of the World*, Vol. VII, pp. 240–242; in his *Discourses and Mathematical Demonstrations* of 1638, Vol. VIII, pp. 329 f.; and in his treatise *On the Force of Percussion*, Vol. VIII, pp. 329–330. See also footnote 149, below.

[121]This particular usage will be discussed later.

[122]*Le Mecaniche*, pp. 159–160. Galileo's definition was inspired by those of Pappus and Guido Ubaldo del Monte. Pappus (*Collections*, VIII, Propositions I and II) had defined the center of gravity as a point in a body such that if the body were suspended from it, it would remain at rest. In practice, the

With the help of these definitions, Galileo was able to introduce the three principles which, he believed, governed the behavior of simple machines. The first of these, presiding over the general conditions of equilibrium, was based directly on the principle of the lever: that is, "unequal weights hanging from unequal distances will weigh equally whenever the said distances are inversely proportional to the weights," for only in these conditions will the moments of the two bodies be equal.[123] Imagine a cylinder *CFDE*, of uniform density and size throughout, and let it be suspended by its end points *C* and *D* from the balance *AB* equal in length to the cylinder. Now dividing *AB* equally at the point *G* and suspending the cylinder from this point, there can be no doubt that it will balance in *G*, because the line drawn from that point to the center of the earth would pass through the center of gravity of the cylinder. And of the latter, parts of equal moment would exist

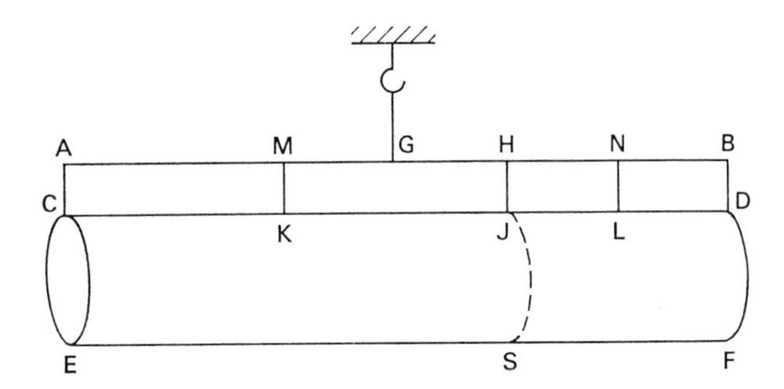

around such a line; it would be the same if from the points *A* and *B* there were suspended the two halves of the cylinder *CFDE*. Next suppose the cylinder to be cut into two equal parts along the line *JS*; it is obvious that the parts *CS* and *SD* would no longer be in equilibrium, so long as the strings *AC* and *BD* continue to constitute their sole supports. But suppose a new string *HJ* be added to support both parts of

center of gravity is the point marking the intersection of all the vertical planes that can be drawn through all the points from which the bodies are suspended successively. Guido Ubaldo del Monte, in his *Paraphrasis*, remarked that the two parts determined by each vertical plane (whatever the point of suspension) were merely "equiponderant," that is, they had the same effect, but separated from each other they would not have the same weight (cf. Duhem, *Les origines*, Vol. II, p. 7). Galileo converted Guido Ubaldo's remark into a definition and removed any remaining uncertainties with his concept of moment.

[123]Ibid., pp. 161–162.

the cylinder; it will sustain them jointly in their pristine state of equilibrium. Now add an additional string at *MK*, equidistant from *AC* and *HJ*; since the center of gravity of *CS* now lies directly below *MK*, it follows that we can cut the two strings *AC* and *HJ* without altering the position of *CS* with respect to the line *AH*. And if we do the same with the other part of the cylinder *JF* (cutting the strings *HJ* and *BD* and adding in the center the sole suspension *NL*), it is likewise apparent that this part will not vary its position or relation with *AB*. Hence, the parts of the whole cylinder being the same with respect to the balance *AB* as they have been all along, there is no doubt that equilibrium will still exist at the same point *G*. All that remains to be proved is that "the ratio existing between the weights *CS* and *SD* exists also between the distances *NG* and *GM*."[124]

Directly inspired by Archimedes, this proof no less than its model was not altogether beyond reproach. Thus, as Mach has noted, Galileo's artifice of restoring equilibrium with the help of the two new strings, *MK* and *NL*, was based on the implicit assumption that the moment of a heavy body depends exclusively on its weight and on the distance of its center of gravity from its point of suspension.[125] Since this assumption takes for granted what Galileo had set out to establish, his "geometrical" proof was, in fact, no more than the explication of a mechanical intuition to which nothing essential was added. However, Galileo's reasoning cannot be completely equated with that of Archimedes. First, it was more economical in that it embraced the case of incommensurable no less than of commensurable weights. But above all it was more general. Galileo expressly attributed the equilibrium of two weights to the equality of their static moments, thereby formulating the basic principle of the analysis of simple machines: In all such machines equilibrium will be maintained or established whenever the moment of the force remains equal to the moment of the resistance. In other words, the further away a force is from the fulcrum, the greater the resistance it can overcome.

Let us now suppose that we had a means of evaluating the spatial component appearing in the respective moments of the force and the resistance of every simple machine; would we not be able to say where the mechanical advantage lies and how great it is? Now this means

[124]The same proof is also found in the *Discourses* (Vol. VIII, pp. 152–154).

[125]Mach, *The Science of Mechanics*, pp. 17 ff.

does exist, and it is none other than the reduction of every simple machine (pulley, windlass, capstan, etc.) to a lever or system of levers. To the Archimedean principle of the equality of the moments of force and the resistance, Galileo added a further principle borrowed from the Peripatetics, thus fusing two traditions and generalizing the results of the later with the help of the earlier.

After he produced this generalization, all Galileo had to do—and this would be his third principle—was to show that the theory of simple machines must indeed fall into the province of statics. Consider a lever, in which a force B balances a resistance A. If we wish to raise the latter, we must communicate an additional impulse to the force B, large enough to surmount the resistance. But by upsetting the equilibrium, are we not, in fact, abandoning statics for dynamics? The solution is to treat the additional impulse as an infinitesimal quantity and to consider the displacement of the resistance by the force as the continuous transformation of one equilibrium position into another—in short, to assume that a system in equilibrium is set in motion by a force small enough to be neglected for all practical purposes. Galileo formulated this preliminary condition in impeccable fashion: "And since to make the weight B descend, any minimal (charge) added to it is sufficient (*ogni minima gravità accresciutali è bastante*), we shall leave out of account this insensible quantity and shall not distinguish between the power of one weight to sustain another and its power to move it."[126] For the first time the theory of simple machines had been fully integrated into statics and, what is more, by rigorous arguments.

All that remained was to make that theory itself as rigorous as possible, and Galileo's efforts in this direction are clearly reflected in the following two examples. Consider, first of all, a lever of the second class

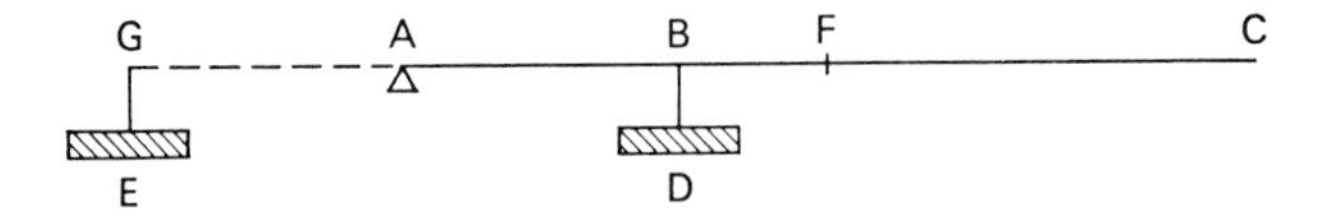

supported at one of its ends, and place the weight to be lifted, that is, the resistance R, anywhere on the line between the point of support

[126]*Le Mecaniche*, p. 164. Guido Ubaldo del Monte, for his part, maintained that a greater force is needed to move a weight than to keep it in equilibrium; see Duhem, *Les origines*, Vol. I, p. 224.

and the force or power P. Let the lever be ABC: the point of support in A, the weight to be raised (that is, D) in B, and let the force be applied at C. Had we placed B in F, the center of the lever, it is clear that the work of supporting the resistance R would have been equally divided between A and C, so that the resistance would have been balanced by the force $P = R/2$; that is, $P/R = AF/AC$ (or AB/AC). The same ratio can easily be shown to hold when B does not coincide with

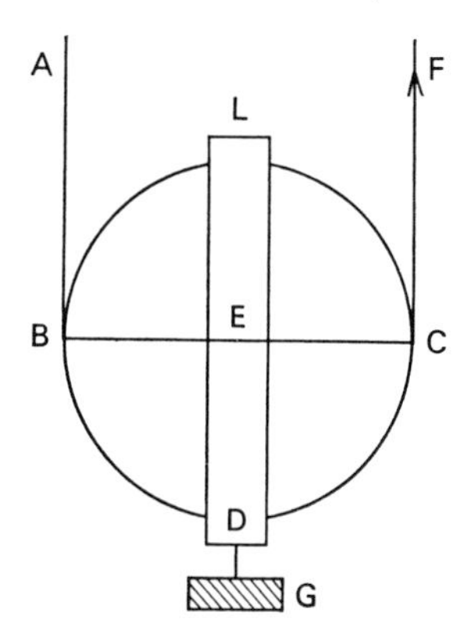

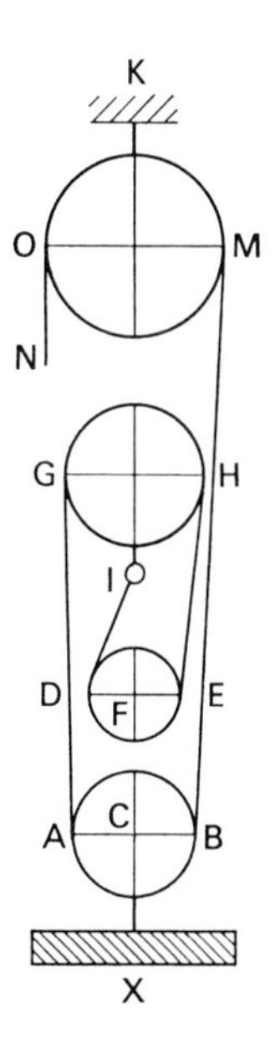

F. To that end let us increase AC by the distance $AG = AB$, remove the weight D, and suspend the weight $E = D$ from G. A force P applied at C will balance E whenever $P/E = GA/AC$. But $GA/AC = AB/AC$, and since $E = D$ by definition, it follows that $P/D = P/R = AB/AC$; in other words, the weight D suspended from B will have the same moment as the weight E suspended from G.[127] Consider next the pulley BDC, turning about its center E, and arranged in a box or housing LD, to which is suspended the weight G. Now pass about the pulley the cord $ABDCF$, whose end A is attached to a fixed point, and let the force act upward in the direction of CF. It suffices to construct the diameter BEC, to show that the arrangement reduces to a lever of the second class, from whose center (E) hangs the weight G (the resistance); the support of the lever is in B, the force is applied at C. From the preceding it follows that $P/R = BE/BC = 1/2$. The effect of the force will therefore have been doubled.[128]

Next, Galileo considered a system of four pulleys in two blocks. It is obvious that only the lower block (consisting of a combination of two levers of the second class) helps to diminish the work, the upper block simply transmitting the motion.[129] Now suppose the weight X to be suspended from the two lower pulleys AB and DE, the cord being wound round these and the upper pulley GH by the line $IDEHGAB$, the whole frame being sustained by the point K. "I say next that the force applied at M will be able to sustain the weight X when it is equal to one quarter of this. For if we imagine the diameters DE and AB and the weight hanging from the midpoints F and C, we shall have two levers, the fulcrums of which correspond to the points D and A; whence the force placed at B, or let us say at M, will be able to sustain the weight X if it is the fourth part of it."

Even though none of these conclusions were completely original, there is no doubt that Galileo had come a very long way since the *De Motu*: he had at last mastered the tradition, or rather he had reconstructed it for his own purposes. From the Archimedean current, he borrowed and made explicit the concept of static moment and the method of geometrical exposition; from the Peripatetic current he took the idea of reducing the behavior of simple machines to that of levers. From the first current he drew a demonstrative, from the second a

---

[127]Ibid., pp. 171–172.
[128]Ibid., pp. 173–174.
[129]Ibid., p. 176.

heuristic principle. Then, proceeding from the simple to the increasingly complex with the help of previously defined concepts and propositions, he was able to produce his first authentically "Archimedean" work, one in which, however, he no longer kept strictly to the letter of Archimedes, as he had done in the *De Motu*, but also used Peripatetic ideas, with the help of which, once he had treated them in the spirit of Archimedes, he was able to extend the Archimedean contribution in the most fruitful manner.

These contributions did not, however, exhaust the role of the *Mecaniche* in the history of statics. For, in that work, Galileo did more than demonstrate that simple machines can be reduced to levers; he also formulated a general principle based on the relative displacements of the moving force and the resistance and thanks to which all machines can be compared to one another and finally considered as being equivalent. To that end, he not only drew on the Archimedean and Peripatetic currents that had presided over his earlier analyses but also had recourse to the medieval tradition.

Imagine a lever *AB* in equilibrium about a point *C*, and apply to one of its ends, say *B*, "the least 'moment of heaviness.' "[130] The lever will immediately occupy the new position *DCE*. Now consider the dis-

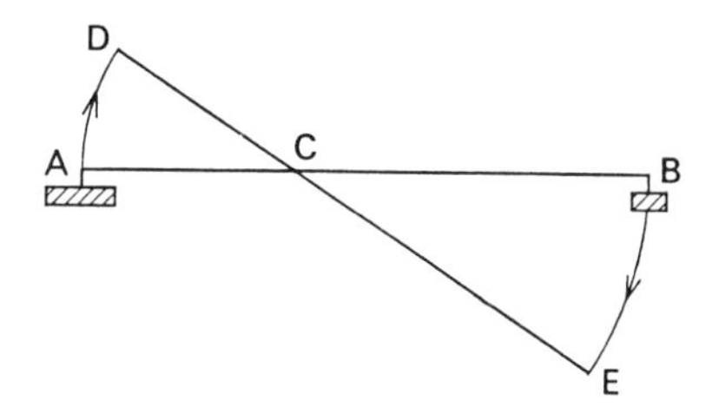

tances traversed by the extremities *A* and *B*, that is, the arcs *AD* and *BE*. Since the angles *ACD* and *BCE* are equal, it follows that the two arcs *AD* and *BE* must be "similar" and hence bear the same proportion as the radii *AC* and *CB*. Consequently the speed of the motion of *B* must be as many times greater than that of *A*, as the heaviness of the latter exceeds that of the former; hence, in evaluating the *moment* of either, it makes no difference whether we multiply its weight

[130]Ibid., p. 163.

by its virtual velocity or by its distance from the fulcrum: the two moving bodies will remain in equilibrium whenever relative increases in weight are compensated by corresponding decreases in speed, that is, whenever the ratio of the two weights is in inverse proportion to the ratio of the virtual velocities.[131] Despite the continued use of an Archimedean vocabulary, it is clear that the argument was inspired by Peripatetic and medieval ideas, elegantly combined into a new entity. From the first source came the concern with velocity and its compensatory role; from the second the concern with virtual displacements; for, as Galileo realized, it is with the help of the latter that the workings of simple machines can be reduced to a single principle. Galileo's own definition of the relationship between the force and the resistance in the behavior of levers illustrates this point most clearly. Let a lever *BCD* be in equilibrium about the point *C*, and let $CD = 5\ BC$;[132] the re-

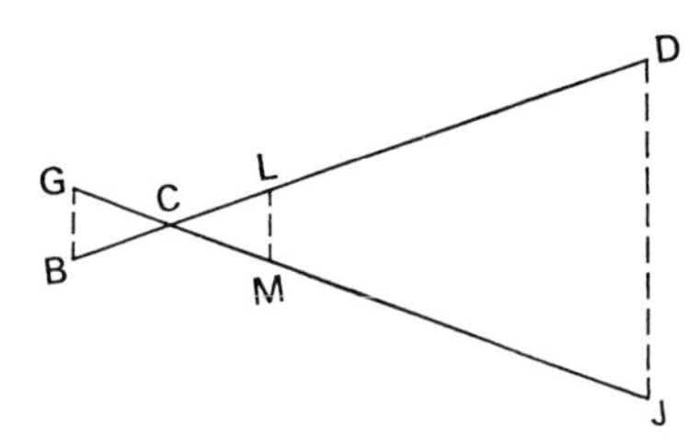

sistance (weight) is applied at the point *B*, the force at the point *D*. Let us increase the force until the lever takes the position *GCJ*; the force will have passed through the space *DJ*, that is, it will have traveled five times as far as the resistance *BG*. Let us now consider a point *L*, such that $LC = BC$; if the same movement is repeated, point *L* will move through the space $LM = BC$; and if we suppose that the force is applied directly at *L*, it will have to be equal to the resistance to produce the same effect. Now it is obvious that the same total force is needed, no matter what its point of application: to displace five times as great a resistance over five times as small a distance demands the same force as displacing five times as small a resistance over five times as great a distance. The two operations are completely equivalent, and the gain is one of convenience only.

---

[131]Ibid., p. 164. Galileo here uses *momento* in its second sense, that is, as the product of the weight and the velocity, and his subsequent analysis of equilibrium was conducted in frankly dynamic terms, quite especially in his development of the theory of the inclined plane.

[132]Ibid., pp. 166–167.

However, it was in his treatment of the inclined plane that Galileo developed this principle of equivalence to its fullest extent.[133] He began by asking himself whether it was not possible, on an inclined plane, to have a resistance *E* cover the same distance as a much smaller force *F*. When *F* descends, does *E* not move up *AC* through an equal distance? In fact, as Galileo went on to explain, "although the movable body *E* will have passed over all the line *AC* in the same time that the other heavy body *F* has fallen through an equal interval, nevertheless the heavy body *E* will not have been removed from the common center of heavy things more than the distance of the perpendicular *CB*, while

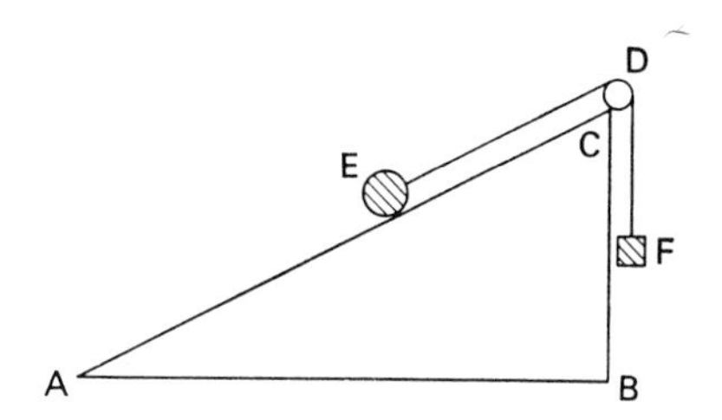

the heavy body *F* descending perpendicularly will have dropped by a space equal to the whole line *AC*. And since heavy bodies do not have any resistance to transverse motions except in proportion to their removal from the center of the earth, then the movable body *E* not being raised more than the distance *CB* in the whole motion on *AC*, while *F* has dropped perpendicularly as much as the whole length *AC*, we may rightly say the travel of the force *F* has the same ratio to the travel of the force *E* as the line *AC* has to the line *CB*, or as the weight *E* has to the weight *F*." In other words, the respective travel of the force and of the resistance must not be evaluated by reference to the actual distances each has traversed, but by reference to the projections of these distances on a vertical axis.

A general principle thus emerges from his analysis, namely, that in all simple machines a gain in force or power is invariably offset by a corresponding loss in travel or, as Galileo himself liked to put it, "whatever is gained by force is lost in time and in speed."[134] Since, moreover, Galileo's proof was based explicitly on the distances traversed, we are entitled to think that he was in fact asserting the absolute equivalence of the work done by the moving force to the work done by the resist-

---

[133]Ibid., pp. 185–186. [134]Ibid., p. 185.

ance.[135] The progress accomplished since the thirteenth century was striking: while the *De ratione ponderis* had merely employed virtual displacements to adduce indirect proofs of particular problems, Galileo made them the basis of a *general*[136] principle applicable to all simple machines. Moreover, the general principle he formulated, for the first time, opened the way for a complete revision of statics; the principle had merely to be treated as an axiom, as Descartes was to do later,[137] for all the basic propositions of statics to be fitted into a coherent and much simpler whole. One last step will then suffice—the substitution of infinitely small for finite displacements—for the principle of equivalence to become the principle of virtual work. And though Galileo foresaw neither the form nor the role his principle was to assume in the hands of Descartes and Jean Bernouilli,[138] there is no doubt that a continuous thread ran through his work and that of his successors.

Thus Galileo's new attitude to the tradition had crystallized out. Having first founded his demonstrations on a fruitful combination of Archimedean and Peripatetic currents, he skillfully completed and unified the results by borrowing from medieval statics both the theory of the inclined plane and also the idea of comparing the respective displacements of the force and the resistance. The thirteenth-century tradition was thereby linked to the other two traditions, and their synthesis in *Le Mecaniche* was a direct prelude to the modern reconstruction of statics on the principle of virtual work. It is remarkable that Galileo,

---

[135] Galileo later extended this principle to hydrostatics when explaining why the surfaces of liquids in communicating vessels of different internal diameters invariably rise to the same horizontal plane. Cf. *Discourse on Bodies in Water*, Vol. IV, pp. 77–78.

[136] Guido Ubaldo del Monte, for example, used the idea of equivalence when treating the pulley, but rejected it in the case of the lever; Duhem, *Les origines*, Vol. I, p. 223.

[137] Descartes may well have drawn his inspiration from Galileo himself; Mersenne's translation of the *Mecaniche* appeared in 1634, and the *Explication des engins par l'ayde desquels on peut, avec une petite force, lever un fardeau fort pesant* followed in 1637. Torricelli's axiom ("Two bodies connected together cannot move spontaneously unless their common center of gravity descends") also was no more than a generalization of Galileo's contention in the *Discourses* (Vol. VIII, p. 215) that a system formed of several bodies cannot move spontaneously unless its center of gravity approaches the common center of gravity to which all heavy bodies tend.

[138] It was Descartes who was the first to realize that virtual displacements must be considered as being of infinitely small magnitudes; see Duhem, *Les origines*, pp. 337–338. The first formulation of a principle of virtual work was given by Jean Bernouilli, ibid., Vol. II, p. 269.

though a staunch Archimedean, should nevertheless have appreciated the great importance of the medieval current and that he should have looked to it for what indications he needed to generalize his demonstrations. Not only did he thus take a view opposite to that of his chief predecessors and contemporaries,[139] but by assimilating and then dominating the tradition he was able to turn his *Le Mecaniche* into the first truly modern treatise on statics.

Le Mecaniche II: Toward a New Dynamics

Although its avowed purpose was to complete the theory of simple machines with an explanation of the workings of the screw, Galileo's analysis of the inclined plane took him far beyond statics. In particular, a careful analysis of the changes in the downward tendency of heavy bodies on inclined planes led him to a description of gravity of which one can justly say that it contained the germs of a completely new dynamics.

When he first approached the subject of inclined planes in the *Mecaniche*, Galileo must already have been familiar with the solution, probably thanks to Tartaglia's new edition of the *De ratione ponderis*. Hence, his originality lay not so much in the results as in his approach, that is, in his manner of formulating the problem. Why, he asked himself, should it be easier to maintain a heavy body in equilibrium on the

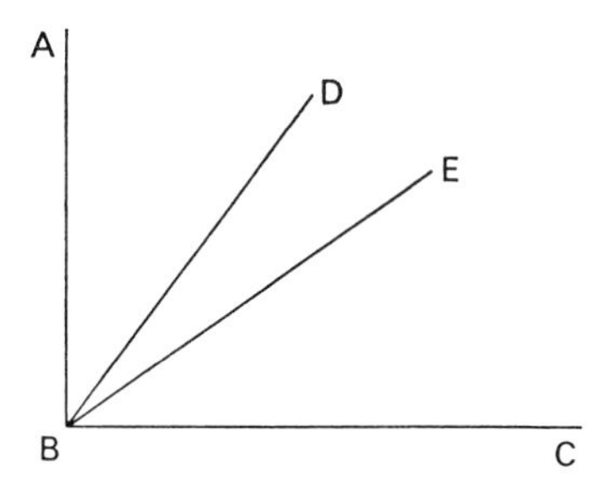

plane *BE* than on the plane *BD* and, a fortiori, than on the vertical *BA*? "This," Galileo explained, "comes from its having a greater *impeto* to go downward along the line *AB* than along *EB*, and along *DB* than along *EB*";[140] or, to use an expression he introduced a few pages later

[139] Among them Guido Ubaldo del Monte and Benedetti; cf. Duhem, *Les origines*, Vol. I, pp. 225–229.

[140] *Le Mecaniche*, p. 180.

and again in the *Discourses*, thus emphasizing the great importance he attached to the matter: the reason for the resistance of a heavy body to motion upon an inclined plane to decrease as the angle of inclination diminishes is that the "moment of descent" (*momento di discendere*) on which its downward motion depends decreases in the same proportion from a maximum value along the vertical *AB* to zero along the horizontal *BC*. Hence, the real problem posed by the inclined plane is not, as Galileo's precursors had thought it to be, to determine the correct equilibrium ratio of the weights of two bodies joined by a string and placed on two contiguous inclined planes, but to determine the ratio in which the moment of descent of a body, which is maximal along the vertical, is reduced by decreases in the inclination of the plane on which the body rests. In his attempt to establish this ratio, Galileo most ingeniously solved the case of the inclined plane by reference to the much simpler case of the lever.

Let *ACJ* be a circle with the diameter *AC* and suspend two weights of equal moment from the extremities *A* and *C*, so that the line *AC* being a lever moving about the center *B*, the weight *C* will be sustained by the weight *A*.[141] Now imagine that the arm *BC* of the lever be in-

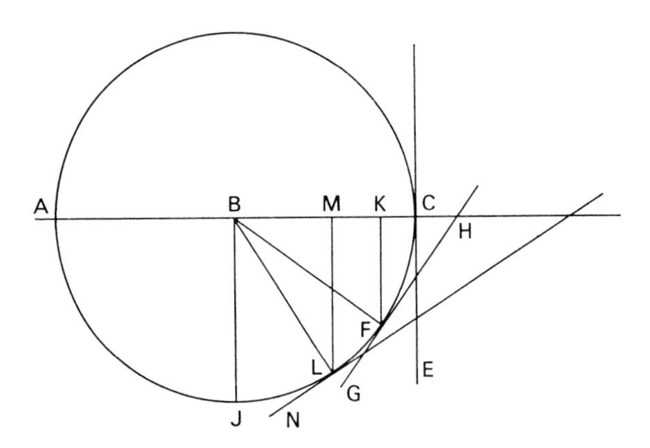

clined downward along the line *BF*, so that we now have a bent lever *ABF* instead of the straight lever *ABC*; then the moment of the weight *C* will no longer be equal to the moment of the weight *A* but will be diminished in precisely the same ratio as the line *BK* (the projection of *BF* on *BC*) is shorter than the line *AB*. The moment of the weight *C*

[141] Ibid., pp. 181–183.

will decrease even more if we bend the arm of the lever until it coincides with the line *BL*. By two successive steps, Galileo then proceeded from the lever to the case of the inclined plane. First of all, he imagined the circumference *CFLJ* to be "a surface of the same curvature" placed beneath the weight *C* so that the weight is constrained to descend along it; in that case, its downward *impeto* will decrease as the inclination of the surface on which it rests. In other words, if it starts from *F* instead of *C* or from *L* instead of *F*, the decrease of its downward *impeto* will be directly proportional to the decrease in its static moment. Galileo's second step was more remarkable still. Consider the points *C*, *F*, and *L*; the "inclination" of the circumference at each of these points is identical with the "inclination" of the tangents *CE*, *FG*, and *LN*; we can therefore assume that the moments of descent of a heavy body at the points *C*, *F*, and *L*, causing it to move along *CE*, *FG*, and *LN* are identical to the moments of descent with which it would start its downward motion from the points *C*, *F*, and *L* if it moved along the circumference. But we have already seen that the tendency of a heavy body to move downward along the circumference decreases with its static moment; its moments of descent along *CE*, *FG*, and *LN* will therefore vary in the same proportion, so that the moment of descent of a heavy body along *CE*, for example, will have the same ratio to its moment of descent along *FG* as its static moment in *C* to its static moment in *F*, that is, as *BC* is to *BK*. The rest follows as a matter of course. Extend *FG* until it cuts the diameter *AC* at point *H* and consider the inclined plane *FH* whose altitude is *FK*; since the tangent *GH* has the same inclination throughout its length, it follows that the moment of descent of a heavy body placed on *FH* will be in the same ratio to its moment of descent along the vertical as *BK* is to *BC*. Now since the triangles *BKF* and *FKH* are similar, it follows that $BK/BF = FK/FH$, and finally, since $BF = BC$, that $BK/BC = FK/FH$. In other words, the moment of descent of a weight on an inclined plane has the same ratio to its moment of descent along the vertical as the height of the inclined plane has to its length.[142]

[142]Ibid., p. 183. The whole proof, as the reader will have noted, rests on the proposition that at *F*, the point of contact between the circumference and the tangent *GH*, a heavy body will have the same downward *impeto* or moment of descent, regardless of whether it moves down the circumference or the tangent. Now this argument presupposes that the slope of a curve in a given point is identical with that of the tangent at this point. Galileo therefore relied on a proposition whose justification would only come with the introduction of the calculus a century later.

A first consequence of this analysis was undoubtedly that it bestowed a new meaning on the concept of moment. By "moment," the reader may remember, Galileo referred to the action of a body on a mechanical system, or more precisely to "that tendency to go downward composed of heaviness, position, and of anything else by which this tendency may be caused." We saw that Galileo used this concept in a sense akin to static moment and also in another (close to the medieval tradition), namely, to refer to a magnitude of the type $W \times L$. To these two senses, the theory of the inclined plane added yet another, since by "moment," in the sense of "moment of descent," it referred to a magnitude conceived as a function of both gravity and the oblique of the plane. Now this third sense differs profoundly from the first two. For when it is used in the sense of static moment, or as the product of the weight and the virtual velocity, a *momento* cannot be treated in isolation from the system of constraint; its value can be determined only by reference to the point of the system to which the body is attached. Not so with the moment of descent on an inclined plane, for that moment remains unchanged throughout the length of that plane; moreover it has a dynamic as well as a static significance, for, once the body has begun to move down the inclined plane, its moment of descent is a measure of the force on which that motion depends. Now it is precisely to the part it can play in dynamic analyses that the concept of moment of descent owes its exceptional importance.

What, in fact, was Galileo's approach to dynamics at the time he wrote the *Mecaniche*? His dominant idea was that a force inherent in all bodies causes them to tend toward the center of the earth; to this force he attributed a twofold role: first, it was responsible for the weight of the body, or rather for the pressure it exerted when at rest, and second, it helped to explain the spontaneous downward motion of all heavy bodies free to approach the center of the earth. His use of the term *gravità* to refer both to the weight of a body and the cause of its motion toward the center clearly reflected this view, that the same physical entity presides over two distinct functions: the gravific and the motive. Thus consider a heavy body whose descent is blocked: the resistance opposing its natural motion, and which is equal to its downward tendency is, in fact, indistinguishable from its weight, which explains why traditional writers on dynamics, ever since Aristotle, had all considered weight as the obvious measure of gravity as a motor force. Their dynamic analyses were, however, based on two distinct conceptions. For Aristotle and his orthodox successors, the motor force impelling a heavy body toward the center was proportional to its individual (absolute) weight; subsequently a greater interest in the experimental evidence together with a return to Archimedean ideas during the second half of the sixteenth century suggested that the speed of fall of a heavy body varied not as its individual but as its specific weight. This view, which Galileo adopted in his *De Motu*, was undoubtedly an improvement on the earlier one: asserting that the downward tendency of a body varies as the quantity of matter contained in a given volume of that body meant, to some extent, introducing the concept of "mass" into dynamics. Moreover, there is no doubt that the replacement of absolute with specific weight opened the way for a correct appreciation of the role of the medium, as witness the close links between the *De Motu* and the *Discourses* of 1638. But even though it was preferable to the old, the new interpretation was but one stage on the road to progress; in no way did it put an end to the fundamental identification of the motor and gravific functions, nor did it allow any independent treatment of the former.

But let us return to the inclined plane. On it a heavy body will, in the absence of friction, have a spontaneous tendency to move downward and to resist any motion in the opposite direction, both effects being the direct result of the action of the natural motor force. Let us decrease the angle of inclination of the plane: the downward tendency will decrease as well, as will the resistance to motion in the opposite direction. But since the specific weight itself has not changed in the process, it cannot serve as the measure of the motor force, so that it is no longer possible to equate gravity *qua* motor force with gravity *qua* weight, and a new concept is needed for the dynamic description of the downward motion of heavy bodies. By introducing his moment of descent, Galileo showed, first of all, that he was fully aware of this need: being proportional to the inclination of the plane, the moment of descent expresses the intensity of the force on which the downward motion depends, and hence may be treated as the proper cause of motion. In one case at least, a new magnitude had taken the place of weight in the dynamic description of natural motions; for the first time in the history of mechanics, the traditional identification of the gravific with the motor function of gravity had been eliminated. Nor was that all. For, in addition to underlining the impossibility of describing the motor function purely in terms of the gravific function, the new concept also made it possible to consider the motor function as such. Thus while it had previously been lumped together with weight under the general heading of gravity, the motor force impelling all heavy bodies downward was transformed into a distinct physical entity—into a distinct object of thought.[143]

True, the *Mecaniche* considered only motion along an inclined plane, not the case of freely falling bodies, so that the scope of the concept of moment of descent remained restricted. It is clear, however, that one of the most essential conditions for the construction of modern dynamics had been met. The moment of descent as we saw, was defined as being constant throughout the length of a given plane; a heavy body rolling down that plane will accordingly be impelled by a constant force. But in asserting that this was the case, was Galileo not, in fact, challenging the fundamental axiom of traditional mechanics? When he described the action of the natural motor force on an inclined plane by the moment of descent, that is, by a constant, was he not preparing a refutation of the established idea that the motor force is proportional to the velocity? In fact, when he eventually succeeded in providing an intrinsic analysis of naturally accelerated motion, the close affinity between moment of descent and moment of speed, whose continuous summation endows every accelerated motion with a special character, became self-apparent. In this way, the moment of descent undoubtedly led Galileo to anticipate the Newtonian definition of force.[144]

An even more immediate proof of the fruitfulness of Galileo's approach is the rejection, in the *Mecaniche*, of one of the most persistent prejudices

---

[143]This is also borne out by the fact that Galileo at no time referred to the medieval concept of *gravitas secundum situm* with which he was certainly familiar, and which his own moment of descent might appear to resemble at first sight: both expressed the effect of the inclined plane on a heavy body, and both were proportional to the obliquity of the plane; moreover, it was on their evaluation of the *gravitas secundum situm* or of the moment of descent that both Jordanus and Galileo based their theories of the inclined plane. Yet there was an essential difference between the two. The sole purpose for which Jordanus introduced the concept of *gravitas secundum situm* was to conceptualize the apparent decrease in the weight of a heavy body on an inclined plane, and the chief problem he attempted to solve was to determine the ratio between the total weight acting along the vertical and the decreased weight acting along the inclined plane. At no time did Jordanus attempt to distinguish the motor force from the weight; instead, as the second supposition of the *De ratione ponderis* shows very plainly, he continued to relate changes in speed directly to the decrease in weight.

[144]This approach also paved the way for the unification of statics and dynamics, of which we shall speak later.

of traditional mechanics. Consider a sphere rolling down a perfectly smooth and polished plane. We know that the motive force takes the form of a moment of descent; we also know that if the inclination of the plane is decreased, the downward tendency will decrease proportionally. Let us now suppose our plane to be parallel to the horizon, as happens, for instance, with the surface of a frozen lake or pond.[145] The disposition to motion which appears when the plane is tilted "only by a hair" is suddenly eliminated, and the body remains immobile. However, the moment of descent is not responsible only for the downward tendency of the sphere; it also accounts for its resistance to motion in all other directions. That being the case, what is the precise significance of the complete disappearance of the moment of descent on a horizontal plane? Would the sphere put up no resistance at all to lateral motion along this plane or would some force, however small, still be needed to disturb its rest, in accordance with the tenets of traditional mechanics? Galileo's answer was quite unequivocal: "It is perfectly clear," he wrote, "that on an exactly balanced surface the ball would remain indifferent and questioning (*dubia*) between motion and rest so that even the least force would be sufficient to move it, just as on the other hand any little resistance, such as that merely of the air that surrounds it, would be capable of holding it still."[146] This was the only logical solution, once it had been assumed that through the moment of descent, the force residing within the moving body is responsible for its motion no less than for its resistance to motion; released from both tendencies on the horizontal plane, the sphere cannot but remain "indifferent and questioning between motion and rest." The importance of this conclusion cannot be overstated. According to traditional mechanics, all matter had an innate tendency to come to rest, and this conception, which Galileo still shared when he wrote the *De Motu*,[147] made it impossible to conceive of the conservation of motion. By contrast, when in his discussion of the inclined plane he suggested that the specific weight, that is, the quantity of matter, could not provide a valid expression of the natural motor force and when he showed that the latter could disappear without the

---

[145] *Le Mecaniche*, p. 179.

[146] Ibid., p. 180.

[147] In the dialogue version of the *De Motu*, Vol. I, p. 373, Galileo wrote: "*Ista autem motus aeternitas ab ipsius terrae natura longe abesse videtur, cui quies jucundior quam motus esse videtur*." The same interpretation, which had vitiated Pappus's analysis of the inclined plane, was also adopted by Guido Ubaldo del Monte. (cf. Duhem, *Les origines*, Vol. I, p. 225).

former being modified in any way, Galileo pointed to at least one situation in which matter does not display that tendency to come to rest which was previously thought to be its essential characteristic. Not that the *Mecaniche* formulated the principle of inertia; the reader will look in vain for any discussion of the behavior of a sphere once it had been set in motion on a horizontal plane. It is nevertheless true that his study of the inclined plane forced Galileo to proceed to a radical revision of the prevailing view on matter, a revision which, as we shall see, eventually led him to the principle of the conservation of motion.[148] Thus while Galileo did not yet describe motion as a state, he no longer debarred it a priori from being described as such.[149]

The *De Motu* and the *Mecaniche* thus enable us to draw up a fairly complete picture of the development of Galileo's early thought: starting with a commentary on the traditional views, he gradually transcended them and in so doing paved the way for the emergence of the "new science." And as he proceeded, he gained an increasingly better appreciation of the meaning of mathematicization in natural philosophy. Mathematicizing, as the *Mecaniche* showed quite plainly, means first of all substituting quantitative for qualitative concepts and introducing into physics the deductive order of geometry. But it also means making a clear break with sense experience, abandoning the complexity and con-

---

[148]To appreciate Galileo's progress, we need merely contrast his conclusions with the dynamic arguments following the analysis of the inclined plane in the *De ratione ponderis*, whose author did little more than repeat the traditional commonplaces (cf. Duhem, *Les origines*, Vol. I, pp. 136 ff.). Thus he assumed that a freely falling body is accelerated not only because its weight increases as it falls but also because, carrying part of the medium along with it, it exerts an ever-increasing pressure on the lower strata. At no time did the medieval author use his analysis of the inclined plane to distinguish between motor force and weight, let alone define the first without reference to the second.

[149]Hence, we cannot agree with those commentators who have argued that it was chiefly as the product of weight and velocity ($W \times V$) that the concept of *momento* played a role in the development of the Galilean science of motion (see particularly Wohlwill, *Galileo Galilei und sein Kampf für die Copernicanische Lehre*, Vol. I, p. 143; and Koyré's *Études galiléennes*, Vol. II, p. 20). In fact, it was in the form of moment of descent, and in this form only, that the concept of *momento* exerted a fruitful influence on Galileo's thought. Not that Galileo did not make any attempt to construct dynamic magnitudes based on *momento* defined as the product of the weight and the velocity, but in every case—whether in his treatment of the centrifugal force or of the force of percussion—all these efforts proved completely abortive (for his treatment of the centrifugal force see Chapter 5; for the force of percussion see Appendix 3).

tingency of concrete situations for the analysis of the most general cases with the help of a small number of factors; in short, mathematicizing means *idealizing*. Galileo himself explained how the "superhuman Archimedes" had guided him along this path.[150] Thus he used the authority of Archimedes to explain that two weights suspended from the arms of a balance may be said to make right angles with the balance, even though, strictly speaking, the verticals passing through their respective centers of gravity are not parallel but converge toward the center of the universe—for this was precisely how Archimedes himself had dealt with the matter in his *Quadrature of the Parabola*. The rigor of Archimedes' demonstration provides formal proof that such abstractions are indeed legitimate and that simplifying does not necessarily mean tampering with the truth. The physicist speaking as a mathematician can safely ignore useless complications, and since it suffices that the weights make equal angles with the balance, he can say equally well that "either the suspended weights actually do make right angles, or else that it is of no importance."[151] Under "the protecting wings" of Archimedes, Galileo thus felt entitled to introduce the essential concept of "geometrical license," that is, the physicist's right to postulate the universal and ideal conditions that reign in geometry.

In the *De Motu*, this methodological principle still remains a hope;

---

[150] *De Motu*, p. 300.

[151] In fact, this is of some importance in physics, if not in geometry. Thus Galileo himself assumed that all bodies had a fixed point or center of gravity, which tends to approach the common center of heavy bodies along a straight line. Now, a fixed center of gravity implies a force that is constant in magnitude and direction; this condition is not satisfied unless the verticals passing through the centers of gravity of heavy bodies are parallel; if they converge toward the center of the earth, the force ceases to be constant in magnitude and direction, and we can no longer attribute fixed centers of gravity to heavy bodies (cf. Duhem, *Les origines*, Vol. II, pp. 152–185). When Galileo asserted that two weights suspended from the arms of a balance make right angles with the latter, he was simply adopting (though for reasons he did not yet fully comprehend) the only physical approach compatible with his conception of the center of gravity. In fact, in his treatment of equilibrium no less than of motion, he never succeeded in discarding the idea of a common center of heavy bodies which he identified with the center of the earth. The consequences for dynamics will be examined at the end of Chapter 7; with regard to statics, we might mention briefly that the breach was filled by Torricelli, who was the first to dispense with references to the common center of heavy bodies, by basing his analysis exclusively on the center of gravity of the moving bodies themselves. The famous axiom bearing his name ("Two heavy bodies connected together cannot move spontaneously unless their common center of gravity descends"), thus marked the end of a long era of theoretical equivocation.

the only example of "geometrical license" found in it, namely, the use of the void to determine the true velocity of freely falling bodies, had no dynamic repercussions. The *Mecaniche*, by contrast, showed to what great extent Galileo had meanwhile learned to apply geometrical license to full advantage first in his analysis of equilibrium conditions and next in his study of the inclined plane. While the *De Motu* never even conceived of the possibility of describing the phenomenon of motion as such, that is, regardless of the moving bodies, the *Mecaniche*, though not ostensibly concerned with the problem of motion, nevertheless pointed the way to a dynamics in which gravity would be defined as an independent force. Again, while the *De Motu* retained many essential features of the traditional cosmology, the *Mecaniche*, though it did not completely strip the center of the earth of its motive role, tended to abolish the privileged position rest had always been thought to enjoy over motion. On a horizontal plane, as we saw, matter was said to be quite indifferent to rest or motion. It is to geometrical license that we must also attribute some of the most fruitful propositions of the *Mecaniche*. Thus it was geometrical license, that is, the postulation of ideal conditions, that enabled Galileo to consider as infinitely small the additional impulsion needed to set a simple machine in motion and hence to show that the study of such machines falls into the province of statics. Also, it was geometrical license that enabled him to identify the decrease in *momento* experienced by a heavy body on a curved surface with the decrease in *momento* which it would experience when placed on the tangent to the curved surface at the precise point on which it is located. Finally, geometrical license alone enabled him to show that, on a horizontal plane, even a minute force can always set even the largest sphere in motion.[152] Without doubt, these were no more than isolated flashes of insight, and Galileo still had to travel a long way before he arrived at the perfection of the *Dialogue* or *Discourses*; for all that, it was in the *Mecaniche* that he first revealed the tremendous advantage of the geometrical method in the analysis of physical problems. The preparatory period was ended; that of the great discoveries could begin.

---

[152]This shows what a vast distance separated Galileo's ideal cases from the divine cases of his fourteenth-century precursors.

II The Copernican Doctrine and the Science of Motion

Introduction

From 1602 to 1642 Galileo focused his attention on two topics: the justification of Copernicanism and the construction of a science of the motion of heavy bodies on mathematical principles. The result was the publication of the *Dialogo supra i due sistemi del mondo (Dialogue Concerning the Two Chief World Systems)* of 1632, and of the *Discorsi e demostrazioni matematiche intorno a due nuove scienze (Discourses on Two New Sciences)* of 1638. Though the same period also saw the publication of his *Geometric and Military Compass* (1606), of his *Siderius Nuncius* (1610), of his *Discourse on Bodies in Water* (1612), of his *Letters on Sunspots* (1613), and of his *Saggiatore* (1623), none of these works contained any fundamentally new themes. Thus the *Geometric and Military Compass* was a practical manual; the *Discourse on Bodies in Water* merely bore witness to Galileo's continued interest in statics and hydrostatics and, apart from important reflections on the forces ensuring the cohesion of bodies, was remarkable chiefly for the mastery with which it expressed physical problems in geometrical terms. The *Siderius Nuncius*, the *Letters on Sunspots*, and the *Saggiatore* were all based on astronomical observations and reflections and, as such, foreshadowed many of the general cosmological arguments used in the *Dialogue*. In the main, these works offered Galileo an opportunity to lay claim to many of his earlier discoveries and to develop some of them in greater detail; and they in no way questioned the two projects to which he devoted his creative energy for more than forty years.

Now it is easy to show that in the foundation of classical mechanics, Galileo's cosmological contribution played as important a role as did the *Discourses*. In fact, the heliocentric idea had set astronomers a difficult poser ever since antiquity: how to reconcile the diurnal motion of the earth with everyday observations and mechanical experience. To solve this problem or rather this series of problems, Galileo took an unprecedented step: he set out the traditional objections to the diurnal motion in the form of mechanical arguments and showed that, in that way, the objections fell away and the rotation of the earth could be reconciled with common experience. Admittedly, his was still a far cry from the creation of a new celestial mechanics, and Galileo, unlike Kepler, adduced no causes capable of rendering the new cosmology intelligible; for all that, his translation of the traditional objections to the earth's diurnal motion into the language of mechanics was a landmark in the history of the science of motion.

In fact, the true problem is to appreciate the real scope of this contribution, as distinct from the contribution of the *Discourses*. Indeed, Galileo's own procedure suggests a solution. As we just saw, his refutation of the objections to the earth's diurnal motion rests on principles and concepts borrowed from the science of motion, so that there are some grounds for supposing that the mechanical part of his cosmology was but the application to a special case of a system he had elaborated beforehand, that is, the system he would develop in full in his *Discourses*. In that case, should we not examine his justification of the earth's diurnal motion in conjunction with his ideas on accelerated motion?

Though not altogether mistaken, this approach nevertheless has grave disadvantages, not least the minimization of the obvious differences between the *Dialogue* and the *Discourses*. Thus the former, unlike the latter, made no direct use of the very precise theorems and propositions of the science of naturally accelerated and projectile motions, and referred to them only incidentally, for instance, when attacking Peripatetic opinions,[1] or when attempting to explain certain tidal phenomena.[2] Moreover, the two works employed quite distinct principles. Thus one that plays a crucial part in the *Dialogue*[3] is not even mentioned in the *Discourses*; moreover, what principles appear in both works differ in the manner of their formulation no less than in their application. Now, the most cursory examination of the problems discussed in the *Dialogue* and the *Discourses* will show that, far from being accidental, these differences are a direct consequence of the disparate character of the two works.

Thus the theory of naturally accelerated motion faced Galileo with a problem that, though not simple, was nevertheless limited in scope: its solution called for a minimum number of concepts and variables, some of which had been developed by Galileo's precursors. By contrast, when he countered the traditional objections to the earth's diurnal motion with mechanical arguments, Galileo had to turn pioneer, defining principles and concepts as he went along. However, his simple mechanical ideas proved quite impotent in the face of problems so complex that they could only be solved satisfactorily with the help of Newtonian mechanics. Hence, far from providing him with an opportunity to apply the general concepts and results embodied in the *Discourses*, the

[1]*Dialogue* II, pp. 248 ff.
[2]Ibid., IV, pp. 474 ff.
[3]The principle of mechanical relativity.

Dialogue, based at it was on the mechanical analysis of problems that still proved intractable to geometrical treatment, was bound to follow a separate path. To fuse the two into a single body of concepts and propositions would be doing violence to Galileo's own thought and lending it a degree of cohesion that it could not possibly have possessed.

Moreover, the danger of this approach is not only that it distorts the contribution of Galileo's cosmology to classical mechanics but also that it completely ignores one of the most important aspects of that contribution. The traditional philosophic objections to the Copernican doctrine were based, inter alia, on three fundamental theses: the theory of natural motions, according to which every body can be associated with one type of motion, and one type only; the theory of the elements, according to which the earth is quite unlike any celestial body; and the treatment of motion as a process, from which it followed that the only ontologically meaningful motion of the earth had to be in a straight line toward the center. Now, besides being obstacles to the Copernican doctrine, these theses also provided the real premises of the traditional science of motion. Clearly, therefore, when he replaced these theses with "Copernican premises," compatible with a motion of the earth, Galileo rendered a double service to mechanics: he laid the foundations for the construction of a general science applicable to celestial no less than to terrestrial bodies; and he freed local motion from its Aristotelian constraints. Though less spectacular, this second contribution was no less far-reaching than the first.

The foregoing remarks may serve to explain the order in which we propose to discuss Galileo's main work. Rather than lump the cosmological contributions of the *Dialogue* quite arbitrarily together with the mathematical arguments of the *Discourses*, we shall look at them consecutively and examine each in its proper context; in that way we hope to arrive at a valid appreciation of their respective significance in the construction of a new science of motion. This procedure strikes us not only as being more faithful to the historical facts but also as the only one capable of reconstructing the living complexity of Galileo's thought in all its fruitfulness—and, if need be, even in its contradictions.

It should, however, be stressed that if we begin with Galileo's cosmological contribution, that is, with his justification of the Copernican doctrine, we wish in no way to imply that it came first in Galileo's own scientific development. On the contrary, all the evidence tends to show that from the year 1602, which we have chosen as the end of his preparatory period, he divided his interest equally between Copernicanism and the motion of heavy bodies. This is borne out by two series of letters, one dating back to the years 1602–1604 and the other to the years 1609–1610. In 1602, writing to Guido Ubaldo del Monte, Galileo announced his discovery of the isochrony of pendular oscillations; at the same time he also propounded a law he would later incorporate in the *Discourses:* "If from the highest or lowest point in a vertical circle there be drawn any inclined planes meeting the circumference, the times of descent along these chords are each equal to the other."[4] Two years later, in a letter to Paolo Sarpi, Galileo announced that the distance traveled by a body falling from rest is proportional to the square of the elapsed time (the "square law").[5] That same year he also made an important contribution to cosmology with his *Lectures on the New Star of October 1604*, in which he showed himself a convinced Copernican and challenged the Peripatetic division between heaven and the earth, and quite particularly the idea that the moon had a smooth and polished surface.[6] Five years later, a new spate of letters shows what great headway he had meanwhile made in cosmology and mechanics. In February 1609 a letter to Antonio de' Medici bore witness to his considerable advances in the study of projectile motion;[7] in July of the same year a letter from Luca Valerio tells us that Galileo had discovered the principle on which he proposed to base the analysis of naturally accelerated motion.[8] Finally, in May 1610 Galileo informed Belisario Vinta, Secretary to the Grand Duke, that he had written three books of propositions on local motion, "a new science I have constructed from first principles."[9] The same letter also tells us that Galileo had not neglected his cosmological studies; indeed, the *Siderius Nuncius* had just been published, and it is

---

[4]*Letter to Guido Ubaldo del Monte*, November 29, 1602, *Opere*, Vol. X, p. 98. Galileo added that he was not yet able to adduce a proof of this law. While we can date most of his astronomical discoveries fairly accurately, we know nothing at all about the dates on which he first perfected his analysis of uniformly accelerated motion, when he established that the velocity of such motion increases in proportion with the time, or when he first formulated the principle of the conservation of motion.

[5]*Letter to Paolus Sarpi*, October 16, 1604, Vol. X, p. 115.

[6]*Opere*, Vol. II, p. 283; for Galileo's practical Copernicanism, see ibid., p. 622.

[7]*Letter to Antonio de' Medici*, February 11, 1609, Vol. X, pp. 228–230.

[8]*Letter from Luca Valerio to Galileo*, July 18, 1609, Vol. X, pp. 248 f.

[9]*Letter to Belisario Vinta*, May 7, 1610, Vol. X, p. 351.

quite certain that at the time a first version of the *Dialogue* had already been drafted; among the other works Galileo hoped to conclude in the near future he mentioned a treatise entitled *De sistemate seu constitutione Universi*, "an immense subject embracing philosophy, astronomy, and geometry."[10] In view of this inextricable mixture of currents, we have thought it best to analyze Galileo's various contributions by following an essentially logical clue, namely by proceeding from the more general, and sometimes the more confused, to the more particular and the more perfect, as we look first at his cosmological theories and then at his theory of motion.

[10]Ibid., p. 351; another reference to this first draft of the *Dialogue* can be found in the *Siderius Nuncius* III, p. 75. Moreover, the *Letters on Sunspots* contained numerous passages that would be repeated almost verbatim in the *Dialogue*.

4 The Construction of a Copernican Cosmology

There is little doubt that, even at the beginning of the seventeenth century, the main objection to the Copernican doctrine continued to be of a philosophical nature. To endow the earth with motion and hence to put it on a par with celestial bodies meant flying in the face of one of the most basic tenets of traditional philosophy, namely, the difference in essence between heaven and earth. And while it challenges the accepted view of the heavens, the Copernican doctrine also involved a physical, as distinct from philosophical, claim, whose justification was an essential preliminary to any proof of the validity of the heliocentric system.

Now, the distinction between heaven and earth held a remarkable place in Aristotle's cosmology: it stood halfway between his basic premises —the orderly structure of the cosmos, the theory of natural motions, the theory of the elements—and his actual conclusions, for instance, that the earth stood still at the center of the universe. It thus provided an essential element of the cohesion of his general doctrine. In other words, if it fell, the whole Aristotelian edifice would come tumbling down with it. By unifying heaven and earth and then presenting the principles of a Copernican cosmology, that is, changing the framework in which mechanical thought was traditionally steeped, Galileo cleared the way for a completely new theory not only of celestial but also of terrestrial mechanics.

The New Universe

Aristotle had defended his distinction between heaven and earth with the help of two distinct arguments: the doctrine of natural motions and the theory of contraries.

Let us first consider the doctrine of natural motions. Combining sense experience with the assumption that only the circle and the straight line represent perfect geometrical lines, that is, lines whose successive parts, when superposed, can be made to fit precisely, Aristotle was led to posit that simple local motion can appear in two forms only: the circular and the rectilinear.[1] Then, harmonizing this assumption with the a priori orderly structure of the cosmos, he was able to infer the existence of

[1]For this argument see Chapter 1.

three simple natural motions: straight upward motion away from the center, straight downward motion toward the center, and circular motion about the center.[2] From the further assumption that each of these motions is appropriate to a different element, Aristotle concluded that the number of elements must be limited. Thus while earth and fire (to which he later added water and air) provided the physical foundations, and accounted for the number of rectilinear motions, a fifth element, the ether, was the basis of simple circular motion. Moreover, simple natural motions held the key not only to the possible number of elements but also to some of their most essential properties. Thus, because all its successive parts are alike, natural circular motion is perfect and hence confined to bodies in actual existence, and as such ingenerable and incorruptible; while rectilinear motion, being indeterminate in itself[3] and definable only by reference to external factors, is typical of imperfect bodies, that is, bodies possessed of gravity, levity, generability, and corruptibility. This explains why terrestrial bodies, which alone are generable, corruptible, and changeable, move in straight lines, while celestial bodies, which are ingenerable, incorruptible, and constant, move in circles.

Now, the soundness of any argument depends on the soundness of its premises, and Galileo thought it was possible to show that several of Aristotle's implicit or explicit assumptions were completely arbitrary or even false. Thus Aristotle had based his classification of simple motions on purely geometric considerations: they were said to be simple because they were made along simple lines—the straight and the circular—and because all their parts were similar. Now, as Galileo pointed out, the same definition also applied to the case of the cylindrical helix, which is anything but a simple geometrical figure.[4] But Aristotle went even further astray in his implicit assumptions—for instance, when identifying the different simple *natural* motions after his definition of simple lines. This classification, as Galileo noted, made sense only on the assumption that there could be a single center in the universe. But if, instead, we assumed that the real universe contained thousands of centers, we should also have to grant the existence of thousands of nat-

[2] Cf. *Dialogue* I, pp. 38–39. (For an excellent English version, the reader is referred to Stillman Drake's translation, published by the University of California Press, 1962).

[3] Ibid., p. 42.

[4] Ibid., p. 40. Aristotle himself believed that a helix was not a uniform figure; cf. *Physics* V, 4, 228 b 24–25.

ural circular motions and thousands of natural motions upward and downward.[5] In ignoring this possibility, Aristotle had simply been "pulling cards out of his sleeve," and likewise when, at almost every step of his argument, he appealed to other, quite arbitrary, considerations. Thus he had based his theory of simple bodies on the alleged correspondence between motions and elements, that is, on the assumption that all simple bodies were mobile by nature. The arbitrary character of his assumption is the more apparent in that Aristotle himself had explained elsewhere that all things constituted by nature had "within themselves a principle of motion and rest,"[6] thus contradicting his own words.[7] Again, if all natural motions were simple, why associate some rather than others with particular bodies? Why, in other words, were those bodies to which Aristotle attributed natural circular motion unable to describe rectilinear motions as well?[8] By ignoring this possibility, Aristotle once again selected a premise that, though essential to his own argument, was completely baseless.

Nor were these logical errors the only flaws in Aristotle's theory: the basic idea that there was a direct correspondence between simple motions and the elements had no physical foundations either. Before we can deduce the simplicity of bodies from the simplicity of their motions we must first establish that such motions are indeed as simple as Aristotle assumed them to be. But there is no justification for the view that every straight motion is simple in itself or that it is confined to simple bodies. Thus when a piece of wood, compounded of the elements earth and air, is set in motion, the earth in it will draw it downward while the air in it will tend to impel it in the opposite direction. Far from being simple, the motion of the wood will therefore have a mixed character, so that "the apparent simplicity of the motion no longer corresponds to the simplicity of the line."[9] But perhaps there is a better criterion of simple motions by which Aristotle's theory of the elements might yet be saved. Can we, for instance, say with Simplicio that the simple rectilinear motion characteristic of simple elements is always swifter than that which comes from their predominance in mixed bodies?[10] But this assumption, too, is erroneous: not only do such mixed bodies as

[5]*Dialogue* I, p. 40
[6]*Physics* II, 1, 192 b 13–15.
[7]*Dialogue* I, p. 39.
[8]Ibid., p. 40.
[9]Ibid., pp. 40–41.
[10]Ibid., p. 41. The three interlocutors presented in the *Dialogue* are: Salviati, Galileo's mouthpiece; Sagredo, the unprejudiced man; and Simplicio, the champion of Aristotle.

lead or bronze fall more swiftly than do simple bodies, but if we argue in that way we are, in fact, identifying the simple motions from the simple bodies, whereas Aristotle's original assumption was that the latter were determined by the former. In short, Aristotle's claim that the natural upward and downward motions of the two fundamental terrestrial elements, fire and earth, could be established a priori was quite untenable. And because that claim was the basis of his separation of the sublunary from the superlunary world, that separation proved to be spurious as well. Thus when Simplicio argued that "the motion of heavy bodies is contrary to that of light ones; but the motion of light ones is seen to be directly upward, that is, toward the circumference of the universe; therefore the motion of heavy bodies goes directly toward the center of the universe,"[11] he, like Aristotle before him, was simply assuming what is in question, namely, that the center of the earth coincides with the center of the universe. This can be illustrated by a simple geometrical example. Let us consider a mobile point in a circle; if the point travels in a straight line in any direction whatsoever, it will doubtless go toward the circumference of the circle and, continuing its motion, will arrive there. But it will not always be true that anything moving by the same line in the opposite direction would go toward the center; this will happen only if the point started out from the center. Thus when Aristotle argued that, because fire rises straight up from the surface of the earth toward the circumference of the universe, heavy bodies must necessarily tend toward its center, he simply took it for granted that the center of the earth coincided with that of the universe.[12]

The importance of these critical remarks cannot be overstated. Aimed though they were at the doctrine of natural motions and the theory of the elements, they also undermined the most crucial argument by which Aristotle had tried to establish the immobility of the earth.[13] For, if it is indeed impossible to show that heavy bodies move toward the center of the universe, and hence that the center of the earth coincides with that of the universe, then nothing can prevent us from placing it elsewhere. Moreover, though heavy bodies undoubtedly tend to approach the center of the earth, they do so simply to return to the whole from

[11] Ibid., p. 59. Cf. Aristotle, *On the Heavens* II, 14, 296 b 8–12.

[12] *Dialogue* I, pp. 60–61. Galileo had used the same argument previously in his *Letter to Ingoli* (1624), Vol. VI, pp. 535–539.

[13] Aristotle, *On the Heavens* II, 14, 296 b 15 ff.

which they have been separated by force. And what is true of terrestrial bodies may equally well be true of celestial bodies. "Why may we not believe that the sun, moon, and other world bodies are also round in shape merely by a concordant instinct and natural tendency of all their component parts?"[14] In other words, there is no reason why rectilinear motion should not be as appropriate to celestial as it is to terrestrial bodies.

Though Aristotle's theory of natural motions was the most weighty philosophical argument against the Copernican doctrine, it was not the only one; another, and perhaps even more ill-founded, objection of Peripatetic physics was based on the theory of contraries, which in turn was based on two central ideas, namely, that generation and corruption are confined to bodies with contrary qualities, because everything that is generated or destroyed comes into being or disappears through the agency of a contrary;[15] and that the only bodies to possess contrary qualities are those "movable in contrary motions."[16] Moreover, contrary motions could be only two in number, toward the center and away from the center, and only the elements earth and fire (to which water and air were added later) could partake of such motions. Since the third type of simple motion (around the center) had no contrary (because the other two motions were contrary and any one thing can have one contrary only), it follows that the element corresponding to that motion could have no contrary itself. Now since circular motion is confined to celestial bodies, these alone can be ingenerable and incorruptible.

Galileo had no difficulty in showing that this argument, too, could be demolished with Aristotle's own weapons.[17] For, according to Aristotle himself, the contrary quality responsible for the generation or corruption of a body had to reside outside the latter. "You must therefore say," Galileo asserted, "that when a body becomes corrupted, this is

---

[14]*Dialogue* I, p. 58.
[15]*On the Heavens* I, 3, 270 a 12 ff. *Generation and Corruption* II, 2, 1 and 3.
[16]*Dialogue* I, p. 63.
[17]"And if you, Simplicio, know how to teach me nature's method of operation in quickly begetting a hundred thousand flies from a small quantity of musty wine fumes, showing me what the contraries are in that case and what things corrupt and how, I should esteem you even more than I do; for I can comprehend these matters not at all. In addition I should very much like to understand how and why these corrupting contraries are so favorable to daws and so cruel to doves, so indulgent to stags and so impatient with horses, as to allow the former many more years of life (that is to say, of incorruptibility) than they give weeks to the latter." Ibid., p. 64.

occasioned by a quality contrary to its own residing in another body."[18] In other words, to make a celestial body corruptible, "it is sufficient that there are in nature bodies having a contrariety to the celestial bodies. And such are the [terrestrial] elements, if it is true that corruptibility is contrary to incorruptibility."[19]

Needless to say, Simplicio, as Aristotle's champion, was ready with his reply: "No, this is not sufficient, my dear sir. The elements become altered and corrupted because they contact and mix with one another, and thus can exercise their contrariety. But celestial bodies are separated from the elemental, by which they are not even touched—though they, indeed, do influence the elements. If you want to establish generation and corruption in celestial bodies, you must show that contrariety exists among them."[20] In fact, this is easily done. Let us begin with the contrariety of the terrestrial elements. We know that the source from which it is derived is the contrariety of natural upward and downward motions, which in turn is derived from the contrary principles of lightness and heaviness. "According to you yourself, levity and gravity occur in consequence of rarity and density; therefore density and rarity will be contraries."[21] But these two qualities exist abundantly in celestial bodies as well, for, as the Peripatetics themselves believe, the stars are merely denser parts of their region of heaven.[22] And being the seat of such contrariety, celestial bodies cannot possibly be free of generation and corruption.

In a final effort to defeat his opponent, Simplicio tried to draw a distinction between rarity and density in celestial bodies and rarity and density in terrestrial bodies: "In celestial bodies density and rarity are not contraries to each other as they are in elemental bodies. For there they do not depend upon the primary qualities, cold and heat, which are contrary, but upon greater or less matter in proportion to size. Now 'much' and 'little' have only a relative opposition, which is the most trifling there is, and has nothing to do with generation and corruption."[23] Without realizing it, the Aristotelian had fallen into his own trap. Not

[18]Ibid., p. 67.
[19]Ibid.
[20]Ibid.
[21]Ibid.
[22]Ibid.
[23]Ibid. This is in effect the correct Aristotelian thesis. Since density and rarity are but the effects of matter "changing from what it is potentially to what it is actually" during the process of generation (cf. *Physics* IV, 9, 217 a 26 – b 33) and since the latter takes place under the influence of contraries (heat-cold; dry-wet), Aristotle doubtless attributed density and rarity in the sublunary world to the action of heat and cold.

only must density and rarity in the heavens differ fundamentally from those prevailing on earth (and this distinction is, to say the least, quite unintelligible), but on earth these causes must be sought in the contrariety of heat and cold. Now the simplest experiment will show that this is not the case: thus iron weighs the same and moves in the same manner whether it is glowing or cold.[24] Taken to its limits, the theory of contraries thus leads to untenable conclusions. In short, it was no more capable of justifying the natural distinction between heaven and earth than was the theory of natural motions.

Though it demolished the traditional cosmology, Galileo's critique offered no positive alternative: the old theory had been refuted, but there was as yet no proof that the new, heliocentric hypothesis was true. It cannot even be said that Galileo's critique was wholly original—a host of others including such ancient philosophers as the Atomists and Plutarch and such modern scholars as Nicholas of Cusa and Giordano Bruno had plainly expressed their opposition to the Aristotelian arguments;[25] and a number of discreet allusions suggest clearly that Galileo was familiar with their views.[26] However, he made no explicit reference to them, possibly because he appreciated the personal danger he would run in repeating the heretical doctrines of the Atomists and of Bruno, but more probably for quite a different reason. Aristotle, as Simplicio remarked, would generally proceed by two steps: he would first prove the necessity of his conclusions by means of a priori assumptions and then try to prove their agreement with reality by appeals to tangible experience.[27] Now, both Nicholas of Cusa and Giordano Bruno, to quote only two of Galileo's more immediate precursors, used precisely the same approach: they too would make a priori pronouncements on the world and the bodies in it, and then appeal to experience to confirm the hypotheses they had framed without the least reference to it; despite the differences between their conclusions and Aristotle's, their methods were therefore identical to his. Galileo, by contrast, preferred to give

[24]*Dialogue* I, p. 68.

[25]For Nicholas of Cusa, see M. de Gandillac, *La Philosophie de Nicolas de Cuse*, pp. 78 ff.; for Giordano Bruno, see P. H. Michel, *La Cosmologie de G. Bruno*, pp. 198 ff.

[26]Thus, in his *Second Letter on Sunspots* (Vol. V, p. 138) he alluded to "those wise philosophers whose opinion of the celestial substance differed from Aristotle's"; as for Bruno, though Galileo did not mention him by name, he did refer to his ideas on the infinite number of worlds during the Third Day of the *Dialogue* (p. 354).

[27]*Dialogue* I, p. 75.

priority to experience, and by arguing *from it* rather than *about it* he made a clean and deliberate break with the traditional method. The result was the dawn of modern science.

In this he was greatly aided by a new device, one that helped to change the very conditions of cosmological thought, namely, the spyglass or telescope.[28] The opening pages of the *Siderius Nuncius* are devoted to this momentous event. During the summer of 1609 news reached Padua, where Galileo was then teaching, that a certain Fleming had constructed a spyglass by means of which distant objects could be seen as distinctly as those nearby. Having received confirmation of this "truly remarkable effect,"[29] Galileo decided at once to construct a similar instrument of his own. To that end, he prepared a tube of lead, fitted two glass lenses to its ends, "both plane on one side, while on the other side one was spherically convex and the other concave," and, using the plano-concave lens as the eyepiece, he perceived objects "three times closer and nine times larger than when seen with the naked eye alone." Possibly Galileo exaggerated the power of his instrument, or perhaps he took greater advantage of other constructions than he cared to admit;[30] in any case, hardly had he built his spyglass than he appreciated its great scientific possibilities and turned it toward the heavens. The results were immediate and spectacular: during the first weeks of 1610, a magnificent series of discoveries—the most important that any man had ever made in so short a time—was to deliver the coup de grace to the old cosmology. Not only was Galileo right to claim he was the first man ever to speak of celestial bodies other than "in the imagination,"[31] but he was also the first to study a problem in natural philosophy by systematically questioning experience. It is to his elaboration of a new cosmology, as the necessary prelude to classical mechanics, that we shall now turn our attention.

Because it is nearest to us and hence the most easily observed celestial object, it was the moon to which Galileo looked first in his attempt to unify heaven and earth. To that end, he adduced three sets of facts

---

[28]In his *Siderius Nuncius* Galileo referred to the new instrument by the name of *cannochiale*, but in his correspondence he described it as the *telescopio*.

[29]*Siderius Nuncius*, Vol. III, 1, pp. 60 ff.

[30]This point has been discussed at length by Vasco Ronchi in *Galileo Galilei e il cannochiale*, especially in Chapter II.

[31]*Letter to Gallenzone Gallenzoni*, July 16, 1611, Vol. XI, p. 142; *Letter to Antonio de' Medici*, January 7, 1610, Vol. X, p. 277.

that, he claimed, were bound to persuade all men capable of "adding reason to observation."[32]

Observing the planet a few days after new moon, when it is seen with "brilliant horns," Galileo was struck by the uneven and wavy lines separating the bright part from that still plunged in darkness; he also observed many luminous excrescences extending beyond the boundary into the darker portion. These proved conclusively that the moon could not possibly be the perfectly polished sphere postulated by the Aristotelians, the less so as the irregularities were in no way diminished in the course of the lunar cycle.[33]

Other observations were to confirm these early conclusions. In the luminous part of the moon, the telescope revealed the existence of relatively small dark zones, quite distinct from the large gray spots that cover the moon's surface at all times and that can be seen with the naked eye. These dark zones always have their blackened parts directed toward the sun, while on the opposite side they are surrounded by a bright rim. They gradually lose their blackness as the illuminated region grows larger and larger, much as the shadows in the hollows of the earth diminish in size as the sun rises higher.[34] Again, and no less surprisingly, many bright points appear within the darkened portion of the moon, completely divided and separated from the illuminated part and at a considerable distance from it. These points gradually increase in size and brightness until, an hour or two later, they fuse with the rest of the lighted part, which has now increased in size. A similar phenomenon can, once again, be observed on earth, just before the rising of the sun, when the mountain peaks are illuminated by the sun's rays while the plains below remain in shadow.[35]

On the basis of these observations, Galileo drew up a remarkable description of the lunar surface. At least two parts can be distinguished with certainty. The first of these corresponds to the large and permanent gray spots we have mentioned: here the small number of bright points preceding the general illumination, and the relative absence of dark zones, suggest an even and uniform relief.[36] Within these large lunar spots, we also find brighter patches, some of them very bright indeed. Since they invariably present the same appearance, they cannot be at-

---

[32] *Letter to Father Grienberger*, September 1, 1611, Vol. XII, p. 183.
[33] *Siderius Nuncius*, p. 63.
[34] Ibid.
[35] Ibid., p. 64.
[36] Ibid., pp. 65–69.

tributed to mere irregularity of shape but rather suggest that the moon's surface is characterized by a diversity quite comparable to that of the earth.[37] On the other hand, the lighter tracts of the moon are in marked contrast to these gray spots: the appearance of isolated bright specks well before the rest of the moon is lit up and the persistence of large pockets of shade prove clearly that this part of the moon is full of cavities and prominences.

Galileo's arguments were so persuasive that in the autumn of 1610 most leading astronomers and mathematicians in Italy and the rest of Europe[38] rallied to his side, agreeing, in particular, with the claim of the *Siderius Nuncius* that with the help of the telescope "one may learn with all the certainty of sense evidence that the moon is not robed in a smooth and polished surface but is in fact rough and uneven, covered everywhere, just like the earth's surface, with huge prominences, deep valleys, and chasms."[39]

With his demonstration of the unevenness of the lunar surface, Galileo had taken the first step in demolishing the "perfection" of celestial bodies. However, since no natural activity appeared to take place on the moon, the traditionalists could still maintain that the latter, unlike the earth, was not the seat of generation and corruption, and hence still essentially different from it. Only by proving them wrong in this respect also and by carrying into the heavens this exclusive property of the earth could Galileo hope to demolish the philosophical foundations of

[37]Ibid., p. 69.

[38]Among them Kepler (cf. his famous *Galilaeae vicisti*, Vol. X, p. 436), Magini (Letter of September 28, 1610, Vol. X, pp. 437–438), and Father Clavius (Letter of December 17, 1610, Vol. X, pp. 484–485). Other discoveries mentioned in the *Siderius Nuncius* (the description of the Milky Way as a congeries of innumerable stars grouped together in clusters, and the identification of the satellites of Jupiter) were greeted with similar enthusiasm.

[39]*Siderius Nuncius*, pp. 59–60. This is perhaps the place to mention that Galileo developed two distinct methods of determining the height of lunar summits: one based on the comparison of the lunar diameter to the distance between an illuminated summit and the boundary of the light; the other based on the lapse of time between the appearance of such a summit and its fusion with the luminous zone. See *Siderius Nuncius*, pp. 71 f., and *Letter to Breugger*, November 8, 1610, Vol. X, pp. 466–473, in which Galileo compared the two methods at some length and concluded in favor of the first. He added that the most reliable measurements were those made at quadrature, for only then is the line between individual bright peaks, and also between them and the general limit of illumination, seen "head on" instead of obliquely. (Ibid., pp. 472–473). He also dwelled at length on the fact that the circumference of the moon, though covered by mountains like the rest, does not resemble a toothed wheel, cf. *Siderius Nuncius*, pp. 69–70; *Letter to Father Christophe Grienberger*, September 1, 1611, Vol. XI, pp. 184 ff.

geocentrism and hence to construct the physical framework required by the Copernican doctrine.

During the latter part of 1610, Galileo observed the presence of dark spots on the surface of the sun.[40] Careful observations throughout the year 1611 convinced him of their relative permanence and permitted him to declare that, far from being optical illusions, the spots were real and material objects, and that their behavior was regular in several respects. Thus daily drawings showed that they traveled across the solar disk from east to west parallel to the equator, that they fell within that narrow zone of the solar globe "corresponding to the space in the celestial sphere that lies within the tropics," and that they appeared in the more southerly parts of the sun's body and disappeared or separated from it in the more northerly regions, that is, that they moved "obliquely to the horizon."[41] Moreover, while some of the spots did not alter their appearance as they crossed the solar disk in just under two weeks (some reappearing briefly after the same interval), others changed their shape continually, splitting up or fusing into larger spots.[42]

Summary though they were, these early observations showed that sunspots were true celestial phenomena, not "mere illusions of the telescope or at best some phenomenon produced by the air."[43] For if they were mere illusions, as many of Galileo's opponents held them to be, why should they be confined to the region between the sun and the earth and appear nowhere else in the sky? For none are perceived against the lighted face of the moon, or picked out by the sunlight at a distance from the sun, as happens with clouds. Again, what power could keep such atmosphere phenomena as clouds or vapors in such perfect order that, in collecting or separating, they invariably accompany the sun?[44]

---

[40]Though Galileo was not the first to have mentioned these spots (that honor belongs to Johann Fabricus and Father Christopher Scheiner), he was the first to study them systematically and to offer an explanation. His observations and comments are contained in three letters he wrote to Marcus Welser, a wealthy merchant of Augsburg, who, in 1612, had published Scheiner's letters to him and had sent a copy to Galileo. Cf. E. Wohlwill, *Galileo Galilei und sein Kampf für die Copernicanische Lehre*, Vol. I, pp. 438 ff.

[41]All these descriptions are taken from the *First Letter on Sunspots* dated May 4, 1612 (Vol. V, pp. 15–16). Galileo later discovered the cause of their oblique motion: the axis around which the sun revolves is not perpendicular to the plane of the ecliptic; cf. *Dialogue* III, pp. 372 ff.

[42]For the method of observation developed by Galileo with the help of his pupil, Benedetto Castelli, see *Second Letter on Sunspots*, August 14, 1612, Vol. V, pp. 136–138.

[43]The view of certain philosophers; cf. *Dialogue* I, p. 77.

[44]*Second Letter on Sunspots*, pp. 128–129.

In short, all the arguments mustered by the traditionalists against the real existence of sunspots were so farfetched that they could be dismissed out of hand. But were the spots part and parcel of the body of the sun, or were they celestial objects in its vicinity? In his first letter to Marcus Welser, Galileo still refused to commit himself,[45] but by the time he wrote the *Second Letter* four months later, he had made up his mind: the spots were either contiguous to the solar surface or else separated from it by an imperceptible distance. In short, the sun, no less than the moon, must be the seat of generation and corruption. Though this was the only logical conclusion, the Peripatetics, not surprisingly, fought it tooth and nail in a final attempt to save the incorruptibility of the heavens. All their arguments were baseless, as Galileo stressed in his second and third letters to Welser; but they did serve some useful purpose in that their refutation not only forced Galileo to base his own solution on more solid foundations but also provided him with the occasion to deliver a remarkable lesson on methodology.[46]

Take the argument of the "philosophers" that the very regularity of their motions proved that sunspots could not move independently of one another or at a great distance from the sun, so that they must be attributed to a single cause. The most likely explanation, according to them, was that the spots were attached to a sphere rotating somewhere between the earth and the sun in such a way as to ensure the regular passage of the spots in front of the solar disk. In that case the sun itself would remain untainted by generation and corruption.[47]

Galileo's reply was both systematic and conclusive. If the sun is spherical and rotates on its own axis,[48] there must be two points at rest called the poles, and all other points on its surface will describe parallel circles that are larger or smaller according to their distance from the poles; the largest of all will be the central or equatorial circle, equidistant from

---

[45]*First Letter on Sunspots*, p. 111. See also the Introduction to the *Discourse on Bodies in Water* (May 1612), in which Galileo first mentioned his work on sunspots.

[46]The Peripatetics had previously advanced the most improbable suppositions in their attempt to save the traditional description of the moon; thus they claimed that while the lunar surface was admittedly uneven and jagged, it was surrounded by a perfectly spherical and transparent layer of crystal; cf. Galileo's *Letter to Marcus Welser*, February 1611, Vol. XI, pp. 38 f., and also his *Letter to G. Gallenzoni*, July 16, 1611, Vol. XI, pp. 142 f.

[47]*Second Letter on Sunspots*, p. 118.

[48]For reasons that will be examined in the next chapter, Galileo believed that the rotation of the sun could be deduced from the motion of the sunspots.

the two poles. Under these circumstances, the length of a spot would be precisely its dimension determined along one of the circles parallel to the equatorial, and its breadth (or width) its dimension along a circle passing through the poles and perpendicular to that which determines their length.[49] Having defined these terms, Galileo found it a simple matter to show "from the special characteristics of the motion of the spots"[50] that the Peripatetic view had no substance.

Thus, observation shows, first of all, that though the spots, both at their first appearance and final disappearance near the edges of the sun, generally have very little length, they have the same breadth as they possess in the central part of the sun's disk. Now, as anyone familiar with the phenomenon of foreshortening on a spherical surface will realize, this would not occur if the spots moved at an appreciable distance from the solar globe. "For the maximum thinning takes place at the point of greatest foreshortening, and it would occur outside the face of the sun if the spots were any perceptible distance away from its surface."[51]

Second, it shows that the distances traversed by a spot in equal times decrease as the spot approaches the edge of the sun. Moreover, the decreases vary as the versed sines of equal arcs, as could happen only in the case of circular motion on the sun itself. "In circles even slightly distant from it, the spaces passed in equal times would appear to differ very little against the sun's surface."[52]

Highly significant, too, is the mode of variation of the distances between one spot and another. Spots close together along the same parallel seem almost to touch when they first emerge in the east, but as they move away from the edge, they separate more and more. They have their maximum separation at the center, and, as they move on from there, they again approach each other. Accurate observation of the ratios of these separations and approaches shows that they must take place on the actual surface of the solar globe. Let *A* and *B* be two spots, for instance, those first observed on July 1, 1611, whose progress could easily be

[49]Ibid., p. 118.
[50]Ibid., p. 117.
[51]Ibid., p. 119. Cf. *Dialogue* I, p. 79.
[52]*Second Letter on Sunspots*, pp. 119–120. By "versed sine," seventeenth-century mathematicians understood the straight line $M'A$ on the accompanying diagram, that is, the segment between the origin *A* of the arc *AM* and the projection of its extremity *M* on 0*A*.

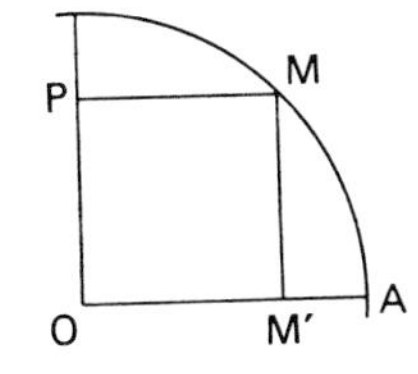

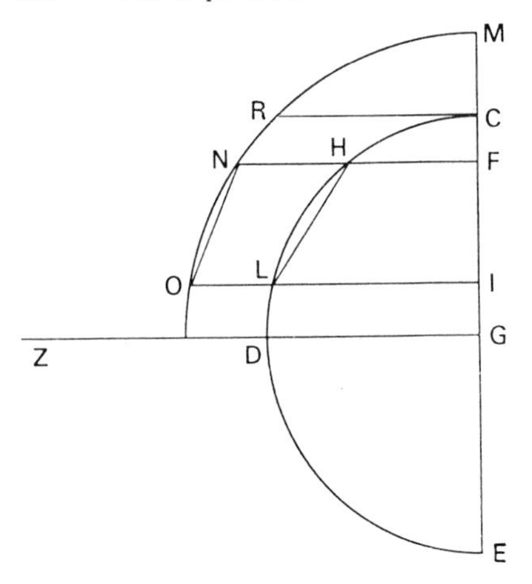

followed during subsequent days.[53] The parallel along which they traveled ran just to the north of the equatorial circle, and their motion was oblique with respect to the plane of the ecliptic. Let *GZ* represent a ray from our eye to the sun.[54] Describe the semicircle *CDE* around the center of the sun, *G*; it will correspond to the real distance traversed by the spots *A* and *B*, while the diameter *CGE* will represent their apparent travel. On July 1, the spot *B* was already close to the center of the solar disk, while the spot *A* was still near the edge, say, at the distance represented by *CF*. From *F* draw a perpendicular to *CG*, and let it meet the circumference at *H*; clearly, when the spot *A* appears in *F*, it actually occupies the position *H* on the solar surface. Let us next project the real distance between the two spots *A* and *B*, that is, the chord *HL*, on to the diameter *CE* to obtain *FI*, and draw *IL* perpendicular to *CG* in *I*; *FI* will represent the apparent distance between the two spots. Now, it is only when the center of *HL* coincides with the point *D* that we can observe the real distance between the two spots directly. Let us finally look at what happened on July 5, by which time the two spots had moved appreciably toward the right of the solar disk, to occupy positions equidistant from the center. Not only had their

[53] Ibid., pp. 120 f. The precise development of the spots is described in Vol. V, pp. 172 f.

[54] Because the distance between the earth and the sun is very much greater than the sun's diameter, the angle subtended by two visual rays drawn to opposite edges of the sun's disk is very small, so that these rays can, for all practical purposes, be treated as parallels; see *Second Letter on Sunspots*, p. 121.

apparent separation increased, as could have been predicted, but an attentive analysis of the data led Galileo to conclude that this increase is in precisely the same ratio as *FI* is to *HL*.

Now, this result is incompatible with the Peripatetic hypothesis, as the following construction will readily show. Draw the arc *MNO* so that its distance from the solar surface is one-twentieth part of the diameter *CE*, and produce the perpendiculars *FH* and *IL* to *N* and *O*; then, clearly, if the spots *A* and *B* would move along the circumference *MNO*, *A* would appear in *F* when in fact it occupies the point *N*, and *B* would appear in *I* when in fact it occupies the point *O*. The real distance between them is therefore the chord subtending the arc *NO*, and the two spots are furthest away from each other when the chord *NO* is perpendicular to the line *GZ*. It follows that the apparent separation would have to vary as *FI* is to *NO*, not as *FI* is to *HL*. Since they in fact vary as the latter,[55] it follows that the spots must be either contiguous with the sun or, at most, at an imperceptible distance from it. Moreover, had we chosen two spots closer to each other than *A* was from *B* and also closer to *C* or *E*, this fact would have been more obvious still. Thus if one spot had been in *C* and the other in *F*, their real separation, the chord *CH*, would have been at least seven or eight times greater than their apparent separation, *CF*; whereas if the same spots were in, say, *R* and *N*, their real separation would be the chord *RN*, which barely measures one third of *CH*; and since, after some time, the apparent distance between the spots would be equal to *CH* (which is at least seven times as long as *CF*) and no longer to *RN* (which is almost twice as long as *CF*), it is clear that the sunspots are indeed contiguous to the surface of the sun.[56]

The philosophers also advanced a second, "saving," hypothesis, whose only advantage over the first was that it provided a better account of the variable nature of sunspots, that is, of their change in shape while combining or separating. In brief, that hypothesis amounted to replacing the stable spots fixed on a concentric sphere surrounding the sun with a large number of small, opaque bodies revolving about the sun like so many tiny planets, and traveling at different velocities; this would explain their combinations, separations, recombinations, and so on.[57] Now this hypothesis, as Galileo was quick to show, was even

---

[55] Ibid., p. 123.
[56] Ibid., pp. 123–124.
[57] *Third Letter on Sunspots*, December 1, 1612, Vol. V, p. 231.

less tenable than the first. In his first letter to Welser he had already remarked that such small planets, if they did indeed exist, could not revolve far from the sun, both because they had never been observed outside the sun's surface and also because they would have to move across that surface much more swiftly than they actually do.[58] Moreover, the spots could never even have appeared under such conditions. Thus, if they were indeed a congeries of many minute celestial bodies revolving uniformly, though at different speeds, around the sun, at least fifty of them would have to unite in order to constitute some of the larger spots. If their velocities varied, there would be little chance for a large enough number of them ever to combine together, let alone as frequently as they would have to do; even if such aggregations should occur fortuitously, their constituent bodies would quickly draw apart—the unequal speeds would most certainly prevent them from staying together for almost two weeks. "Hence anyone who wished to maintain that the spots were a congeries of minute stars would have to introduce into the sky innumerable movements, tumultuous, uneven, and without any regularity. But this does not harmonize with any plausible philosophy."[59]

Thus the heavens themselves were refuting their incorruptibility, confounding all those who attempted "to measure the whole universe by means of their tiny scale" and who forgot that "nature has no obligation toward man and has signed no contract with him." The discovery of sunspots was therefore the end of a long road, begun with the acrimonious debate about the nature of comets and novae at the end of the sixteenth century, when disagreement between the astronomers still cushioned the traditional cosmology against too rude a jolt.[60] The position of the sunspots, by contrast, could be determined with some accuracy, and their continual reappearance was quite incompatible with the tenets of the old system. Hence it was with good reason that Galileo wrote to Paolo Gualdo on June 16, 1612, that he was greatly consoled by the fact that "the spots and my other discoveries are not things that will pass away with time as did the novae of 1572 and 1604 or the comets, whose disappearance from the skies did so much to calm those whose minds were in such great anguish during their presence. The

---

[58] *First Letter on Sunspots*, pp. 111–112.
[59] *Third Letter on Sunspots*, pp. 232 f.
[60] Thus Tycho Brahe placed comets and novae alike beyond the lunar orbit, while other astronomers placed them within it.

spots, by contrast, will torment them ceaselessly, because they refuse to vanish, and it is only fitting that nature should thus punish the ingratitude of men who have maltreated her for so long by foolishly shutting their eyes before the very light she provided for their edification."[61]

Sunspots thus attested to the physical unification of heaven and earth, without which it would have been quite impossible to test the truth of the Copernican doctrine. Not that all celestial bodies had been transformed into identical objects; Galileo himself never maintained that they were;[62] nevertheless, the same principles could now be applied to all of them, with the result that the Aristotelian distinction between celestial and terrestrial science could be dropped once and for all. Was this the ultimate conclusion of Galileo's analysis? In fact, a moment's thought will show that the justification of Copernicanism called for yet another unification, based on the demonstration that, in a heliocentric system, *all* the planets (and their satellites) constitute a system of bodies whose identity in status must go hand in hand with a special type of physical resemblance. Traditional cosmology had always based its refusal to place the earth among the planets on the fact that while celestial bodies were "luminous and resplendent," the earth alone was "obscure and without light."[63] Would the telescope be able to do away with that distinction as well?

Two separate arguments can be distinguished in Galileo's very precise answer to this question. The first, based on methodical observations of Venus enabled him to demonstrate that the planets were not luminous in themselves. In October 1610, Venus was an "evening star" and had just passed out of opposition with the sun; to Galileo, who had closely followed the course of the planet as it traveled toward the east, it appeared in the shape of a perfectly round disk of small dimensions.[64] It retained that shape for several weeks, but as it approached the first quarter, its disk seemed to shrink on the side opposite the sun until it

---

[61] *Letter to Paolo Gualdo*, June 16, 1612, Vol. XI, pp. 326–327.

[62] Thus he always rejected the idea of life on the moon; cf. *Dialogue* I, pp. 86–87; and *Letter to G. Muti*, February 28, 1616, Vol. XII, pp. 240–241.

[63] *Dialogue* I, p. 72.

[64] Cf. *Letter to Father Clavius*, December 30, 1610, Vol. X, p. 500, and *Letter to Paulus Sarpi*, February 12, 1611, Vol. XI, pp. 48 ff. In fact, when he wrote these two letters, Galileo had not yet observed the complete cycle of the planet, and hence based much of his argument on an extrapolation that in the end turned out to be perfectly correct.

became semicircular at the moment of maximum elongation. A few weeks later, Venus once again approached the sun; although its apparent diameter had increased,[65] its visible surface had shrunk further into a crescent and continued to thin out, until the next conjunction, when Venus passed in front of the sun and became invisible. Then the same cycle was repeated, but this time in reverse, until, several weeks later, the planet resumed its circular shape. Now these phases showed beyond the least doubt that Venus revolves about the sun, as most late sixteenth-century astronomers had in fact believed it did. But at the same time—and this was Galileo's own conclusion—the phases also proved that Venus had no luminosity of its own: no luminous star could hide large parts of its surface for months at a time. "And in this way we are assured," Galileo wrote to Father Clavius, "that Venus, like Mercury, revolves about the sun; and hence that the planets have no light of their own but owe their brilliance to the sun, which, I believe, is not the case with fixed stars." For what was true of Venus was equally true of the superior planets: the observed decrease in brightness as the telescope was moved on from Mars to Jupiter and Saturn could be due only to a decrease in the light these planets receive from the sun, while no fixed star, however far from the sun, has ever been known to lose any part of its luminosity.[66]

Nor was that all, for Galileo went on to show that the earth, too, is capable of reflecting the sunlight and of scattering it into space. Let us attend to Salviati:[67]

Salviati: Now tell me, do you believe that the moon is really brighter by night than by day, or does it just by some accident look that way?
Simplicio: I believe that it shines intrinsically as much by day as by night, but that its light looks brighter at night because we see it in the dark field of the sky. In the daytime, because everything around it is very bright, by its small addition of light it appears much less bright.
Salviati: Now tell me, have you ever seen the terrestrial globe lit up by the sun in the middle of the night? . . .

So you have never chanced to see the earth illuminated except by day, but you see the moon shining in the sky on the darkest night as well. And that, Simplicio, is the reason for your believing that the earth does not shine like the moon; for if you could see the earth illuminated

[65]Galileo noted that the increase in the apparent diameter of Venus between superior and inferior conjunction was roughly from 1 to 6.

[66]*Letter to Giuliano de' Medici*, February 1611, Vol. XI, p. 62.

[67]*Dialogue* I, pp. 113–114.

while you were in a place as dark as night, it would look to you more splendid than the moon. Now if you want to proceed properly with the comparison, we must draw our parallel between the earth's light and that of the moon as seen in daytime; not the nocturnal moon, because there is no chance of our seeing the earth illuminated except by day. Is that satisfactory?

Simplicio: So it must be.

Salviati: Now you yourself have already admitted having seen the moon by day among little whitish clouds, and similar in appearance to one of them. This amounts to granting at the outset that these little clouds, though made of elemental matter, are just as fit to receive light as the moon is. More so, if you will recall in memory having seen some very large clouds at times white as snow. It cannot be doubted that if such a one could remain equally luminous on the darkest night, it would light up the surrounding regions more than a hundred moons. If you were sure, then, that the earth is as much lighted by the sun as one of these clouds, no question would remain about its being no less brilliant than the moon. Now all doubt upon this point ceases when we see those same clouds, in the absence of the sun, remaining as dark as the earth all night long. And what is more, there is not one of us who has not seen such a cloud low and far off, and wondered whether it was a cloud or a mountain; a clear indication that mountains are no less luminous than those clouds.

But while it is fairly simple to demonstrate the earth's ability to reflect the light of the sun, can it also be shown that it scatters the reflected light into space, in the manner of the other planets? Once again the "courtesy"[68] that nature had always shown him came to Galileo's aid. If we observe the moon when it approaches the sun, that is, just before or after conjunction, we shall see not only a crescent illuminated by the sun but also a fainter light covering the rest of its surface and making it stand out from the dark sky. Remarkably enough, this secondary light diminishes as the moon moves away from the sun, to wane very rapidly after the first quarter.[69] Now, traditional cosmology could account for this peculiar phenomenon only with the help of ad hoc hypotheses; by refuting the latter, Galileo gave further proof not only of the falsehood of that cosmology but also of the consistency of his own solution.

[68]This expression is found in the *Second Letter on Sunspots*, p. 137.

[69]*Siderius Nuncius*, pp. 72 f. Leonardo da Vinci had previously given a correct account of the existence of this secondary or "earth" light, and so had Maestlin, Kepler's teacher. It is difficult to tell whether Galileo was acquainted with Maestlin's account, but we do know that he was unfamiliar with Leonardo's; cf. Wohlwill, *Galileo Galilei und sein Kampf*, Vol. I, pp. 260–261.

The first of these hypotheses, namely, that the secondary light was of lunar origin, could be dismissed out of hand, for in that case the light would be particularly noticeable during eclipses, when in fact it is not.[70] Nor could the secondary light be derived from the sun, for in that case, its intensity could not vary except during eclipses,[71] and the moon would have to be transparent, when in fact the lunar mountains can be seen to obstruct the rays of the sun and to cast deep shadows.[72] More absurd still was the suggestion that the secondary light might be derived from the fixed stars[73] or from Venus.[74] The only one of the traditional explanations to which Galileo attached any importance, namely, that the moon is surrounded by a relatively dense envelope of ether, which reflects part of the sunlight in which it is continuously bathed to the dark lunar surface, had to be rejected as well: the only known phenomenon of this kind—the scattering of sunlight after sunset by the terrestrial atmosphere—is of short duration and, in any case, does not extend over the whole hemisphere facing the sun.[75]

All these more or less gratuitous suppositions can be safely discarded once it is granted that the earth is capable of reflecting the sun's rays into space. Thus if we place ourselves on the moon in our mind's eye, we shall see that, at conjunction, that is, when the moon occupies a position between the sun and the earth, its side which is facing us receives no sunlight, while there lies before it the entire surface of that hemisphere of the earth which is vividly illuminated by the rays of the sun. The moon therefore receives the light which this hemisphere reflects, and it is thanks to this "earthlight" that we can see the moon even during the day. Then, as the moon moves on, it faces an ever darker earth until, when it is ninety degrees away from the sun, it sees but half of the earth illuminated. Finally, when the moon is in opposition to the sun, it faces a hemisphere of the earth steeped in darkness and receives no light from it at all. And the relation between the two globes is such that whenever the earth is most brightly lighted by the moon, the moon is least lighted by the earth, and vice versa.[76]

---

[70] *Siderius Nuncius*, p. 73.
[71] Ibid.
[72] *Third Letter on Sunspots*, pp. 224–225.
[73] *Letter to Prince Leopold de' Medici*, May 1640, Vol. VIII, pp. 516 f.
[74] *Siderius Nuncius*, p. 73.
[75] *Letter to Prince Leopold de' Medici*, pp. 507–523. Galileo did not, however, deny the possible existence of an envelope of ether around the moon and thought it might explain the coppery light in which the moon is bathed during eclipses; ibid., pp. 514–515.
[76] Ibid., pp. 502–503; *Siderius Nuncius*, pp. 74–75.

This theory not only was extremely simple but offered a very precise explanation of why the earthlight is much stronger two or three days before conjunction than two or three days after, in other words, why the moon is brighter when we see it before dawn in the east than in the evening after the setting of the sun in the west. "The reason for this difference is that the terrestrial hemisphere opposite the moon when it is in the east has fewer seas and more land, containing all of Asia; but when the moon is in the west it faces great seas—the whole Atlantic clear to America—a very plausible argument for the surface of the water showing itself less brilliantly than that of the land."[77] Thus, as Galileo concluded, "the earth, in fair and grateful exchange, pays back to the moon an illumination similar to that which it receives from her throughout nearly all the darkest gloom of night."[78]

All these discoveries and theories opened up completely new vistas to astronomy. In particular, once it had been shown that celestial bodies were of the same nature as the earth, there was no longer the least reason why the results of terrestrial studies should not be applied to the celestial realm. This new approach, which was a direct consequence of the physical unification of the universe, was the more significant as it marked out the path along which classical science was to forge ahead. This is a point worth dwelling upon.

Let us return to the sunspots: granted that they were a part of the sun's body, what material processes were responsible for them? In this connection, Galileo noted that the surface of the solar disk lying between spots is often distinct "all the way out to the very limit of the sun," from which he deduced that the spots are very thin in comparison with their length and breadth.[79] This might have been due to the fact that the spots were like "lakes or caverns in the body of the sun," but Galileo was quick to reject this explanation[80] on the grounds that all the evidence seemed to point to a similarity between the spots and terrestrial clouds. The appearance and disappearance of the spots during relatively short periods, their expansion, division, recombination, and

[77]*Dialogue* I, p. 124.
[78]*Siderius Nuncius*, p. 74. Elsewhere, Galileo explained that the earth seen from the moon would have different phases, just like its satellite, though in the reverse order: "full earth" would take place at new moon, and "new earth" at full moon. Cf. *Dialogue* I, pp. 89–90.
[79]*Second Letter on Sunspots*, pp. 124–125.
[80]*Third Letter on Sunspots*, pp. 202–203.

assumption of the most irregular shapes, as well as their opacity, were all so many properties they shared with the clouds. Moreover, if we could look at the earth from an external vantage point, the clouds would surely offer us the same aspect as sunspots, changing all the time and thinning out when they reached the edges of the terrestrial disk.[81] In the *Third Letter*, however, Galileo put forward an alternative explanation, namely, that the spots were made up of smoke. Thus, if one wished to imitate them with the help of terrestrial matter, one could do no better than put a few drops of incombustible bitumen on a red-hot iron plate. "From the black spot thus impressed on the iron, there will arise a black smoke that will disperse in strange and changing shapes"—a comparison that was the more remarkable in that it suggested a completely novel idea, namely, that the sun might well be the seat of tremendous combustion phenomena.[82]

Perhaps this example will be thought somewhat farfetched, although it in no way detracts from the value of Galileo's general procedure. In any case, there is nothing less arbitrary than the analysis by which he deduced the marked unevenness of the moon's surface from simple optical experiments.[83] The fact that the moon is covered with mountains had already been established in the *Siderius Nuncius*. The reader may recall the main arguments: the wavy boundary of shadow and light, and the presence of bright points in the darkened zone and of dark patches in the bright. However, as Galileo explained to one of his correspondents, the mountainous nature of the lunar surface remained a mere deduction,[84] and still had to be confirmed by simple experiment. To that end, he suggested the following test: Affix a plane mirror to a wall illuminated by the sun; upon retiring into the shadow, you will notice that the mirror looks darker than the surrounding area. True, a Peripatetic philosopher may object: on the opposite wall, which receives the reflection from the illuminated side along with that of the mirror, the reflection of the mirror is much the brighter. But far from "saving" the traditional interpretation, this observation merely serves

[81]This idea was first formulated in the *First Letter on Sunspots* (p. 106), developed in the *Third Letter on Sunspots* (p. 230), and repeated in the *Dialogue* (I, p. 76).

[82]*Third Letter on Sunspots*, p. 230.

[83]The importance he attached to this kind of analysis is clearly attested by a passage in the *Discourses*, in which Sagredo praises Salviati for solving questions that are difficult not merely in appearance but also in reality with the help of "reasons, observations, and experiments that are common and familiar to everyone"; *Discourses* II, Vol. VIII, p. 131.

[84]*Letter to Father Christophe Grienberger*, September 1, 1611, Vol. XI, p. 183.

to emphasize its weakness, for it shows that the light reflected by the mirror is not diffused over the whole opposite wall but goes to a single place no larger than the mirror itself. Now take the moon: the fact that it appears bathed in an even and soft light instead of blinding the beholder shows clearly that, like the wall, it has an irregular and rough surface, made up of a succession of valleys and mountains.[85]

It might, of course, still be objected that the moon is not a plane surface, and that its reflection must therefore be compared to that of a spherical mirror, which, by virtue of its curvature, scatters light evenly in all directions. Now this objection would undoubtedly be valid if the point of the comparison was to show how a given object reflects light. In fact, however, its purpose was to show how the light is perceived by an observer, so that what matters is not the total amount of light sent back by the subject but the amount of light received by the subject. Now in this respect, what applies to a plane mirror applies a fortiori to a spherical mirror: since only a small portion of its surface reflects the incident light in a fixed direction, we may take it that only a very small part of that light will manage to reach our eye, and therefore that only a tiny fragment of the mirror will seem bright to us. Thus if we attach a spherical mirror to a wall illuminated by the sun and allow the reflected light to fall on the opposite wall, we shall be able to notice the mirror only as a bright spot if we stand in the direct path of its reflected light, but shall observe no change in the illumination of the opposite wall. Hence, there is indeed a sharp contrast between the way in which a wall and a mirror—be it plane or spherical—reflect the light: while the former scatters it evenly, the latter concentrates it in a small point. The reason is obvious: because the wall is made up of a host of very small surfaces placed in an innumerable diversity of small slopes, it will disperse the incident rays in all directions so that the wall will look evenly illuminated over its entire surface, while the mirror will appear as a single bright point against a darker background.[86]

---

[85] *Dialogue* I, pp. 96–98.

[86] Ibid., pp. 98–101. In the preceding analysis, Galileo ignored the "adventitious irradiation of the eyes through the reflection made in the moisture at the edges of the eyelids," which causes luminous bodies to look much larger than they would otherwise appear. However, this effect in no way supports the traditional thesis: the resulting luminosity is produced by a very small part of the mirror. Hence, if the moon were indeed a highly polished surface, it would be undetectable, since the very small part of it which can reflect the image of the sun to the eyes of any individual would remain invisible because of the great distance (ibid., pp. 101–102). For the theory of irradiation, see *Dialogue* III, pp. 364–365.

In his attempt to justify the Copernican doctrine Galileo therefore not only deprived the earth of its unique position in the universe but turned it into a laboratory for testing celestial phenomena, and lunar phenomena in particular. Thus, in a sense, he may be said to have embraced geocentrism even more wholeheartedly than any supporter of the old system. The paradox was admittedly superficial, but it nevertheless showed that physical (as distinct from positional) geocentrism is the natural complement of the heliocentric doctrine.

The Premises of a Copernican Cosmology

With the physical unification of the universe, all the old philosophical arguments in support of geocentrism were swept away, and so were the main objections to the heliocentric alternative, which, moreover, was corroborated by every new observation and every advance in astronomical knowledge. One of the main objections to the Copernican doctrine was that, if the earth revolved about the sun and the moon about the earth, the moon would be describing two circles at once, whereas Aristotle had declared explicitly that such compound motions were impossible. But the discovery of Jupiter and its satellites made short shrift of this argument,[87] and numerous other new findings also lent weight to the Copernican doctrine. Thanks to its ability to eliminate the "irradiation effect," the telescope showed that the apparent diameter of Mars was some fifty times greater at opposition, when it is closest to the earth, than it is at conjunction; similarly, the apparent diameter of Venus could be shown to increase from one to six between upper and lower conjunction.[88] All these facts, together with the observed phases of Venus and the retrograde motion of the superior planets, were so many necessary consequences of Copernicanism; they followed directly from the position of the sun at the center of the planetary revolutions. Thus, while Ptolemy's system had been constructed piecemeal to save the appearances, that of Copernicus was able to explain and predict all the apparent anomalies in the motions of the planets.[89] The conviction that Copernicanism did indeed reflect the "true constitution of the world"[90] would remain with Galileo for the rest of his life.[91]

---

[87] *Dialogue* III, p. 368.
[88] Ibid., pp. 366–367.
[89] Ibid., pp. 370–372.
[90] *First Letter on Sunspots*, p. 99.
[91] It was in his correspondence that Galileo felt most free to express this conviction. See particularly *Letter to Paolo Sarpi*, February 12, 1611 (Vol. XI, pp.

But such a conviction was of course far from constituting a proof, and Galileo was much too good a logician to assume that the truth of a conclusion was a guarantee of the validity of its premises. To justify the Copernican doctrine, he followed three distinct paths, two of which are generally known if not always fully appreciated: a methodical attempt to show that the motion of the earth is compatible with everyday experience; and a search for phenomena that could be explained only in terms of this motion.[92] But though they were the most important when viewed from the vantage point of modern science, they were not the only ones. Galileo's third approach is generally passed over in silence, though the fact that it came first in the *Dialogue* shows how much importance he himself attached to it. Let us therefore take a closer look at Galileo's own train of thought.

A mechanical justification of Copernicanism must involve a definition of "mass" and "attraction," and a correct analysis of circular motion. Since Galileo had not yet taken either of these steps, he was unable to show that the sun was prevented from revolving about the earth for purely mechanical reasons and was thus thrown back on observation and philosophical speculation. Now, the Copernican doctrine hinged on two basic assumptions, namely, the diurnal and annual motions of the earth, two assumptions that were largely descriptive in character; to transform the model they defined into a true cosmological system, in the absence of mechanical deduction, therefore called for a set of more general and rationally justifiable suppositions concerning the universe at large and the regular motions of its constituents.[93] Thus suppose it were possible to prove that these general suppositions, and they alone, were reasonable cosmological premises; would geocentrism not appear to be as "irrational" as difficult to reconcile with observation? And would Copernicanism not assume that intelligible necessity that is the hallmark of all science? Moreover, demonstrating that Copernicanism is the only doctrine befitting those who stick to the facts and who ignore

48 ff.); *Letter to G. Gallenzoni*, July 16, 1611 (Vol. XI, p. 152). *The Letters on Sunspots* were also almost overtly Copernican; thus he referred to the Copernican doctrine in the very first letter as a "vera e buona filosofia."

[92]The first procedure will be examined in the next chapter; for the second, which has no direct bearing on the genesis of the new mechanics, the reader is referred to Appendix 4.

[93]We may therefore put it that in the pre-Newtonian context of Galileo's thought these general assumptions bore the same relationship to the descriptive assumptions as Newton's laws bore to Kepler's.

the authority and prejudices of the old texts would therefore mean not only defining the premises of a new cosmology but also changing the basis of all physical arguments, especially in what concerns motion. Hence by looking at the new cosmological premises which Galileo substituted for the old we not only shall be examining his justification of Copernicanism but shall also come to appreciate one of the most important reasons for the transformation of mechanics that took place at the beginning of the seventeenth century.

Galileo's first cosmological premise seems, on cursory examination, to fit into the traditional frame. Thus he declared quite openly that he was adopting one of the fundamental theses of Aristotle's *On the Heavens*, namely, that the parts of the world formed an orderly whole. In fact, his argument differed little from Aristotle's, though he substituted a precise geometric definition of the three-dimensional nature of celestial bodies for Aristotle's vague formulations.[94]

"I admit that the world is a body endowed with all the dimensions, and therefore most perfect," Galileo asserted. "And I add that as such it is of necessity most orderly, having its parts disposed in the highest and most perfect order among themselves."[95] Two thousand years after Aristotle, the concept of an orderly cosmos thus continued to dominate cosmology and to supply it with its most basic theme.

However, this is where the similarity between Galileo and Aristotle ends. In particular, Aristotle's idea of six a priori directions had completely disappeared from Galileo's account. Moreover, the theory of gravity showed that the world could not have a center determined a priori; gravity was simply the inner tendency of bodies to rejoin, by the shortest route, the whole from which they had been separated. And being a constituent property of matter, gravity could not possibly affect the future location of material bodies in the general scheme of the uni-

[94]"Therefore if you assign any point for the point of origin of your measurements, and from that produce a straight line as a determinant of the first measurement (that is, of the length), it will necessarily follow that the one which is to define the breadth leaves the first at a right angle. That which is to denote the altitude, which is the third dimension, going out from the same point, also forms right angles and not oblique angles with the other two. And thus by three perpendiculars you will have determined the three dimensions by three unique, definite, and shortest lines." *Dialogue* I, p. 37.

[95]Ibid., p. 43

verse.[96] Now, by denying that the orderly structure of the universe implied any particular arrangement of the bodies contained in it, Galileo also demolished another basic concept of traditional cosmology: the idea that every body has a natural place; at best, he occasionally used the "natural place" to refer to the links between a whole and its parts.[97]

But Galileo not only rid cosmology of its a priori directions and natural places; he showed *eo ipso* that the orderly structure of the universe lacked an a priori basis and that it can be deduced only from astronomical observations. This point is worth stressing because this interpretation set him apart from any of his contemporaries, including even the greatest. Kepler, in particular, thought that there must be a divine reason for the distances and velocities of celestial bodies being what they are.[98] This explains his attempt, in the *Mysterium Cosmographicum*, to describe the distances of the planets from the sun and the dimensions of their orbits by reference to five regular polyhedra. Thus, while he had transformed the traditional essences into mathematical "harmonies," he fully retained the idea of a formal cause responsible for the a priori order of the world. Galileo, for his part, though no less persuaded of the importance of that order, discarded this "mathemythical" approach and in so doing greatly simplified his cosmological premises.

Now the traditional idea of a cosmological order was, as the reader may recall, bound up with the idea of a finite universe. This theory, too, Galileo threw overboard. There was nothing to support the view that an orderly world must be "finite and bounded rather than infinite and unbounded,"[99] and since observation could not decide the issue, either solution might apply from case to case. However, by greatly increasing the number of fixed stars and by showing that they are at immense distances from one another, the telescope clearly suggested, at the very least, that the skies must have "a very great depth."[100] This explains why Galileo so often adopted the "infinitist" view, for instance,

---

[96]At the same time, Kepler described gravity as the mutual tendency of cognate bodies to join each other; this conception of gravity as a form of magnetism was, however, no closer to Newton's view than was Galileo's theory. Cf. J. L. E. Dreyer, *A History of Astronomy from Thales to Kepler*, p. 399. See also Chapter 5.

[97]See, for instance, *Letter to Ingoli*, 1624, Vol. VI, p. 557.

[98]Dreyer, *A History of Astronomy*, pp. 374–375. Kepler also believed that divine reasons alone could explained why the planets were only five in number.

[99]*Dialogue* III, p. 347.

[100]*Letter to Ingoli*, Vol. VI, p. 525.

when, like Bruno, he compared the stars to so many suns[101] or when he mentioned that point in "infinite space" from which Ptolemy's sphere of fixed stars would look no larger than a single star seen from the earth.[102] Unfortunately, as we saw, observation was unable to settle this issue. Thus Tycho Brahe, considering the diameter of the fixed stars and the absence of the least parallax in them, concluded that, if the Copernican hypothesis was correct, the smaller stars must have dimensions equal to or greater than the earth's orbit, which meant that the limits of the universe were an infinite distance away from us, a point he refused to concede. Now the telescope, by eliminating the "irradiation" effect and hence by reducing the apparent diameter of the fixed stars, suggested that the boundaries of the universe must indeed be finite. This may explain why Galileo considered the problem insoluble, a view he first expressed in 1624 and to which he still adhered in 1641.[103] His attitude was not only prudent but also justified by the fact that, in principle, an orderly world could equally well be finite or infinite.[104]

Though the picture of the cosmos drawn in the *Dialogue* was thus far more flexible than that of the traditionalists, it would be quite wrong to dismiss it as mere speculation without any real bearing on the progress of science. In particular, the principle of an orderly universe led Galileo to conclude that natural bodies have a clear preference for certain states. Take a universe constructed of a system of bodies: every

---

[101]*Dialogue* III, p. 354. In his letter to Ingoli, he called this assumption "most reasonable." For Bruno's views, the reader is referred to Michel's *La Cosmologie de G. Bruno* (pp. 165 f.). Bruno's arguments were actually confused and rather unconvincing. For instance, he countered Aristotle's claim that the infinite exists only in the imagination with the argument that, if the imagination did not correspond to reality, a natural phenomenon (such as the imagination) would transcend nature, which would be absurd (p. 172). Hence Michel was right in remarking that, as far as Bruno was concerned, infinity was a certainty, not a problem: "He provided the answer even before he posed the question" (p. 177). In this connection, we might also remark that Nicholas of Cusa, while contending that infinity was an exclusive attribute of God, proclaimed the impossibility of assigning a limit to the universe (ibid., p. 178).

[102]*Dialogue* II, pp. 397.

[103]*Letter to Ingoli*, p. 529 (1624); *Letter to Liceti*, January 1641, Vol. XVIII, pp. 293–294. See also A. Koyré, *Du monde clos à l'univers infini*, p. 96. Koyré has also shown that because Kepler refused to grant the possibility of an infinite world he went to great lengths in lending a finitist interpretation to Galileo's astronomic discoveries.

[104]This is quite obvious in a heliocentric system, though a geocentric system must of necessity be finite (cf. *On the Heavens* I, 5).

one, whether moving or at rest, must be at a certain distance from the center, and every part of every body will be united to its whole thanks to its natural gravity. Now, Aristotle to the contrary, the characteristic order of that system can never be maintained by rectilinear motion, and this is so for two main reasons: first, because rectilinear motion must lead bodies farther and farther away from their starting points, and second, because, since a straight line is infinite and indeterminate, rectilinear motion must itself be indeterminate and hence tend nowhere in particular. Such motion is quite incompatible with an orderly universe because, far from maintaining its order, it tends continuously to destroy it. To consider it a natural motion of bodies, without first imposing restrictions that run counter to all reason, is therefore tantamount to denying the very possibility of a cosmological system. The order of the universe is, in fact, compatible with two states only: the state of rest and a state of motion that is "terminate" and that, while carrying bodies from one place to the next, nevertheless keeps them at an equal distance from the central point with respect to which their arrangement can be described. Thus the only kind of motion to meet this condition is circular motion, that is, "the one made by the moving body upon itself," and "the one that carries the moving body along the circumference of a circle about a fixed center."[105] The first keeps the body always in the same place, and the second puts neither the body nor those about it in disorder.[106] Did Galileo refuse to grant any role to rectilinear motion in the structure of the universe? Actually he granted it two: that of restoring order by returning the parts to the whole along the shortest route and that by which nature normally ensures the acceleration and retardation of bodies. For the rest, "it seems to me that one may reasonably conclude that for the maintenance of perfect order among the parts of the universe it is nec-

---

[105] *Dialogue* I, pp. 55 f.

[106] Without clearly developing this implicit premise of his cosmology, Copernicus had previously put forward the idea that circular motion was the only true natural motion; cf. *De revolutionibus orbium coelestium* I, 8. As for his view of gravity, though it differed from the traditional, it was still a long way from Galileo's, for while Copernicus also held gravity responsible for the return of the parts to the whole, he believed its main function was to ensure the combination of these parts into spheres, which not only were the most perfect forms in the universe but also engendered circular motions by themselves. Galileo, for his part, viewed gravity exclusively as an attribute of matter. For the Copernican theory of gravity, see *De revolutionibus* I, 9, p. 101.

essary to say that movable bodies are movable only circularly; if there are any that do not move circularly, these are necessarily immovable, nothing but rest and circular motion being suitable to the preservation of order."[107] The reasons that persuaded Galileo to consider circular motion the only form of motion worthy of inclusion among the premises of his cosmology are therefore perfectly clear, and there is little need to introduce aesthetic motives[108] or the intrinsic perfection of circular motion. "For my own part," Galileo wrote in the *Saggiatore*, "never having read the pedigrees and patents of nobility of shapes, I do not know which are more and which are less noble, nor do I know their rank in perfection. I believe that in a way all shapes are ancient and noble; or to put it better, that none of them are noble and perfect, or ignoble and imperfect, except insofar as for building walls a square shape is more perfect than a circular, and for wagon wheels, the circle is more perfect than the triangle."[109] In short, the ability of circular motion to keep the order of the universe intact sufficed Galileo to promote it to the rank of a cosmological premise, the more so as in the absence of the idea of universal attraction the order of the universe was bound to strike any positive mind as a prime datum of cosmology.

The preservation of an orderly world and the primacy of circular motion—these were the two general premises that reason itself placed at the head of the new cosmology. Now, while they corroborated the Copernican doctrine, these premises also had a bearing on the behavior of terrestrial bodies and hence imposed certain limitations on Galileo's thought, which we shall now examine in brief.

First of all, by ascribing a special role to rest and to circular motion in the preservation of the order of the universe, Galileo reintroduced an

---

[107] *Dialogue* I, p. 56. After putting forward the same view in his *Letter to Ingoli*, Galileo went on to say: "Hence I conclude that if the earth has a natural tendency to move, that tendency can only be toward circular motion, rectilinear motion being relegated to its parts; this applies not only to the earth but also to the moon, the sun, and all the other bodies in the whole universe whose parts, when violently removed and hence disordered, will rejoin the whole along the shortest route." Vol. VI, p. 559.

[108] The view of M. Panofsky in "Galileo as a Critic of the Arts" (*Isis*, March 1956, Vol. 147); cf. A. Koyré, "Attitude scientifique et pensée esthetique," in *Critique*, September–October 1955).

[109] *Saggiatore*, Vol. VI, p. 319. Cf. *Letter to G. Gallenzoni* July 16, 1611, Vol. XI, pp. 145 f. In the *Dialogue* (pp. 110 f.), Galileo remarked that if a spherical shape were a sign of intrinsic perfection, then all material bodies would have to be perfect, since by appropriate division it is possible to reduce all of them to an infinite number of spheres.

important idea of traditional philosophy, namely, the existence of natural states. Though this left him free to decide the order of celestial bodies by observation, it robbed him of the power to choose the appropriate motions; in this respect and despite its anti-Peripatetic contents, Galileo's doctrine thus remained pre-Newtonian. Moreover, accepting the idea of natural states also meant preserving the distinction between natural and violent motions. Admittedly, Galileo, unlike Aristotle, defined "natural" motions in terms of the disposition of bodies and not in terms of their essence, but because he also held that this disposition is best maintained by circular motion, he was prevented from treating the latter as a form of violent motion and thus from analyzing it by purely mechanical methods. Finally, though it avoided a number of pitfalls, Galileo's theory of gravity was also not entirely beyond reproach. It is true that his conception of gravity as the inner compulsion of parts to rejoin the wholes from which they have been separated allowed him to organize the world as he pleased, but it nevertheless turned gravity into an inalienable attribute of these parts, thereby making it impossible to attribute to bodies, considered by themselves, a general neutrality toward motion.[110]

However, though these ideas encumbered Galileo in more than one respect, they did not hamper his attempt to refute the objections against the earth's diurnal motion by valid mechanical arguments. In fact, the real obstacle lay elsewhere, namely, in the alleged supremacy of circular motion; when he made it the sole guarantor of the universal order, he impeded not only the mechanical interpretation of planetary motions but also progress in astronomy as such. To appreciate this point, we need merely recall his part in the controversy of the comets, as a result of which circular motion, by virtue of its ordering function, was transformed into the *sine qua non of the existence of celestial bodies.*

The story began with the appearance of three comets toward the end of the year 1618. In 1619 a Roman Jesuit, Father Orazio Grassi, published a short work in which he treated these comets in much the same way as Tycho Brahe had treated the comet of 1577.[111] In particular, he argued that, because of their very small parallaxes, comets must be true

[110]In this part of the present work we are concerned mainly with the mechanical repercussions of Galileo's cosmology, not with his mechanical thought as such, so that we need not dwell on the fact that various passages in the *Discourses* run counter to some of the arguments he put forward in defense of the heliocentric hypothesis.

[111]Cf. Dreyer, *A History of Astronomy*, p. 366.

celestial bodies, situated at immense distances from the earth and moving in circular orbits.[112] Galileo, who considered this interpretation completely unfounded, mounted a series of attacks against it, particularly in his *Discorso delle Comete* and his *Saggiatore*, in which he mustered a whole battery of arguments to show that comets were not real objects but mere appearances, issuing from the reflection of the sun's rays by certain substances probably composed of terrestrial vapors,[113] and hence comparable to rainbows, halos, and sundogs.[114]

Now, not only was Galileo completely mistaken, but his view was the more surprising in that it led him to brush aside the same parallax argument which during the Third Day of the *Dialogue* he would fully accept when agreeing with Tycho Brahe that new stars were authentic celestial bodies. Thus, oddly enough, while his Peripatetic adversary adopted the "modernist" view of comets, Galileo himself defended what can only be called an Aristotelian position.

Not, of course, that his arguments were without any substance. Thus he pointed out quite correctly that if comets were celestial bodies, they must nevertheless differ from the rest, since all the observations tended to show that their matter was exceedingly thin and rarefied,[115] that is, precisely of the same nature as terrestrial reflection phenomena. Also, in the case of the latter, no one has ever been able to detect the least parallax,[116] so that Tycho's hypothesis lost much of its weight. Moreover, many of Grassi's arguments were blatantly false, for instance, his claim that the further away an object is from the earth, the less it is magnified by the telescope, and precisely because the telescope fails to enlarge the apparent size of comets, the latter must be further away than the moon and must therefore be genuine celestial bodies.[117] Galileo had no difficulty in demolishing this line of reasoning: the telescope, in fact, magnifies all objects in precisely the same way, no matter what their distance; the reason why very distant objects appear to be less highly magnified by

---

[112]A refutation of Father Grassi's *Disputatio astronomica de tribus cometis anni MDCXVIII* was published in June 1619. It was entitled *Discorso delle comete* and, though it bore the signature of Mario Guiducci, was written largely by Galileo himself (see *Opere*, Vol. VI, Introduction, pp. 9 ff). At the end of 1620, Father Grassi (using the pseudonym of Lothario Sarsi Sigensano) published a counterattack, the *Libra astronomica ac philosophica*, which Galileo criticized page by page in his *Saggiatore* of 1623. The final chapter in this protracted controversy was Father Grassi's *Ratio ponderum Librae ac Simballae*, published in Paris in 1626. This work went almost completely unnoticed.

[113]*Saggiatore*, p. 278; cf. *Dialogue* I, p. 77.

[114]*Saggiatore*, p. 238.

[115]Ibid., pp. 274–275.

[116]Ibid., p. 300.

[117]Ibid., pp. 271 f.

it than those nearer to it must be sought in the elimination of the irradiation effect, which causes an artificial magnification of celestial objects when viewed with the naked eye.[118]

But, however pertinent his objections may have been, they did not justify his own position. This he tried to do by more general arguments in both the *Discorso delle Comete* and the *Saggiatore*. First of all, Galileo, unlike Father Grassi, realized full well that if the comets were celestial bodies, they would have to travel in enormously large orbits—for how else explain that the comet of 1577 did not return to view until 1618?[119] In the forty days of 1618 during which it stayed in sight, this comet covered more than a quarter of the great circle of the celestial sphere; hence its orbit in forty-one years must be "exorbitant and monstrous."[120] Nor could the comet possibly have traveled in a circular orbit; its long disappearance meant that it must follow an oval or else a completely irregular trajectory. And while the introduction of such trajectories may have helped Father Grassi "to save not only this appearance but any other," such trajectories or lines "have no determinacy whatsoever" and are therefore indefinable. "Hence to say 'such events take place by reason of an irregular line' is the same as saying, 'I do not know why they occur'; and the introduction of such lines is in no way superior to the sympathy, antipathy, occult properties, influences, and other terms employed by some philosophers as a cloak for the correct reply, which would be 'I do not know.' "[121]

This may help explain Galileo's own view: comets cannot be true celestial objects because, if they were, they would have to follow oval trajectories, and these ran counter to the basic tenets of the Copernican doctrine. It may also explain why, though Galileo expressly numbered the ellipse among the "regular lines"[122] and hence might very well have recognized comets as true celestial bodies, he did not even examine this possibility: to admit comets as permanent features of the heavens was to put an end to the supremacy of circular motion and thus to deny the simplicity and order of the Copernican universe.[123]

For all that, Galileo's new cosmological premises remained of positive value. While retaining the ideas of an a priori world order and of the

[118]Ibid., pp. 273 ff.

[119]*Discorso delle Comete*, Vol. VI, pp. 51–52. If the two objects sighted were not the same comet, the orbits would of course have to be even more enormous.

[120]*Saggiatore*, pp. 243 f.

[121]Ibid., p. 244.

[122]Ibid.

[123]Ibid., pp. 87–89.

contrast between natural and forced motions (though in modified form), they nevertheless freed the study of local motion of most of its old constraints. The reader may recall that in the Peripatetic view local motion derived its fundamental categories no less than its true significance from such basic cosmological postulates as the orderly structure of the cosmos and the associated theories of natural motions and the elements. In other words, once natural places had been assigned to all elementary bodies, and with them the appropriate natural motions and directions, it fell to local motion to maintain these bodies in, or restore them to, these places, where alone they could exist in actuality. Hence, though local motion was assigned a clear role of its own, it was also subordinated to an ontological function, without which it could not even be conceived.

This view was one that Galileo changed completely. Aristotle's orderly structure, though inseparable from natural motions, was nevertheless logically anterior to them: it introduced the center and periphery of the world even before specifying the six directions that occurred in it and moreover, by validating the theory of the elements, endowed local motion with its true significance. Now it was the very attributes (places and directions) by which Aristotle had defined an orderly universe even before defining motion that disappeared completely from Galileo's universe. As a result, order became identical with the states through which it manifests itself, that is, with circular motion and rest. From being a means of achieving (or conserving) a preexisting order, circular motion and rest were thus *transformed into that order itself*, and moreover—and this is of primary importance—ensured its very existence and stability.

The consequences were far-reaching. For Aristotle, motion was the transformation of the potential into the actually existing and "was not even conceivable in the absence of a subject."[124] Galileo, having identified motion with order, could dispense with its ontological function. At the same time he severed the link between local motion and changes in respect of quality and quantity, and thus robbed the concept of *κίνησισ* of all meaning. In Galileo's world, local motion, at least in its perfect, or circular, form was thus endowed with a goal that for the first time was identical with it.[125]

---

[124] *Dialogue* II, p. 147.

[125] In a strict sense, this new interpretation applied exclusively to uniform circular motion. Rectilinear motion, which in Galileo's view served the restoration and not the maintenance of order,

Nor was that all. The concept of order, in fact, concerns only the relations between bodies, not the bodies themselves; hence local motion, identified with order, could in no way affect the nature of moving bodies. But in that case, was it not true to say that all bodies as such were completely indifferent to motion and immobility? Now this was quite contrary to the view of traditional mechanics, for which motion could never be equivalent to rest. In the sublunary world, for example, rest alone had a positive sense, while motion, no matter whether natural or violent, represented an abnormal condition for bodies. In Galileo's view, on the other hand, not only were motion and rest no longer described in terms of the actual or the potential, but motion had ceased to have the least effect on the nature of bodies and in that sense had become akin to rest. "If I had all the time and space I need to explain my view, I would go so far as to say that motion, inasmuch as it is simple, cannot make a moving body hot or cold or alter it in any other way except to change its place; it hence produces nothing that would not have occurred had the body remained at rest."[126]

But ridding motion of its ontological function also meant severing the links between motion and the essence of bodies. It was with the help of these links that Peripatetic mechanics had explained the immobility of certain bodies and the mobility of others. Thus Aristotle had deduced the immobility of the earth not so much from direct experience as from the assumption that its own gravity must cause it to remain immobile at the center of the universe; conversely, the celestial bodies which revolved about the center because they were neither heavy nor light were endowed with local motion, a sign of their perpetual existence in actuality.[127] Galileo, on the other hand, saw no purpose in linking motion or rest to the essence of bodies, for as he saw it, the arrangement of materially like bodies in a stable system tells us nothing at all about their mobility or immobility; but above all, motion and rest could have no bearing on the formal nature of bodies: once the ontological function had disappeared, the idea that bodies were inherently mobile or im-

---

remained so to speak inferior to circular motion. Anticipating some of our later findings, we would, however, like to add that rectilinear motion suffered no other limitations; it was, as Galileo himself put it, "infinite by nature." Hence we are not misrepresenting him when we assert that by local motion in general he referred to all the changes his cosmological premises entailed for kinematics.

[126] *Discorso delle comete*, p. 55.

[127] Cf. Chapter 1.

mobile had to be dropped as well, for it was no longer possible to decide a priori which were in motion and which were at rest. The cosmological premises introduced in the *Dialogue* thus implied the absolute relativity of motion. In other words, the Peripatetic definition of motion as "the functioning of what is potential as potential" was completely meaningless. From a letter by Benedetto Castelli we gather that Galileo arrived at this conclusion in 1607: "Motion," he remarked at that time, "is nothing other than the change of one thing with respect to another,"[128] a view he repeated several times during the Second Day of the *Dialogue*. Admittedly, the principle of visual relativity underlying that definition was still a far cry from a new science of motion, but when, as we shall see, it became linked to other factors, it gave rise to a much broader principle of relativity.

By linking motion to the essence of bodies, Aristotle tried to show not only which bodies were mobile and which immobile but also why some forms of motion must be reserved for certain bodies. Thus he contended that the earth, which is impelled toward the center of the world by its own gravity, is bound to experience circular motion as a violent and transitory form, whereas celestial bodies cannot be in rectilinear motion, however briefly, for such motion is incompatible with their privileged status and hence with the order of the world. Galileo, on the other hand, by making that order his starting point and identifying it with circular motion and rest, had no difficulty in attributing both rectilinear and circular motion to all bodies in the universe: when separated forcibly from their whole, fragments of any celestial body tend to return to it in a straight line, while circular motion is the natural attribute of all celestial bodies including the moving earth.[129] Not only did all these bodies have the capacity of moving either rectilinearly or circularly, but all of them did so at some time or another. What simpler way, for instance, of imagining the formation of the solar system than by a combination of rectilinear and circular motions? That the "divine architect," after creating all the planets in a particular region of space, caused them to drop in a straight line until, having gained their "proper and permanent speed," their rectilinear acceleration was converted into a circular one.[130] Though

---

[128]"Il moto non altro che una mutatione d'una cosa in relatione a un'altra"; Letter of April 1, 1607, Vol. X, p. 170.
[129]*Letter to Ingoli*, Vol. VI, p. 559.

[130]This view, which Galileo mistakenly attributed to Plato, was propounded in the *Dialogue* (I, p. 53) and repeated in the *Discourses* (IV, pp. 283–284). For its

no more than a "fantasy," this theory nevertheless reflected the view that, though circular and rectilinear motion are not equivalent, they appertain to all celestial bodies. For that reason, while Aristotle had argued that the irreducible difference between circular and rectilinear motion reflected the irreducible differences between heaven and the earth, Galileo's conception was due only to a provisional defect in mechanics. Thus, during the cosmological discussions that filled the Third Day of the *Dialogue* he was at great pains to show that the same models and the same explanatory schemata apply to all bodies, no matter where they are located, and to all phenomena no matter where they are produced.

However, Galileo's cosmological premise did more than remove some of the most cumbersome restrictions imposed on mechanics by the old school; it also bestowed an entirely new status upon local motion. Thus if bodies are really indifferent to any particular form of motion, there is nothing to stop them from moving with several motions at once. Now, the idea of compound motions was basic to the Copernican doctrine, so that we need not be surprised that Galileo dwelled on this point at such length.[131] In particular, he tried to show that the earth can easily describe the three motions demanded by Copernicus, namely, rectilinear motion (by the parts toward the whole), a daily whirling motion upon itself, and another whirling motion about the sun completed in a year, and that, in an orderly world, circular motions (from west to east) are compatible not only with rectilinear motion but also with each other. The arguments he put forward during the Second Day of the *Dialogue* were meant to consolidate this supposition by a definition of the nature and consequences of the composition of motions, but for the moment we shall simply recall Galileo's concern to demonstrate that the latter is fully borne out by concrete experience. Thus he remarked that the fact that a ball can roll down an inclined plane while turning upon itself shows that there is nothing unnatural in the composition of motions.[132] The behavior of the compass was no less significant, for it in-

refutation by Newton, see *Isaac Newton's Papers and Letters on Natural Philosophy*, edited by I. B. Cohen, pp. 284, 296 ff., and 304 ff.; Cohen demonstrated its untenability on mechanical grounds. On this point see also A. Koyré and I. B. Cohen, "Newton, Galileo and Plato" in *Proceedings of the Ninth International History of Science Congress* (Barcelona, 1959), pp. 165–197.

[131]For the complete analysis see *Dialogue* III, p. 424.

[132]Ibid.

volves three types of motion: a rectilinear motion toward the center of the earth as a heavy object; a horizontal circular motion by which it conserves "its axis in the direction of certain parts of the universe"; and a third motion, discovered by Gilbert, of "dipping its axis in the meridian plane toward the surface of the earth, in greater or less degree proportionate to its distance from the equator."[133] It then suffices to recall how Aristotle's denial of the possibility of compound motions had vitiated any attempt to analyze the trajectories of projectiles to perceive immediately how the new cosmology went hand in hand with a fresh reflection about motion.

However, this was by no means its chief importance. When he identified motion with the cosmological order, Galileo also dispensed with the need for an external goal, and this, coupled to his elimination of the ontological function, made it possible to treat motion as a *state*, and as such capable of indefinite self-conservation. In the case of circular motion, this consequence was only too obvious, for such motion has the special property that any point along its trajectory is at one and the same time a beginning and an end, the moving body leaving and approaching it continuously and hence displaying no particular tendency to reach it and no particular reluctance to escape from it, so that it alone can be said to be essentially uniform. "From this uniformity and from the motion being finite there follows its perpetual continuation by a successive reiteration of the circulations which cannot exist naturally along an unbounded line or in a motion continually accelerated or retarded." Hence circular motion is the only form of motion capable of preserving order and, what is more, of conserving it indefinitely.[134]

The case of motion in a straight line is much more complicated, for, as we saw, it serves merely to eliminate a temporary state of disorder and hence lacks the power of self-conservation. Moreover, though it also is finite in fact, it is by no means finite in principle, since nature has not "arbitrarily assigned to it some terminus."[135] Now this could mean only that rectilinear motion, too, might in the absence of all external impediments, continue indefinitely, so that in this sense it could be said to resemble circular motion. In short, if we dispense with the demand for order, the de facto finitude of rectilinear motion will disappear, and it will become a state as natural to bodies as rest or circular motion.

[133]Ibid., pp. 426 and 437.
[134]*Dialogue* I, p. 56.
[135]Ibid., p. 43.

Admittedly, Galileo's arguments still lacked a precise mechanical meaning since, instead of attributing the conservation of motion to the indifference of matter to motion or rest, he attributed it to certain properties of the trajectory. Hence it would be idle to look in the discussions that make up the First Day of the *Dialogue* for an anticipation of the law of inertia. Moreover, it should be stressed that, despite the important transformation that his ideas underwent during the Second Day of the *Dialogue* and later in the *Discourses*, Galileo never abandoned the view that circular motion alone had an actual and real power of indefinite self-conservation. However, and just as remarkably, he never felt the slightest need to attribute the conservation of motion to a motor or formal cause,[136] thus demonstrating beyond all doubt that, despite the limitations which they served to perpetuate, his new premises did indeed tend to represent motion as a state. And the last two comments attest, perhaps better than anything else, to the ambiguous but nonetheless fruitful role these premises played in Galileo's development of mechanics.

* * *

The common view that the Copernican contribution to mechanics was confined to the study of the motion of bodies on a moving earth is therefore quite mistaken; in particular, it neglects the effects of Copernicanism on the cosmological premises, which from Galileo's and probably until Newton's day presided, at least in part, over the general interpretation of motion. The real problem is the precise scope of this basic transformation.

Now there is no better way of demonstrating the role of the new cosmological premises than to recall some of the most important fourteenth-century contributions to mechanics. Take, for instance, the theory of *latitudines:* because it made a clear distinction between speed and acceleration, the modern reader might well consider it a direct refutation of the Aristotelian view that there could be no such thing as the motion of a motion, and hence an invitation to treat motion as a physical state.[137] Now neither the Mertonians nor the Parisians, not surprisingly, saw

---

[136]In this respect he was well ahead of Copernicus, who attributed the circular motion of celestial bodies to their spherical form which, he believed, was capable of engendering such motion quite spontaneously: "The mobility of a sphere consists in moving in a circle and thus expressing its form" (*De revolutionibus* I, 4). Galileo was unquestionably the first to assume that motion does not have to be described by reference to a formal cause, whether of a substantialist or a geometric type.

[137]Cf. Chapter 2.

matters in this light: because they retained the idea of an a priori world order in which all elementary bodies are assigned definite places, they were unable to divorce local motion from its ontological function and hence to treat it as a state instead of a process. Other examples are equally revealing—for instance, Oresme's paradoxical view of the motion of the earth. Thus, though he realized full well that the appearances would not be changed in the slightest if the earth moved about the sun instead of standing still, and though he was the first to refute the objections to the earth's diurnal motion, he nevertheless ended up by rejecting the whole idea, claiming, as the reader may remember, that he had merely suggested it "by way of diversion." And though he may have added this rider for purely prudential reasons, as Duhem has suggested, this cannot be the full explanation. In fact, because Oresme continued to accept the Aristotelian distinction between heaven and earth, the theory of the elements and the doctrine of places and natural motions, he was bound to defend the physical arguments of the geocentrists; hence, however intrigued he may have been by the idea of the earth's diurnal motion, the temptation was not great enough to help him overcome all the philosophical prejudices that stood in its way. Oresme's case is all the more interesting in that on another occasion he was also unable to carry an anti-Aristotelian idea to its logical conclusion. Thus, when dealing with the plurality of worlds, he rejected the Aristotelian thesis that bodies are heavy or light in themselves and hence the idea of natural places. Nevertheless, in his commentary on Aristotle's *On the Heavens* he saw fit to defend the Aristotelian view. Once again this inability to generalize findings gathered in a particular sphere cannot be blamed on the medieval habit of compartmentalizing what for us are related problems; the only true explanation is that Aristotle's cosmology still seemed solid and unassailable. This also explains why Bishop Tempier's "divine cases" (and notably the rectilinear motion of the world) were not pursued to their logical conclusion. So close was the link between the traditional conception of motion as an active process and the orderly structure of the world that it was impossible to reject the first without discarding the second. The revolutionary contribution of Copernicanism was precisely that it did reject it, thus making feasible what fourteenth-century physicists, however bold, could not even imagine.

Naturally we do not wish to suggest that the replacement of Aristotelian with Copernican premises explains the genesis of Galileo's rational mechanics, which was, in fact, based on concepts and studies of quite a different kind. Moreover, the critique of Aristotle's mechanical ideas was no innovation, as our analysis of fourteenth-century studies must have shown abundantly. It nevertheless remains a fact that in dismantling the major bastions of the traditional cosmology, Galileo eliminated a host of principles whose tacit acceptance had impeded the development of the science of motion. By demonstrating their lack of substance, or showing that they were rooted in defective experiments, he not only did away with a great many unnecessary limitations but also opened the way for the most fruitful application of the new ideas and the formulation of new problems. In other words, having first eliminated the major philosophical obstacles in the path of mechanics, Galileo was able to introduce a new conceptual system that, although transitory, nevertheless rid that science of its age-old shackles.

The importance of this contribution may also be appreciated from the support it provided for the Copernican doctrine. Oresme's great weakness had been his inability to justify the earth's diurnal motion by other than optical arguments or by appealing to other than reasons of simplicity. Having discarded the traditional restrictions, Galileo, for his part, found himself in possession of incomparably more effective weapons. By severing the link between local motion and the nature of moving bodies, by presenting motion as a neutral state, by asserting its relativity and equivalence to rest, and by granting the existence of compound motions, he was able to construct a new framework in which the debate with those who denied the earth's diurnal motion could be conducted with the help of entirely new arguments. If we add that these arguments, which we are about to examine at greater length, represent one of the two chief contributions of Galileo to classical mechanics, we may have helped the reader appreciate the full importance of the Copernican premises he presented in the *Dialogue*.

5 The Problem of the Earth's Diurnal Motion

While the new cosmological premises did away with the idea that the motion of the earth was impossible in principle, they said nothing about whether it was possible in fact. Hence, before he could establish the soundness of the Copernican doctrine, Galileo had first to refute the argument of all those philosophers and astronomers who had maintained since antiquity that the annual and diurnal motions of the earth ran counter to everyday experience.

The case of the annual motion is quite different from that of the diurnal one. To begin with, poor observation or errors in interpretation accounted for most of the objections to the former. Tycho Brahe, for example, had deduced from the apparent diameters of the fixed stars that according to the Copernican doctrine even the smallest of them must be much larger than the orbit of the earth. Moreover, since despite these "incomprehensible" dimensions the fixed stars showed no detectable parallax, the limits of the universe would have to be placed at an infinite distance from us.[1] In fact, apart from evincing a bias in favor of a finite universe, Tycho's argument merely reflected his inability to suppress the irradiation effect. The telescope, by contrast, did just that, for with its aid the apparent diameter of a fixed star of the first magnitude was reduced from two minutes to five seconds and that of a star of the sixth magnitude from twenty seconds to fifty thirds; in other words, their dimensions ceased to be "incomprehensible." Moreover, as Galileo showed during the Third Day of the *Dialogue*, the Copernican doctrine in no way implied an infinite universe; thus if one assumes that a fixed star of the sixth magnitude is no larger than the sun, its distance from us would be 2160 radii of the earth's orbit,[2] a perfectly "conceivable" figure and one that also explained why no parallax could be observed.[3]

The earth's diurnal motion, while introducing no modification in the universe, posed a quite distinct problem, namely that of its compatibility with our experience. Would not the natural order of things be destroyed on a rapidly spinning earth? Would not the ordinary

---

[1] *Dialogue* III, pp. 385 ff.

[2] Galileo's estimate was based on the assumption that the distance of the earth from the sun was 1208 terrestrial radii, when in fact it is 46,000. His general conclusion, however, was correct. Ibid., pp. 386–387.

[3] Galileo added, however, that careful observations of the most suitable stars might well reveal a parallactic displacement on their part. Ibid., pp. 408–411.

movements of the earth's surface be altered by such a motion beyond recognition or even prevented altogether? Unlike the annual motion, which posed optical problems, the diurnal motion thus raised genuine mechanical difficulties which no amount of observation could hope to overcome. The only alternative was to introduce a new conceptual system based on a complete redefinition of motion and its properties. This was precisely the task Galileo set himself.

The Objections and Their Refutation

There were four main objections to the earth's diurnal motion; some of these went back to antiquity. The first was based on the simplest of all natural motions, that is, the perpendicular fall of heavy bodies;[4] it alleged that if the earth was indeed a rotating body, a tower from whose top a rock was dropped would travel many hundreds of yards during the time the rock consumed its fall, so that the rock would strike the earth that distance behind the tower.[5] This effect, they claimed, could be confirmed by another experiment, which is to drop a lead ball from the top of a ship's mast. If the ship is at rest, the ball will hit the foot of the mast, but if the ship moves the ball will strike that distance from the foot of the mast which the ship will have run during the time of the fall of the lead.[6]

The second objection was raised by Tycho Brahe and was based on projectile rather than on natural motion. To begin with, it is almost certain that the infinitely more rapid motion of the earth would "absorb" the motion of cannonballs, so that all of them would go in the same direction.[7] But, above all, a ball shot toward the west could never have the same range as one shot toward the east; for when the ball goes toward the west, and the cannon, carried by the earth, goes east, the ball ought to strike the earth at a distance from the cannon equal to the sum of the two motions; while from the trip of the ball shot toward the east it would be necessary to subtract that which was made by the cannon following it.[8] But since experiment shows the shots to fall equally,

---

[4]This argument was first presented in Ptolemy's *Almagest* I, 7.; it was restated by Tycho (cf. J. L. E. Dreyer, *A History of Astronomy from Thales to Kepler*, p. 360) and criticized by Copernicus (*De revolutionibus orbium coelestium* I, 7).

[5]*Dialogue* II, pp. 151–152.

[6]Needless to say, this experiment was never performed.

[7]Tycho Brahe, *Astronomicarum epistolarum liber*, ed. Dreyer, pp. 218 ff.; Tycho's argument had previously been criticized in Kepler's *In commentaria de motibus Martis*, *Opera*, Vol. III, pp. 458 ff. (Frisch edition).

[8]*Dialogue* II, pp. 152–153 and 194.

it follows that the cannon is motionless and consequently the earth as well.

The third objection was based on everyday experience, namely, the flight of birds and the movement of clouds;[9] since neither one nor the other adheres to the earth, how could they possibly keep up with the speed of a revolving globe? Does it not seem obvious that they could only be borne very swiftly toward the west? "If we, carried by the earth, pass along our parallel (which is at least sixteen thousand miles long) in twenty-four hours, how could the birds keep up on such a course, whereas we see them flying east just as much as west or any other direction without any detectable difference?"[10] Finally, if the earth did indeed spin on its axis, we should be perpetually lashed by a tremendously strong east wind.

The fourth and probably the most ingenious objection of all was framed by Ptolemy, who based it on a property of motion about a center,[11] namely, that it tends to cast off, scatter, and drive away the parts of the moving body whenever the motion is swift enough or the body has been insufficiently secured. Since in the case of the earth's rotation the motion would undoubtedly be swifter than any circular motion known to man, all heavy bodies, men, and animals upon it would be hurled into space, and "what tenacity of lime or mortar would hold rocks, buildings, and whole cities" to the ground? But, as with the birds and the clouds, we need only open our eyes to behold that this grain of sand and that leaf repose peacefully in their proper places instead of rushing headlong into space, so that the earth must indeed be immobile.[12]

Though each of them dealt with a particular case, the first three objections nevertheless posed a single problem: how to reconcile the rotation of the earth with everyday experience. A single set of principles and concepts could therefore be applied to their refutation.

We saw earlier that Galileo had rid motion of its ontological function.[13] This enabled him to establish the legitimacy (and not merely the possibility) of the principle of the visual or mathematical relativity of mo-

[9]This argument, which was used by Ptolemy (*Almagest* I, 7), was later criticized by Copernicus (*De revolutionibus* I, 7).

[10]*Dialogue* II, p. 158.

[11]Ptolemy, *Almagest* I, 7. The argument was quoted and rejected by Copernicus (*De revolutionibus* I, 7 and I, 8).

[12]*Dialogue* II, pp. 158–159 and 214.

[13]See Chapter 4.

tion. Galileo put it all most forcefully when he asserted that the motions of the celestial bodies are only "in relation between the latter and the earth" and that precisely the same effect follows whether the earth is made to move and the rest of the universe to stay still, or the earth alone remains fixed while the whole universe shares a single motion.[14] "Motion," he wrote elsewhere, "does no more than make variations in appearances, which take place in the same way by making the earth move and holding the sun still as they would by the opposite."[15] No doubt, this principle sweeps away some difficulties. Thus Aristotle had argued that to account for the appearances of diurnal motion while the heavens stood still, the earth must be moved with at least two motions, in which case the risings and settings of the fixed stars would have to vary.[16] He simply does not see that a diurnal rotation of the earth would produce the same changes in the sky, in the same time, as the revolution of all the heavens. And since the latter have no need for more than a single motion, neither has a rotating earth.[17] Just as ill-founded was the objection that on a rotating earth objects near the equator would have to travel more swiftly than those nearer the poles; for a rotating earth would in no way differ from a rotating "stellar sphere in which those things closest to the equinoctial plane move in larger circles than those more distant from it."[18]

But useful though all these remarks were, they did not go very far, and it is easy to see that this first principle of relativity could not stand up against the traditional objections. For the problem they raised was no longer a problem of *description*; by arguing that if the earth were indeed a moving body, it must produce detectable changes in all its parts, they posed the quite distinct problem of the *mechanical effects* of motion. In other words, they challenged Galileo to treat motion not as a simple form of displacement, but in terms of its action on moving bodies, or rather to prove that on a rotating earth no effects would appear that did not also appear on an earth at rest. To meet this challenge, Galileo was forced to generalize his conception of relativity and in so doing to bestow upon it a radically new significance.

Consider a system of bodies joined together in any way, and let an observer placed among them detect a "variation in their appearances"

---

[14] *Dialogue* II, p. 143.
[15] Ibid., pp. 121–122.
[16] *On the Heavens* II, 14, 296 a 34 f.
[17] *Dialogue* II, 163–164.
[18] Ibid., p. 267.

with respect to a number of external reference points. Clearly, he can account for these variations equally well by the motion of the reference points as by that of the system with which he himself is associated. But let us now suppose that the possible motion of the system is in fact the true motion, and let us by way of illustration consider the case of a cargo ship on its way from Venice to Aleppo.[19] While the goods with which it is laden are undoubtedly being transported from one port to another, they do not change their relative position inside the hold, so that an observer locked up with them would have no reason to think they were not at rest. "It is obvious then that the motion common to many moving things is idle and inconsequential to the relation of these movables among themselves, nothing being changed among them, and that it is operative only in the relation that they have with other bodies that lack that motion."[20] But though an observer shut up in a ship's hold finds it impossible to tell whether he is at rest or in motion, this impossibility is of a different kind from that associated with the appearances; it springs from the fact that motion in no way alters the relations of a system of bodies, once all of them share equally in it. We saw earlier that, from a purely optical point of view, motion and rest can be attributed indifferently to the observer or the object observed; we have now also learned that there is no way of telling from within a system (at least under certain conditions)[21] whether the system is in motion or at rest. From the principle of visual relativity we have thus passed on to the principle of mechanical relativity, discovering our first weapon for the refutation of the traditional objections: "Then let the beginning of our reflections be the consideration that whatever motion comes to be attributed to the earth must necessarily remain imperceptible to us and as if nonexistent so long as we look only at terrestrial objects."[22]

As a result, the very basis of the discussion was transformed. For if it is true that a system can move without any inner changes, then a body apparently at rest might easily be in motion with the system of which it constitutes a part; moreover, a body just set in natural or violent motion might not have left a state of rest but might equally

[19] Ibid., pp. 141–142.
[20] Ibid., p. 142.
[21] Thus the system must be in uniform motion, a point with which we shall be dealing later on, but which we can safely ignore for the moment.
[22] *Dialogue* II, pp. 139–140.

well have left a state of motion. Now, as Galileo rightly remarked, the traditional physicists completely ignored this fundamental fact; on the pretext that they were comparing the normal behavior of bodies to their possible behavior on a rotating earth, they were all making the same mistake, namely, to assume that a moving body must start from absolute rest and can never start from a primary state of motion. Thus they all supposed that an arrow shot vertically into the air, and indeed all freely falling bodies, must have been immobile immediately prior to starting.[23] On the contrary, if we assume that the arrow started from an initial state of motion, then the Peripatetic objections must fall by the wayside, particularly if it could also be shown that all projectiles leaving the ground conserve the earth's diurnal motion within them even while describing their own motion. Let us now see in what way Galileo developed this crucial proof.

The simplest solution would of course have been to argue without further ado that, once it is granted that the diurnal rotation is the earth's own and natural motion, then it must also be granted that this motion is "indelibly impressed" on all parts of the earth.[24] Thus a rock at the top of a tower must have as "its primary tendency a revolution about the center of the whole in twenty-four hours, and it eternally exercises this natural propensity no matter where it is placed."[25] In short, he might easily have replaced the Aristotelian principle of natural rest with another principle based on the fact that "if the natural tendency of the earth were to go around its center in twenty-four hours, each of its particles would also have an inherent and natural inclination not to stand still but to follow that same course."[26] However, Galileo realized full well that this purely dialectical argument would have brought the mechanical solution of the problem not a whit nearer—much the same argument had been advanced unsuccessfully by Copernicus and the great majority of his earlier disciples. By telling their adversaries either, like Copernicus, that freely falling bodies have a mixed rectilinear and circular motion[27] or, like Rothmann, that the falling stone participates in the motion of the earth or again, like Gilbert, that because the diur-

[23]Ibid., pp. 196 and 200.
[24]Ibid., p. 168.
[25]Ibid.
[26]Ibid.
[27]"As for the things that fall and rise we avow that their movement must be double with respect to the world and generally composed of the rectilinear and the circular." *De revolutionibus* I, 8.

nal rotation is the natural motion of the earth it cannot produce any perturbations on it[28]—all of them relied on the traditional conception of natural motion.

Galileo's merit was precisely to have transcended this simple *ad hominem* argument and to have offered a solution in strict keeping with the idea of mechanical relativity. That solution he based largely on the systematic study of the behavior of bodies on an inclined plane. Thus, he asserted that if a ball was placed on a plane surface "as smooth as a mirror and made of some hard material like steel," and the surface was tilted however slightly, the ball would roll down it spontaneously and with a continually accelerated motion; conversely, if the ball was drawn or thrown against the gradient, its motion would constantly slow down until it came to a halt and began to roll down again.[29] Does this mean that Galileo considered the weight of the body, that is, what he himself had described as inalienable property of all matter, as the dynamic factor responsible for its propensity or resistance to motion? Not in the least, for he realized that whenever the tilt of the plane is decreased, so is the propensity or resistance to motion, and yet the weight of the body remains quite unchanged.[30] Hence he considered it far more reasonable to assume that the body follows a moment of descent that increases with the tilt of the plane and to relate its propensity or resistance to motion to that moment. Now, we saw that the concept of a moment of descent enabled Galileo to represent the motor function of gravity as such and hence to distinguish it from the purely gravific function.[31] Beyond that, it also allowed him to analyze the behavior of bodies in which the propensity (and hence the resistance) to motion had been completely suppressed. Thus, if our inclined plane has no tilt at all and therefore imparts no tendency to the ball to approach or withdraw from the common center of heavy bodies, so that the motor function of gravity is eliminated, the ball will no longer have the least moment of descent and will consequently be "indifferent to the propensity and resistance to motion."[32] Hence, if it is at rest, it will remain so indefinitely. Now, though Galileo had advanced the same argument in

---

[28]For Rothmann, see Dreyer, *A History of Astronomy*, p. 361; for W. Gilbert, see *De Magnette* VI, 3, p. 326; VI, 5, p. 340. The case of Bruno will be considered later.

[29]*Dialogue* II, p. 172.

[30]Having severed the traditional link between the velocity of a falling body and its total weight, Galileo was of course referring to the specific weight of the moving body.

[31]See Chapter 3.

[32]"Indifferente tra la propensione e la resistenza al moto," *Dialogue* II, p. 173.

the *Mecaniche*, where he had even asserted that under these conditions the slightest force would suffice to set any body into motion, in the new context it paved the way for a decisive advance. For let us imagine that thanks to an impulsion a certain speed or *impeto* has been conferred on a ball resting on a horizontal plane; since the entire plane is equidistant from the common center of heavy bodies, any moment of descent will have disappeared and, with it, the ability of the motor force to increase or decrease the ball's velocity. With admirable skill, Galileo was then able to introduce the idea of the conservation of impressed motion: on a horizontal plane equidistant from the center of heavy bodies a motion once started will continue indefinitely and uniformly and will thus be inertial in the precise sense of that term.[33]

The final step could now be taken. For if uniform motion around a center has the power of continuing indefinitely, this can mean only that the moving body has neither a "resistance nor a propensity" to such motion as distinct from motion toward or away from the center.[34] Thus if the circumstances under which the motion began happen to change, especially by the addition of new motions, there is no reason why it should abolish itself. Being an inertial form of motion by definition, motion around a center is thus "incompatible"[35] with no other form of motion, whether spontaneous or otherwise: in all cases the *impeto* of each of these motions will combine with that of uniform circular motion, which, being constant in magnitude and direction, will conserve the original velocity impressed upon the body.

The way was now clear for a simple and direct refutation of the traditional arguments. Take first of all the Peripatetic assertion that all projectiles on a rotating earth must be deflected toward the west. Galileo's answer was in two parts: he first demonstrated the absence of any deflection in a case to which the principle of conservation could be applied with particular ease and then went on to show that precisely the same argument applied to the case of freely falling bodies on a rotating earth.[36]

---

[33]Ibid., p. 174. Though we cannot say precisely when Galileo discovered this principle, there is good reason to think that he did so at a fairly early stage. Thus in 1607, Castelli wrote in a letter that, according to Galileo, though a mover is needed to initiate a motion, once the motion has begun, it will continue indefinitely in the absence of external impediments. (Vol. X, p. 170).

[34]*Dialogue* II, p. 175.

[35]Ibid.

[36]Galileo's methodology will be examined in Chapter 8.

Thus, supposing the earth were a vast ocean, then, if the conservation principle holds, a ship once set in motion will, in the absence of any resistance, keep traveling around the globe, because its motion neither brings it nearer nor removes it farther from the center of heavy bodies.[37] Now suppose a stone has been placed on top of its mast; it, too, will move with a circular motion about the center and, being carried along by the ship, will move at the same speed as the latter.[38] Now let the stone drop from the mast; according to our principle, it will conserve this inertial motion, which will combine with the downward motion. Instead of leaving the ship it will thus follow it and finally land in the same place where it would have landed had the ship been motionless. Now the case of a heavy body on a rotating earth—for example, a rock on top of a tower—would be similar in all respects. From the diurnal motion that carries it along together with the tower, the rock will receive an *impeto*, which will be impressed upon it indelibly. Hence, when released from the top of the tower, it will follow the motion of the latter despite the loss of direct contact and, keeping to the vertical, will finally land in the same spot as it would if the earth were at rest. When applied correctly, the principle of conservation thus suffices to dispose of one of the most weighty of the objections to the motion of the earth.[39]

Tycho Brahe's objection, that projectiles would have unequal ranges on a moving earth, was disposed of with equal ease. Again Galileo's argument proceeded by two stages. Take an open carriage, place in it a crossbow with the bolt at half elevation so as to obtain the maximum range, and then, while the horses are still running, shoot once in the direction of their motion and again in the opposite way, taking careful note where the carriage is at the moment the arrow strikes the ground in each case.[40] According to Tycho, the distance between the arrow and the spot where the carriage was when the arrow struck the ground

[37]*Dialogue* II, p. 174.

[38]Ibid.

[39]It is a remarkable fact that Galileo saw fit to establish the principle of conservation by reference to forced motions (a ball on a horizontal plane, a rock on a moving ship) before applying it to the natural motion of the earth. This could mean only that, as far as Galileo was concerned, all distinction between forced and natural motions had vanished. But it is also very likely that only forced or "artificial" motions, which could not be accounted for by invoking some conformity to the natural order of things, made possible genuine conceptual advances. The reader may recall that the wish to account for a forced motion, namely, that of projectiles, was responsible for the chief innovation in fourteenth-century dynamics, that is, the impetus concept.

[40]*Dialogue* II, p. 194.

would be much less when the shot went in the direction of the carriage than when it went the other way. Thus if the crossbow had a range of 300 yards and the carriage traveled 100 yards while the arrow was in the air, then the arrow fired in the direction of the course would strike the ground 200 yards from the carriage, while if it were fired in the opposite direction, the distance between the two would be 400 yards.[41] To make these shots cover equal distances, the bow would have to be bent harder with the course and more weakly against the course; to put it more simply, if the carriage moved with, say, one degree of speed, then in order to shoot the arrow in one direction as well as the other and have it depart equally from the moving carriage, it would be necessary that if on the first shot it left with, say, four degrees of speed, then on the other shot it must leave with only two.[42]

Now let us apply the principle of conservation of motion. When the carriage is moving, it impresses its own *impeto* of one degree of speed on all the objects inside, not least on the crossbow and the arrow it carries. Hence, when the arrow is fired in the direction of the course, the bow impresses its three degrees of speed upon a bolt which already possesses one degree thanks to the carriage. And if the arrow is fired in the opposite direction, the same bow will confer its three degrees upon a bolt moving with one degree in the opposite direction. "But you yourself have already declared that in order to make the shots equal, it is required that the bolt leave with four degrees in the one case, and with two in the other. Hence, without changing the [tension of the] bow, the course of the carriage itself regulates the flights."[43] Now what applies to the crossbow and arrow applies equally well to a cannonball following the earth's diurnal motion: shots made with the same force will always carry equally far no matter whether they are sent east or west. In other words, if V is the velocity that the crossbow or cannon impresses upon the arrow (or ball), t the duration of the flight, S or Vt the distance the shot would have covered had it been fired from a stationary vehicle (or

---

[41]Galileo used the term *braccio* (21 to 22 inches), which, for convenience, some translators have rendered as "yard" and others as "cubit."

[42]*Dialogue* II, p. 145. Galileo's analysis was not completely accurate, for the velocities must be compounded as vectors, not as scalars. Nevertheless the general conclusion is valid. Galileo gave a correct account of the composition of velocities in his *Discourses* (p. 280).

[43]Ibid., p. 196. The reader will have noted that Galileo applied the principle of conservation directly, that is, without first reducing the initial motion of the arrow to a circular inertial motion.

earth), v the velocity of the carriage (or moving earth), and s or vt the distance the latter would have covered during the flight of the arrow (or ball), then Tycho and Galileo would have expressed the situation as follows:

Tycho Brahe

a. Shot fired in the direction of the course.
   With S_1 as the distance between the vehicle (or earth) and the arrow (or ball) when it strikes the ground:
   $s_1 = (S - s) = (V - v)t.$

b. Shot fired in the opposite direction.
   With s_2 as the distance between the vehicle (or earth) and the arrow (or ball) when it strikes the ground:
   $s_2 = (S + s) = (V + v)t.$

Galileo

a. Shot fired in the direction of the course.
   $s_1 = (V + v)t - vt = Vt$
   (vt representing the distance the vehicle has covered during the flight of the shot).

b. Shot fired in the opposite direction.
   $s_2 = (V - v)t + vt = Vt.$

As for the third objection, its refutation followed even more directly from the new principle. Since birds and clouds can move for long periods at considerable distances from the ground, it seems hard to believe that, even while they are in the air, they continue to be part of the same system as are objects on the earth's actual surface. However, this difficulty disappears the moment we reflect on the behavior of the air: because it is mechanically bound to the earth, it will have the same uniform circular *impeto* as the more solid parts. Hence it, as well as the birds in it, will describe the same motion as the terrestrial globe; they "do not have to worry about following the earth, and so far as that is concerned, they could remain forever asleep." Much as a stone will drop straight down by virtue of the motion impressed upon it, so a bird or a cloud can, without any effort, keep vertically above a fixed point on the earth.[44]

The argument is interesting for another reason as well. At first sight

[44]Ibid., p. 209.

it might seem that since birds are living creatures and hence capable of moving contrary to the diurnal motion, they are in no way comparable to projectiles, so that the general explanation would not apply to their case. In fact, this view of the matter is easily disposed of. Thus if we drop a dead bird from the top of a tower, it will do the same as a stone; it will continue its vertical downward motion with the uniform circular *impeto* impressed upon it by the earth's rotation. Next, let us release a live bird; like the dead bird it will participate in the diurnal motion and conserve the corresponding velocity.[45] But, precisely because it is alive, it is free to send itself by beating its wings to whatever point of the compass it pleases, and this second motion, in which we do not participate, is for that very reason the only one we can observe. "For after all, its leaving toward the west in flight was nothing but the subtraction of a single degree from, say, ten degrees of diurnal motion, so that nine degrees remain to it while flying. And if it alighted on the earth, the common ten would return to it; to this it could add one [degree] by flying toward the east, and with the eleven it could return to the tower."[46] Now, not only does this argument extend the principle of conservation to the air, but it also shows that this principle applies at all times to all parts of a system in which spontaneous motions occur. And precisely because it applies to natural and forced no less than to spontaneous motions, the principle of conservation has the widest possible scope.

The fourth objection was of quite a different nature. Unlike the first three, which were meant to show that the earth's diurnal motion was incompatible with everyday experience, it contended that the centrifugal effects of a rotating earth would be such as to vitiate the coherence and unity of all terrestrial bodies and processes.

Let us first try to describe the centrifugal effect of circular motions in general. If we tie one end of a cord to a bottle of water and, holding the other end firmly by the hand and making our shoulder joint the center of the rotation, cause the bottle to go around swiftly so that it describes the circumference of a circle, we shall find that, no matter whether that circle is parallel to the horizon or vertical or slanted in any way, the water will never spill out of the bottle, rather "he who swings it will always feel the cord pull forcibly to get farther away from

---

[45] Ibid., p. 212.

[46] Ibid.

his shoulder."[47] Bodies whirled about a fixed center therefore acquire a force or *impeto* that tends to remove them from the center, even though they have a natural tendency to approach it. Now since this force is unlike any other, what kind of motion does it communicate to bodies subject to its influence? A long discussion between Simplicio and Salviati conducted in the Platonic manner made it clear that, in Galileo's view, the resulting motion could be neither circular nor perpendicular to the plane of the circumference. A projectile rapidly rotated will, upon being separated from the thrower, retain an *impeto* to continue its motion along the tangent to the circle described by the projectile at the point of separation.[48] This particular property, as Galileo clearly realized, made the fourth objection to the earth's diurnal motion the most crucial of all. Thus, while the first three had been based on the view that the diurnal motion would lead to disorder within the system—an argument that was easily refuted by the demonstration that the possible perturbations were nullified thanks to the conservation of the diurnal motion within terrestrial bodies—the fourth was based on the argument that a force resulting directly from a circular motion must lead to the ejection of bodies from the system to which they belong. Hence, the objection could be met, not by reference to the general properties of motion, but only by reference to the causes and effects of circular motion as such.

Two observations, Galileo believed, pointed to the correct answer. It is clear, first of all, that if a rotating body tends to escape along a tangent to its circular trajectory, then its escape can be prevented only by a balancing force. Now on a moving earth, that balancing force exists in the form of gravity, that is, in the intrinsic property of all bodies to move toward the center.[49] Let us next examine how a body moving along the tangent parts from its circular trajectory: it will clearly not move away uniformly, but its distance from the circle will increase in an increasing ratio. Thus in a circle with a diameter of ten yards, a point on the tangent two or three feet away from the contact will be three or four times as far from the circumference as a point one foot away.[50] This fact is particularly important because it shows that the

---

[47]Ibid., p. 216.

[48]Ibid., pp. 217–220. In Galileo's own words, the centrifugal force was the "impeto di muoversi par la retta tangente il cerchio del moto nel punto della separazione."

[49]Ibid., p. 220. In the analysis that follows, "gravity" refers to the motor force, not to the body's weight.

[50]Ibid., The Italian *piede*, which we have rendered as "foot," measured some eight inches.

tangential escape of the body is extremely small at first. Gravity, by contrast, acting toward the center, will make its full force felt the moment the separation has been effected, that is, when the separation due to the centrifugal force is still so insignificant that the slightest impulsion will suffice to maintain the body in its circular trajectory. Now, precisely the same thing happens in the case of a rotating earth: the curvature of its surface is so small that a tangent will not recede from that surface by "an inch in a thousand yards," so that the centrifugal tendency conferred upon all heavy bodies will always be great enough to offset what, at the beginning, is a "very minute motion away" from the center.[51]

Strange though this conclusion may have been, the argument as a whole was not uninteresting. In particular, its starting premise was absolutely correct: it is certain, in fact, that only the action of a greater force can offset the centrifugal effects of the earth's rotation and hence prevent the destructive results envisaged by the geocentrists. In more modern terms (but also in accordance with Galileo's own treatment of gravity in the *Discourses*) we might put it that the motion of the earth produces no detectable centrifugal effects because its dispersive action is not large enough to overcome the accelerative effects of gravity. But this argument is not far off from the correct interpretation of circular motion. If the earth spins on its own axis, all bodies on its surface must describe a circular trajectory in space; and of the two forces acting upon them simultaneously, the one (the centripetal force) will always neutralize the rectilinear motion which the other (the centrifugal force) tends to impress upon them. It follows that the motion of such bodies must be compounded of two other motions, and two rectilinear motions at that. Thus for the first time (and perhaps also for the last) Galileo realized that the curvature of a circular motion is due to the action of one force on another.[52] This is why, despite its shortcomings, his solution was greatly superior to that of Copernicus, who simply invoked the *natural* character of the earth's rotation to explain the absence of any centrifugal effects. "If someone thought that the earth moved," Copernicus asserted, "he would certainly say that this motion is natural and not violent. Now whatever happens in conformity with nature produces effects contrary to those engendered by violence. For the things to which

---

[51]Ibid., p. 221.

[52]Ibid., p. 220.

force or violence are applied are necessarily destroyed and cannot subsist for long; but those which are made by nature are done so in a suitable way and remain in their best disposition. Ptolemy therefore need not be afraid that the earth and all the terrestrial bodies would be destroyed by a rotation due to the action of nature."[53] Once again, Copernicus' solution was devoid of any mechanical significance.

However, Galileo's interpretation, though born of the correct insight, also ended in confusion. For hardly had he stressed the compensatory role of gravity in preventing bodies on earth from flying off at a tangent than his physical insight made way for an oversimplified geometrical representation. Thus, instead of proceeding to a determination of the factors to which the centrifugal force is proportional and hence of its physical structure, which would have been the only fruitful approach, he passed on from the centrifugal force, which he had originally introduced as a physical reality, to the distance through which it removed bodies from the earth's surface. From the fact that, on a geometric model, this distance is almost undetectable near the point of separation, Galileo concluded that no object would be thrown off from a moving earth, thus stripping the centrifugal force of all its physical attributes.

And because he found the geometrical approach so simple, he quickly extended its application. Thus, while he had opposed a distance to a force in the first part of his refutation, he simply opposed two distances in the second part. For the preceding argument did not take into account a very important fact, namely, that different bodies on the surface of the earth manifest distinct centripetal propensities, the lighter among them having a particularly "weak tendency to descend toward the center."[54] Under these circumstances, could it still be maintained that their gravity was great enough to offset the effect of the earth's rotation, so that very "light" bodies, in particular, would not be shot into the sky? Galileo's answer deserves being quoted in full: "If, in order for the stone or feather resting on the surface of the earth to be retained, it were necessary that its descent should be greater than or equal to its motion made along the tangent, then you would be right in saying that it would have to move as fast or faster along the secant downward than

[53]*De revolutionibus* I, 8.

[54]*Dialogue* II, pp. 221–222. This tendency, as Galileo had already realized, depends exclusively on external causes, since gravity has the same effects on all bodies, at least if we ignore the effects of the medium. Hence his argument in no way implied the slightest link between the speed of descent and the absolute weight of the falling body.

along the tangent eastward. But didn't you tell me a little while ago that a thousand yards along the tangent from the point of contact would be scarcely an inch away from the circumference? So it is not enough for the tangential motion (which is that of the diurnal rotation) to be simply faster than the motion along the secant (which is that of the feather downward). The former must be so much faster that the time required to carry the feather a thousand yards along the tangent shall be less than that of its moving a single inch downward along the secant; which I tell you will never be, though you make the latter motion as fast and the former as slow as you please."[55]

Now, though Galileo based his argument expressly on the slight curvature of the earth, that property was not its real crux, as witness Salviati's reply to one of Simplicio's objections. Thus when Simplicio, rightly disconcerted by what had gone before, refused to grant that, even if its speed along the tangent were "a million times faster" than its speed along the secant, a feather or a stone would not be extruded from the earth,[56] Salviati replied: "Saying this, you say what is false; not from a deficiency in logic or physics or metaphysics, but merely in geometry. For if you were aware of only its first principles, you would know that a straight line may be drawn from the center of a circle to a tangent, cutting this in such a way that the portion of the tangent lying between the contact and the secant will be a million, or two, or three million times greater than that portion of the secant which remains between the tangent and the circumference."[57] Having thrown physical intuition to the wind, Galileo was thus thrown back on a geometrical proposition which he could easily prove by the following argument: Given the ratio of *BA* to *C*, *BA* being as much greater than *C* as you please, and let there be a circle with center *D* from which it is required to draw a secant so that the tangent shall have the same ratio to this secant as *BA* has to *C*. With respect to *BA* and *C* take the third proportional *AI*; as *BI* is to *IA* make the diameter *FE* to *EG*. From the point *G* draw the tangent *GH*. "I say that this is what was required, and that as *BA* is to *C*, so *HG* is to *GE*. For *FE* being to *EG* as *BI* is to *IA*, by composition *FG* is to *EG* as *BA* is to *AI*; and since *C* is the mean proportional between *BA* and *AI*, *GH* is the mean between *FG* and *GE*. Therefore as *BA* is to *C*, so *FG* is to *GH*; that is, *HG* is to *GE*; which is what was re-

---

[55]Ibid., p. 223.

[56]Ibid.

[57]Ibid., p. 224.

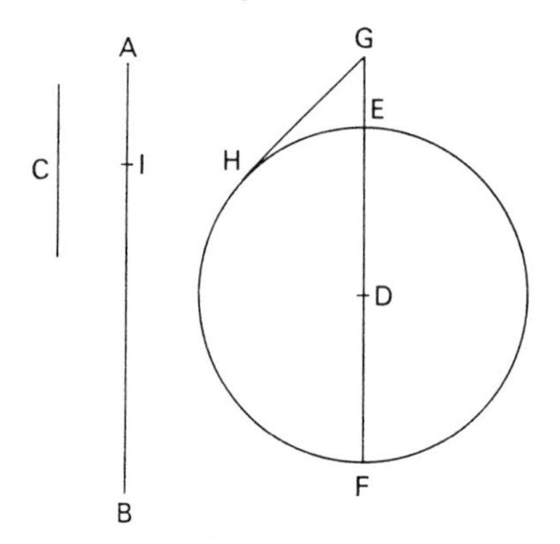

quired to be done."[58] In other words, even if the centrifugal force is incomparably greater than that of gravity, it is unable, during the first few moments of the motion, to remove a body from the earth's surface by any appreciable distance. Hence there is no danger, "however fast the whirling and however slow the downward motion, that the feather (or even something lighter) will begin to rise up, for the tendency downward always exceeds the speed of projection."[59]

This conclusion did not, however, represent Galileo's last word on the subject of the centrifugal force. For, taken literally, it implied that no rotational motion of any kind can have any detectable consequences. Galileo knew perfectly well that such consequences could be discerned; thus a heavy body placed on a wheel spinning rapidly in a horizontal plane will invariably be flung off, and yet a brief calculation will show that its linear (or tangential) velocity is by no means smaller than that impressed on bodies on the earth's surface by the diurnal motion. Aware of this difficulty, Galileo felt compelled to substitute a new proof, which marked a return to the correct physical intuition and paved the way for a valid description of the centrifugal effect.

Take the case of a rotating wheel with a stone placed near its rim. Clearly, the greater the speed of the wheel, the greater the *impeto* with

[58]Ibid.

[59]Ibid. On pp. 228–229, Galileo added further strength to this argument by asserting that even if the downward speed decreased in a much greater ratio than the force of gravity, it would nevertheless always remain large enough to restore the moving body back to the circumference of the earth, "from which it is distant [at the beginning] by the minimum distance, which is none at all." For another example of excessive geometrization, though by Descartes and not by Galileo, see A. Koyré: *Études galiléennes* II, pp. 37–38.

which it will hurl off the stone: "when the speed increases, the cause of projection will increase also in the same ratio."[60] In other words, the angular velocity of the wheel is proportional to the magnitude of the centrifugal force. However, the angular velocity cannot be the only determinant factor. Thus if we increase the diameter of the wheel but leave its angular velocity unchanged, a stone placed on the larger circumference will undoubtedly possess a greater linear velocity than it would have on a smaller wheel.[61] But will the centrifugal force still vary in the same way? Will it increase as the radius of the wheel, or will it perhaps decrease and thus dispose of Ptolemy's objection? Oddly framed though it may appear, this question nevertheless marked Galileo's return from a purely geometric approach to a mechanical one.

This may also be inferred more directly from the static basis of Galileo's new proof, namely, that two equal weights in a balance remain in equilibrium because "the gravity of one weight resists being raised by the gravity with which the other, pressing down, seeks to raise it."[62] However, gravity alone cannot always explain a body's resistance to motion; thus on a balance with unequal arms (that is, a steelyard) a large weight may be insufficient to raise a very much smaller one. This effect, as we saw, can be explained in terms of virtual motions: the reason why a small weight can balance a larger one is that it moves through a larger distance or, as Galileo himself put it, because "the smaller weight overcomes the resistance of the greater by moving much when the other moves little."[63] Now, there is nothing to prevent us from applying the special case of the steelyard to any mechanical situation involving the motion of heavy bodies: "Fix it well in mind as a true and well-known principle that the resistance coming from the speed of motion compensates that which depends on the weight of another moving body, and consequently that a body weighing one pound and moving with a speed of one hundred units resists restraint as much as another of one hundred pounds whose speed is but a single unit. And two equal movable bodies would equally resist being moved if they were to be made to move with equal speed."[64]

For instance, let there be two unequal wheels around a center *A*, *BG* being on the circumference of the smaller and *CEH* on that of the

[60] *Dialogue* II, p. 238.
[61] Ibid., pp. 238–239.
[62] Ibid., p. 240.
[63] Ibid., p. 241; cf. Chapter 3.
[64] *Dialogue* II, p. 242.

larger, the radius *ABC* being vertical to the horizon. Through the points *B* and *C* draw the tangents *BF* and *CD*, and in the arcs *BG* and *CH* take two arcs of equal length, *BG* and *CE*. The wheels are to be understood as rotating about their center in such a way that two moving bodies will be carried along the circumferences *BG* and *CE* with equal speed. Let the bodies be two stones placed at *B* and *C*, so that in the same time during which stone *B* travels over the arc *BG* stone *C* will pass the arc *CE*. "Now I say that the whirling of the smaller wheel is much more powerful at projecting the stone *B* than is the whirling of

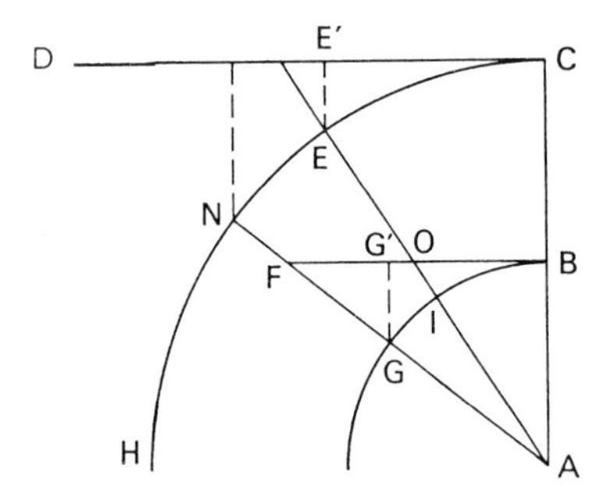

the larger wheel at projecting the stone *C*."[65] For clearly, since the projection must be along the tangent, the two stones will be impelled along *BF* and *CD* by the *impeto* impressed upon them by the rotation; moreover, to keep the stone on the smaller wheel, that is, to offset the motion of projection along the tangent *BF*, its weight would have to be pulled back as far as the secant *FG*, or rather the perpendicular drawn from the point *G* to the line *BF*, whereas on the larger wheel the withdrawal would need to be no more than the secant *DE*, or rather the perpendicular drawn from the point *E* to the tangent *DC*. But *EE′* is less than *GG′*, and decreases more and more as the large wheel is increased in size. Therefore, the force needed to prevent the projection will have to compensate for a much greater withdrawal in one case than in the other.[66] Let us now apply the principle of virtual velocities. Since the distances *GG′* and *EE′* correspond to motions made in equal times, the small wheel will impress on the stone *B* a greater virtual velocity along the tangent than the large wheel impresses on the stone *C*. If we multiply these velocities by the identical weights of the two stones, we

---

[65]Ibid.

[66]Ibid., p. 243.

see at once that the small wheel impresses a greater *impeto* on the stone *B* than the larger wheel impresses upon the stone *C*. In other words, the centrifugal force varies inversely as the dimensions of the wheel, so that the greater the radius, the smaller the centrifugal effect. Hence, on a wheel the size of a terrestrial parallel, the effect would be quite undetectable.[67]

Once again Galileo's conclusion was wrong, though this time not because of excessive geometrization but because of an apparently inexplicable confusion: he had based his proof on the assumption that the two bodies, *C* and *D*, traveling on the circumferences of their respective wheels, will traverse equal distances *CE* and *BG* in equal times. However, the circumference on which the body *C* travels is much greater than that of the body *B*; hence when he accorded them equal *linear* velocities, Galileo failed to realize that he was conferring a very much lower *angular* velocity on *C*; this prevented him from comparing the centrifugal effects of the two rotations. In fact, his argument proved one thing, and one thing only: when the angular velocity decreases, so does the centrifugal effect. As for the part played by the length of the radius, all he had to do was to note that, if both wheels had the same angular velocity, the body *C* will reach the point *N* at the same time as the body *B* reaches the point *G*, from this it follows immediately that the distance through which the centrifugal force will fling the body *C* from the big wheel must be very much greater than the distance through which the body *B* will be flung from its wheel in the same time. Why did Galileo close his eyes to something that seems so obvious to us? Quite apart from an evident error in judgment, his failure to arrive at a genuine "physical" interpretation of the centrifugal force had a more basic reason: because it served to conserve order, circular motion tended to be "natural" and hence by definition incapable of producing the least perturbation. Moreover, by treating circular motion as a natural one, Galileo was led to neglect the centripetal force, which alone could prevent the appearance of the centrifugal effect. As a result, he failed to consider the latter as a force sui generis, whose magnitude must increase with the velocity and the mass of the

---

[67] "And thus it might be supposed that the whirling of the earth would no more suffice to throw off stones than would any other wheel, as small as you please, which rotated so slowly as to make but one revolution every twenty-four hours." Ibid., p. 244.

moving body. Huyghens would be the first to treat centrifugal force as the physical result of circular motion and to show to what factors it is proportional.[68] But circular motion would no longer be for him a natural form of motion, and recognizing its specificity, he would be able to endow it with specific effects.

New Principles and New Concepts

Though the *Dialogue* clearly failed to produce a correct account of the centrifugal force, it nevertheless defined several authentic mechanical principles and concepts. Let us now examine the precise role the latter played in Galileo's thought, and also the remaining gap between them and the principles and concepts of classical mechanics.

The first new concept introduced by Galileo was that of an inertial system, which, in the form of the principle of mechanical relativity, presided over his refutation of the first three traditional objections: the reason why the fall of heavy bodies, the travel of cannonballs, and the flight of birds do not experience the least alteration on a moving earth is that each part of the earth's surface can be likened to an inertial system. Now it is remarkable to see how perfectly Galileo was able to define several of the essential properties and conditions needed for the definition of such a system. Thus he showed that only uniform motions could be inertial motions—capable of ensuring the displacement of a system without disturbing any of the mechanical phenomena that normally occur in it. Conversely, every change in speed must cause reactions in the system, and these reactions will, in fact, indicate the direction of the general motion. Thus when a basin containing water is moved without being tilted, so as to advance with a changing velocity "being sometimes accelerated and sometimes retarded," the water, which does not adhere as firmly to the basin as its solid parts, will not be compelled to follow all its displacements. When the vessel is slowed down, the water will retain a part of the *impeto* already received and run to the forward end, where it will necessarily rise. If, on the other

[68]Cf. Christian Huyghens, *Horologium oscillatorium sive de motu pendulorum ad horologia aptato demonstrationes geometricae* (Paris 1673), Part V. In this work Huyghens merely stated, but did not prove, the relevant theorems. He returned to the subject more fully in the posthumously published *De Motu et vi centrifuga* of 1703. Let us recall that the centrifugal force, while proportional to the length of the radius, is also proportional to the mass of the moving body and to the square of its angular velocity, in such a way that $Fc = mr\omega^2$.

hand, the motion of the vessel is speeded up, the water will retain part of its "slowness" *(tardità)* and will fall behind "while becoming accustomed to the new *impeto*."[69] In other words, Galileo considered that a system in nonuniform motion cannot be an inertial system, and it was only because he took too broad a view of uniformity that he differed from Newton in this respect.

However, the full implications of Galileo's great step in likening every part of the earth's surface to an inertial system cannot be fully appreciated unless it is remembered that, to his mind, the uniform circular motion of the parts continued to be a natural form of motion. Thus, if circular motion is indeed a *natural* form of motion, we might expect to find that, in the direction opposite to that in which it draws bodies, there would appear a resistance no less marked than, say, the resistance to upward motion. From the fact that it does not obstruct displacements in the same direction, does it really follow that it will not obstruct displacements in the opposite direction? By claiming that this was indeed the case, Galileo showed once again that, even when apparently adopting the traditional standpoint, he still introduced concepts of a distinctly modern stamp. In his refutation of Tycho's argument, he had previously shown that violent motions could occur freely within a system in uniform motion; it was by reference to one of his favorite examples, that of a ship, that he took this idea a step further and demonstrated its general validity. Thus, shut yourself up with a friend in the main cabin below decks on a large ship and have with you some butterflies and other small insects, together with "some fish in a bowl, and a suspended bottle that empties drop by drop into a wide vessel beneath it." When the ship stands still, the insects will fly with equal speed to all sides of the cabin, the fish will swim indifferently in all directions; the drops will fall into the vessel beneath, and if you throw something to your friend, you need throw it no more strongly in one direction than in another. Now have the ship proceed in any direction you like; then, "so long as the motion is uniform and not fluctuating this way and that, you will discover no change in all the effects named." In jumping, you will cover the same spaces on the floor as before, and you will make no larger jumps toward the stern than toward the prow even though the ship is moving quite rapidly while you are

[69]*Dialogue* IV, p. 450. It was on the perturbing effects of nonuniform motion within a given mechanical system that Galileo also based his tidal theory.

in the air. The same object can be thrown forward and backward to cover the same distance with the same force, and the droplets will continue to fall into the vessel "without dropping toward the stern";[70] the fish in their bowl and the insects will continue to swim or fly indifferently in all directions, "nor will it ever happen that they are concentrated toward the stern, as if tired out from keeping up with the course of the ship."[71] True, up on the bridge noticeable differences can be observed, especially in the behavior of the insects, which, even at a slight distance from the ship, would be held back by the air and hence prevented from following the ship's motion. "But keeping themselves near it, they could follow it, without effort or hindrance, for the ship, being an unbroken structure, carries with it part of the nearby air."

Now, if a system in uniform motion in no way alters the relative motions of its parts, it follows that no mechanical experiment, no observation, however careful, performed within that system will be able to indicate whether the system as a whole is moving or at rest. For instance, if the ship were traveling from Venice to Aleppo, and if there were a magic pen on the bridge which had the property of leaving visible traces of the trip in the air, it would mark an arc of a circle extending from Venice to Aleppo. Yet what happens with a normal pen? If an artist had begun drawing with it on a sheet of paper as the ship began to make sail and had continued doing so throughout the trip, he might cover the sheet with landscapes, buildings, animals, and other things by means of lines insignificantly small in comparison with the line from Venice to Aleppo; yet only these tiny marks would appear on the paper while not the least trace at all would appear of the pen's own motion across the Mediterranean.[72] In other words, there is nothing within a system to tell us whether it is at rest or in uniform motion—the only detectable motions are those with respect to which the system is at relative rest.

Although Galileo held that only systems in uniform motion could be inertial, he did not go so far as to assume that this motion must also be rectilinear; ingoring this fact, the *Dialogue* argued about systems that, properly speaking, were not inertial at all. Indeed, the very success of his analysis prevented Galileo from making a clear distinction between systems in uniform circular and uniform rectilinear motion. For if it is true that the earth's diurnal motion suffices to offset all the

[70]*Dialogue* II, p. 212.
[71]Ibid.
[72]Ibid., p. 198.

perturbations envisaged by the geocentrists, how could a system in such motion be anything but inertial? To say otherwise, Galileo would have had to realize that circular motion is a mixed motion, maintained by an accelerative force acting toward the center; viewing it, on the contrary, as a natural form of motion, he had no theoretical reason not to treat systems in such a motion as inertial systems, and the more so as the evidence of his senses told him that on the surface of a revolving earth any systems in uniform, though not strictly in rectilinear, motion can for all practical purposes be likened to inertial systems.

But while there are no practical disadvantages in treating systems in uniform circular motion as inertial systems (at least if the radius of their orbit is sufficiently great), it is quite impossible to extend this conclusion to the case of rotatory motions and to claim, as Galileo did, that no motion of a body north or south, east or west, or up or down on the surface of a rotating earth can ever show whether or not the system as a whole is in motion. This conclusion was not only false but also in manifest conflict with the very principles on which Galileo had based his refutation of the traditional objections. In fact, a systematic development of the implications of the principle of conservation would have quickly convinced him that the earth's rotation must necessarily affect the observable behavior of bodies on it. Take the case of a rock dropped from the top of a tower. Galileo, as we saw, proved conclusively that it would not be left behind as the tower followed the earth's diurnal motion. But if we imagine a tower as an extension of a terrestrial radius, we must grant that, since its top is further away from the center of the earth than its base and hence describes a larger arc, it must also have a greater linear (or tangential) velocity than the base. Now this greater velocity will impress a greater circular *impeto* on the rock and, according to the principle of conservation, this *impeto* must be preserved in toto; from this it follows that the rock, far from falling behind, will in fact land slightly to the front of the tower. Galileo's failure to appreciate this fact is the more surprising in that, a few pages later in the *Dialogue*, he implicitly offered the correct solution. The occasion was a problem posed by the philosopher Locher—to determine how long it would take a ball falling from the moon, and keeping an *impeto* equal to diurnal motion, to reach the earth and along what path it would travel.[73] According to Locher, the ball would fall for at least six days

---

[73]Ibid., pp. 245 ff.

and, except for a fall just above a pole, it would describe a spiral, staying perpendicularly above that point of the earth that was beneath it when the descent began. We need not dwell on the first part of Galileo's critique,[74] for all that concerns us here is that, in his refutation of the proposition, Galileo accused Locher of having failed to grasp the principle of the conservation of motion. Thus he argued that if the ball really left the moon's orbit with an *impeto* equal to the diurnal motion, then, according to the principle of conservation, this *impeto* would undergo no diminution during the entire descent. And as the ball approached the earth, its "rotational motion would have to be made in ever smaller circles, so that if the same speed were conserved in it which it had within the orbit, it ought to run ahead of the whirling of the earth."[75] Now, this was precisely the argument that Borelli, probably basing himself directly on Galileo's analysis, was to use in 1668

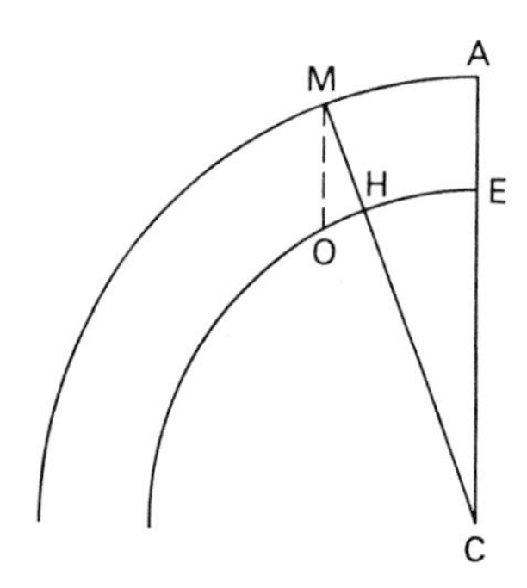

when he asserted that all heavy bodies in free fall are deflected toward the east. Thus if *EA* is a tower built on the terrestrial equator, and *HM* the position it occupies after its foot has traversed the arc *EH*, then a stone dropped from it when it is in *EA* and which hits the ground when it reaches *HM* will fall ahead of the point *M*: since its circular *impeto* (or "transverse impetus," as Borelli called it) is completely conserved, the stone must cover, relatively to the terrestrial surface, the same distance that the top of the tower traverses during the descent, so that the

---

[74]Galileo showed, in particular, that the ball would take only 3 hours, 22 minutes, and 44 seconds to reach the earth (ibid., pp. 249–250); his computation was, however, based on an estimate of the natural acceleration of heavy bodies at half the correct value.

[75]Ibid., pp. 259–260.

stone will hit the ground at the point *O*, obtained by dropping a perpendicular from *M* onto the arc *EH*.[76]

Just as he ignored the increase in linear velocity associated with increases in distance from the center, so Galileo also ignored the fact that on a rotating earth the linear velocity of a moving body cannot be the same at the equator as it is elsewhere on the earth. Yet later, when he analyzed the path of a cannonball fired at a southerly target, he noted that since the cannon is situated on a smaller parallel, it would move eastward more slowly than the target. If Galileo had but developed this idea, he would have been forced to grant the existence of a westward deflection, instead of rejecting this possibility on the ground that any such deflection would be quite imperceptible,[77] and with it the conclusion that a rotating system cannot possibly be an inertial one.[78] Whence the paradoxical nature of Galileo's final position: while his view of inertial systems allowed him to establish the possibility of the earth's diurnal motion, it also led him to deny the existence of phenomena in which he ought to have discovered the proof of such motion. The paradox was an irony of history: the very perturbations that the Peripatetics thought necessary but whose absence precluded for them the earth's diurnal motion were dismissed by Galileo when in fact they not only existed but proved that the earth was not at rest. Thus even while he turned the earth's diurnal motion into a real possibility, Galileo debarred himself from proving its existence.[79]

---

[76]Borelli's analysis is quoted in A. Koyré, "A Documentary History of the Problem of Fall from Kepler to Newton," *Transactions of the American Philosophical Society*, New Series, Vol. 45, Part 4, pp. 373–374. Borelli made an attempt to measure the deviation from the Torre degli Asinelli in Bologna, but with the primitive equipment at his disposal he was of course bound to fail.

[77]As is the case with normal shots.

[78]*Dialogue* II, p. 205. We know that the problem can be solved only with the aid of complementary inertial, or Coriolis, forces. Galileo encountered the same difficulties when he tried to account for the trade winds; cf. Appendix 4.

[79]It is certain that G. Bruno came very close to the idea of an inertial system well before Galileo (cf. *La Cena de le Ceneri* II, 5, *Opere italiene*, Wagner edition, Leipzig 1830, pp. 169 ff.). In his attempts to refute the geocentric arguments, Bruno affirmed that the earth's motion would cause no perturbation because "all things situated on the earth follow its motion." Thus a stone dropped from the top of the mast of a moving ship will land at its foot because the motion of the vessel lends the stone a *virtus impressa* thanks to which it can follow the ship's motion even after it has lost direct contact with the mast. Since Galileo never quoted Bruno's ideas, undoubtedly for prudential reasons, it is impossible to tell whether he was familiar with them. In any case, we wish in no way to detract from Bruno's originality when we point out that his account of inertial systems was

No less important than the concept of inertial systems, whose full consequences it revealed, the principle of the conservation of motion led to the permanent abandonment of the traditional approach. It rid motion, definitively, of its ontological function and of its subjection to the physical nature of bodies and transformed it into a state that, at least under certain conditions, was capable of indefinite self-perpetuation. At the same time it led to the rejection of both the Aristotelian view of the role of the medium and of the medieval impetus concept. Regarding the alleged motor role of the medium, the *Dialogue* simply contented itself with repeating the arguments of the *De Motu*,[80] but while the latter had criticized the Aristotelian doctrine chiefly in order to pave the way for the impressed-impetus concept, the critique of the *Dialogue* was mainly concerned to show that no contact or motor was needed for the conservation of uniform motions. This is how Salviati summed it all up for Simplicio's benefit: "There is an enormous difference between this experience of yours and our example. You make the wind arrive upon this rock placed at rest, and we are exposing to the already moving air a rock which is also moving with the same speed, so that the air need not confer upon it some new motion but merely maintain—or rather, not impede—what it already has. You want to drive the rock with a motion foreign and unnatural to it; we, to conserve its natural motion in it."[81] As for the medieval impetus theory, its rejection is best illustrated by the shift in meaning Galileo bestowed upon the word *impeto*. For Buridan, no less than for the *De Motu*, impetus had been the sine qua non of continued motion, and though the *Dialogue* also continued to treat *impeto* as a manifestation of the motor in the moving body, it granted it an entirely different status. Let us look once more at our moving ship: its (supposedly uniform) motion is engraved indelibly on all the bodies it carries along, including the stone on top of its mast. The stone is dropped: the motion that has been

---

greatly inferior to Galileo's. Not only did he fail to realize that an inertial system must move in a straight line, but he never even mentioned the fundamental condition of its uniformity. Moreover, when explaining the trajectory of the falling stone by reference to its participation in the ship's motion, he failed to introduce the principle of conservation. Now, in the absence of that principle, the idea of participation was bound to be confused and much closer to the traditional views than to the modern interpretation of inertial motion. It nevertheless remains a fact that Bruno's reply to the geocentrists was greatly superior to that of his precursors, including Copernicus.

[80]*Dialogue* II, pp. 177–179; cf. Chapter 3.

[81]*Dialogue* II, p. 169.

impressed upon it acts as an *impeto* and hence enables it to follow the ship's motion even while it has lost contact with the mast. However, this *impeto* has clearly ceased to be a motor cause: it has become the motion of the ship "engraved" in the stone, or rather the quantity of motion that the stone derives from its participation in the ship's motion. Reduced from the rank of cause to that of effect, the *impeto* thus no longer refers to anything but the indefinite conservation of the motion originally acquired by the moving body.[82]

While the case of a ship, like that of the earth, can mislead one into thinking that the *impeto* is a transitory phenomenon, since once the stone has landed at the foot of the mast or the tower, it stays there immobile, the truth of the matter is that, thanks to its original *impeto* which was identical with the ship's motion, the stone, when it ceases its descent and once again makes direct contact with the ship, continues to move, though no longer *separately* from its support. It is, moreover, perfectly feasible that a falling body should change from one inertial system whose *impeto* has been engraved in it to another system at relative rest with respect to the first; in that case, the motion acquired in the first system must also be conserved in the second. Thus a ball released by a galloping horseman will not come to rest as soon as it hits the ground but will roll along with the horse except as the roughness and unevenness of the ground impedes it. Were the ground perfectly smooth, there would be no reason to think that the ball would not follow the horse indefinitely.[83]

The links of the principle of conservation with motion as a state do not, however, exhaust its importance; in Galileo's formulation it is also inseparable from a specific interpretation of gravity and one, moreover, whose historical significance cannot be overstated. Considering it as a cause of weight and of downward motion, Galileo looked upon gravity as a force inherent in all matter. As a motor force, gravity was that internal tendency by which all bodies tend to approach the center of the earth as closely as they can. It follows that no material body can be neutral to motion or rest; whenever it is free to approach the center it will spontaneously begin to move or, if it is already moving, to ac-

[82]Galileo also used the expression *impeto* to describe the way in which a constant force engenders a uniformly accelerated motion, and then to define the dynamic properties which that motion confers upon bodies.

[83]For this analysis, see *Dialogue* II, p. 182.

celerate its motion. Conversely, if the body is forcibly separated from the center, the impressed motion will gradually slow down under the action of gravity and finally become exhausted. Hence, inertial motion can take place only in the absence of an accelerative or retarding cause, that is, after the elimination of the motor function of gravity. This elimination cannot be achieved by a vacuum which, in Galileo's physics, in no way alters the force by which a heavy body is impelled downward but demands a uniformly accelerated motion of an infinitely increasing velocity—a somewhat gratuitous supposition that would deprive gravity of its *ultima ratio.*[84] Galileo contended therefore that it was only when they moved on a plane all of whose points were equidistant from the center[85] that bodies were immune to the action of gravity; and it is then perfectly clear that if such motions, once begun, were capable of continuing indefinitely, they were also bound to be circular motions.

This erroneous conclusion on Galileo's part raises an important problem. At the time he wrote his *Dialogue* it was common to look upon gravity, not as an internal motor, but as the result of an external force. Such was the view particularly of Gilbert and Kepler and hence could not have been unknown to Galileo. Now, had he followed in their footsteps and imagined an isolated body moving in a vacuum, that is, in a situation from which all motor force had been eliminated, and with it the cause of acceleration and retardation, then would he not perforce have come to appreciate that the resulting motion must be inertial? Moreover, since no one direction in a vacuum is more important than any other, would he not have come to appreciate that the inertial motion must also be rectilinear? In short, had he, too, conceived of gravity as the effect of an external cause, would he not have framed the principle of inertia in its correct form? The reader may perhaps object that the attribution of gravity to an external source was precisely one of those highly speculative hypotheses that he so deeply mistrusted. But though he expressly stated that the idea of an attractive force was not only unverifiable but contrary to the scientific spirit,[86] and however recalcitrant he may have been to frame hypotheses, he never denied their necessity. Moreover, since the attractionist hypothesis was later to prove so fruitful, it is reasonable to speculate about the services it might have rendered to the author of the

[84]Namely, that it impels bodies toward the center.

[85]In other words, on a spherical surface.

[86]See Chapter 8, footnote 18.

Dialogue. For all that, it is possible to show that this hypothesis, unlike the Galilean interpretation, was then quite incapable of leading to the principle of conservation of motion. There is no better way of making this point than an examination of Kepler's refutation of the geocentric thesis.[87]

One principle or "axiom" presided over Kepler's entire theory: a body at a large enough distance from others will remain permanently at rest,[88] that is, it will have no tendency to move toward any particular point.[89] It follows that the only possible source of motion must reside in the "mutual affection" of bodies. Thus, if it were possible to place two stones at a great distance not only from all celestial bodies but also from all other bodies, their "magnetic virtue" would cause them to travel toward each other, and once each of them had covered a distance inversely proportional to its mass, the two would join together.[90] Let us now replace one of these stones with the earth: although each of the two bodies continues to attract the other, the relative "attractive virtue" of the stone has become so small that it can be ignored for all practical purposes; hence the only manifest attractive force will reside in the incomparably larger mass of the terrestrial globe and can be measured precisely by the "gravity" of the stone. Thus every body in the vicinity of the earth experiences a "traction" or *raptus* thanks to which it is bound to the center "as if by very thin chains or ties."[91]

It was on this basis that Kepler proceeded to refute the traditional objections. Since all bodies are linked perpendicularly to the earth by "very thin chains or ties," they cannot possibly be deflected to the west, unless the *raptus* has vanished. Now, much as the tractive action of the earth decreases with distance, so it must increase when a falling body approaches its surface; hence there is no chance that we can ever observe even the slightest deflection;[92] a freely falling body will invariably keep to the vertical line joining its point of departure to the earth.[93] Kepler's refutation of Tycho Brahe's argument that two cannonballs will not fall equally if one is fired to the west and the other to the east

---

[87]All of our Kepler quotations are taken from Koyré, *Études galiléennes* III, pp. 26 ff.

[88]Ibid., pp. 28 and 31.

[89]"Heavy bodies only tend to approach the center of the earth inasmuch as it is a center of attraction; if the earth were not a sphere, they would move toward a lateral point," ibid., p. 28.

[90]Ibid., p. 32.

[91]Ibid., p. 28.

[92]At least, any perceptible deflection, as Kepler specifies; ibid., p. 34.

[93]Ibid., pp. 28–29 and 33–34.

was conducted on much the same lines. A cannonball traveling through the air may be imagined as being "tied to the earth by an infinite number of chains or elastic ties,"[94] which form a kind of cone with the cannonball as its apex. In the case of a simple upward projection, the tension of all these chains or nerves will prevent the ball from leaving the earth. Similarly, if the ball had been fired toward the west or the east, it would be hauled back by the "western chains" or the "eastern chains." Now since the number of active chains is identical in both cases, the resistance that the earth's traction offers to the two projectile motions will also be identical, as will the travel of the ball in either direction. To pretend otherwise would mean pretending that the earth's traction differs from one direction to the other.

Not only did this refutation fail to invoke the principle of conservation, but that principle was quite foreign to Kepler's entire argument—the reason why a stone dropped down vertically was that "chains" from the earth kept it constantly beneath its point of departure, and not that the diurnal motion was engraved upon it. Only the constant action of a force therefore prevented its deflection toward the west, or, as Koyré has put it so well, it was thanks to a real physical action and not to a mechanical state, that is, the state of motion, that a stone was thought to fall to the foot of a tower, or a cannonball fired vertically into the air was thought to return to its starting point.[95] And since this interpretation was quite incompatible with the idea of an inertial system, it is not surprising that the fall of a heavy body from the mast of a ship should have faced Kepler with an insoluble enigma. How can a stone possibly follow the motion of a ship when the latter, unlike the earth, exerts no traction upon it? Kepler himself stated it all quite bluntly when he said that "a ship does not attract the objects it transports with a magnetic virtue, but carries them along by mere contact, while the earth continues to attract them by virtue of its gravity which it fails to communicate to the motion of the ship."[96] Under these circumstances, a stone dropped from the top of a mast could not possibly fall to its foot while the ship was moving; once it had lost contact with the ship, the traction of the earth would necessarily deflect it toward the stern.

In fact, the very "axiom" on which Kepler based his analysis explains his inability to frame the principle of conservation, and hence

[94]We are following Koyré's excellent analysis; ibid., p. 37.

[95]Ibid., p. 34.

[96]Ibid., p. 39.

to offer an effective refutation of the traditional objections. For when he asserted that a body remote from all others can have no natural state other than perfect immobility, Kepler was continuing to characterize matter by its persistent tendency to come to rest. This tendency, which Kepler designated by the term "inertia"[97] and which, according to him, conferred upon every body "the power to remain in its place,"[98] was bound to offer a perpetual resistance to motion, with the result that motion had to be re-created from one moment to the next and could never enjoy the same status as rest. Kepler was merely repeating the Peripatetic cry when he asserted that "every material body is by nature immobile and destined to come to rest in whatever place it happens to be."—"Rest, like a shadow, is a form of privation needing no special creation, but appertaining to created things as a trace of nothingness; motion, by contrast, is positive like light."[99] And if the spontaneous attraction of bodies alone explains their motions toward or around each other, there is nothing to distinguish that attraction from the *conjoined motor*, whose essential and constant presence had been affirmed by Aristotle and all his disciples. Explaining as it did motion in the traditional terms or, as it would later, in Newtonian terms, the concept of attraction by itself could not lead to physical progress; before it could be put to fruitful use, a radically new conception was needed, namely, that matter is inherently indifferent to rest and uniform motion. Short of this new conception, which Kepler's attractionist interpretation of gravity was quite incapable of introducing, there was little chance of arriving at the principle of conservation.

Galileo, by contrast, did succeed in formulating this principle, having first established that the tendency of bodies to come to rest, far from being a basic attribute of matter, could disappear as completely as the tendency toward motion. How precisely did he arrive at this conclusion, even while he continued to treat gravity as an immanent force? Let us return to his study of motions on an inclined plane. We saw that the first conclusion he drew was that a clear distinction must be made between the gravific and motor functions of gravity, the first being characterized by its constancy, the second by its possible variability. This distinction was best reflected in the concept of moment of descent; by

---

[97]Paradoxically enough, Kepler was thus the first to employ this term; ibid., p. 26.

[98]Ibid., p. 33.

[99]Ibid., pp. 31–32.

providing him with a direct expression of the motor function of gravity, it enabled Galileo to study variations of the latter resulting from variations in the inclination of the plane and to show that the moment of descent decreases from its maximum value when the angle of the plane is greatest to its minimum value when the plane is horizontal. However, this early conclusion could not yet lead him to the idea of the conservation of motion; it put an end to the traditional identification between weight and the motor function of gravity, but did not assure that, when this function vanishes, matter would not manifest a spontaneous propensity for rest. Let us then take a closer look at the moment of descent. Situated as it is at the frontiers of statics and dynamics, it can serve equally well to define equilibrium conditions as to describe the motor function of gravity; in fact, we can say that if a body on an inclined plane is joined by a string to another body hanging freely, the two will balance each other whenever their moments of descent are equal. Now if this is the case, it is so because the tendency of each to move in a downward direction is precisely offset by the resistance of the other to upward motion: in other words the moment of descent measures not only the propensity to downward motions but also the resistance to upward motion. Now this remark holds the key to Galileo's decisive advance, namely, his identification of the apparent tendency of heavy bodies to come to rest with their resistance to upward motion. In making that identification, Galileo took a momentous step: he now held one and the same force responsible for motion and for rest, or rather the propensity to rest whenever the latter appeared. Associated with a variable magnitude that could vanish in certain cases, the propensity to rest thus ceased to be the direct attribute of matter it had been for the traditional physicists no less than for Kepler; there were no longer any theoretical reasons why it should not be discarded from the repertoire of physics. To do so, we need merely envisage a situation in which the moment of descent has vanished completely, as happens, for instance, on a plane equidistant from the center. Here all bodies will have lost their apparent tendency to move, and the motor force, for its part, will be as good as abolished. With it, both the propensity to rest and that to motion will have vanished as well, and matter, left to its own devices, will persist indefinitely in either of these two states, not showing any less repugnance to leave a state of motion once acquired than to leave a state of rest.

The reason, therefore, why Galileo was the first to formulate the idea of inertial motion was that he took full theoretical advantage of the inclined-plane experiment, whereas Kepler, for instance, was prevented by his conception of gravity from engaging in a fruitful confrontation of theory with experience. In the attractionist hypothesis, the changes in speed which a body experiences on planes of different inclinations can be explained only in terms of the accelerative effects of the earth's attractive force, which means that every decrease in speed is due to a decrease in acceleration. This interpretation called, inter alia, for a conception of force close to the modern one, so the inclined-plane experiment could not of itself have led Kepler to a transformation of the prevailing dynamic ideas; also that transformation was for him a sine qua non of the correct interpretation of the inclined-plane experiment. Galileo's approach avoided this trap; even before the essential steps we have mentioned were taken, it helped to open mechanics, albeit in a rudimentary fashion, to a series of basic experimental facts. Thanks, no doubt, to its great simplicity, it was able to assimilate these facts in an intelligible way, thereby leading mechanics out of its age-old impasse. Kepler's solution, by contrast, was much too abstract and much too general to have modified the traditional approach; being of a purely speculative nature and hence divorced from the facts, it could at best present the old doctrine in a new guise. There are thus rather good reasons for the claim that the postulate of an immanent motor force alone could, at the beginning of the seventeenth century, have disposed of the mistaken view that matter has an innate propensity to rest; and though the principle which Galileo elaborated under these conditions was necessarily incorrect, it nevertheless introduced the essential fusion of the concepts of motion and indefinite conservation.

A good example of the advances this principle ushered in is provided by the way in which his *Second Letter on Sunspots* succeeded in introducing the idea of the sun's rotation. As the reader will remember, Galileo's observations had shown him that sunspots travel across the solar globe with a regular motion in about two weeks. Now, observation cannot, of course, by itself tell us whether sunspots revolve independently about the sun following the motion of its ambient "or whether they are part of the sun's [rotating] body."[100] Purely physical reasons

---

[100] *Second Letter on Sunspots*, Vol. V, pp. 133 ff.

made it highly probable that the second hypothesis was the right one; thus the sun's alleged circumambient substance would have to be "very fluid and yielding" to account for the fact that the spots contained in it changed their shapes so easily and so often. An orderly motion, such as that of the spots, seems incapable of having its basis in a fluid substance whose parts do not cohere and are therefore subject to disturbances and other accidental movements.[101] But it suffices to put the problem in mechanical terms to show that, even if originally the motion belonged to its ambient, the sun must necessarily rotate on itself. Let us take again our principle that, in the absence of any forces tending to remove a body from or to draw it toward the center, it will be indifferent to rest and to motion. From it, we can deduce that no part of the sun will have a propensity or aversion to circular motion about a fixed center and that, being the sum of its parts, the sun as a whole cannot have any internal repugnance to a rotational motion. And, in fact, "by such rotation it is neither removed from its place nor are its parts permuted among themselves. Their natural arrangement is not changed in any way, so that as far as the constitution of its parts is concerned, such movement is as if it did not exist."[102] Now if the sun's ambient, be it ever so fluid and rare, were indeed capable of independent motion, it would exert some pressure on the solar body, and since the latter is indifferent to motion and rest, the smallest impulse would suffice to set it in motion and cause it to spin on its own axis. "This," Galileo added, "may be further confirmed, as it does not appear that any movable body can have a repugnance to a movement without having a natural propensity to the opposite motion, for in indifference no repugnance exists; hence anyone who wants to give the sun a resistance to the circular motion of its ambient would be putting in it a natural propensity for circular motion opposite to that. But this cannot appeal to any balanced mind."[103] As for this rotation of the sun, because it was neither helped nor hindered by gravity, it was bound to last indefinitely; indeed, it was a necessary consequence of Galileo's conception of gravity

---

[101]Ibid., p. 134.
[102]Ibid., p. 135.
[103]Ibid., p. 135. It should be pointed out that, though Kepler and Bruno had also expressed the view that the sun rotates on its axis, neither of them had tried to offer a mechanical proof. Thus Bruno merely put the matter forward as a pure surmise, and Kepler because it provided him with a dynamic explanation of the elliptical trajectory of the planets (for Kepler, see Dreyer, *A History of Astronomy*, pp. 393 ff.).

and the principle of conservation that, once a rotation has started, it must, in the absence of external impediments, continue for all time. Admittedly, in all this analysis, he relied largely on intuition and never referred explicitly to the conservation of angular momentum,[104] but his conclusion was perfectly clear: just like uniform circular motion on a spherical surface, an isolated rotational motion needs no external agents to ensure its indefinite continuation.

To the principles of relativity and conservation of motion, the *Dialogue* also added a third, namely, that it is possible to combine two motions without suppressing either. This affirmation was by no means new; the Greek geometers, for instance, had distinguished "regular" from "mechanical" curves, the latter referring to curves resulting from the composition of two motions;[105] moreover, the constructions by which astronomers had tried to "save the appearances of celestial phenomena" also involved the composition of certain motions. However, the astronomers had considered this composition as a pure mathematical artifice without any true physical significance and in any case had confined their attention to uniform circular motions, while Galileo not only posed the problem in strictly physical terms but also generalized its scope.

In fact, there is no physical reason why circular and rectilinear motions, though enjoying a difference in status on the cosmological plane, should not be compounded. Traditional mechanics saw an insurmountable obstacle to such compositions in the inherent tendency of all bodies to move in a specific way, but once the ontological function of natural motions had been discarded, there was no longer any reason why bodies should not be equally indifferent to motions of all kinds. Gravity, though acting from within the moving body, in no way invalidates this argument: the downward thrust by which it manifests itself causes heavy bodies to describe a motion that does not differ essentially from other motions and hence does not stand in the way of their combination. But though the rectilinear motion due to gravity can be combined with the uniform circular motion resulting from the earth's diurnal rotation,[106]

---

[104] In classical mechanics, an isolated system in rotation must conserve its angular momentum, that is, the product $I\omega$ of the moment of inertia I and the angular velocity ω.

[105] Mechanical curves included the conchoid, the cissoid, the quadratix, and the spiral; see J. Vuillemin, *Mathématiques et métaphysique chez Descartes*, pp. 82 ff.

[106] *Dialogue* II, p. 175.

this composition nevertheless has a number of apparently disconcerting consequences. Thus while a stone dropped from the mast of a ship at rest will touch deck, say, after two pulse beats, it will also touch it after two pulse beats if the ship is pursuing a steady course. But during these two pulse beats the ship will have traveled, say, twenty yards, so that the actual motion of the stone will have been a "diagonal" line much longer than the first straight and perpendicular one, and the difference between the two will be the greater, the faster the speed of the ship.[107] Similarly, if a cannon mounted on a tower were fired parallel to the horizon, it would not matter if a small or great charge was put in it—the ball might fall a thousand yards away, or four thousand, or even ten thousand, but all these shots would take the same time as the ball would have taken had it been dropped straight down from the tower. In other words, a motion can be combined with another motion, and the initial trajectory extended considerably, without its duration being altered in any way.[108] In accepting this strange conclusion, imposed by reason[109] even before it could be verified by experiment, Galileo showed his complete grasp of the principle of the composition of motions and hence the special role that must be assigned to inertial motion.

Oddly enough, when it came to the particular case of the earth, he was not altogether successful in combining its diurnal, that is, its uniform, circular motion, with the rectilinear motion of its parts. In particular, his treatment of the projectile motion of a cannonball fired vertically upward and the motion of heavy bodies toward the center of a rotating earth showed clearly how far ahead his theoretical analysis still was of the practical applications. Take first the case of a cannonball. The principle of conservation made it easy to explain why it should stay vertically above the mouth of the cannon throughout its motion: since the earth has impressed its diurnal motion upon the cannon, the ball will not start from rest but will join this impressed motion about the center to the upward projection in such a way that, following the eastward motion of the earth, it will keep continuously over the gun during both its rise and its fall.[110] The real or absolute motion of the ball will therefore correspond to a "diagonal" trajectory, that is, to a curve whose nature Galileo did not, moreover, define any further. But how precisely did Galileo prove that this trajectory was diagonal from

---

[107]Ibid., pp. 180–181.
[108]Ibid., p. 197.
[109]Ibid., p. 181.
[110]Ibid., p. 200.

the very outset, as he believed it to be? He considered the case of an erect cannon *AC*, and assumed that on an immobile earth the ball *B* would leave the mouth *A* to describe the perpendicular line *BA*.[111] But if the earth rotates and carries the cannon with it, then during the time in which the ball is moving through the gun, the cannon will have passed to the new position *DE*: hence the ball will leave the gun from the point *D*, having first been hurled by the charge along the slanted line *BD*, "both motions being toward the east."[112] In other words, Galileo attributed the diagonal trajectory of the ball to the motion of the gun and hence to the earth's diurnal motion, thus substituting a simple displacement for the principle of composition.

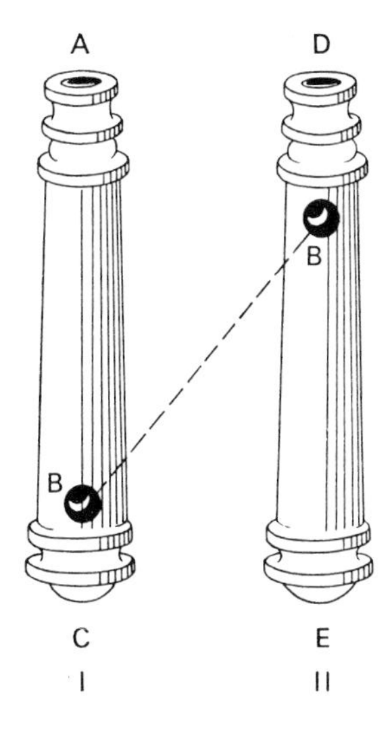

Galileo's analysis of the curve along which a heavy body tends to approach the center of the earth provides an even better illustration of his incapacity to proceed to a de facto composition of the inertial motion engendered by the earth's rotation with a rectilinear motion. On mechanical grounds, Galileo realized clearly that the problem came back to combining a uniform with an accelerated motion, even if he did not consider it useful to define the nature of the latter.[113] He also realized that if both the straight motion toward, and the circular motion about,

---

[111] Ibid., p. 201.
[112] Ibid., p. 202.
[113] Ibid., pp. 189–190. The context, however, suggests that he was familiar with its characteristics.

the center were uniform, then the two could be compounded into a spiral line as defined by Archimedes in his book on the spirals bearing his name, that is, on spirals generated when a point moves uniformly along a straight line that is being uniformly rotated about a fixed point at one of its extremities. But since the motion toward the center is continually accelerated, the real trajectory of the heavy body must have an ever increasing ratio of successive distances from the circumference of the circle which it would have described about the center had it never left its starting point. As a result, the two lines (the real trajectory and the circumference), which are nearly coincident at first, will draw farther and farther apart as the downward speed of the heavy body increases. Having introduced these definitions, Galileo went on to determine the precise trajectory of the falling body. To that end, he drew the circle *BI* (representing the terrestrial globe) with *A* as the center, and then,

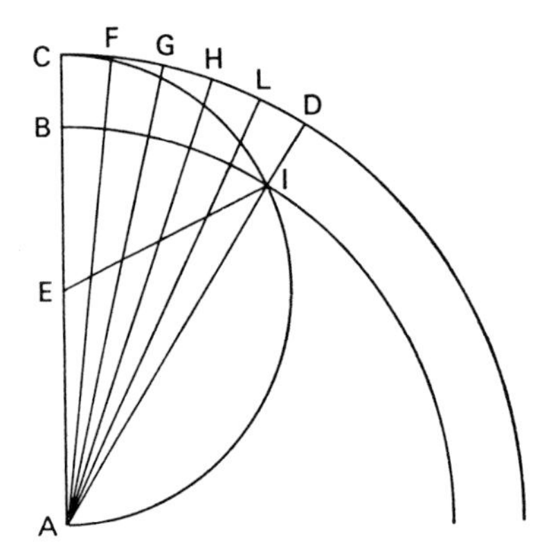

prolonging the radius *AB*, he drew in the height of the tower *BC*, which, carried by the earth along the circumference *BI*, marks out the arc *CD* with its top. "Now dividing the line *CA* at its midpoint *E*, and taking *E* as a center and *EC* as radius, the semicircle *CIA* is described, along which I think it very probable that a stone dropped from the top of the tower *C* will move, with a motion composed of the general circular movement and its own straight one."[114] For if equal sections are marked on the circumference *CD* and straight lines are drawn to the center *A* from the points *F*, *G*, *H*, and *L*, the parts of these intercepted between the two circles represent always the same tower *CB* carried by the earth

[114]Ibid., p. 191.

toward *DI*. The points where these lines are cut by the arc of the semicircle *CIA* are the places at which the falling stone will be found at the various times. Now these points become more distant from the top of the tower in an ever increasing proportion, and that is what makes the stone's straight motion along the side of the tower increasingly rapid.

Galileo's conclusion was not nearly as inexplicable as it might seem. We know that the inertial *impeto* due to the earth's rotation is characterized by a fixed curvature as well as by a fixed magnitude; the motion conserved in heavy bodies after they have lost contact with the earth's surface is therefore a qualitatively determined motion, that is, one whose form is given a priori. Rectilinear downward motion, for its part, has only one role to play: to bring bodies nearer to the center of the earth. That being so, and since falling bodies must in any case describe a curve, are we not forced—once we have granted the inertial character of uniform circular motion—to evaluate their real trajectory too by reference to its form? At the nonmathematical, purely representational level on which this discussion took place, it is therefore not in the least astonishing that Galileo should have based the final determination of the motion of a heavy body toward the center on that of its clearly determined component. Just how predominant circular motion was in his thought appears even more clearly from one of his other arguments. Consider the arcs that the heavy body describes during its fall, that is, those intercepted by the radii *AC* and *AF*, *AF* and *AG*, *AG* and *AH*, *AH* and *AL*, *AL* and *AD*, and so on, and compare them to the corresponding arcs *CF*, *FG*, *GH*, *HL*, and *LD*, which the heavy body would have passed in the same intervals of time had it remained above the tower. A very simple demonstration enabled Galileo to show that the length of the first is precisely equal to that of the second.[115] "From this there follows a third marvel—that the true and veritable motion of the stone is in no way an accelerated motion but is always equable and uniform." Moreover, its speed is identical to that of the earth's diurnal motion.[116] As a consequence, the motion of the heavy

[115]Ibid., pp. 192–193. This argument shows once again to what extent geometrization, when it is not based on adequate physical concepts, can distort a problem; cf. footnote 117, below.

[116]The time it takes a heavy body to reach the center of the earth would therefore be equal to the time in which the tower would have covered a quarter of the terrestrial circumference, that is, six hours. In 1635, in a letter to Fermat (cf. *Œuvres de Fermat*, edited by P. Tannery and C. de Waard, Vol. IV, pp. 15 ff.), Mersenne noted that this

body toward the center of the earth had lost the least resemblance to a compound motion: excluded from the genesis of the real trajectory, rectilinear motion was reduced to the deflection of the inertial circular motion and deprived of the means of changing the latter in any way.[117]

* * *

Clearly, therefore, the principles and concepts Galileo employed to establish the mechanical possibility of the earth's diurnal motion were by no means the principles and concepts of classical science. Neither his view of inertial systems nor his principle of conservation—nor a fortiori his attempts to compound motions—was truly satisfactory. Must we therefore dismiss his contribution during the Second Day of the *Dialogue* as totally irrelevant to the birth of classical science? Nothing could be more unfair or mistaken. The uncertainties and limitations in his path were the typical obstacles encountered by every creative thinker who finds that each new principle and each new concept must be wrested from diametrically opposite principles and concepts and

---

solution was incompatible with Galileo's own "square law" (that the spaces passed over by a body starting from rest have to each other the ratios of the squares of the times in which such spaces were traversed) (*Dialogue* II, p. 248). Giving g the approximate value of 7.89 m/sec^2, Mersenne concluded that it would take the body no more than twenty minutes to reach the center of the earth.

[117]Galileo not only failed to produce a geometric composition of rectilinear downward and inertial circular motion but also failed to offer a valid mechanical solution. Koyré has retraced the whole history of this problem in his "Documentary History of the Problem of Fall from Kepler to Newton." Here we need merely mention the contributions of Fermat and Borelli. The former informed Mersenne that a heavy body in uniform rotational plus uniformly accelerated motion must describe a spiral of the second degree (*Œuvres de Fermat*, Vol. IV, pp. 15 ff.). When Galileo was told of this solution by Carcavi, he declared that he had been entertaining the same idea for a long time (*Letter to Carcavi*, June 5, 1637, Vol. XVII, p. 89). However, though geometrically sounder, Fermat's solution was mechanically no more correct than Galileo's. In particular, neither of them realized that the substitution of the uniform motion of a radius vector for the uniform circular impetus of the heavy body ran counter to the principle of conservation. The first to point this out (even though he was unable to determine the shape of the curve) was Borelli, possibly influenced by Galileo's analysis. Imagine a heavy body falling with a uniformly accelerated motion along a radius that revolves with a constant angular velocity: clearly, the *linear* velocity communicated to the body by the motion of the radius will decrease as the body approaches the center of the circular motion. This means that the original uniform circular *impeto* must keep decreasing, thus violating the principle of conservation. Hence, if we wish to argue in mechanical terms and not in terms of a geometrical transposition that totally deforms the problem, we must endow the vector radius with a constantly increasing angular velocity (cf. Koyré, "Documentary History," pp. 359 ff.).

that each new step not only introduces a new content but also marks an advance from one intellectual universe to another. How difficult that advance really is may be gathered not only from the rough and ready nature of the principles and concepts we have been examining but also from the contradictions and even regressions that punctuate the *Dialogue*. Thus to refute the first of the geocentrists' objections, Galileo takes for granted that the earth's diurnal rotation communicates an inertial motion to all terrestrial bodies, namely a motion deprived of any effect. Yet a few pages later, when dealing with the centrifugal force, he declared that a rotational motion may well produce a tendency in heavy bodies to escape along the tangent. Another remarkable contradiction was that while he likened the moving earth to an inertial system during the Second Day, he attributed the tides to the double motion of the earth during the Fourth Day, thus contradicting all that had gone before.[118] And what can be more disconcerting than to find that, despite the great transformations his thought had undergone in the meantime, he nevertheless introduced one of the oddest themes of the *De Motu* into the *Dialogue* when he asked himself whether the acceleration commonly attributed to freely falling bodies might not, after all, be an optical illusion? However, neither these contradictions nor these regressions must be allowed to detract from the value of the radical innovations of the *Dialogue*, of which they even supplied the best proof. For, when all is said and done, anyone who had read the *Dialogue* or the *Discourses* needed far less effort to formulate the principle of inertia or to apply the principle of composition correctly than did Galileo when, with great intellectual courage, he concluded in at least one case in favor of the conservation of an impressed motion, or when he asserted that a system in uniform motion in no way alters the normal course of phenomena within that system. Only Newton showed the same intellectual courage when, fifty years later, he propounded the theory of universal attraction.

But just as it would be quite wrong to underestimate the importance of Galileo's reflections on the earth's diurnal motion, so it would be incorrect to take these reflections out of their context and to put them on a par with the arguments presented in the *Discourses*. To appreciate the truth of this remark, we need merely examine the subsequent fate

---

[118] See Appendix 4.

of the principles and concepts we have been examining. If we do so, we shall find that, while the *Dialogue* failed to apply the principle of the composition of motions correctly, and while its account of the curved trajectory of projectiles or of falling bodies on a rotating earth led more or less to the suppression of the idea of compound motions, the *Discourses* rightly explained the nature of the trajectory by the simple composition of two rectilinear motions. Again, while the *Dialogue*, however admirable its construction of the principle of the conservation may have been, reserved this principle exclusively for the case of uniform circular motion, the *Discourses*, though not formulating the principle in its modern form, nevertheless offered what in practice amounted to an identification of inertial motion with rectilinear uniform motion. Finally, it is altogether remarkable that the principle of mechanical relativity, on which the refutation of the first three objections to the diurnal motion was ultimately based, should have been ignored in the *Discourses*, which totally dispensed with the idea of inertial systems. The reasons for these differences are easily explained by the distinct nature of the problems considered in the two works. When he dealt with the motion of the earth, Galileo encountered a complex problem whose solution called not only for an analysis of circular motion in general and of rotatory motion in particular but also for a clear definition of the concepts of mass and force; but the *Discourses*, which deal exclusively and in purely abstract terms with the natural and projectile motions of heavy bodies, did not in any way provide him with the necessary conceptual resources. Moreover, the contrast was equally marked on the methodological plane, for mathematics played a purely illustrative part in the *Dialogue*, but not in the *Discourses*, In particular, the *Dialogue* used mathematics to drive home or refine specific arguments, but never as a language for formulating general problems or for developing their analysis, thus eschewing the demonstrative rigor of mathematics in favor of substitutes that, however ingenious they may have been, could not rival the efficacy of the former.

This, we repeat, does not mean that there are no links at all between the mechanical ideas expounded in the two works, as shown, for example, in a common debt to statics. To go beyond that, however, would mean overlooking the fact that, while a large part of Galileo's cosmological work admittedly paved the way for classical mechanics, it did so in a context quite distinct from that of the *Discourses*.[119] Because he was analyzing two sets of mechanical problems—those raised by the diurnal motion and those raised by the motion of heavy bodies—Galileo was led to two sets of conclusions which, though he did not consider them incompatible, he nevertheless failed to combine. Hence his contribution to classical mechanics must be evaluated on two distinct planes, lest the importance of either be lost in the mists of an artificial continuity.

---

[119]This is why we need not be astonished to find that Galileo failed to apply his geometrical analysis of projectile motion to the case of freely falling bodies on a rotating earth. He lacked the conceptual means needed to reduce the problem to that which the *Discourses* solved so elegantly.

III The Birth of Classical Mechanics

Introduction

While the defense of the Copernican doctrine enabled Galileo to outline some of the most significant ideas and methods of classical mechanics, his *Discourses* must be considered an integral part of that science. Hence Galileo could declare proudly at the beginning of the Third Day: "My purpose is to set forth a very new science dealing with a very ancient subject," and never was a declaration less presumptuous, nor, for that matter, has there ever been a better summary of the marked contrast between the *Discourses*, the peak of Galileo's studies, and those books, "neither few nor small," that traditional philosophers had devoted to the subject of motion. At the heart of this contrast lay an event of momentous importance: the successful geometrization of the motion of heavy bodies, or rather the first successful geometrization of a physical problem since Archimedes' work on statics and hydrostatics.

What precisely do we mean by the geometrization of motion? First of all, that motion, instead of being simply described, is interpreted in terms of quantitative laws ensuring accurate predictions and hence the ascendancy of reason over nature. But this first aspect, though the more spectacular, is by no means the most important. For once it has been turned into an autonomous phenomenon, in the manner of a mathematical object, motion can be defined genetically and its fundamental properties studied methodically. The traditional, ontological interpretation of motion made way for an entirely new approach, in which the proportions between two primitive magnitudes, space and time, held the key to its true essence. In other words, Galileo took motion out of the province of philosophical speculation and carried it into the rational domain provided by geometry; while in the *De Motu* and the *Dialogue* he had still used geometry as a mere aid to what was essentially a philosophical view, he now turned it into the very language of science.

Nor was that all. Geometrizing motion also means presenting the whole body of established theorems and propositions as a *coherent* whole. This Galileo succeeded in doing in at least two ways: first of all by introducing theorems and propositions in such a way that the proof or justification of each followed from results previously established, and second, by ordering his material in three main categories—uniform motion, uniformly accelerated motion, and projectile motion—each subdivided so as to range from the simplest to the most complex. This procedure found its clearest expression during the Third Day of the *Discourses*, which was devoted exclusively to the natural motion of falling bodies; all the propositions fit neatly into five groups of increasing complexity: free fall, motion along an inclined plane, combination of the two, motion along a horizontal or inclined plane after free fall, and determination of the swiftest line of descent. All this, as we have said, was the result of the successful geometrization of the motion of heavy bodies; it alone enabled Galileo to deduce an intelligible and coherent body of theorems and propositions from the ratios of two primitive magnitudes, and hence to turn motion into an object of thought sui generis.

A few examples taken from the Third Day of the *Discourses* may serve to illustrate this point. That day began with a discussion of the natural motion of bodies in free fall in terms of Galileo's "square law," which states that the distances traversed by a moving body are proportional to the square of the time intervals. Because he lacked the appropriate mathematical language, Galileo's original formulation of that theorem proved rather impractical,[1] and he accordingly tried to improve it by applying the operational resources of traditional geometry.[2] To that end, he considered a line SV traversed by a freely falling body, and a segment ST of the same line, and then applied Theorem II to demonstrate that the time the body needs to traverse SV is to the time it needs to traverse ST as SV is to the proportional mean SX: "Now since it has been shown that the spaces traversed are in the same ratio as the squares of the times; and since moreover the ratio of the space SY to the space ST is the square of the ratio SV to SX, it follows that the ratio of the times of fall through SV and ST is the ratio of the respective distances SV and SX."[3] In this slightly changed form, the theorem became easy to utilize and hence could play a leading part in the systematic study of the motion of heavy bodies.

---

[1]In fact, he applied the original theorem on one occasion only (*Discourses* VIII, p. 273), when he established the parabolic trajectory of a projectile.

[2]Corollary II to Theorem II (*Discourses* III, p. 214).

[3]By definition, $SX^2 = SV \times ST$. If T_{ST} is the time of descent along ST, and T_{SY} the time of descent along SY, it follows from Theorem II that $T^2_{SV}/T^2_{ST} = SV/ST$; now, $SV/ST = SV/SX^2/SV = SV^2/SX^2$; therefore, $T_{SV}/T_{ST} = SV/SX$.

S
T
X
V

Passing on from the motion of freely falling bodies to that of bodies on an inclined plane (in the absence of all friction), Galileo established in a remarkable series of theorems how changes in the length and/or inclination of the plane lead to determinate changes in the time of descent. Thus Theorem III stated that "if one and the same body starting from rest falls along an inclined plane and also along a vertical, each having the same height, the times of descent will be to each other as the length of the inclined plane and the vertical";[4] and Theorem IV, that "the times of descent along planes of the same length but of different inclinations are to each other in the inverse ratio of the square roots of their heights."[5] Perhaps the most remarkable of all was Theorem VI:[6] "If from the highest or lowest point in a vertical circle there be drawn any inclined planes meeting the circumference, the times of descent along these chords are equal to each other." Galileo's proof was as follows: On the horizontal line *AB* construct a vertical circle, and from its point of tangency with the horizontal draw the diameter *CD*. From the highest point *D* draw inclined planes to *F* and *E*, which are arbitrary points on the circumference; from *C* draw *CF* and *CE*: the times of descent along *DF*, *FC*, and *EC* will be equal to the times of descent along the diameter *DC* and hence equal to one another. To prove that the time of descent along *DF* is equal to that along *DC*, we need merely draw *FG* perpendicular to the diameter *DC*, and join *FC*. Since the time of descent along *DC* has the same ratio to that along *DG* as the mean proportional between *CD* and *DG* has to *DG* itself; since that mean proportional is *DF* (the angle *DFC* being a right angle and *FG* being perpendicular to *DC*), it follows that the times of descent

[4]*Discourses* III, p. 215.
[5]Ibid., p. 219.
[6]Ibid., pp. 221 ff.

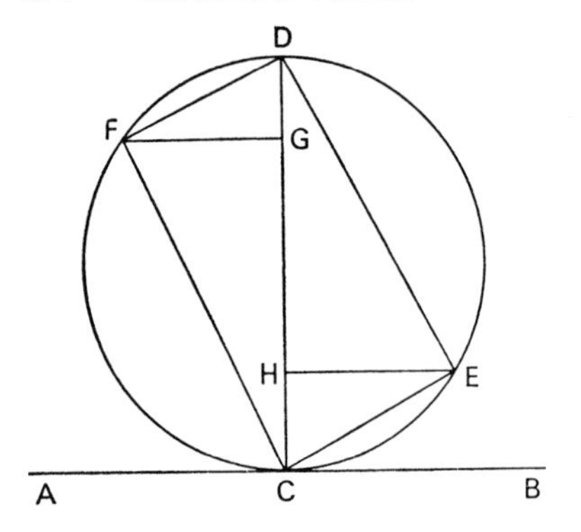

along DC and DG will have the same ratio as DF has to DG. But it has previously been demonstrated that the times of descent along DF and DG also have the same ratio as DF has to DG; therefore, the times of descent along DF and DC must be equal.[7]

From these preliminary results Galileo could readily pass on to the analysis of increasingly complex mechanical problems, all of which he solved by methods whose rigor was in no way inferior to that of geometry. Proposition XV provides a particularly good illustration.[8] In it, Galileo set himself the following problem: "Given a vertical line and a plane inclined to it, find a length on the vertical line below its point of intersection which will be traversed in the same time as the inclined plane, each of these motions having been preceded by a fall through the given vertical line." A construction followed by an argument that was as elegant as it was brief provided the solution: "Let AB represent the vertical line and BC the inclined plane; it is required to find a length on the perpendicular below its point of intersection, which after a fall from A will be traversed in the same time that is needed for BC after an identical fall from A. Draw the horizontal AD, intersecting the prolongation of CB at D; let DE be a mean proportional between CD and DB; make BF equal to BE; also let AG be a third proportional to BA and AF. Then, I say, BG is the distance which a body, after falling through AB, will traverse in the same time which is needed for the

---

[7] Ibid., pp. 222–223. It is required to prove that $T_{DF} = T_{DC}$. Now, $T_{DC}/T_{DG} = (\sqrt{CD} \times DG)/DG$ (Theorem II, Corollary II); but $DF^2 = CD \times DG$; therefore, $T_{DC}/T_{DG} = DF/DG$. Now, since $T_{DF}/T_{DG} = DF/DG$ (Theorem III), we have $T_{DF}/T_{DC} = DF/DF = 1$. Q.E.D.

[8] Ibid., pp. 232–233.

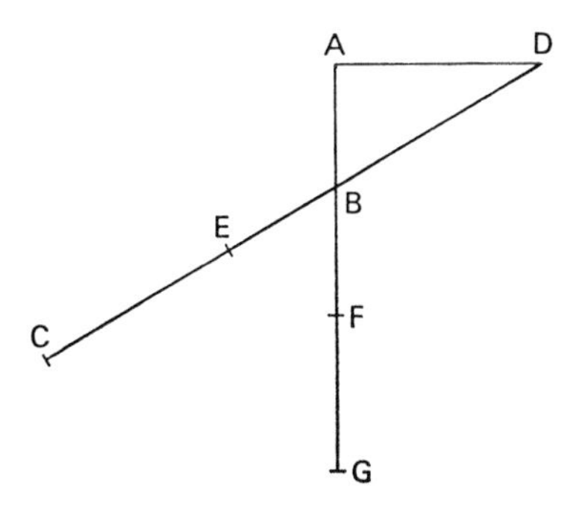

plane BC after the same preliminary fall. For if we assume that the time of fall along AB is represented by AB, then the time for DB will be represented by DB. And since DE is a mean proportional between BD and DC, this same DE will represent the time of descent along the entire distance DC, while BE will represent the time required for the difference of these paths, namely, BC, provided in each case the fall is from rest at D or at A. In like manner we may infer that BF represents the time of descent through the distance BG after the same preliminary fall; but BF is equal to BE. Hence the problem is solved."[9] There is no need to quote further examples to show that Galileo had indeed succeeded in constructing a geometrical model of uniformly accelerated motion from which he could derive a host of propositions by rigorous deductions. In short, by interpreting motion in terms of temporal and spatial relationships, he had transformed it into an object of rational science, removing it from the sphere of intuition and sensible experience, so much so that during the Third and Fourth Days of the *Discourses* he was able to lay the foundations of what came to be known as point mechanics.

In saying this, we have not, however, told the whole story. For though we have shown that he did indeed contribute "a very new science dealing with a very ancient subject," we must also try to explain how he came to be the first to take the great leap forward into mathematical

---

[9] *Hypothesis:* AB is the perpendicular, BC the inclined plane. Draw the horizontal AD, intersecting CB produced in D. Let $DE^2 = CD \times DB$, $BF = BE$, and $AG = AF^2/AB$; then BG is the required distance.
Proof: If $T_{AB} = AB$, then $T_{DB} = DB$ (Theorem III); since $DE^2 = BD \times DC$, $T_{DC} = DE$ (Theorem II, Corollary II), and for a motion starting from D or A, $T_{BC} = BE$. Similarly, $T_{BG} = BF$ (since $T_{AG} = AF$ and $T_{AB} = AB$); but $BF = BE$; hence the proposition.

physics. In fact, the direct basis of Galilean science was no more pure mathematics than it was pure observation; perfect knowledge of traditional geometry or closer attention to the observable phenomena do not suffice for the elaboration of a mathematical theory of motion. Without pretending that we are explaining the inexplicable genius of Galileo, we nevertheless feel certain that the real roots of that theory must be sought in a number of interpretative ideas, with the aid of which alone motion as a physical object could be opened up to mathematical reason and to its tremendous resources. Now these interpretative ideas, without which Galilean science could never have taken shape, can be divided into two main categories. First of all came the ideas on which the representation of motion can be based directly—among them speed, acceleration, and gravity as a motor force; these may be said to have constituted the *conceptual* mainsprings of Galileo's mechanics. In addition, there were several more general ideas, serving less as aids to the intrinsic analysis of motion than in defining its physical status. This category included, inter alia, the idea that under certain conditions uniform motion is capable of indefinite self-conservation. As complements to the concepts, this category represented the *principles* of Galilean mechanics. Was there a direct true continuity between these concepts and principles and Newton's? And if not, what are the characteristics of Galileo's own system? The answers to these questions will help us to a deeper understanding of the *Discourses;* it alone can show us in which context Galileo developed his thought and hence help us determine its precise scope.

6 The Geometrization of the Motion of Heavy Bodies (Part 1)

The Two Representations of Velocity in the Analysis of Naturally Accelerated Motion

The properties belonging to uniform motion have been discussed in the preceding section; but accelerated motion remains to be considered.

And first of all it seems desirable to find and explain a definition best fitting this motion as it occurs in nature. For anyone may invent an arbitrary type of motion and discuss its properties; thus, for instance, some have imagined helices and conchoids as described by certain motions which are not met with in nature, and have very commendably established the properties which these curves possess by virtue of their definitions; but we have decided to consider the phenomena of bodies falling with an acceleration such as actually occurs in nature and to make this definition of accelerated motion exhibit the essential features of observed accelerated motions. And this, at last, after repeated efforts we trust we have succeeded in doing. In this belief we are confirmed mainly by the consideration that experimental results are seen to agree with and exactly correspond with those properties which have been, one after another, demonstrated by us. Finally, in the investigation of naturally accelerated motion we were led, by hand as it were, in following the habit and custom of nature herself, in all her various other processes, to employ only those means which are most common, simple, and easy.

For I think no one believes that swimming or flying can be accomplished in a manner simpler or easier than that instinctively employed by fishes and birds.

When, therefore, I observe a stone initially at rest falling from an elevated position and continually acquiring new increments of speed, why should I not believe that such increases take place in a manner which is exceedingly simple and rather obvious to everybody? If now we examine the matter carefully, we find no addition or increment more simple than that which repeats itself always in the same manner. This we readily understand when we consider the intimate relationship between time and motion; for just as uniformity of motion is defined by and conceived through equal times and equal spaces (thus we call a motion uniform when equal distances are traversed during equal time intervals), so we may also imagine that, in an interval of time similarly divided into equal parts, increases in speed will take place just as simply. Thus we may picture to our mind a motion as uniformly and continuously accelerated when, during any equal intervals of time whatever, equal increments of speed are given to it. Thus if any equal intervals of time whatever have elapsed, counting from the time at which the moving body left its position of rest and began to descend, the degree of speed acquired during the first two time intervals will be double that acquired during the first time interval alone; so the degree acquired during three of these time intervals will be treble; and that in four, quadruple that of the first time interval. To put the matter more clearly, if a body were to continue its motion with the degree of speed which it had acquired during the first time interval and were to retain this same speed uniformly, then its motion would be twice as slow as that which it would have had with the degree acquired during two time intervals.

And thus, it seems, we shall not be far wrong if we admit that the intensification of speed is proportional to the extension of time; hence the definition of motion we are about to discuss may be stated as follows: a motion is said to be equally, or uniformly, accelerated when, starting from rest, it acquires equal moments *(momenta)* of speed in equal times.[1]

An admirable passage, no doubt, but an extremely complicated one as well, with the new barely distinguished from the old, so that nearly every phrase has to be evaluated with great care.

Let us first consider, not so much the process of increases in speed as such, as the general manner in which Galileo fitted speed into his new framework. We are struck, first of all, by the nonmathematical nature of his representation; far from treating speed as the ratio of the distance traversed to the time, as we might have expected him to do,[2] he presented it as a quasi-physical magnitude increasing by the successive additions of parts, a process that could apparently be grasped by some sort of intuitive perception. Nor was that all. Speed was also said to have the remarkable property of being characterized at every instant by a *degree* and being so much greater as we take it farther from the beginning of the motion. In other words, speed was thought to possess, at every moment, a fixed intensity or, which comes back to the same thing, its growth was assumed to take the form of an intensification (or *intensio*, as Galileo called it). The conclusion is obvious: by presenting speed as a quasi-physical object to which a given degree of intensity can be assigned at any one moment, Galileo showed clearly that, three centuries after the Mertonians and Parisians, the treatment of speed as an intensive magnitude was still one of the mainstays of mechanical thought.

This claim, which may perhaps surprise the reader, is not difficult to justify, for it is easily demonstrated that even in his most advanced

[1] *Discourses* III, pp. 197 f.
[2] The more so as he did just that a few pages earlier when discussing uniform motion.

studies Galileo never ceased to employ the medieval approach to speed. This may be gathered above from his continued use of the traditional vocabulary; such expressions as *gradus velocitatis* (or *grado di velocità*), *intensio velocitatis*, and *motus difformis* had far too precise a meaning for their retention to signify anything but the survival of the ideas they had originally been coined to reflect.[3] Nor were the new terms Galileo introduced any less revealing. One of the most typical of these was the concept of *impeto*, which we have met so frequently. Because Galileo attached so many distinct meanings to it, it cannot be easily defined, though we may say quite generally that it referred to a dynamic magnitude associated with motion, to a direct function of speed made manifest particularly in the force of percussion; in particular, he employed it to represent the quantity of motion and the *vis viva* of classical mechanics. It is then remarkable to see how easily he substituted *impeto* for *velocità*, even in the course of a single argument, as he did when dealing with the composition of speeds[4]—a substitution that can be explained only by the fact that he continued to treat speed implicitly as a quasi-physical magnitude coextensive with motion.[5]

Galileo's terminology, however, was not the only indication of the persistence of fourteenth-century ideas in his work; thus his attempts both in the *Dialogue*[6] and in the *Discourses* to demonstrate the continuous nature of speed increases do little more than develop one of the most interesting implications of the traditional view.

Unlike *impeto*, the continuous nature of speed increases is easily described; it results quite simply from the fact that "a moving body, departing from rest and entering into the motion for which it has a natural inclination, passes through all the antecedent gradations of

[3]The term *grado di velocità* was used throughout the *Discourses* as it was throughout the *Dialogue* and as it would be in one of the last texts written by Galileo, the *Essay on the Force of Percussion* (cf. Vol. VIII, pp. 337 ff.); the term *intensio velocitatis* occurred in the very passage we are analyzing; and the term *motus difformis* appeared, inter alia, in the *Dialogue* VII, p. 451.

[4]*Discourses* IV, pp. 288–289.

[5]The following examples of this synonymous use are chosen at random: *Dialogue* I: *Impeto e velocità* (p. 46); *l'impeto, cioè il grado di velocità* (p. 52); in *Discourses* III, *impeto* and *velocità* are used interchangeably throughout the argument set out on pages 205–208; similarly in *Discourses* IV, pp. 280 ff., which includes among others the expressions *impeto o momento di velocità* (p. 280) and *velocità e impeto* (p. 287).

[6]In a passage containing a discreet allusion to his own studies of the motion of heavy bodies; since similar passages in the *Dialogue* often explain and complement the analyses presented in the *Discourses*, we shall have occasion to refer to them more than once.

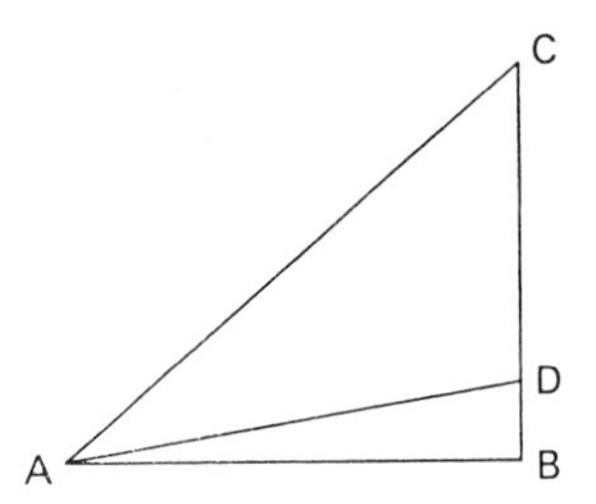

slowness that exist between a state of rest and any assigned degree of speed.[7] This conclusion, which at first sight seems to run counter to experience,[8] could, in Galileo's view, be confirmed by reference to a singularly important proposition in mechanics, namely, that the speeds acquired by a heavy body descending along the inclined plane *CA* and the perpendicular *CB* to the terminal points *A* and *B* on the horizontal *AB* are identical.[9] Without entering into the merits of this proposition, we shall merely note that thanks to it we can allocate to every fall along the perpendicular a corresponding descent along an inclined plane having the same height. Imagine then a plane such as *DA*: a body descending along this plane will arrive at the terminal point *A* with less speed than it would have attained after descending along the line *CA*, and also with a speed equal to that with which it would have attained on reaching the terminal point *B* after descending along the perpendicular *DB*. Moreover, it is obvious that an unlimited number of planes of increasingly smaller inclinations can be constructed between *DA* and *BA*, and that the speed of our moving body along these planes will grow smaller and smaller although never decreasing to zero. Now these increasingly small degrees of speed are precisely those with which the body would descend from increasingly low points along the perpendicular, or, rather, they are the degrees with which the body will start its motion along the perpendicular as we consider states closer and closer to rest. It follows that a heavy body, as it begins to move, will indeed pass through all the "infinite gradations of slowness" comprised between any degree

[7] *Dialogue* I, p. 45; the same idea was repeated in *Discourses* III, p. 199.

[8] Experience seems to show that after a heavy body starts from rest it immediately acquires certain speed (*Discourses* III, p. 199). This may explain why Cavalieri (see Galileo's *Opere*, Vol. XIII, pp. 311 ff.) did not at first agree with the argument and why Descartes thought that "this does not happen in the ordinary course of events but may happen sometimes" (*Letter to Mersenne*, October 11, 1638; *Œvres*, published by Adam and Tannery, Paris, 1897–1913, Vol. II, p. 399.

[9] *Dialogue* I, pp. 117 ff., and *Discourses* III, p. 205. In both passages, Galileo used *impeto* and *velocità* interchangeably.

of speed and the state of rest, and this without pausing in any of them.[10] The same conclusion can also be reached by combining the principle of sufficient reason with an argument much closer to common experience than the preceding one, namely, that transitions from one state to another are "more naturally and readily made to a closer than to a more remote one."[11] To illustrate this point, Galileo, considering the upward motion of a shot, had Sagredo ask Simplicio: "Does not this cannonball, sent perpendicularly upward by the force of the charge, continually decelerate in its motion until it finally reaches its ultimate height, where it comes to rest?[12] And in diminishing its velocity—or, I mean, in increasing its slowness—is it not reasonable that it makes the change from 10 degrees to 11 sooner than from 10 to 12? And from 1000 to 1001 sooner than to 1002? And, in short, from any degree to a closer one rather than to one more remote?" The answer was obvious: "No doubt can remain that the ball, before reaching the point of rest, passes through all the greater and greater gradations of slowness, and consequently through that one at which it would not traverse the distance of one inch in a thousand years. Such being the case, as it certainly is, it should not seem improbable to you, Simplicio, that the same ball, in returning downward, leaving rest, recovers the velocity of its motion by returning through those same degrees of slowness through which it passed going up; nor should it, on leaving the larger degrees of slowness which are closer to the state of rest, pass by a jump to those farther away."[13]

Though the continuous nature of motion must thus be granted by anyone placing reason above the evidence of his senses, it nevertheless raises a number of problems. Thus we saw that Galileo thought it possible to fit an "unlimited" (read: "infinite") number of planes between the points C and B on the accompanying diagram; now since each of these planes represents a different degree of speed along the perpendic-

---

[10]*Dialogue* I, p. 52; cf. *Discourses*, III, p. 199: "We must infer that as the instant of starting is more and more nearly approached, the body moves so slowly that if it kept on moving at this rate it would not traverse a mile in an hour or in a day or in a year or in a thousand years; indeed it would not traverse a span in an even greater time."

[11]*Dialogue* I, p. 55.

[12]This awkward formulation might suggest that Galileo continued to subscribe to the traditional thesis of the *quies media*, which he had expressly rejected in the *De Motu* (cf. Chapter 3).

[13]*Dialogue* I, pp. 54–55; cf. *Discourses* III, p. 200. Galileo made much the same point with the help of a geometrical example in a letter to Carcavy (June 5, 1637, Vol. XVII, pp. 92–93).

ular *CB*, and since speed increases are continuous, are we not entitled to conclude that it is always possible to fit an infinite number of degrees between the state of rest and any given degree of speed or, more generally, between any two degrees of speed however close?[14] And under these conditions, would it not follow that any motion, no matter how brief, must be of infinite duration? Hence, does not the idea of a continuous scale of speed, logically unexceptional though it may be, render the concept of motion completely unintelligible?[15] In fact, this question was merely a reformulation of Zeno's *Dichotomy*, and Galileo resolved it by following Aristotle (*Physics* VI) almost to the letter.

The problem is to show that motion is possible even if length and time are infinitely divisible. To that end, Aristotle had argued that a finite distance does not embrace the infinite in the sense that it is composed of an infinite number of points which the moving body must traverse in turn, because a point comes into being solely after a division or, in the case of motion, after a pause, so that only the possible reiteration of these processes can be said to be infinite. Definable as "that which is divisible into parts that are themselves divisible," length is infinite in a potential sense only, and a body in continuous motion will, as it were, slide over this potentiality without ever actualizing it. And what applies to length applies equally well to time; much as the former is not composed of a multiplicity of points, so the latter is not constituted of a multiplicity of instants; being a limit without duration, the instant like the point—a limit without extension—has no reality as such. And precisely because time contains infinity in the same way as length, all the objections to the reality of motion disappear.

Galileo's solution took much the same form, except that he substituted speed for distance. Motion would become inconceivable only if the moving body "paused" in each degree of successive speed it traversed, whereas in fact, the body glides uninterruptedly through all of them "so that even if the passage requires but a single instant of time, still, since every small time contains infinite instants, we shall not lack a sufficiency of them to assign to each its own part of the infinite degrees of slowness, though the time be as short as you please."[16] This kind of argument, which neither Bradwardine nor Oresme would have

[14] *Dialogue* I, p. 46.
[15] Cf. *Discourses* III, p. 200.
[16] *Dialogue* I, p. 46; cf. *Discourses* III, pp. 200–201. See also Aristotle's refutation of Zeno, Chapter 1.

disavowed, shows perhaps better than anything else to what extent fourteenth-century ideas continued to hold sway in Galileo's thought.

In all fairness, however, we ought to add a reservation. For while it is true that Galileo, like the Mertonians and the Parisians, considered speed an intensive magnitude characterized by its continuous growth, he also gave it a much broader significance and scope. This is best appreciated from his conception of "degree of velocity." The reader will recall that when Oresme tried to determine the magnitude of the instantaneous speed in a uniformly difform motion, he assumed that it was proportional to the distance that a body in which it was wholly conserved would traverse on a horizontal plane in a given time. Having failed to define the length of that time and the distance actually traversed by the body, Oresme's conclusion was at best a mere suggestion. Galileo, on the contrary, showed that it was possible to assign a precise value to every degree of speed (*Discourses*, Proposition XXIII): a body passing from a uniformly accelerated motion to a uniform motion on a horizontal plane will, with the degree of speed it possessed when it reached that plane, traverse twice the distance it covered in an equal time during the first part of its motion.[17] To the qualitative aspect of the "degree of speed" Galileo therefore added an authentic quantitative dimension, thus greatly broadening the scope of that concept. In addition he felt free to rely on his treatment of speed as an intensive magnitude to introduce several conclusions in formal opposition to the traditional theories. We just saw that since speed is subject to continuous increases, it is always possible to interpose a further degree of speed between any two, including the state of rest. In that case how can we ever decide where the motion commences and where the state of rest ends? In other words, would the difference between motion and rest as two fundamentally distinct states of bodies not completely disappear? Now, far from being disconcerted by this conclusion, Galileo welcomed it warmly, as witness his definition of the degree from which the movable body began to move as "that of the most extreme slowness, namely rest."[18] Fourteenth-century physicists had never arrived at similar conclusions; far from considering the state of rest as one of an infinite degree of slowness, they resolutely refused to renounce its privileged

[17] If v_i is the instantaneous speed of the moving body after traversing the distance s with a uniformly accelerated motion in the time t, then $v_i = 2s/t$; cf. *Discourses* III, pp. 242–243.

[18] *Dialogue* I, p. 44.

status: rest, being the goal of all motions either as their *terminus a quo* or else as their *terminus ad quem*, was said to be a state naturally and irreducibly distinct from them. Galileo, on the other hand, having rejected the idea that motion is a process (partly for cosmological reasons), used his identification of velocity with an intensive magnitude to support the argument that motion and rest were two equivalent states.

Nor did he simply apply the traditional representation of speed in a new way; by correlating it with a thesis of his own, he lent it quite an unsuspected breadth. The *Mecaniche* had already stated that in the absence of friction even the smallest force could set the largest body in motion on a horizontal plane. Of course, the body could be so large and the force so small that the effect could become imperceptible; must we therefore conclude that it has vanished altogether? The continuous nature of speed increases enables us to state categorically that it must persist even under these circumstances. Between rest and any degree of speed there is always room for an intermediate degree, and especially for a degree so small that the moving body to which it has been communicated will not cover "a mile in an hour, nor in a day, nor in a year, nor even in a thousand years." The transmitted speed can therefore approach as close as you like to immobility without ever vanishing altogether, and this agrees perfectly with the idea that no lower limit can be assigned to the effect one body can have on another. Now this conclusion was remarkable in that it meant that a proposition on the causes of motion was linked to and confirmed by a proposition on its effects; for the first time, therefore, the dynamic interpretation of motion had been brought into harmony with the kinematic, or, to use the traditional language, the analysis *quoad causas* had been reconciled with the analysis *quoad effectus*. True, this reconciliation failed to explain the precise way in which a force transmits its effects; moreover, it became meaningless just as soon as classical mechanics had been fully constructed; for all that, it paved the way for a general science of motion, so that we may justly claim that, even while he continued to use the older ideas, Galileo exploited them in a way that his precursors, blinded as they were by philosophical prejudice, could never even have imagined.

Important though they were, these changes in no way exhausted Galileo's personal contribution. Thus, while his theory of motion still involved medieval elements, it also reflected a conception of motion, of speed, and of speed increases that classical mechanics would later adopt for its own. A closer look at his definition of naturally accelerated motion will show us, perhaps better than anything else, how Galileo succeeded in developing the first modern analysis of motion within the framework we have just described.

To begin with, this definition established the complete autonomy of local motion. In particular, it dispensed with the need to relate local motion to the actualization of forms, that is, to the presence of a fixed substratum, and more generally to changes in quality or quantity; once the concept of κίνησισ had been dropped, local motion was opened up to direct examination, and there was no longer the least need to introduce such ontological factors as the *terminus motus*, that important concept of ancient physics. In short, the mode of speed increases, and this alone, now presided over the theory of motion. And once termini had been eliminated from the definition of motion or even from the proof of its reality, the possibility of unlimited motion had to be granted as well, so that the conception of motion as a state, first adumbrated in the *Dialogue*, became a tacit assumption permeating the entire *Discourses*. As a result it became possible to treat naturally accelerated motion in a purely rational way, that is, in the manner in which we normally treat mathematical objects. Like them, it was constructed before our very eyes and its definition was seen to engender its object even while describing it, thus revealing that its status was that of a real or genetic definition. Admittedly, the underlying assumption that speed was an intensive magnitude involved a quite considerable appeal to intuition; for all that, Galileo's construction bore a close resemblance to Euclid's. The very form of Galileo's analysis thus reflected the mathematical basis of his approach.

But Galileo's definition did not simply justify the substitution of one general idea of motion for another. For while a general framework close to that of classical mechanics was defined, the theory of naturally accelerated motion underwent its first great revolution since the fourteenth century. We saw earlier what conclusions most medieval physicists were forced to adopt; thus Albert of Saxony, perhaps the most lucid of them all on this subject, had asserted that speed increases formed an arithmetic progression or, more precisely, that if a motion is divided into equal parts, "the speed is doubled after the second part, tripled after the third, and so on." In other words, he identified naturally accelerated motion with uniformly accelerated motion. He also showed why speed increases cannot form a geometric progression. However, the problem was not fully resolved. What precisely are we to make of the "parts" of a motion? Are they divisions of the time or of the distance? Albert of Saxony himself decided in favor of the latter: "When a certain space has been traversed, the motion has a certain intensity; when double that space has been traversed, the speed is doubled; when the space is tripled, the speed is tripled, and so forth."[19] Now this solution, which was adopted by all Renaissance physicists, was also the one which Galileo adopted when in 1604 he tried to deduce from "a completely indubitable principle" his first results. Various writings, includ-

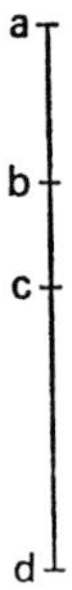

a
b
c
d

ing a letter to Paolo Sarpi, refer to his choice of space as the independent variable; thus, considering the case of a stone descending from the point *a* along a line *abcd*, he explained that "the degree of speed it possesses in *c* "is to the degree of speed it possessed in *b* as the distance *ca* is to the distance *ba*; consequently the degree of speed it possesses in *d* is as many times greater than that which it possessed in *c* as the distance *da* is greater than the distance *ca*."[20] Yet, having assumed that the moving body increases its speed in the same proportion as it moves away from its starting point, Galileo tried to proceed to a direct proof of his square law, without noting then the contradictions this step

[19]P. Duhem, *Le système du monde*, Vol. VIII, p. 295.

[20]Letter to Paolo Sarpi, October 16, 1604, Vol. X, p. 115. Benedetti had offered the same solution in his *Diversarum speculationum mathematicarum liber* (Turin, 1585), as quoted in A. Carugo and L. Geymonat, *Discorsi e dimostrazioni matematiche intorno a due nuove scienze*, p. 768.

entailed.[21] Some thirty-four years later—the decisive transformation cannot be precisely dated[22]—he rightly substituted time as the independent variable appropriate to the analysis of naturally accelerated motion.

Two reasons, so Galileo himself would have us believe, accounted for his substitution of time for space. The first of these was the discovery of an inherent contradiction in the original supposition. To begin with, consider several speeds having between them the same "proportion as the spaces traversed or to be traversed": it is clear that a body moving with these speeds would need equal time intervals to cover these spaces. Suppose now that speed increases as space in naturally accelerated motion; the speed with which a body would traverse a space of eight feet would be double that with which it covered the first four feet, and the time intervals needed for these passages would have to be equal. "Yet for one and the same body to fall eight feet and four feet is possible only in the case of instantaneous motion; but observation shows us that the motion of a falling body occupies time, and less of it in covering a distance of four feet than of eight feet; therefore it is not true that its speed increases in proportion to the space."[23] Did this mean that it increases in proportion to the time? Physical and philosophical reasons combined to persuade Galileo that it did. Thus, the uniformity of motion which "is defined by and conceived through equal times and equal spaces" pointed clearly to the "intimate connection between time and motion"[24]; moreover, it is certain that nature aims to "employ only those means which are most common, simple, and easy" and that there is "no increase more simple than that which repeats itself always in

---

[21]Duhem, *Système du monde*, Vol. VIII, pp. 373 ff. A. Koyré (*Études galiléennes* II, pp. 14–25, passim) has drawn attention to the main cases of Galileo's failure; particularly, "the reason why Galileo geometrized to excess and conferred upon space what appertains to time" (p. 24) may be found in "the predominant role of geometrical considerations in modern science and in the relatively greater intelligibility of spatial relationships" (p. 15).

[22]A valuable hint, however, is contained in a letter by Daniello Antonini, dated April 9, 1611 (*Opere*, Vol. XI, p. 85); it includes a definition of naturally accelerated motion that is couched in precisely the same terms as Galileo would later use in the *Discourses*. Hence it seems most likely that it was during 1610–1611 that Galileo introduced the definite substitution of time for space as the independent variable, and that Favaro and Koyré were wrong when they gave the date of this substitution as 1604 and 1620, respectively.

[23]*Discourses* III, pp. 203–204.

[24]Ibid., p. 197.

the same manner." If, then, we consider an interval of time divided into equal parts, the preceding remarks lead us by the hand, as it were, to the correct definition of naturally, that is, uniformly, accelerated motion, namely, that a motion is uniformly accelerated if and when "during any equal intervals of time equal increments of speeds are assigned to it."[25] This might suggest that Galileo followed a perfectly clear path: having discovered a contradiction in his earlier argument, he was forced to return to the fundamentals, and, applying the idea of natural simplicity, he was led almost inexorably to replace space with time, the more so as the principle of simplicity played so important a role in his cosmology and especially in his defense of the Copernican doctrine.

While this interpretation is not without merit, it does not reflect the full complexity of the problem Galileo set out to solve, nor does it even present an adequate account of the reasons that persuaded him to turn time into the independent variable. Let us look again at the contradiction he discovered in the alternative hypothesis. We readily grant that a body moving successively with speeds "in proportion to the spaces traversed or to be traversed" must traverse these spaces in equal intervals of time. However, the manner in which Galileo used this principle to disprove the idea that speed is proportional to the distance traversed was utterly invalid. For his entire argument was based on the assumption that, if a body traversing a space eight feet long acquired twice the speed it had upon traversing a space of four feet, it would also have to traverse eight feet in the same span of time as it traversed four. Now it is quite mistaken to suppose that the overall motion over the entire eight feet must be twice as fast as the motion over the first four feet; or rather before we can suppose this to be the case, we must first have replaced the *uniformly* accelerated motion over four and over eight feet, respectively, with two simultaneous uniform motions, the second of which takes the body over twice as great a distance with twice as great a speed. In other words, Galileo neglected an essential point, namely, the continuous increase in velocity; his argument was therefore unsubstantiated and in no way undermined the hypothesis it was meant to replace. However, it would be mistaken to attach too much importance to this logical error, for it was exceedingly difficult to refute the initial hypothesis directly without the help of the infinitesimal calculus,[26]

[25]Ibid., p. 198.

[26]The original hypothesis can be expressed by the differential equation $ds = gs\,dt$, where s represents the dis-

and yet a host of reasons persuaded Galileo that it must be wrong.[27]

But to prove that one principle is wrong does not, of course, automatically prove that another is right. Galileo's demonstration that speed could not increase *directly* as distance did not prove that space could not possibly be the independent variable. In fact, after the failure of his first attempt, he had only to postulate that the speed increases, not as the space traversed, but as the square root of that space, in order to obtain a definition of naturally accelerated motion that was in full accord with his square law and hence with all his other laws. Thus, if we divide a line AE into four equal parts, and assume that the degree

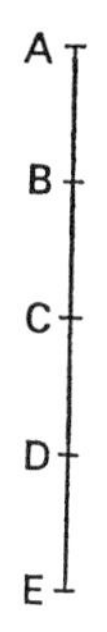

of speed acquired in E will be to the degree acquired in B as $\sqrt{AE}$ / $\sqrt{AB}$, then, because $AE = 4AB$, the speed in E will be twice the speed in B, as the square law would have it be. Hence, Galileo could have defined naturally accelerated motion as follows: "By naturally or uniformly accelerated motion I mean that motion whose speed increases as the square root of the space traversed" or, to use the language of the

---

tance, t the time, and g the acceleration. For purposes of integration we must put ds/s equal to $g\ dt$, whence $\log s = gt$, or $s = Ae^{gt}$, where A is a constant of integration. To say that the velocity increases as the distance is therefore tantamount to turning space into an exponential function of time, a fact that was first pointed out by Ernst Mach, *La Mécanique et son développment*, pp. 123 and 243; English edition, *The Science of Mechanics*, translated by T. J. McCormack.

[27]He may have realized in the following way that the original hypothesis was incompatible with his "square law." Thus consider a line AE divided into four equal parts, AB, BC, CD, DE. If speed increases are proportional to the distances traversed, then a body that has acquired one degree of velocity in B will have acquired two in C, three in D, and four in E. Yet according to the square law, the body can only have twice the speed in E that it has in B! This argument, incidentally, was but an application of the analytical method whose importance for science Galileo had previously been at great pains to stress (cf. *Dialogue* I, p. 75).

time, "whose speed increases in subdouble ratio of the space traversed." But, in that case, why did Galileo not stick to this solution, instead choosing time as the independent variable?

There are three possible answers to this question, all of them in full accord with the avowed explanatory ideal of the *Discourses*. First, if we look more closely at the formulation Galileo might have adopted had he stuck to distance as the independent variable, namely, that "by naturally or uniformly accelerated motion I refer to a motion whose velocity increases as the square root of the space traversed," we find that it lacks the characteristics of a real or genetic definition as demanded by a geometrized science of motion. Not only does it fail to bring out the process of growth by which the uniformly accelerated motion is engendered, but it also fails to explain why a motion whose speed increases as the square root of the distance traversed must necessarily be a *uniformly* accelerated motion. The definition therefore did no more than characterize the natural motion of heavy bodies by one of its properties and in no way led up to the intelligible principle from which all the other properties could be deduced; in short, it was not so much a definition as a description. All these disadvantages disappear once time is chosen as the independent variable: we can take any time interval whatsoever and divide it into equal parts; then, if one degree of speed is acquired in each of these parts, it follows that the speed will be doubled after the second part, tripled after the third, quadrupled after the fourth, and so on, whence "it ought to be conceded without question"[28] that the resulting motion is uniformly accelerated. Moreover, the great advantage of the original hypothesis—that it reflects the genesis of the motion—is preserved, this time without any contradictions. Last, but by no means least, if Galileo had continued to maintain that the speed is proportional to the square root of the distance traversed, he could only have proved his square law indirectly, by a reduction to absurdity;[29] now, in all his scientific writings (and this was a remarkable feature of his approach to which we shall be returning at some length) he invariably tried to adduce direct, ostensive demonstrations, apparently in the belief that in physics, unlike in mathematics,

[28]*Discourses* III, p. 205.

[29]But only because he lacked the resources of the infinitesimal calculus. Thus if he could have put ds equal to $\lambda\sqrt{s}dt$, λ being a constant, he would have had $ds/2\sqrt{s} = \frac{1}{2}\lambda\, dt$; then, integrating $\sqrt{s} = \frac{1}{2}\lambda(t-t_o)$, he would have obtained $s = (\lambda^2/4)t^2$ (if $t_o = 0$). The solution follows immediately if we put $\lambda^2/4 = g/2$ or $\lambda = \sqrt{2g}$.

a proposition cannot be established by the mere proof that its contradictory is impossible. In short, there were solid reasons why he should have substituted time for space, and even though these reasons were less simple than he himself cared to admit, there is little doubt that they reflected his desire to arrive at the simplest and most constructive definition—one, moreover, that agreed best with his ideal of mathematicized physics. Admittedly, other factors may also have influenced his choice,[30] but the true explanation can be found only in Galileo's own mathematical ideal.

For a full appreciation of this point, we must also remember what tremendous repercussions Galileo's definition of naturally accelerated motion had on the traditional links between time and motion. For Aristotle, motion was the prime reality, and time no more than its reflection in the human mind; being the "calculable measure or dimension of motion with respect to before-and-afterness,"[31] time was merely a means of introducing order into our observations and quite particularly into our astronomical observations. While Galileo's analysis of naturally accelerated motion did not reject this view explicitly, it nevertheless laid the foundations for a different interpretation. The fact that equal degrees of speed are added during each successive and equal time interval can mean only that time has become a continuous magnitude and hence the physical basis of motion and its evolution. In other words, time had come to enjoy physical primacy over motion, and because it allows the intelligible reconstruction of the latter, it also had to enjoy de jure precedence over it. The Aristotelian representation of time as an abstraction from motion thus made way for a new view reflecting the demands of mathematical physics, namely, that time is logically anterior to motion. Admittedly, Galileo did not yet draw any general conclusions from this new interpretation, and the description of the motion of the universe and of its constituent bodies in terms of the unconditional and independent flow of time had hardly started; for all

---

[30]Thus Koyré suggested that Galileo's choice was based on a causal analysis inspired by Gilbert's attractionist theory; in that case he would have been led to assume that a freely falling body experiences a constant pull by the earth and that time was the obvious variable (*Études galiléennes* II, p. 25). We think that this interpretation is erroneous, first of all, because Galileo would have had to be familiar with the fundamental law of Newtonian dynamics and second, because it ignores his own very precise interpretation of gravity as a force inherent in the moving body itself; cf. Chapter 8.

[31]*Physics* IV, 11, 219 b 1–2.

that, his firm assertion that speed increases as time may without exaggeration be considered an early expression of the Newtonian concept of absolute time.

We are therefore on safe ground when we claim that Galileo's definition of naturally accelerated motion did away with a host of traditional prejudices. But did it do more than that as well? Did the same progress also mark his analysis of the essential object of that definition, namely, the manner in which speed increases take place? No impartial reader, we are convinced, can really affirm it at once, so obvious again is the parallel with some of the main results in the fourteenth century. We need merely recall the approach of the Mertonians and Parisians to change in speed: they divided a time interval into equal parts, and contended that if the motion was difform, then the speed must experience an increase (*intensio*) or a decrease (*remissio*) during each of these parts. If these changes were regular, the resulting motion was said to be uniformly difform; if they were irregular, the resulting motion was said to be difformly difform. Now, when Galileo wrote that "the intensification of speed is proportional to the extension of time,"[32] was he not merely echoing these fourteenth-century views? And when he advanced his final definition—"a motion is said to be uniformly accelerated when, starting from rest, it acquires during equal time intervals equal moments of speed"—was he not repeating, almost literally, Heytesbury's formulation that "any motion is uniformly intensified if in each of any equal parts of the time whatsoever it acquires an equal increment of speed"?[33] In short, did Galileo do more (and this would already constitute a considerable advance) than produce a formal identification of naturally accelerated with uniformly difform motion as defined by his fourteenth-century precursors?[34]

In fact, so radical was Galileo's transformation of those changes in

---

[32]"Intensionem velocitatis fieri juxta temporis extensionem," *Discourses* III, p. 198.

[33]Cf. Chapter 2, footnote 17.

[34]According to Duhem, a similar identification had been made by Domingo de Soto in the fourteenth century (cf. P. Duhem, *Études sur Léonardo de Vinci*, Vol. 3, p. 558; see also Chapter 2 above). It is, however, difficult to attribute a precise meaning to Soto's text. In particular, his identification of motion occurs in a somewhat confusing passage and, moreover, in a purely philosophical context, so that it must be considered a fleeting reference rather than a carefully constructed basis of an objective study of the motion of heavy bodies. No reference to Soto is found in Galileo's *Juvenilia*, which mentions most leading fourteenth-century authors by name.

speed to which the Mertonians had referred as *latitudines* that, despite the close resemblance between his formulation and theirs, there is no justification at all for the view that Galileo's definitions had "their almost exact Merton counterparts."[35] To begin with, there was a clear change in attitude. Thus, fourteenth-century writers considered speed, defined as an intensive magnitude, to be a quality of motion *(qualitas motus)* with a fixed intensity at any one moment, and a difform motion to be one that could have no two successive intensities of the same value. To render the underlying processes somewhat less elusive, they first distinguished the possible modes of variation in the *qualitas motus* and then defined the motions to which these modes could be applied; if the changes in each equal part of the time were uniform, the motion itself was said to be uniformly difform; if the changes themselves were difform, the variation was said to be difformly difform. Now this meant that, far from trying to reconstruct the composition of speeds in non-uniform motions, the theory of *latitudines* served simply to describe and classify the chief changes capable of affecting the *qualitas motus* and then applied these changes to the description and classification of the chief types of motion. Hence the Mertonians' *latitudo* or Oresme's *velocitatio* referred in no way to the difference between two successive and instantaneous degrees of speed but simply to the way in which the intensity varies over equal time intervals, on the fundamental assumption that this variation must be either regular or irregular. The medieval explanation of uniformly difform and difformly difform motions was therefore based on a classification of their modes of variation; this type of analysis was the only one known in the fourteenth century.

Galileo's definition was of quite another type. When a heavy body starts to fall from rest, it acquires a certain *momentum velocitatis* during the first time interval, which is repeated in each successive interval, thus conferring a continuous speed increase upon the moving body. As Galileo himself expressed it: "If any equal intervals of time whatever have elapsed, counting from the time at which the moving body left its position of rest and began to descend, the amount of speed acquired during the first two time intervals will be double that acquired during the first time interval alone; so the amount added during three of these time intervals will treble; and that in four, quadruple that of the first

---

[35]The view of M. Clagett; cf. his *Science of Mechanics in the Middle Ages*, pp. 251–253.

time interval. To put the matter more clearly, if a body were to continue its motion with the same speed which it had acquired during the first time interval and were to retain the same uniform speed, then its motion would be twice as slow as that which it would have if its velocity had been acquired during two time intervals."[36] In other words, whereas medieval physicists simply classified qualitative changes in speed, Galileo introduced the quantitative concept of an elementary increase in speed during each equal part of the time. This meant that he had ceased to explain uniformly accelerated motion by regular intensifications in speed, and instead tried to discover what composition of elementary increments renders such motion intelligible. In short, he had stood the traditional principle of explanation on its head: while Oresme's *velocitatio* at best anticipated the modern concept of acceleration, Galileo based his definition on the *momentum velocitatis*, thus isolating the increase of speed as the chief analytical factor. For the descriptive and classificatory endeavors of traditional science he substituted the principle of a genetic explanation, borrowed from geometry.

This transformation went hand in hand with another. Not only did Galileo turn the concept of *momentum velocitatis*, that is, of instantaneous increases in speed, into the basis of his definition of uniformly (or naturally) accelerated motions, but he also gave that concept a degree of precision whose importance cannot be overstated. Let us return to the theory of *latitudines* and consider a speed subject to a uniform *intensio*. Its regular increase can be explained by the fact that an equal variation takes place in each equal part of the time. However, medieval physicists never bothered to turn this variation into an autonomous concept; it helped them explain increases in the *qualitas motus* but did not enjoy logical anteriority over the latter. Moreover, it could not be expressed in other than purely qualitative terms; it simply represented a change in quality in equal time intervals. Thus when Heytesbury defined uniformly difform motion as that which, *in quacumque equali parte temporis aequalem acquirit latitudinem*, he simply meant "that speed which, in any equal parts of the time whatsoever, acquires an equal tendency to increase its intensity." To give his definition a quantitative sense would be quite wrong, for not only is there nothing in the text itself to warrant that interpretation, but the philosophic concerns it presupposes were quite alien to the Mertonians as well as to Oresme.

[36]*Discourses* III, p. 198.

Galileo's definition of uniformly accelerated motion, by contrast, was based directly on the idea of elementary speed increases. In order to grasp what he meant by these, we can do no better than look once again at his own formulation: "When, therefore, I observe a stone initially at rest falling from an elevated position and continually acquiring new increments of speed, why should I not believe that such increases take place in a manner which is exceedingly simple and rather obvious to everybody? If now we examine the matter carefully, we find no addition *(additamentum)* or increment *(incrementum)* more simple than that which repeats itself always in the same manner."[37] Defined in this way, each new increment is therefore superimposed upon all the previous increments by an additive process. Now this was a definition of the greatest importance, because it implied that speed increases were the sum of elementary increases in the amount of speed. Moreover, since time is a continuous magnitude, there is nothing to prevent us from dividing it into parts as small as we like and hence from considering the amount by which the speed increases in each equal and successive part of the time, as a measure of its instantaneous change. We are thus not reading more into Galileo's text than we are entitled to when we claim that his *momentum velocitatis* referred precisely to what classical mechanics would call the rate of change in speed. In other words, Galileo did no less than express speed in its differential form.

In so doing, he accomplished what his fourteenth-century precursors could never have done, namely, to evaluate the magnitude of the speed at any given instant directly from its mode of increase. For the Mertonians and the Parisians the theory of *latitudines* and the concept of instantaneous speed had always remained two distinct consequences of their interpretation of speed as an intensive magnitude. To illustrate this paradoxical situation, we need only recall the way in which Oresme determined the distance traversed by a body in uniformly difform motion. Unable to deduce the magnitude of the speed at a given moment from its mode of variation, he was forced to introduce the concept of *quantitas velocitatis* as the sum of all the instantaneous speeds through which the motion had passed successively; not only did he fail to define "instantaneous speed" in terms of the mode of increase, but by simply substituting it for that mode he assigned it an independent role. With Galileo, on the other hand, the link between the mode of the increase in

---

[37]Ibid., p. 197.

speed and the value of the speed at any moment t followed quite automatically, because if γ is the "moment of speed" or the acceleration, we see at once that γt represents the amount of speed at that instant. For the first time the speed of a body in nonuniform motion had thus been reconstructed analytically from its mode of variation. This is borne out by the great ease with which Galileo's definition can be expressed in the language of infinitesimal calculus. Since the interval of time during which a new "moment of speed" is added to the moments previously acquired can be chosen to be as small as we like,[38] we can express this *momentum* in the form of the differential coefficient $\gamma = dv/dt$. But the speed at a given instant is simply the summation of all the moments acquired successively since the beginning of the motion; it is therefore perfectly possible to treat this summation as a form of integration, and to express the speed attained at the moment t by

$$v_t = \int_{t_0}^{t} \left(\frac{dv}{dt}\right) dt = \int_{t_0}^{t} \gamma dt = \gamma t + C,$$

where C is a constant of integration defined by the initial conditions of the motion. Despite the fact that he lacked adequate mathematical techniques, Galileo therefore arrived, by purely conceptual methods, at an analysis that modern kinematics would unreservedly have accepted as its own. Traditional though his terminology still was, his definition of naturally accelerated motion may therefore be said to have ushered in the beginnings of classical mechanics.

There is, however, one possible objection to this interpretation of the historical facts. Thus the reader may recall that the theory of *latitudines* represented no more than a part of the deliberations of medieval authors on the subject of changes in speed; that even while the Mertonians and Parisians described and classified the principal types of variation, other thinkers were trying to discover the intrinsic principles underlying them all. Of the several solutions they offered, the most remarkable was that of the Scotists, who favored a frankly quantitative interpretation: according to them, a process of *intensio* involved the addition of distinct parts, each new part being added to the preceding in the manner of extensive magnitudes. This poses the question of whether fourteenth-

---

[38] "Thus if *any equal intervals of time whatever have elapsed*, counting from the time at which the moving body left its position of rest and began to descend. . . ." Ibid., p. 198.

century writers—had they but applied the results of their analysis to the description of *latitudines*, as we know that none of them did in fact—would not have been led to treat the equal change in speed that takes place in each equal part of time during uniformly difform motions as an elementary quantity, and hence to explain the increase in speed characteristic of such motions in Galilean terms. Now this question is the more important in that Galileo was familiar with the Scotist theory of *intensio* and *remissio*, so much so that when he examined the ontological problem of increases in intensive magnitudes in his *Juvenilia*, he pronounced in favor of an augmentation by the addition of parts. In short, was Galileo's undoubted advance not the result, in large part, of a fusion of what had been two distinct branches of traditional philosophy?

There are several reasons for rejecting this interpretation without the least hesitation. Apart from its highly conjectural character, it turns Galileo's original mind into that of a mere commentator who, looking back over three centuries, sees what advantages might have sprung from a synthesis of past ideas and then acts accordingly. Now there is nothing to show that Galileo looked upon the theory of *latitudines* and the Scotist analysis of *intensio* and *remissio* as two related, and hence easily combined sets of ideas. In fact, the contrary was just as likely, and Galileo's position in this respect was perhaps no different from that of his fourteenth-century precursors. Moreover, the alternative view has the tremendous disadvantage that, in trying to reconstruct the genesis of a concept with its help, we totally destroy the meaning of that concept. The essential point to remember is that Galileo, in his definition of naturally accelerated motion, was the first to arrive at a clear idea of acceleration and the first to interpret it as the rate of change of speed. Any attempt to account for this step in purely historical terms means reducing one of the cornerstones of modern science to mere hindsight and book learning. The new step called for the kind of intellectual initiative which Galileo alone was able to provide when he introduced the idea of an elementary amount of speed and then based his reconstruction of naturally accelerated motion on that idea. To detract from the importance of this step is not to add greater verisimilitude to the presentation of the facts but to attempt the impossible task of constructing a concept on a totally artificial basis and in the final analysis to place all ideas on a par. This is precisely the attitude Galileo ridiculed in the *Dialogue*, when he referred to those Peripatetics who twisted quotations in so ingenious a way as to claim as their own all modern inventions, including the telescope.

We may sum it all up by saying that Galileo's theory of naturally accelerated motion marked a tremendous advance over the traditional theory of uniformly difform motion in that it substituted a genetic and quantitative approach for a purely classificatory and qualitative one. In particular, it replaced the idea that intensifications in speed resulted from successive changes (be they regular or irregular) in quality with the idea of increases due to the continuous summation of an elementary quantity. Under these circumstances, does it not seem exceedingly odd that Galileo should have retained the medieval interpretation of speed as an intensive magnitude,[39] apparently seeing no incompatibility between this interpretation and his own conception? Why, one wonders, did he persist in treating speed as a quasi-physical magnitude even while developing a theory of motion untrammeled by the traditional restrictions? And was his combined use of two distinct representations of speed a mere accident, or was it rather a basic element of his geometrization of naturally accelerated motion? The answer can be discovered only in the way Galileo proceeded from his original definition to the enunciation of his theorems and propositions.

The Geometrization of Naturally Accelerated Motion

One law, the "square law," presided over Galileo's treatment of the motion of heavy bodies. It figured prominently in his proof of most of the subsequent theorems (notably in the form of Corollary II)[40] and may thus be called the cornerstone of the entire Galilean edifice. Hence, there is no better way of illustrating and appreciating the geometrization Galileo attempted in the *Discourses* than to examine the proof of the very proposition on which it was based. The square law was deduced from two other laws: Theorem IV on uniform motion, which states that "if two particles are carried with uniform motion but each with a different speed, the distances covered by them during unequal intervals of time bear to each other the compound ratio of the speeds

[39]In the first extant version of his definition of naturally accelerated motion (which probably dates back to the years 1610–1611; cf. Vol. II, p. 263) no less than some twenty-five years later in the *Discourses*.

[40]See footnote 3 of this chapter.

of the time intervals,"[41] and Theorem I on naturally accelerated motion, which states that "the time in which any space is traversed by a body starting from rest and uniformly accelerated is equal to that in which that same space would be traversed by the same body at a uniform speed whose degree of velocity is one-half of the highest and last degree of velocity of the original uniformly accelerated motion." The first of these two laws poses no special problems, and it is to the second that we must look if we wish to discover the precise way in which Galileo introduced the geometrization of naturally accelerated motion.[42]

THEOREM I—PROPOSITION I

The time in which any space is traversed by a body starting from rest and uniformly accelerated is equal to the time in which that same space would be traversed by the same body moving at a uniform speed whose degree of velocity is one-half of the highest and last degree of velocity of the original uniformly accelerated motion.

Let us represent by the line *AB* the time in which the space *CD* is traversed by a body which starts from rest at *C* and is uniformly ac-

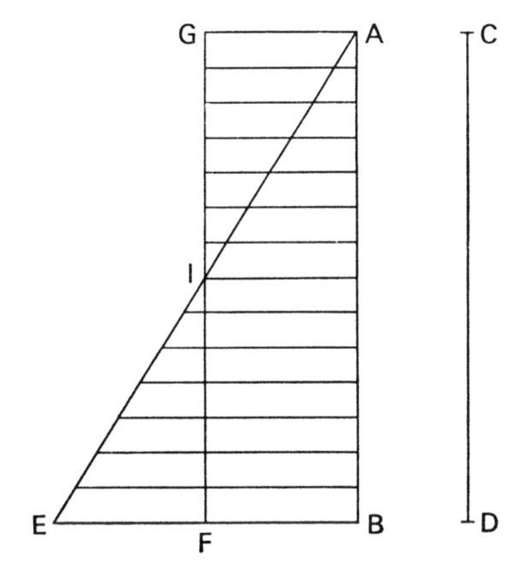

celerated; let the final and highest degree of velocity gained during the interval *AB* be represented by the line *EB* drawn at right angles to *AB*; draw the line *AE*, then all lines drawn from particular points on *AB* and parallel to *BE* will represent the increasing degrees of velocity after the initial instant *A*. Let the point *F* bisect the line *EB*; draw *FG* parallel

[41] *Discourses* III, p. 194; if S_1, V_1, T_1 represent the distance, velocity, and time, respectively, of the first motion, and S_2, V_2, T_2 those of the second motion, we have $S_1/S_2 = V_1/V_2 \times T_1/T_2$.

[42] Ibid., pp. 208–209.

to *BA*, and *GA* parallel to *FB*, thus forming a parallelogram *AGFB* which will be equal in area to the triangle *AEB*, since the side *GF* bisects the side *AE* at the point *I*; for if the parallel lines in the triangle *AEB* are extended to *GI*, then the sum of all the parallels contained in the quadrilateral is equal to the sum of those contained in the triangle *AEB*; for those in the triangle *IEF* are equal to those contained in the triangle *GIA*, while those included in the trapezium *AIFB* are common. Since each and every instant of time in the time interval *AB* has its corresponding point on the line *AB*, from which points parallels drawn in and limited by the triangle *AEB* represent the increasing degrees of the growing velocity, and since parallels contained within the rectangle represent as many degrees of nonincreasing but constant velocity, it appears that there are as many moments of velocity consumed in the accelerated motion represented by the increasing parallels of the triangle *AEB* as in the uniform motion represented by the parallels of the rectangle *GB*. For, what the moments may lack in the first part of the accelerated motion (the deficiency of the moments being represented by the parallels of the triangle *AGI*) is made up by the moments represented by the parallels of the triangle *IEF*.

Hence it is clear that equal spaces will be traversed in equal times by two bodies, one of which, starting from rest, moves with a uniform accelerated motion, while the other, moving with a uniform motion, has a moment one-half of the highest velocity of accelerated motion. Q.E.D.[43]

---

[43]On the basis of this and of Theorem IV (on uniform motion) Galileo had no difficulty in proving the square law.

Let the line *AB* represent the lapse of time measured from the initial instant *A* and the lines *AD*, *DE* represent any

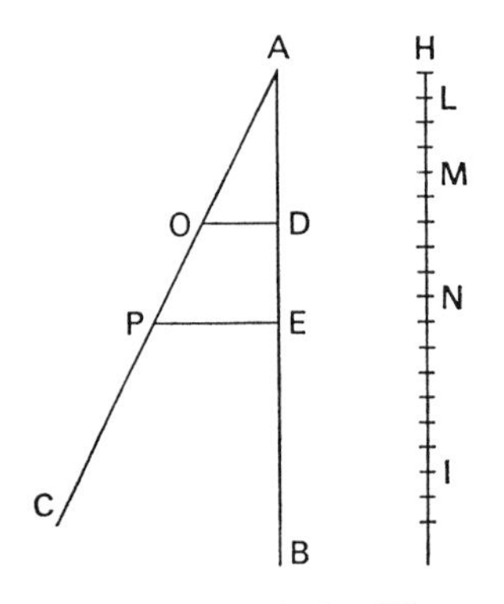

two time intervals; let *HI* represent the distance through which the body, starting from rest at *H*, falls with uniform acceleration. Let *HL* represent the space traversed during the time interval *AD*, and *HM* that covered during the interval *AE*. Draw the line *AC* at any arbitrary angle from the line *AB*; from the points *D* and *E*, draw the parallel lines *DO* and *EP;* of these two lines, *DO* represents the greatest velocity attained during the interval *AD*, that is, after the fall through the distance *HL*, and *EP* the greatest velocity attained during the interval *AE*, that is, after the fall through the distance *HM*.

Let V_1 be the velocity of the uniform motion that can be substituted for the accelerated motion along *HL;* it follows from Theorem I that $V_1 = DO/2$. Similarly, if V_2 is the velocity of the uniform motion that can be substituted for the uniformly accelerated motion along *HM*, then $V_2 = EP/2$. Therefore $V_2/V_1 = EP/2 : DO/2 = EP/DO$; and since $EP/DO = AE/AD$ (per construction), we have $V_2/V_1 = AE/AD$.

Let us now apply Theorem IV on uniform motion: Space *HM*/Space *HL* $= V_2/V_1 \times T_2/T_1$ (where T_2 is the time interval *AE*, and T_1 the time interval *AD*). Hence Space *HM*/Space *HL* $= EP/DO \times AE/AD = AE/AD \times AE/AD = T_2^2/T_1^2$. Q.E.D.

This type of proof poses a conceptual problem: precisely on what representation of speed did Galileo base his argument? More precisely, by what concepts did he succeed in linking speed to the space traversed?

To answer these questions, we must look more closely at the way in which he described speed increases. He assumed, first of all, that from different points on the line *AB* it is possible to draw lines parallel to *AE* and that, since each of these lines represents an increasing degree of speed, it follows that at each point on *AB* an instantaneous magnitude or intensity can be associated with the speed. But the number of points on *AB* and hence the number of parallel lines that can be constructed is infinite; moreover, a new degree can always be interposed between two adjoining degrees however close to each other; hence the speed must increase in a continuous fashion. In other words, the representation of speed Galileo used in Theorem I on naturally accelerated motion was identical with the fourteenth-century conception of speed as an intensive magnitude. Let us now examine how Galileo passed on from the speed to the space traversed. From *F*, the center of *BE*, draw *FG* parallel to *BA*, and from *A* draw *GA* parallel to *FB*: "The parallelogram *AGFB* will be equal in area to the triangle *AEB* since the side *GP* bisects the side *AE* at the point *I;* for if the parallel lines in the triangle *AEB* are extended to *GI*, then the sum of all the parallels contained in the quadrilateral is equal to the sum of those contained in the triangle *AEB*." Now the sum of all the lines contained in the triangle *AEB* represents the increasing degrees of an increasing speed, whereas that of all the lines contained in the parallelogram *AGFB* represents the values of a speed that is not increasing. It follows that the sum of the degrees in either case will be identical and that, since the distances traversed are proportional to the speeds, equal spaces will be traversed in equal times by a body moving with a uniform acceleration and another moving with uniform speed whose degree is half the maximum degree attained by the first. Only the summation of the successive intensities thus enabled Galileo to correlate the two motions and to conclude as to the equality of the spaces traversed. In other words, in his proof of Theorem I, Galileo not only treated speed as an intensive magnitude but also made use of the Oresmian concept of *quantitas velocitatis*.

One might be tempted to reduce the importance of this conclusion with the observation that in Theorem I Galileo was merely paying lip service to a traditional principle, the "law of the median degree" known to all fourteenth-century physicists. A careful perusal of the *Discourses*, however, will show that he used the Oresmian approach and hence continued to represent speed as an intensive magnitude on several other occasions as well. Thus in Corollary I to Theorem II, having first established the square law and having shown that "the increments in the distances traversed during equal time intervals are to one another as the odd numbers beginning with unity," Galileo had Sagredo explain matters in a way that Oresme would have endorsed without the least reservation:

Sagredo: Please suspend the discussion for a moment since there just occurs to me an idea which I want to illustrate by means of a diagram in order that it may be clearer both to you and to me.

Let the line *AI* represent the lapse of time measured from the initial instant *A*; through *A* draw the straight line *AF* making any angle whatever; join the terminal points *I* and *F*; divide the time *AI* in half at *C*; draw *CB* parallel to *IF*. Let us consider *CB* as the maximum degree of the velocity which increases from zero at the beginning, in simple proportionality to the intercepts on the triangle *ABC* of lines drawn parallel to *BC*; or what is the same thing, let us suppose the velocity to increase in proportion to the time; then I admit without question, in view of the preceding argument, that the space described by a body falling in the

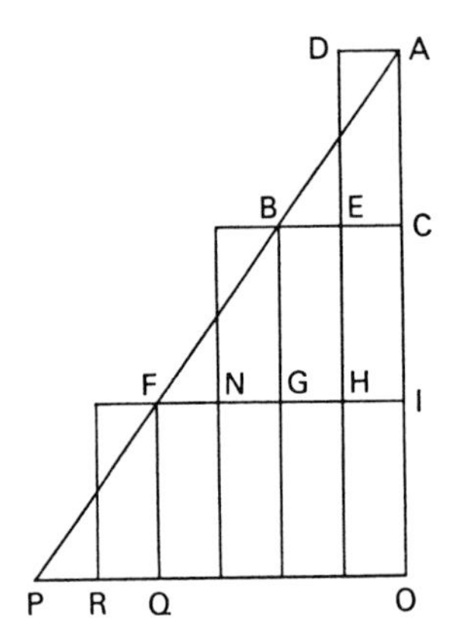

aforesaid manner will be equal to the space traversed by the same body during the same length of time traveling with a uniform speed equal to *EC*, the half of *BC*. Further let us imagine that the body has fallen with accelerated motion so that at the instant *C* it has the degree of velocity *BC*. It is clear that if the body continued to descend with the same degree *BC* without acceleration, it would in the next time interval *CI* traverse double the distance covered during the interval *AC*, with the uniform degree *EC* which is half of *BC*; but since the falling body acquires equal increments of speed during equal increments of time, it follows that the degree *BC* during the next time interval *CI* will be increased by these moments of velocity increasing as the parallels of the triangle *BFG* which is equal to the triangle *ABC*. If, then, one adds to the degree of velocity *GI* half of the degree of velocity *FG*, the highest degree acquired by the accelerated motion and determined by the parallels of the triangle *BFG*, he will have the uniform degree of velocity with which the same space would have been described in the time *CI*; and since this degree *IN* is three times as great as *EC*, it follows that the space described during the interval *CI* is three times as great as that described during the interval *AC*. Let us imagine the motion extended over another equal time interval *IO*, and the triangle extended to *APO*; it is then evident that if the motion continues during the interval *IO* with the degree *IF* acquired by acceleration during the time *AI*, the space traversed during the interval *IO* will be four times that traversed during the first interval *AC*, because the degree *IF* is four times the degree *EC*. But if we enlarge our triangle so as to include *FPQ*, which is equal to *ABC*, still assuming the acceleration to be constant, we shall add to the uniform speed an increment *RQ*, equal to *EC*; then the value of the equivalent uniform speed during the time interval *IO* will be five times that during the first time interval *AC*; therefore the space traversed will be quintuple that during the first interval *AC*.

Clearly, it was only by identifying the increase in speed during the time intervals *CI* and *IO* with the areas of the triangles *BFG* and *FPQ* that Sagredo was able to compare the magnitude of the spaces traversed during either time interval with the space traversed during the interval *AC*. Now this identification presupposed that the successive increases in speed during the intervals *CI* and *IO* could be represented by lines drawn parallel to the base in the triangles *BFG* and *FPQ*. In other words, each of these triangles was thought to represent the sum of the increases in speed during one of the time intervals, so that the concept of *quantitas velocitatis* did indeed constitute the crux of the whole argument. Let us now examine how the same concept was applied in Theorem III.[44]

THEOREM III—PROPOSITION III

If one and the same body, starting from rest, falls along an inclined plane and also along a vertical, each having the same height, the times of descent will be to each other as the lengths of the inclined plane and the vertical.

Let *AC* be the inclined plane and *AB* the perpendicular, each having the same vertical height above the horizontal, namely, *BA*; then I say,

---

[44]*Discourses* III, pp. 215–217.

the time of descent of one and the same body along the plane *AC* bears a ratio to the time of fall along the perpendicular *AB*, which is the same as the ratio of the length *AC* to the length *AB*. Let *DG*, *EI*, and *LF* be any lines parallel to the horizontal *CB*; then it follows from what has preceded[45] that a body starting from *A* will acquire the same degree of velocity at the point *G* as at *D*, since in each case the vertical fall is the same; in like manner the degrees at *I* and *E* will be the same; so also those at *L* and *F*. And in general the moments or degrees of velocity at the two extremities of any parallel drawn from all the points on *AB* to the line *AC* will be equal.

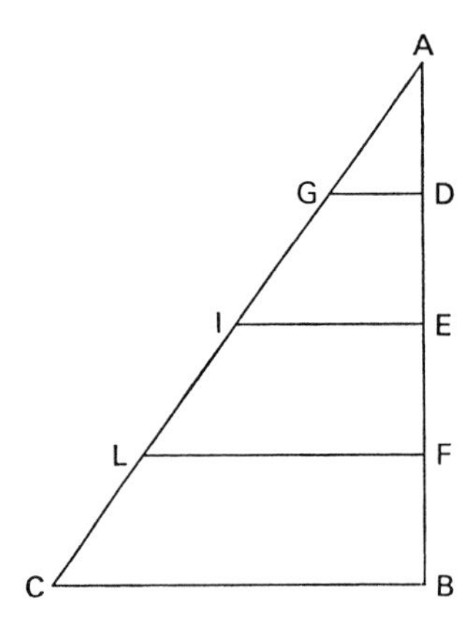

Thus the two distances *AC* and *AB* are traversed with the same degrees of velocity. But it has already been proved that if two distances are traversed by a body moving with the same degrees of velocity, then the ratio of the times of descent will be the ratio of the distances themselves; therefore, the time of descent along *AC* is to that along *AB* as the length of the plane *AC* is to the vertical distance *AB*. Q.E.D.

The aim of this proof was to show that the two motions can be treated as two uniform motions with the same velocity but over two distances of unequal length. To that end, Galileo considered first the set of the degrees of velocity through which each moving body had to pass. These degrees could be represented by the set of the parallels drawn "from all the points on *AB* to the line *AC*." But the speeds at the extremities of each of these parallels are equal (by virtue of the general principle stated at the beginning of the Third Day); so that the two distances *AB* and *AC*

[45]Namely, the principle to which we have already alluded and on which Galileo claimed he had constructed his entire theory of the motion of heavy bodies: "The speeds acquired by one and the same body moving down planes of different inclinations are equal when the heights of these planes are equal." This principle will be examined at length in Chapter 7.

are traversed *with the same degrees of speed.* And precisely because they possess the same quantity of speed, the two motions along the lines on *AB* and *AC* can be likened to two uniform motions with the same speed, whose respective times of descent are in the same ratio as the distances.

It is only fair to point out that neither Corollary I to Theorem II nor Theorem III was deduced exclusively from the assumption that speed is an intensive magnitude or from the concept of *quantitas velocitatis;* thus the former was first based directly on the square law,[46] and, as Galileo himself justly remarked, Theorem III "could have been proved clearly and briefly on the basis of a proposition already demonstrated, namely, that the distance traversed in the case of accelerated motion on *AC* or *AB* is the same as that covered by the uniform speed whose value is one half the maximum speed *CB*; the two distances *AB* and *AC* having been traversed at the same uniform speed, it is evident from Proposition I that the times of descent will be to each other as the distances."[47] In fact, it is most probable that the second version of Corollary I, no less than the proof of Theorem III, goes back to the early stages of Galileo's analysis of the motion of heavy bodies; in that case both would tend to show that during this early phase, which called for maximum inventiveness, Galileo made free use of fourteenth-century concepts to shore up and develop his own ideas—clear proof that these concepts retained much of their heuristic value three centuries after they were first elaborated.

But no matter what view we take of the matter, one thing is quite certain: the intervention of medieval concepts in these early attempts at geometrization forces our attention on the problem of Galileo's double approach to speed. We saw that in his definition of naturally accelerated motions he reconstructed the increase in speed with the help of an elementary quantity or *momentum velocitatis*, thus treating acceleration as a rate of change in speed. Yet when he based his geometrization on concepts framed by the Mertonians and Parisians, he had perforce to abandon his earlier definition, and with it the new ideas it introduced. It would therefore seem that his analysis of changes in speed must have been a mere prelude to—but not the direct basis of—his study of naturally accelerated motion. In particular, while his definition of accelerated motion was already of a piece with modern mechanics, the representa-

---

[46]*Discourses* III, p. 211.

[47]Ibid., pp. 218–219.

tion of speed constituting the veritable foundations of his geometrization had no place in that mechanics. And once we see matters in this light, are we not posing the problem of the relationship between Galileo's contribution and the fourteenth-century tradition in the sharpest possible way? True, we have seen that Galileo's approach to increases in speed was anything but a mere extension of the medieval theory of uniformly difform motion, but this observation does not remove all the difficulties. For are not the undeniable links between Theorem I and the fourteenth-century procedures an invitation to restore through the techniques a patently lacking continuity in the general analytic approach? In other words, and though Galileo had a much deeper understanding of the processes of variation than either Heytesbury or Oresme, did the filiation of ideas not manifest itself quite spontaneously as soon as it came to the exposition of the relevant laws and propositions?

That Galileo did not deduce his kinematics from his own skillful definition of naturally accelerated motion is an indisputable fact. Does that mean that his approach was self-contradictory? In fact, a moment's thought will show that he could not have reacted otherwise. We saw that Galileo's idea of acceleration can be expressed by the differential coefficient $\gamma = dv/dt$, so that the speed v of a freely falling body at any given instant can be represented by

$$v = \int_{t_0}^{t} \gamma dt = \gamma t + C.$$

However, a double step was needed to proceed from this representation to the square law. First of all, the speed at any given instant had to be expressed in the form of the differential coefficient $v = ds/dt$, that is, as the ratio of an infinitely small distance to an infinitely small time interval; second, the differential equation $ds = v\,dt$ had to be integrated.[48] In other words, in order to take the double step that alone would have brought out the full consequences of his definition, Galileo would have had to be familiar with the differential calculus. Now, while he undoubtedly felt the need for this type of mathematical representation (we

[48]That is,

$$S = \int_{t_0}^{t} V\,dt = \int_{t_0}^{t} \gamma t\,dt = \tfrac{1}{2}\gamma t^2 + C_1 t + C_2,$$

where C_1 and C_2 are two constants of integration defined by the initial conditions of the motion.

shall see later that the mathematical method he applied in Theorem I meant just that), he neither constructed it nor even suspected that such a construction was possible. In its absence he had to rely on the tools of conceptual analysis; to go beyond that and to relate the distance traversed to changes in speed, he would have had to be a contemporary of Newton and not of Cavalieri.

Having made this point, we can now return to Galileo's geometrization of the motion of heavy bodies. Though he failed to deduce the laws of naturally accelerated motion directly from his definition, the latter nevertheless had an important part to play, since one of Galileo's most fruitful ideas, that speed increases as time, provided a crucial element in the proof of Theorem I. Hence we are entitled to claim that his definition entered into his analysis, though of course not as effectively as it would have had he been able to apply the appropriate mathematical techniques. But the real problem resides elsewhere. Fourteenth-century studies of difform motion had two main aims: first of all, to construct an accurate description of variations in speed, and second, to evaluate the effects of these variations on the distance traversed or, as they themselves put it, on the quantity of perfection attained by the moving body. While all of them used the same artifice (the comparison of difform with uniform motion), most of them proceeded by trial and error and above all considered the result an end in itself. Galileo, by contrast, immediately applied the fundamental law, which in uniformly accelerated motion connects space with time, to the case of an unlimited extension of that motion; second—and this was the crucial difference between him and his precursors—instead of considering this law to be the sole aim of his analysis, he treated it as a mere stepping stone. For hardly had he succeeded in mathematizing the motion of heavy bodies than he proceeded to the systematic study of that motion in the most diverse situations: descent along inclined planes of different lengths or inclinations; descent along inclined planes drawn from the highest or lowest point in a vertical circle constructed on a horizontal line; free fall combined with descent along an inclined plane; descent along curved trajectories; and projectile motion. In each case the formulation of theorems and propositions and the rigorously deductive organization of the exposition make it clear beyond all doubt that Galileo's approach was quite unlike the medieval one. Instead of focusing attention on a single problem, he engaged in a methodic and authentically mathematical study of that rational object into which he had transformed local motion. A difference in kind and not simply in degree thus separated the *Discourses* from all fourteenth-century writings.

We are now in a much better position to appreciate why Galileo used two representations of speed at once: first, a definition of naturally accelerated motion containing virtually all the laws of that motion; second, an unprecedentedly clear analysis of the properties of motion, conducted *more geometrico*. The gap between the two could not be closed until mathematics caught up with the resources of conceptual analysis. In other words, the main reason why Galileo adopted certain of the traditional concepts was that he still found it impossible to couch his own definition in adequate mathematical terms.[49] This failure largely explains why he continued to treat speed as an intensive magnitude and why he was led back to the medieval conception of instantaneous speed; it also explains why he reintroduced the concept of *quantitas velocitatis*, which served him in much the same way as infinitesimal analysis would serve his successors. In brief, the representation of speed as an intensive magnitude no less than the concept of *quantitas velocitatis* provided Galileo with the means of applying on the intuitive plane what his lack of mathematical resources prevented him from doing on the plane of ra-

---

[49]In particular, dependent as he was on the resources of traditional geometry, Galileo was unable to express speed with the help of algebraic symbols. True, he arrived at a rigorous definition with the help of proportional terms, for which purpose it is not enough to define the equality of speeds by saying that two bodies traverse equal distances in equal time; the correct definition must specify that the distances traversed have to be in the same ratio as the times it takes to traverse them (in other words, it is not enough to put s_1 equal to s_2, and t_1 equal to t_2, but we must also put s_1/s_2 equal to t_1/t_2, for s_1 does not necessarily have to be equal to s_2; cf. *Discourses* III, p. 191, and *Dialogue* I, p. 48); but that definition failed to turn speed into a mathematical object that could be expressed directly by a mathematical symbol such as v and by means of which new propositions and formulas could be expressed (when Galileo speaks of the ratio of two speeds, he is, in fact, referring to the ratio of two times or two distances). That is, though he could define speed in the language of traditional geometry, he was not yet able to turn it into a distinct mathematical entity, and this was so, first of all, because, like the Greeks, he did not consider that continuous magnitudes were comparable to numbers except in the special case of magnitudes that were multiples of one and the same aliquot part. We know that the Greeks succeeded in developing rigorous geometrical methods of proof only by taking the same view and that Euclid in particular asserted that two incommensurable magnitudes cannot have the ratio of one number to another (*Elements* X, 7). It was Descartes who first dispensed with this restriction, thus paving the way for a new mathematical language suited to the needs of modern physics (cf. P. Boutroux, *Les principes de l'analyse mathématique*, Vol. I, pp. 121 ff.).

tional analysis. Hence, if traditional ideas continued to play an important part in Galileo's work, they did so largely for technical reasons.[50]

However, this is not the complete explanation, for the mathematical procedure on which Galileo based the proof of his first theorem shows that the use of medieval concepts also reflected the wish to present the properties of naturally accelerated motion in a manner that was both as direct as possible and also in the fullest accord with the spirit of his definition. Let us look once more at the guiding idea of that proof: the reason why a body in uniformly accelerated motion and a second body in uniform motion whose speed is half the maximum speed of the first cover the same distances in equal time is that the amounts of speed involved in either of the two motions are equal. To prove this equality, Galileo had to assume that the sum of the instantaneous speeds associ-

---

[50]A comparison of Galileo's approach with that of Leibniz is highly illuminating. In his later work on the "natural" motion of heavy bodies (that is, those written from 1674 onward) Leibniz made direct use of Galileo's definition of naturally accelerated motion. According to M. Guéroult (*Dynamique et métaphysique leibniziennes*), he based his analysis on two main concepts: *conatus* and *impetus*. Representing as it did the quantity by which the acquired speed varies from one instant to the next, the term *conatus* referred to the elementary acceleration or increment of speed impressed upon a heavy body in an infinitesimally small time interval (Guéroult, p. 35). On this assumption, the speed acquired at the end of a given time interval can clearly be represented by the summation of the increments of speed "impressed successively on the moving body" (ibid). Now this summation also defines the body's *impetus*, that is, its real speed after a given time. In analytical language, if we express the *conatus* by $\gamma\, dt$, then the *impetus* becomes

$$\int_{t_0}^{t} \gamma\, dt = \gamma t = V + C$$

(C being a constant specified by the initial conditions of the motion). Leibniz's concepts were therefore none other than Galileo's, but revised with the help of infinitesimal calculus. The elementary increment of speed, whose continuous addition ensured the uniform acceleration of a motion in Galileo's definition, was given the status of a differential in Leibniz's; similarly, the latter's treatment of *impetus* as an integral merely gave formal expression to what Galileo had referred to as the degree of speed at each instant of the motion. Infinitesimal calculus enabled Leibniz to round off Galileo's analysis of uniformly accelerated motion, while Galileo, lacking the appropriate mathematical tools, was unable to pass directly from his analysis of the mode of speed increases to the determination of the distances traversed. Leibniz, for his part, had no difficulty in taking this step; to that end he had merely to treat the distance traversed during a given time interval as the integral of the *impeti*, that is, to state that

$$S = \int_{t_0}^{t} \gamma t\, dt = \tfrac{1}{2}\gamma t^2 + C_1 t + C_2.$$

ated with the accelerated motion could be represented by a triangle, and the sum of the instantaneous speeds associated with the uniform motion by a rectangle. In other words, he had to assume that an area can be treated as a sum of lines, and a line as a sum of points. An interpretation of continuous magnitudes involving the hypothesis of indivisibles, together with the treatment of speed as an intensive magnitude and the concept of *quantitas velocitatis*, thus formed the core of his demonstration. Since that approach was by no means new, we consider it useful to recall its role in the traditional analysis of continuous magnitudes and to explain briefly why, despite the problems it posed, Galileo considered it a possible basis for a logically satisfactory argument.

While the term "indivisible" seems to have been coined by Bradwardine, the concept itself probably made its appearance at the beginning of the fifth century B.C. The postulate that every continuous magnitude consists of denumerably finite parts, however small, had two aims: first, to restore the harmony between number and continuous magnitudes which had been destroyed by the discovery of irrational magnitudes; and second, to justify such procedures as the determination of limiting values which at first sight seemed to have no basis in reality.[51] But while it was useful (which explains its survival until the seventeenth century), the hypothesis of indivisibles was also beset with clear contradictions, as Zeno's paradoxes proved only too plainly. In fact, the idea of indivisibles can be defended only on the grounds that it is possible to put an end to the infinite divisibility of the continuum at a certain moment; this assumption poses a dilemma whose two horns are equally sharp. Since intuition tells us that every continuum can be divided ad infinitum, the indivisible parts of which it is allegedly composed must necessarily be without magnitude: but in that case the continuum, conceived as the sum of indivisible parts, would have to be lacking in magnitude as well. It may perhaps be objected that indivisibles do have a certain magnitude, but in that case all magnitudes must be infinitely large, since an infinite number of finite magnitudes, no matter how small, cannot be anything but infinite. Hence Zeno's conclusion that, if the addition of an indivisible element to, or its subtraction from, a magnitude neither increases nor diminishes it in any way, then that infinitely small element

[51]Democritus, who is considered one of the founders of this type of mathematical atomism, may have applied it to the determination of the volumes of various solids; cf. T. L. Heath, *A History of Greek Mathematics*, Vol. I, p. 180.

must be absolutely nothing. Greek mathematicians fully accepted this conclusion; we know that Eudoxus, and later Euclid, endorsed the fact that continuous magnitudes cannot be reduced to discontinuous ones, Euclid making it the subject of the famous proposition with which he opened Book X of his *Elements*. Aristotle, too, used the same approach when he argued that the continuum is divisible into divisible parts, and that the indivisible can refer only to the nonextended limit of a finite magnitude. Hence, if the continuum embraces infinity, it does so potentially, not actually, and though this interpretation abandoned the continuum to intuition, it was nevertheless perfectly coherent.

Now there is no reason to suppose that Galileo was not familiar with these arguments, for not only had all the great medieval writers accepted it and written commentaries upon it,[52] but Galileo himself had an intimate and direct knowledge of Greek geometry and of its fundamental ideas.[53] Moreover, the manner in which he posed the problem during the First Day of the *Discourses* shows that he was fully aware of the difficulties inherent in the concept of indivisibles. Thus, if the continuum is indeed made up of indivisible parts, then a continuous magnitude, which by definition is divisible, cannot possibly be engendered from a finite number of indivisibles. "If two indivisibles, say, two points, can be united to form a quantity, say, a divisible line, then an even more divisible line might be formed by the union of three, five, seven, or any other odd number of points. Since, however, these lines can be cut into two equal parts, it becomes possible to cut the indivisible which lies exactly in the middle of the line."[54] And since the parts composing a continuum are necessarily infinite in number, they cannot be finite in size "because an infinite number of finite quantities would give an infinite magnitude."[55] The conclusion is therefore quite obvious: to assume that the continuum is built up of indivisibles is tantamount to assuming that every continu-

[52]Thus they claimed that a syncategorematic but no categorematic meaning could be assigned to infinity. It should, however, be mentioned that the idea of indivisibility and of an actual or completed infinity had several champions in the Middle Ages (especially Gregory of Rimini). See Duhem, *Système du monde*, Vol. VII, pp. 111 ff., and particularly A. Maier's more objective account in *Die Vorläufer Galileis*, pp. 196 ff.

[53]Thus he explained in his *Saggiatore* (p. 327) that "omnis ratio est duarum magnitudinum ejusdem generis"; this is not compatible with the hypothesis of indivisibles because a point does not belong to the same genus as a line, nor a line to the same genus as an area.

[54]*Discourses* I, p. 77; much the same argument was used by Pascal (*Discours sur l'esprit géométrique*).

[55]Ibid., p. 80.

ous magnitude consists of an infinite number of elements without size. But this brings us back to Zeno's paradox, and one wonders why Galileo saw fit to retain a hypothesis with whose drawbacks he was as familiar as anyone else.

The question is the more significant in that Galileo's definition of the continuum seemed to lead to absurd consequences almost as soon as it was formulated. Thus the assertion that every continuum is composed of an infinite number of parts without magnitude seems to make nonsense of the fact that one and the same class can comprise continuums of different size. "Since it is clear that we may have one line greater than another, each containing an infinite number of points, we are forced to admit that within one and the same class we may have something greater than infinity, because the infinity of points in the long line is greater than the infinity of points in the short line."[56] The assumption thus led to an impasse, for either we try to explain the existence, within one and the same class, of continuums of different size, in which case we are led to the absurd conclusion that "there is something greater than infinity," or we renounce but are unable to account for the continuum *sub specie quantitatis*. "This is one of the difficulties," Galileo continued, "which arises when, with our finite minds, we attempt to discuss the infinite, assigning to it those properties which we give to the finite and limited; but this, I think, is wrong, for we cannot speak of infinite quantities being the one greater or less than or equal to another."[57]

However, this objection to the hypothesis of indivisibles is perhaps less crucial than it seems. For if we consider the case of integers, we might similarly be tempted to assume that all numbers, including both squares and nonsquares, are more than the squares alone.[58] Yet on further examination, we shall discover that these are as many as the corresponding number of roots, since every square has its own root and every root its own square, while no square has more than one root and no root more than one square.[59] If we go on to inquire how many roots there are, we shall find that there are as many as there are numbers, because every number is the root of a square. Contrary to our first as-

[56]Ibid., p. 77.
[57]Ibid., pp. 77–78.
[58]Ibid.
[59]Ibid. This result is even more surprising since, as Galileo pointed out, the proportionate number of squares diminishes as we pass to higher numbers; thus up to 100 the squares constitute one-tenth of all the numbers; up to 10,000 we find only one in 100 to be squares, and up to 1,000,000 only one in 1000.

sumption, we must therefore admit that there are as many squares as there are numbers because they are just as numerous as their roots, and all numbers are roots. Thus the idea of the whole and its parts ceases to apply to the system of numbers as soon as we pass from the finite to the infinite; there is no difference in power between the set of integers and its subsets.[60] Now the case of integers helps us surmount the first difficulty we encountered when attempting to adduce a valid definition of the continuum. We see now that to imply that every continuum consists of an infinite number of nonfinite parts does not necessarily condemn the idea of indivisibles, since the paradoxical consequences of this assertion are no different from those inherent in the system of numbers we use every day and with absolute certainty. Hence, if we are told that a greater line must contain a greater number of indivisibles than a shorter line, we can point out that there are as many square numbers as there are integers, although all integers are, of course, not squares. Is the difference between a longer line and two shorter lines really any more difficult to grasp than the difference between the set of integers, the set of squares, and the set of cubes? Hence, the objection, real enough though it is, by no means invalidates the general argument. "When therefore Simplicio introduces several lines of different lengths and asks me how it is possible that the longer ones do not contain more points than the shorter, I answer him that one line does not contain more, or less, or just as many points as another, but that each line contains an infinite number. Or if I had replied to him that points in one line were equal in number to the squares; in another greater than the totality of numbers; and in the little one, as many as the number of cubes, might I not, indeed, have satisfied him by thus placing more points in one line than in another and yet maintaining an infinite number in each? So much for the first difficulty."[61]

[60]Such is the modern interpretation of Galileo's remarks. This remarkable passage in the *Discourses* was one of the first valid contributions to the mathematical study of infinity, cf. A. A. Fraenkel, *Abstract Set Theory*, p. 40. Galileo himself was, of course, unable to distinguish between a denumerably infinite or countable set, such as the set of integers, and an uncountable set, such as the set of real numbers or the set of points on a line. One might say, on the basis of Galileo's text, that the error of the hypothesis of indivisibles lay chiefly in its attempt to treat the continuum as a countable set.

[61]*Discourses* I, p. 79. Galileo also tried to justify the existence of indivisibles with pure sophisms, including "Aristotle's wheel" and "the paradox of the bowl." In Aristotle's wheel the hypothesis of indivisibles is used to explain how a smaller and a larger circle can cover the same distance in the same time

But did Galileo really do more than "diminish one improbability by introducing a similar or a greater one"?[62] For however pertinent his arguments may have been, they did not yet justify the concept of indivisibles or prove that his conception of the continuum was superior to Aristotle's. Indeed, that proof was exceedingly hard to adduce, since the most plausible answer to the question "Are the finite parts of a continuum finite or infinite in number?" seems to be that "their number is both infinite and finite; potentially infinite but actually finite; that is, potentially infinite before division and actually finite after division; because parts cannot be said to exist in a body which is not yet divided or at least marked out; if this is not done, we say that they exist potentially."[63] That being the case, why did Galileo see fit to improve upon that solution?

Consider a continuous magnitude of finite dimensions, say, a line twenty spans long; according to Aristotle and the Peripatetics, the line does not actually contain twenty lines, each one span long, except after division; before division it contains them only potentially. However, once the division has been made, the size of the original line is neither increased nor diminished—"the finite parts in a continuum, whether actually or potentially present, do not make the quantity either larger or smaller."[64] If, for a finite number of finite parts, we now substitute an infinite number of finite parts actually contained in the whole, it is obvious that only an infinite magnitude can contain an actually infinite number of parts. But what if the infinite number of finite parts is con-

(ibid., pp. 68–72); the paradox of the bowl was the basis of Galileo's demonstration that if two unequal areas are diminished continuously and uniformly they will eventually be reduced, one to a circumference, that is, to an infinite number of points, and the other to a single point, and that this reduction could be explained only by the hypothesis of indivisibles (pp. 73–76). For these two examples the reader is referred to our "Le Problème du continu et les paradoxes de l'infini chez Galilée," in *Thales*, 1959. It is important to distinguish Galileo's indivisibles from those of Cavalieri, for while the former used them in the construction of continuous magnitudes, the latter completely rejected this view, inter alia, in a letter to Galileo on the subject of the two paradoxes just mentioned (cf. *Opere di Galilei*, Vol. XVI, *Letter to Galileo*, October 2, 1634, pp. 136–138), and he applied the term "indivisibles" exclusively to fixed elements of a given figure, characteristic of its properties and hence particularly suited to its analysis; see A. Koyré's excellent essay "B. Cavalieri et la géometrie des indivisibles" in *Hommage à Lucien Febvre* (Paris, A. Colin, 1954), pp. 319–340 (reprinted in *Études d'histoire de la pensée scientifique*, pp. 297 ff.

[62] *Discourses* I, p. 73.

[63] Ibid., p. 80.

[64] Ibid.

tained only potentially, not actually, in the whole? Shall we then agree with Aristotle that a magnitude can contain a potentially infinite number of parts and yet be finite itself, in other words, that the introduction of potential infinity in no way alters a magnitude considered *sub specie quantitatis*? Galileo, for one, rejected this interpretation, and moreover with the help of a purely Peripatetic argument. For, according to Aristotle, whatever exists in actuality must first have existed potentially; and there is nothing to prevent what exists potentially from becoming actual in due course, so that there is no justification for extending this privilege to certain properties but not to others. Hence, if a magnitude contains infinity as a potential, every Aristotelian must agree that it contains it in actuality as well, that is, that a magnitude can never contain infinity, whether or not in potential only, unless it is infinite itself.[65]

Let us now return to our finite continuum. The mere fact that we can assign to it "a hundred finite parts, a thousand, a hundred thousand, or indeed any number we may please"[66] shows that the number of finite parts contained in it is neither finite nor infinite; that between Aristotle's finite and infinite magnitudes "there exists a third intermediate term" which corresponds to every assigned number. In order that this may be possible, every continuum must thus consist of an infinite number of parts without magnitude, and this is the only reasonable explanation of the infinite divisibility of continuous magnitudes. Admittedly, we can never hope to encounter, let alone prove, the existence of such sizeless parts, no matter how long we continue the division,[67] but it nevertheless remains a fact that the hypothesis of indivisibles is a logical consequence of the infinite divisibility of the continuum: once the latter has been granted, the former cannot be denied. The whole argument was summed up perfectly by Salviati, when he declared that "the very fact that one is able to continue, without end, the division into finite parts makes it necessary to regard the quantity as composed of an infinite number of immeasurably small elements."[68] The way in which Galileo

---

[65]Ibid., pp. 80–81. Galileo's entire criticism of Aristotle's view may have been based on a passage in the *Physics* in which Aristotle contended, during a discussion of infinity, that "any magnitude that can exist potentially can also exist actually" (*Physics* III, 7, 207 b 17–18).

[66]Ibid., p. 81.

[67]Ibid., p. 82.

[68]Ibid., p. 80.

arrived at this conclusion was quite remarkable. Traditionally, as we saw, the concept of indivisibles served two ends: to restore the links between continuous magnitudes and numbers and to justify the use of limits in reducing an unresolved problem to one that had already been resolved (for instance, in the case of quadratures or cubatures). Galileo used quite a different approach. He considered the infinite divisibility of a continuous magnitude and tried to adduce a rational (as opposed to an intuitive) explanation of this phenomenon. Now, posing that problem was as good as solving it: a division and a subdivision that can be carried on indefinitely "presupposes that the parts are infinite in number, otherwise the subdivision would reach an end," and also that the parts are not finite in size, because "an infinite number of finite quantities would give an infinite magnitude."[69] We have already seen that some of the paradoxes to which the idea of indivisibles gave rise were no more baffling than those appearing in the system of integers; we can now add that the hypothesis of indivisibles, far from being irrational, is imposed by the very demands of reason.

We have tried to show how Galileo arrived at this conclusion; more important still, however, his use of indivisibles provided the clearest possible illustration of the demonstrative ideal of his *Discourses*. For it is questionable whether the resources of traditional geometry would have sufficed to prove Theorem I or, indeed, whether a mathematical transcription of the physical situation to which that theorem applied was even possible. Hence, when Galileo represented the distance traversed by a uniformly accelerated body by the area of the triangle *ABC*,[70] and the distance the same body would have traversed moving uniformly with half the terminal velocity of the first motion by the rectangle *GABF*, thus applying the method of indivisibles, may he not have been using what prior to the advent of infinitesimal calculus was the sole method of formulating and then offering a mathematical solution to a problem of this type? By far the best answer to this question, of paramount importance to anyone anxious to appreciate Galilean science in its original form, was Huyghens' attempt to show that Galileo could have proved Theorem I with the resources of traditional conventions of geometry, for that attempt revealed much more clearly than any commentary could have hoped to do what reasons persuaded Gal-

---

[69]Ibid.

[70]See footnote 43.

ileo to have recourse to the method of indivisibles and hence to rely on a fourteenth-century concept.[71]

PROPOSITION V

The space traversed in a given period of time by a body starting from rest is half the space traversed by the same body moving uniformly with the speed of the fall at the end of that period.

Let *AH* be the time of the whole fall, and let the body have traversed a distance whose length is represented by the plane *P*. Draw *HL* at right angles to *AH* and of any desired length, and assume that this length represents the speed at the end of the fall. Complete the rectangle *AHLM* and let it represent the space which the body would have traversed in the time *AH* with the speed *HL*. Prove that the plane *P* has half the area of the rectangle *MH*, that is, that it is equal in area to the triangle *AHL*.

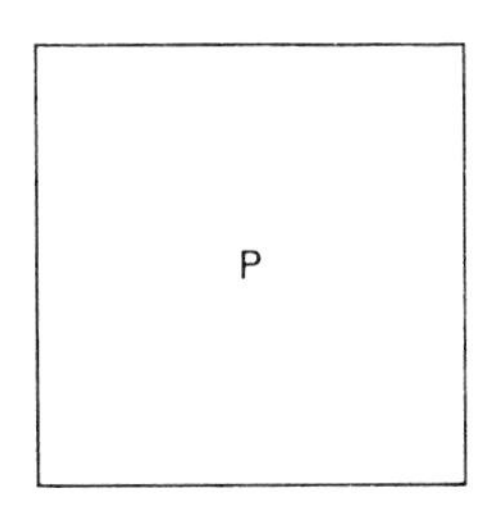

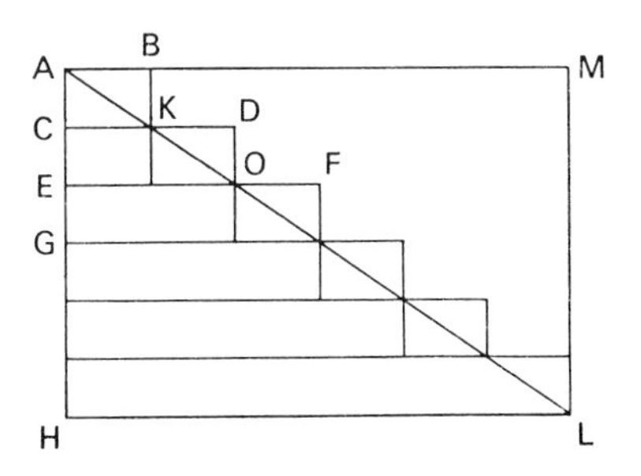

If the plane *P* is not equal to the triangle *AHL*, it must be either smaller or greater. Assume first that it is smaller. Divide *AH* into equal parts *AC*, *CE*, *CG*, etc., in such manner that—when we describe about the triangle *AHL* a figure composed of rectangles whose height is equal to the length of each of these parts of *AH*, that is, the rectangles *BC*, *DE*, and *FG*, and when we inscribe in the same triangle a second figure composed of rectangles of the same height such as *KE*, *OG*, etc.—the excess of the first figure above the second is less than that of the triangle *AHL* above the plane P. This can easily be done, since the total excess of the circumscribed figure above the inscribed figure is equal to a very small rectangle, with *HL* as its base. Hence the excess of the triangle *AHL* above the inscribed figure will certainly be less than its excess above the plane *P*; the figure inscribed in the triangle will therefore be

[71]See Christian Huyghens, *Horologium Oscillatorium*, Part II, Proposition V, *Œuvres*, Vol. XVIII, pp. 137 ff. Huyghens' own, much briefer, proof was presented in Proposition II. Though he disliked lengthy demonstrations, he nevertheless thought them preferable to the approximation involved in the method of indivisibles; cf. *Œuvres*, Vol. XI, p. 158; Vol. XIV, p. 337; Vol. XVI, p. 349.

greater than the plane *P*. Moreover, since the line *AH* represents the time of the whole fall, its equal parts *AC*, *CE*, and *EG* must represent equal parts of time. Now since the speed of the moving body increases in the same ratio as the time of descent,[72] and since the speed acquired at the end of the whole fall is *HL*, the speed acquired at the end of the first part, *AC*, of the time, will be *CK*, because $AH/AC = HL/CK$. Similarly the speed acquired at the end of the second part of the time, *CE*, will be *EO*, and so on. It is clear that during the first part of the time, *AC*, the moving body will have traversed a space greater than zero, and that during the second part of the time, *CE*, it will have traversed a space greater than *KE*, since the space *KE* would have been traversed in the time *CE* by a body moving uniformly with the speed *CK*. In fact, the spaces traversed at a uniform speed are to each other in a ratio composed of the times to the speeds; consequently, it follows from the assumption that a body moving with the constant speed *HL* traverses the space *MH* in the time *AH*, that it will also traverse the space *KE* in the time *CE* with the speed *CK*, considering that the ratio of the rectangle *MH* to the rectangle *KE* is composed of the ratios AH/CE and HL/CK.

And since, as I have said, the space *KE* is that space which would have been traversed in the time *CE* by a body moving with the uniform speed *CK*, and because during the time *CE* the moving body has been traveling with an accelerated motion whose initial speed was *CK*, it is clear that it will traverse a space greater than *KE* in the time *CE*. For the same reason it will traverse a distance greater than *OG* during the third part of the time, *EG*; because in that time it would have traversed *OG* had it moved with the uniform velocity, *EO*, and so on. In each successive part of the time *AH* the moving body will therefore traverse a space greater than the corresponding rectangle of the inscribed figure. Hence the whole space traversed by a body traveling with an accelerated motion must be greater than the area of the inscribed figure. Now that area was supposed to be equal to *P*. Consequently, the inscribed figure would have to be smaller in area than the plane *P*, which is absurd, for it has been shown that it is greater. The plane *P* cannot therefore be smaller than the triangle *AHL*. Nor, as we shall see, can it be greater.

For assume that it could be. Divide *AH* into equal parts and again describe about, and inscribe in, the triangle *AHL* two figures made up of rectangles, such that the excess of the one above the other is smaller than the excess of the plane *P* above the triangle *AHL*. The circumscribed figure will therefore be smaller than the plane *P*. Clearly, then, the moving body will traverse a space smaller than *BC* during the first part of the time *BC*, for in the same time it would have traversed *AC* with the uniform speed *CK*, that is, with a speed the moving body does not attain until the end of the time *AC*. Similarly, during the second part of the time *CE*, the body will traverse a space smaller than *DE*, because it would have traversed the same space in the same time with

[72]Proposition I (note by Huyghens).

a constant speed *EO*, which latter it does not attain until the end of the time *CE*. And so, in each part of the time *AH*, the moving body will traverse a space smaller than the corresponding rectangle of the circumscribed figure. Hence the whole space traversed by the body traveling with an accelerated motion is less than the total space of the circumscribed figure. Now that space was assumed to be equal to the plane *P*; hence the plane *P* will also be smaller than the circumscribed figure, which is absurd seeing that this figure was assumed to be smaller than the plane *P*. Consequently the plane *P* cannot be greater than the triangle *AHL*. But it was previously shown that it cannot be smaller. It must therefore be equal to it in area. Q.E.D.

Now, there was nothing to prevent Galileo from constructing a similar proof; indeed, he had applied the same demonstrative method well before he wrote the *Discourses*, namely, in a small work devoted to the determination of the centers of gravity of various solids.[73] Moreover, it is obvious that this procedure would have led him automatically to eschew medieval concepts and hence to confer upon the science of naturally accelerated motion a much greater coherence and conceptual unity. That being the case, why should Galileo, concerned as he normally was to give his arguments the greatest possible rigor, have preferred to resort to an imperfect geometrization? The answer, we believe, lies in the doubly indirect character of Huyghens' procedure. It was indirect, first of all, because it substituted distances for speeds and, second, because its final conclusion was based on two successive reductions to absurdity. In other words, although it introduced the proportionality of speed to time, it deliberately refrained from engendering the distance traversed from the speed increase—Huyghens simply posed that a given distance had been traversed and then showed that it could not have been otherwise, thus failing to present the result as a consequence of the most essential properties of uniformly accelerated motion. This explains why Galileo preferred to base his own proof on the method of indivisibles. He was determined to keep as closely as possible to the characteristics of the objects he analyzed—in short, to base his conclusions step by step on the phenomena themselves and not on such artifices as a double reduction to absurdity.

But before he could apply the method of indivisibles—the only means of arriving at a direct demonstration—he had first to satisfy one essen-

[73] *Theoremata circum centrum gravitatis solidorum*, *Opere* I, pp. 179 ff. A similar proof, based on the method of exhaustion, can be found in *Discourses* II, pp. 181–184.

tial condition, namely, to construct a representation of a clearly determined physical phenomenon, the increase in speed associated with naturally accelerated motion, to which the method of indivisibles could be applied. Now, only the concept of *quantitas velocitatis* could fulfill this role; describing the natural motion of a mobile between two points as the sum of its instantaneous velocities, it made possible an interpretation akin to that by which the method of indivisibles describes an area as a sum of lines. As a result, he felt free to proceed to and to justify the geometrical expression of a physical phenomenon in its very development and hence to pass on from the continuous increase in speed to the effect of that increase on the distance traversed. Compared with the definition that opened his theory of naturally accelerated motion, the proof of Theorem I may appear a retrogressive step, but we now understand that this retreat was an essential means of developing the demonstrative method best suited to the study of such "fluid" magnitudes as uniformly increasing velocities. In short, what Galileo tried to achieve by the combined use of traditional concepts and the method of indivisibles was what Newton achieved with his method of fluxions.

There is yet another and very telling example of the way in which his demonstrative ideal, as it were, condemned Galileo to employ traditional concepts and the method of indivisibles, namely, Proposition XXIII of the *Discourses*.[74] This proposition took the form of a problem: "Given the time employed by a body in falling through a certain distance along a vertical line, it is required to pass through the lower terminus of this vertical fall, a plane so inclined that this body will, after its vertical fall, traverse on this plane during a time interval equal to that of the vertical fall a distance equal to any assigned distance, provided this assigned distance is more than twice and less than three times the vertical fall." Let AC be the vertical distance covered by the fall in the time AC, and let IR be a distance more than twice and less than three times AC. It is required to pass a plane through the point C, so inclined that a body, after falling through AC, will duirng the time AC traverse a distance equal to IR. On IR, mark two points, M and N, such that $RN = NM = AC$; lay off CE such that $AC/CE = IM/NM$; then on CE prolonged to O lay off $CF = RN$, $FG = NM$, $GO = MI$. By a simple proof, too long to be included here, Galileo then showed

[74] *Discourses* III, pp. 240 ff.

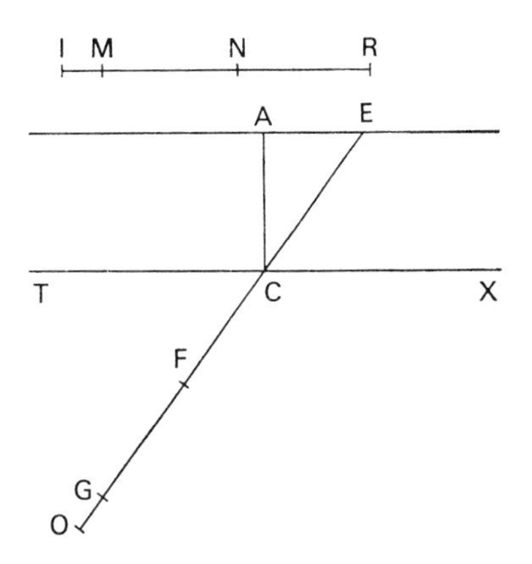

that *CO* is the required plane. If next we decrease the distance *IR* until it is nearly equal to 2*AC*, that is, until *I* is as close as we wish to *M*, and if we again construct *CE* such that $AC/CE = IM/NM$, we shall find that *AC* is very small in comparison with *CE*, which is now so long that it almost coincides with the horizontal line drawn through *C*. The distance *CO*, which the moving body traverses after its fall along *AC* in the time *AC*, will therefore be nearly equal to *CT*, and also to *CG*, that is, to 2*AC*. Consequently we may take it that if, after its descent along a vertical plane such as *AC*, a moving body is deflected onto a horizontal plane such as *CT*, the distance it will traverse on the latter in the same time interval *AC* will be 2*AC*.

Modern kinematics, as we know, can deduce this proposition directly and simply from the fundamental properties of uniformly accelerated motion,[75] but Galileo, who lacked the appropriate analytical tools, could not, of course, steer so simple a course, and the most obvious solution for him, too, would have been to seek the proof in a double reduction to absurdity. Instead, he extended his first proposition to the case in which *IR* is almost equal to 2*AC* by an argument based on limits. Now, what happens when *I* coincides with *M* and the plane *CG* almost coincides with *CT*? Quite simply this: the length previously represented by *OE* is transformed into the infinite length *TX*, and the lengths *FC* and *CE* into the infinite lengths *VX* and *CX*. "The argument here employed," Galileo wrote, "is the same as the preceding. For it is clear,

[75] If we designate *AC* by s_1 and *CT* by s_2, we have $s_1 = gt^2/2$ and $s_2 = vt$, where v is the velocity in *C*, and t the time of descent along *AC*. But $v = gt$, and consequently $s_2 = gt^2$, that is, $2s_1$.

since $OE/EF = EF/EC$, that FC measures the time of descent along CO. But, if the horizontal line TC, which is twice as long as CA, be divided into two equal parts at V, then this line must be extended indefinitely in the direction of X before it will intersect the line AE produced; accordingly, the ratio of the infinite length TX to the infinite length VX is the same as the ratio of the infinite distance VX to the infinite distance CX."[76] This argument clearly reflected Galileo's demonstrative ideal, his wish to establish the required result by keeping as closely as possible to the variations of the magnitude under consideration. However, the difficulties were only too obvious this time: how is it possible to construct a ratio between infinite magnitudes when all such magnitudes are equivalent by definition and when, moreover, all quantitative distinctions and with them any basis for comparison are as good as abolished? This explains why Galileo felt compelled to add that the same result could also be obtained by another method, once again based on the method of indivisibles.[77]

In that proof, he again considered a triangle ABC, and went on to say:

> The lines drawn parallel to its base BC represent [the degrees of] a velocity increasing in proportion to the time, and since these lines are infinite in number, just as the points on the line AC are infinite or as the number of instants in any interval of time is infinite, they will form the area of the triangle. Let us now suppose that the motion is continued, through an equal interval of time without new acceleration, and with the highest degree of velocity attained (that represented by BC); from these degrees will be built up, in a similar manner, by aggregation, the area of the parallelogram $ADBC$ which is twice that of the triangle ABC; accordingly, the distance traversed with these degrees of velocity during any given interval of time will be twice that traversed with the degrees of velocity represented by the triangle during an equal interval of time. But along a horizontal plane the motion is uniform since here it experiences neither acceleration nor retardation; therefore we conclude that the distance CD traversed during a time interval equal to AC is twice the distance AC; for the latter is covered by a motion, starting from rest and increasing in speed in proportion to the parallel lines in the triangle, while the former is traversed by a motion

[76] *Discourses* III, p. 242.

[77] Ibid., p. 243. A proof of the same proposition also appeared in the *Dialogue* (p. 254), where Galileo already made use of the concept of "quantity of velocity," but in an original way because he added the velocities as scalars. He thus obtained a direct but rather unsatisfactory proof, which failed to account for the continuous nature of the speed increases or decreases associated with nonuniform motions. This explains why he offered a further proof in the *Discourses*, and a third based on indivisibles which, to his mind, was by far the most satisfactory.

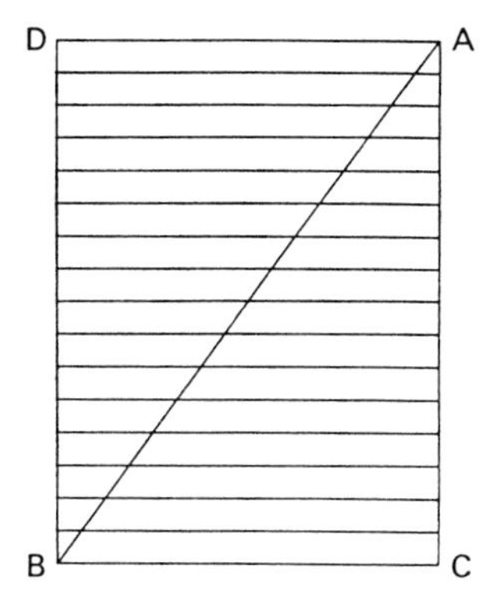

represented by the parallel lines of the parallelogram which, being also infinite in number, yield an area twice that of the triangle.

If we were asked to sum up our preceding remarks on Galileo's attempt to geometrize the motion of heavy bodies, we would say that his study of naturally accelerated motion was guided by a definition which, despite its ties with the medieval conception of speed, introduced a truly rational science of motion: motion was immediately treated as a state of indefinite duration, and acceleration as the rate of change in speed. As a result, the speed of a moving body could for the first time be evaluated from its mode of increase. However, because he lacked adequate mathematical tools, Galileo was prevented from making his definition the basis of the appropriate laws and propositions and had perforce to rely on a number of traditional notions—the description of speed as an intensive magnitude, the concept of *quantitas velocitatis*, and the method of indivisibles—which accordingly played a crucial role in the very formation of the new science. To conclude, however, that Galileo borrowed as much from the past as he contributed to the future would be to ignore that his desire to arrive at a direct determination of the effects of increases in speed on the distance traversed—the main reason why he saw fit to return to the past—was fed by the same source as his successful reconstruction of speed from its mode of increase. Even when he made appeal to the tradition, Galileo continued to be guided by the same demonstrative ideal as had inspired his definition of naturally accelerated motion. The apparent heterogeneity of his concepts and his methods thus hides the incontestable unity of his mechanics, and though the letter of that science may sometimes have conflicted with its spirit, the spirit invariably determined the choice of the letter.

7 The Geometrization of the Motion of Heavy Bodies (Part 2)

Gravity and Differences in the Speed of Freely Falling Bodies

The science of accelerated motion had now been put on its feet, but even before its soundness could be established by experiment, a theoretical difficulty had to be disposed of. Throughout the last chapter it was assumed that the laws and propositions we have examined were applicable to all bodies, no matter what their nature or their weight. Now this assumption, though essential to the construction of a geometrized science, whose laws must by definition have a general validity, is apparently vitiated by a simple observational fact, namely, that freely falling bodies travel at different speeds. How can we postulate that the motion of all bodies obeys the same geometrical laws when there are evident differences in their speeds of descent? How, in particular, can we assume that the motion of a very light body, such as a leaf or an inflated bladder, is governed by the same principle as the motion of a very heavy body such as an ebony or ivory ball? In other words, what guarantee do we have that the increase in speed is independent of the falling body's weight? How can we assume that no intrinsic link exists between the concrete phenomenon of the differences in speeds and the respective weights of bodies? Unless Galileo could answer these questions, the geometrical analyses he advanced during the Third and Fourth Days of the *Discourses* had to remain a mathematical fiction.

The problem was one with which he had long been familiar. In the *De Motu* he had made short shrift of most of the Aristotelian theses on the subject; twenty-three years later, in the *Discourse on Bodies in Water*,[1] he was still holding to much the same view after a much more detailed examination of the influence of the medium on the motion of freely falling bodies. It was therefore only to be expected that he should have prefaced his definitive account in the *Discourses* with a brief résumé of his earlier findings.

One fact, to begin with, was quite certain: the different speeds with which heavy bodies descend toward the center of the earth are related

---

[1]*Discorso intorno alle cose che stanno in su l'acqua*, Vol. IV (1612).

to their specific, not their absolute, weight. This had been established in his *De Motu*;[2] when the matter was reviewed in the *Discourses*, one question dominated the entire discussion—whether or not bodies of the same specific, but of distinct individual, weights descend with identical speeds through identical media. The answer was contained in a "brief and conclusive" argument:[3] Take a body known to move through a given medium with a given speed; tie to it another but smaller body of the same specific weight and release the two. If the speed of descent was in any way dependent on the individual weight, the motion of the larger and ostensibly the swifter, would be retarded by the smaller and ostensibly the slower. Hence if the larger moved with, say, eight degrees of speed and the smaller with four, the two together would move with a speed of less than eight degrees. But the two bodies joined together must be heavier than the larger by itself. "Thus you see how, from your assumption that the heavier body moves more rapidly than the lighter one, I infer that the heavier body moves more slowly." In other words, all those who assume that the speed depends on the individual weights are making an elementary mistake: they fail to see that "it is necessary to distinguish between heavy bodies in motion and the same bodies at rest."[4] True, a large stone placed in a balance "not only acquires additional weight if another stone is placed upon it, but even by the addition of a handful of hemp its weight is augmented by six to ten ounces depending on the quantity of hemp." But once the larger stone has been set in motion, the smaller ceases to exert the least pressure upon it. Similarly, "we always feel the pressure upon our shoulders when we prevent the motion of a load resting upon us; but if we descend just as rapidly as the load would fall, how can it gravitate or press upon us? Do you not see that this would be the same as trying to strike a man with a lance when he is running away from you with a speed that is equal to, or even greater than, that with which you are following him? You must therefore conclude that during free and natural fall the small stone does not press upon the larger and consequently does not increase its weight as it does when at rest."[5] But we have already shown that the smaller body cannot retard the motion of the larger, from which we must infer that "large and small bodies move with the same speed, provided they are of the

[2]See Chapter 3.
[3]*Discourses* I, pp. 107 ff.
[4]Ibid., p. 108.
[5]Ibid.

same specific gravity."[6] It follows that a "bird shot falls (through the air) as swiftly as a cannonball" and "a grain of sand as rapidly as a grindstone."

However, the speed of a falling body cannot depend on its specific weight alone, for the medium too must necessarily affect its motion. In his analysis of this effect, Galileo once again rose up against the traditional view, against the assumption that the speed of freely falling bodies is inversely proportional to the density of the medium, or rather that the medium divides the motive force in direct proportion to its own density. Now, since it is always possible to establish a ratio between two media of different densities, we may put it that the ratio between the densities of water and air is, say, as 10:1. Hence, if a wooden ball descends through air with, say, twenty degrees of velocity, it would have to descend through water with a velocity of two degrees, when in fact it floats near the surface.[7] Aristotle's theory could be faulted in yet another way. For let us replace the wooden ball, which refuses to sink, with some other body of greater specific gravity, such that it will sink through water with a speed of two degrees. What will its speed be in air, assuming Aristotle was right in claiming that the medium divides the speed in direct proportion to its density? The answer is obvious. We shall have to say that it will descend through the air with a speed of twenty degrees. Now this leads to the absurd conclusion that one body moving through the air with twenty degrees of speed will sink in the water with a speed of two degrees, while another body with the same speed in air, will float on the surface of the water. "But without going into the matter more deeply, how have these common and obvious properties escaped your notice? Have you not observed that two bodies that fall in water, one with a speed a hundred times as great as that of the other, will fall in air with speeds so nearly equal that one will not surpass the other with so much as one hundredth part? Thus, for example, an egg made of marble will descend in water one hundred times more rapidly than a hen's egg, while in air, falling from a height of twenty cubits,[8] the one will fall short of the other by less than four finger breadths. In short, a heavy body that sinks through ten cubits of water in three hours will traverse ten cubits of air in one or two pulsebeats; and if the heavy body be a ball of lead, it will easily traverse the ten cubits of water in less than double the time required for ten cubits of

[6] Ibid., p. 109.
[7] Ibid., p. 111.
[8] The Florentine cubit measured 21 to 22 inches

air."[9] Equally false is the view that "one and the same body moving through differently resisting media acquires speeds that are inversely proportional to the resistances of these media" or that, in the same medium, bodies acquire speeds proportional to their weight.[10]

But if the traditional theses had to be rejected, what precisely was the role of the medium? Galileo's general reply in the *Discourses* was no different from that which he had given in the *De Motu* and above all in the *Discourse on Bodies in Water*: when a body is plunged into a medium, it experiences a loss of weight or an upthrust that counteracts and decreases its propensity for downward motion; and it is in this upthrust that the effect of the medium primarily resides.[11] But the loss of weight that a body suffers in a given medium is equal to the weight of the medium it has displaced; it is the greater, the rarer the body and the denser the medium. A body whose density is greater than that of the medium will retain some of its weight in it and sink; one whose density is equal to that of the medium will be weightless; and one whose density is less than that of the medium not only will be weightless but will be "extruded."[12] Hence, the motion of a body in a given medium is governed by two factors and two factors only: its own specific weight and that of the medium.

It was on these conclusions that Galileo had based his first explanation of the difference in speeds. His fundamental idea was very simple: if the downward motion of a body depends primarily on the excess of its specific weight over that of the medium, this can mean only that its downward speed must be directly proportional to its specific weight, due allowance having been made for the upthrust. Hence, if we simply subtract the specific weight of the medium from that of the moving body, we shall arrive directly, if not at the real speed, at least at its order of magnitude; in short, everything hinges on "the excess of the weight of the moving body above that of the medium, and vice versa."[13]

---

[9] *Discourses* I, pp. 111–112.
[10] Ibid., p. 113.
[11] Primarily, but not exclusively, as we shall see later.
[12] Cf. Chapter 3. It follows that two bodies of the same specific weight (that is, of the same density) will react identically in a given medium: since their volume increases as their weight, the upthrust will be proportional to their dimensions. The argument by which Galileo had previously established that the velocity cannot be proportional to the body's absolute weight thus became redundant.
[13] *Discourse on Bodies in Water*, Vol. IV, p. 79. Mathematically expressed, $V = W - M$ (where W is the specific weight of the body and M the specific weight of the medium) or, in the case of two bodies of different specific weights in one and the same medium, $V_1/V_2 = (W_1 - M)/(W_2 - M)$.

Now this explanation was incompatible with the professed geometrical ideals of the *Discourses*, for once the speed of a falling body is assumed to depend directly on its specific weight, there is no certainty that the increase in speed will occur in the same way for all bodies—namely, in simple proportion to time—and hence that the laws of uniformly accelerated motion are also the laws of naturally accelerated motion. But Galileo's problem is now quite clear: How can the observed difference in speeds be explained in terms of the specific weights of the moving bodies and the media without coming back to the first theory?

An appeal to observation not only convinced him of the untenability of his earlier solution but also led him to make one of his boldest extrapolations. Neither the *De Motu* nor the *Discourse on Bodies in Water* had neglected experimental evidence; curiously enough, however, in these early works Galileo had confined his attention to the motion of one and the same body in two different media or of two different bodies in one and the same medium. In so doing he had overlooked what was perhaps the most interesting question, namely, the variation in the speed differences of bodies of different specific weights in media of "different resistances."[14]

When we pass from a fairly rare medium, such as air, to a denser medium, such as water or mercury, we discover that the differences in the speeds of two bodies of different weights increase so markedly that, though their speeds "differed scarcely at all in air," they would in water "fall the one with a speed ten times as great as that of the other. Further, there are bodies that will fall rapidly in air, whereas if placed in water they not only will not sink but will remain at rest or will even rise to the top. . . ." Again, "in a medium of mercury, gold not only sinks to the bottom more rapidly than lead but it is the only susbtance that will descend at all; all other metals and stones rise to the surface and float. On the other hand, the variation of speed in air between balls of gold, lead, copper, porphyry, and other heavy materials is so slight that in a fall of a hundred cubits a ball of gold would surely not outstrip one of copper by as much as four fingers."[15] Hence, it would ap-

[14]*Discourses* I, p. 113. The term "resistance" was rather ambiguous, for in using it Galileo confused (perhaps unwittingly) the two effects of the medium, upthrust and friction. Since only the first concerns us in what follows, we shall, for greater clarity, dispense with Galileo's term.

[15]*Discourses* III, pp. 113 and 116.

pear that the differences in the speeds of freely falling bodies depend primarily on the density of the media through which they travel; while the differences are very great in media of very high density, they decrease as the density of the media decreases, so much so that "in a medium of extreme tenuity, although not a perfect vacuum, we find that, in spite of great diversity of specific gravity (*peso*), the differences in speed are very small and almost inappreciable."[16]

This observation, or rather this double series of observations, had several important consequences. First of all, it showed that the original explanation offered in the *De Motu* had to be abandoned. Let us suppose that lead is 10,000 times, ebony 1000 times, and water 800 times heavier than air. Since the effect of the medium on the moving body makes itself felt primarily as a loss of weight equal to the weight the medium displaced, we can express the specific weight of the body in a given medium by the formula $W_{sb} - W_{sm}$ where W_{sb} is the specific absolute weight of the body[17] and W_{sm} the specific weight of the medium. Hence, the specific weight of ebony in water is to its absolute specific weight as 200 is to 1000, whereas the specific weight of lead in water is to its absolute specific weight as 9200 is to 10,000; the corresponding figures for ebony in air would be 999 to 1000, and for lead in air 9999 to 10,000. While the difference between the specific weights of ebony and lead weighed in air and their absolute specific weights is extremely small, the difference in water becomes considerable in the case of ebony and, though proportionally smaller, anything but negligible in the case of lead. Let us now return to observation. Since the specific weights of bodies weighed in air are nearly identical with their absolute weights, we might expect to find that the ratio of their speeds through air will correspond more or less closely to the ratio of their specific weights. But what do we really find? The precise opposite, for not only does the speed ratio of different bodies not reflect the ratio of their specific weights, but the speed differences tend to become imperceptible. Hence, the speed of a falling body cannot possibly depend on its intrinsic properties. In short, the specific weight cannot play a direct role in the observed speed differences.

But our experiment also has other than purely negative consequences.

---

[16]Ibid., p. 117.

[17]That is, its specific weight in a vacuum. The specific weight of air was also supposed to have been determined in a vacuum.

Using the same example, let us now consider the ratio between the specific weights of each of the media and that of our moving body; that ratio, W_{sm}/W_{sb}, will then express in what proportion the absolute specific weight of the moving body is decreased by that of the medium or, if you will, by what amount the specific weight of the moving body evaluated in that medium will be less than its absolute specific weight.[18] For ebony this ratio will be 800/1000 in water, and 1/1000 in air; for lead the corresponding figures would be 800/10,000 and 1/10,000. But it is obvious that the speed of ebony, for example, determined first in water and then in air, is inversely proportional to the ratio W_{sm}/W_{sb}; that is, the greater the ratio (the closer its value approaches 1), the slower the motion; and the smaller the ratio (the closer its value approaches zero), the swifter the motion. Now, since the same remark applies to the motion of lead, we are clearly dealing with a factor whose variations correspond closely to the observed speed changes. That being so, let us next compare the values of the ratio W_{sm}/W_{sb} for ebony and lead moving first through water and then through air. When the ebony is transferred from water to air, the upthrust decreases dramatically, for instead of representing 0.8 of the body's absolute specific weight, it now represents a mere 0.001; as for lead, the corresponding figures are 0.08 and 0.0001. While the relative gain in weight is thus comparable,[19] it has quite a different significance for each of the two bodies: as a result of the upthrust, the specific weight of ebony in water is reduced to one-fifth of its absolute specific weight; in air, on the other hand, the decrease becomes almost imperceptible and, relatively speaking, no more important than it is in the case of the lead. This fact shows clearly that the difference in the speeds of freely falling bodies is due primarily to the very disproportionate manner in which the lifting effect of a given medium is experienced by bodies of different specific weights; the effect is considerable in the case of bodies whose specific weight is not very much greater than that of the medium, but it drops off sharply as the specific weight of the moving body increases. The drop is reflected in an increase in speed; conversely, in a relatively rare medium with a small lifting effect, the speeds of all bodies will be so close to one another as to become almost indistinguishable. Far from having a single and permanent cause, the observed differences in the

[18] If the ratio is greater than 1 the moving body will rise through the medium.

[19] The upthrust of water in both cases is 800 times greater than that of air.

speed of freely falling bodies thus result primarily from the unequal values which, in the case of different bodies moving in one and the same medium, attach to a *certain ratio*. And though the specific weight of moving bodies continues to play some part in it, it is but one of several factors to be taken into account.

These conclusions, though of great importance, do not, however, take us beyond the threshold of the extraordinary hypothesis on which Galileo based his definitive theory of the difference in speeds. We have just seen that this difference reflects the disproportionate manner in which the lifting effect of a given medium is experienced by bodies of different specific weights; conversely, because this inequality tends to vanish in increasingly rarer media, the speeds too will tend to become equal. Now, all that has changed is the specific weight of the medium, not that of the moving bodies, so that the former plays a determining role in the ratio W_{sm}/W_{sb}. Hence, it is not unreasonable to wonder "what happens to bodies of different weights moving in a medium devoid of resistance, so that the only difference in speed is that which would arise from inequality of weight."[20] Experiment cannot settle the issue, since "no medium except one entirely free from air and other bodies, be it ever so tenuous and yielding, could furnish our senses with the evidence we are looking for."[21] However, nothing prevents us from pressing our results and replacing the media of decreasing densities we have been considering with a medium whose density is zero. If we do so, one thing is certain beyond all doubt: no body will any longer experience the least upthrust, so that the ratios whose proportionally unequal values accounted for the difference in speeds cannot even be constructed. "Having observed this," Galileo declared, "I came to the conclusion that in a medium totally devoid of resistance all bodies would fall with the same speed."[22]

All the elements needed for an acceptable theory had at last been assembled. Observation had suggested that the specific weights of the bodies and the medium could at best explain the *differences* in speed of freely falling bodies. The "highly probable"[23] hypothesis that all bodies must fall with equal speed through a vacuum now turns this suggestion into the very principle of explanation. For let us assume that all bodies do, in fact, travel with the same speed through a vacuum, and let V_0 be that speed; the ratio relating the specific gravity of the medium to that

[20] *Discourses* III, p. 117.
[21] Ibid.
[22] Ibid., p. 116.
[23] Ibid., p. 117.

of various bodies then assumes its full importance: because it measures the effect of the medium, it tells us in what proportion we must reduce the theoretical velocity V_0 to obtain the real speed of these bodies.[24] Let us now return to our previous example, in which we assumed that lead weighs ten thousand times as much as air, but ebony only a thousand times as much: "Here we have two substances whose speeds of fall in a medium devoid of resistance are equal: but when air is the medium, it will subtract from the speed of the lead one part in ten thousand, and from the speed of the ebony one part in one thousand, that is, ten parts in ten thousand. While therefore lead and ebony would fall from any given height in the same interval of time, providing the retarding effect of the air were removed, the lead will, in air, lose in speed one part in ten thousand; and the ebony ten parts in ten thousand."[25] In other words, "if the elevation from which the bodies start be divided into ten thousand parts, the lead will reach the ground leaving the ebony behind by as much as ten, or at least nine, of these parts."[26] Or take another example: "Ebony weighs a thousand times as much as air but this inflated bladder only four times as much; therefore air diminishes the inherent and natural speed of ebony by one part in a thousand; while that of the bladder, which, if free from hindrance, would be the same, experiences a diminution in air amounting to one part in four. So that when the ebony ball, falling from the tower, has reached the ground, the bladder will have traversed only three-quarters of this distance."[27] Now the same principle also helps us to an accurate determination of "the ratio of the speeds of one and the same body in different fluid media."[28] Thus, suppose that tin is one thousand times heavier than air and ten times heavier than water; in that case, the ratio between its own specific weight and that of either of these two media will be, respectively, as 1/1000 and 1/10. Hence, if we divide its unhindered speed into a thousand parts, it is clear that the air will "rob" it of one of these parts in a thousand, and water of one part in ten; in the first

[24]"And since it is known that the effect of the medium is to diminish the weight of the body by the weight of the medium displaced, we may accomplish our purpose by diminishing in just this proportion the speeds of the falling bodies which in a nonresisting medium we have assumed to be equal" (ibid., p. 119).

[25]Ibid.

[26]Ibid.

[27]Ibid.

[28]Ibid.

case it will fall with a speed of 999, in the second with a speed of only 900.[29] Or again, "take a solid slightly heavier than water, such as oak, a ball of which would weigh, let us say, 1000 drachms; suppose an equal volume of water to weigh 950, and an equal volume of air 2; then it is clear that if the unhindered speed of the ball is 1000, its speed in air will be 998, but in water only 50, seeing that the water removes 950 of the 1000 parts which the body weighs, leaving only 50."[30] Hence, no matter whether we compare the motions of two bodies in one medium or the motions of one body in two different media, we must apply the same principle, namely, to determine the amount by which the theoretical speed in a vacuum will be diminished, the diminution being proportional to the ratio of the specific weight of the medium to that of the moving body.[31]

While no one can deny that these computations were in far better agreement with observation than those presented in the *De Motu* and a fortiori than those of Aristotelian physics, it would be idle to pretend that in the form and context in which they were presented they were of more than illustrative value. Before the examples that we have quoted could really support the new theory, Galileo had first to reach a correct evaluation of the absolute specific weights of the media and the moving bodies: but, despite repeated efforts in that direction, he never gained more than a highly approximate idea of their value. Not that he failed to see that the central problem was the determination of the specific weight of the medium, since in order to obtain the absolute specific weight of the moving body, it sufficed to add its specific weight in a given medium to the specific weight of the latter. Several pages of the *Discourses* (First Day) preserve the memory of Galileo's many attempts to determine the specific weight of air as accurately as possible; but, ingenious though they undoubtedly were, none of the attempts was truly conclusive, and the resulting lack of accuracy is clear proof of

[29] Ibid.

[30] Ibid., p. 121.

[31] All the examples we have mentioned can be represented by a single formula. Thus if V_0 is the speed of descent in a vacuum and V the speed with which a body of specific weight W_{sb} descends in a medium of specific weight W_{sm}, then the relation of V to V_0 is given by $V = V_0[(W_{sc} - W_{sm})/W_{sb}]$. It goes without saying that we have been considering mean speeds, not instantaneous increases in speed.

Galileo's inability to verify his theoretical conclusions by direct experiments.[32] Under these circumstances, he had just two means of testing his hypothesis: either he could prove by purely logical means that the differences in speeds could always be reduced to the action of the medium, or else he had to discover a method of experimentation that did not require any prior knowledge of the different specific weights.

To that end he subjected the observations from which he had started to a careful scrutiny, especially, the fact that the differences in the speeds of descent of bodies with distinct specific weights decrease very rapidly in very rare media. It appeared to Galileo that this observation might be far less important than he had at first assumed it to be, for though a lump of lead does not travel more than two or three times more swiftly "over a distance of four or six cubits" than a bladder of the same volume, it is also very true that, over very much larger distances, "the lead might cover one hundred miles while the bladder was traversing but one."[33] In other words, he suspected that a more extensive study of the motion of bodies in rare media might well reveal the existence of considerable differences in speed. Could they once again be attributed to the action of the medium, or were they rather the result of an intrinsic link between the speed of descent and the body's specific weight?

The second alternative could be immediately discarded. For what would an experiment conducted over long enough distances be able to prove? Merely that the speed of the lead increases far more quickly than that of the bladder, in other words, that the ratio of the distances covered by the lead and the bladder during the same time interval keeps increasing. However, the ratio between the specific weights of the two moving bodies has remained constant throughout the descent: their speed can thus in no way have been determined by their specific weight, for if it had, "the ratio between the distances traversed ought to have remained constant."[34] Hence, the continuous increase in the differences between the speeds of relatively heavy and light bodies which would

---

[32]*Discourses* III, pp. 121–128. Galileo concluded in particular that the specific weight of air is 400 times less than that of water, when, in fact, the correct figure is nearer 800. The experiment on which he based that conclusion was, however, brilliantly conceived; its central idea was that air must be weighed in a vacuum, not in air (pp. 125–126).

[33]Ibid., pp. 117 f. The Florentine mile measured 1653.607 meters.

[34]Ibid., p. 118.

be observed if these bodies traveled over large enough distances through a rare medium could in no way restore the leading role to the specific weight. But if the difference between the specific weights is incapable of explaining the continuous change in the speed ratio, how can the medium, "which for its part is assumed to remain constant,"[35] account for this difference any better? And if it cannot, does not the continuous increase in the speed ratio become quite unintelligible in the light of Galileo's new perspective?

This would undoubtedly be the case if it were not possible to demonstrate that in certain respects the medium fails to remain "constant." Let us reexamine the argument by which we have just established that, once the medium had ceased to influence their motion, all bodies would descend with the same velocity; to arrive at that conclusion, we considered the weight-reducing effect of the medium but completely ignored another factor, namely, the medium's resistance to penetration, or the friction effect. Now, unlike the upthrust, the friction effect is variable and, moreover, variable in two respects. First, it "depends directly on the change in rapidity with which it must yield and give way laterally to the passage of the falling body which is being constantly accelerated,"[36] and second, the need to "open a path for themselves and to push aside the parts of the medium does not impede all moving bodies in the same way."[37] Thus, while the friction effect makes itself barely felt in the case of bodies with a very high density, it will severely impede the motion of bodies with a low density and hence already robbed of part of their absolute speed by the upthrust or weight-reducing effect of the medium. The problem is therefore as good as solved: all falling bodies will, in equal and successive intervals of time, experience the same acceleration and hence a continuous increase in speed, but since the friction effect increases as the latter, an ever larger part of the speed must be devoted to overcoming the friction effect, "until the speed finally reaches such a point and the resistance of the medium becomes so great that, balancing each other, they prevent any further acceleration and reduce the motion of the body to one that is uniform and one that will thereafter maintain a constant value."[38] And this will happen much more quickly with rarer bodies whose "absolute speed" has already been reduced by the upthrust than it does

---

[35] Ibid.
[36] Ibid., p. 119.
[37] Ibid.
[38] Ibid.

with denser bodies. The conclusion is obvious: "Now seeing how great is the resistance which the air offers to the slight moment of the bladder and how small that which it offers to the large weight of the lead, I am convinced that, if the medium were entirely removed, the advantage received by the bladder would be so great and that coming to the lead so small that their speeds would be equalized."[39] Once we distinguish the friction from the weight-reducing effect and then examine how the former complements the latter, we can attribute to the medium not only the differences in speed that appear right from the start of the descent but also those that do not appear until after some lapse of time.[40]

The main purpose of this analysis was to fill a gap; but in revealing the great importance of the friction effect in such rare media as air, it also showed under what conditions the general principle of explanation lends itself to experimental verification. We saw that it was only due to the disproportionate manner in which the upthrust is experienced by bodies of different specific weights that the differences in speed were engendered in the first place: hence if in the case of a body traveling through air we should succeed in isolating the friction effect and could show that the differences in velocity were due to it alone, then we should also have shown that, whenever the upthrusts tend to become proportionally identical, the speeds will also become equal. Now, as Galileo was quick to point out, naturally accelerated motions cannot help us here because, to measure their speeds, we should have to use impossibly large heights and above all because it would be almost impossible to distinguish between the respective roles of upthrust and friction. Ideally, we should have to compare the simultaneous motions of a light and a heavy body over a succession of "small heights"[41] in such a way that the retardation due to the friction effect would be brought to light step by step. Now there is one experimental arrange-

---

[39]Ibid.

[40]Not only do the expressions "upthrust" and "friction" not appear in the *Discourses*, but Galileo employed no special terminology to distinguish between them. For all that, the general exposition is quite clear, since the context usually shows to which of the two effects Galileo is referring. The only difficult passages occur during the First Day (pp. 136–137), where Galileo seems to have forgotten the importance of the upthrust when discussing the rapid decrease in speed of a bullet traveling from air into water; and during the Fourth Day (pp. 276–278), when he briefly recalls the conclusions of the First Day.

[41]*Discourses* III, p. 128.

ment that meets these conditions admirably. Take two balls, one of lead and one of cork, "the former more than a hundred times heavier than the latter,"[42] suspend them from equally long fine threads, pull them aside an equal distance from the perpendicular and let them go at the same instant. You will then discover two things: first, that the two balls have identical periods of oscillation and that they keep step so perfectly that "neither in a hundred swings nor even in a thousand will the former anticipate the latter by as much as a single moment;"[43] and second, that the resistance (friction) of the medium decreases the amplitude of the cork's swing more than it does that of the lead. The way in which Galileo used these two facts to justify his general principle of explanation shows clearly to what extent he had already come to appreciate the close connection between reason and experiment, a connection that Newton considered a characteristic feature of the new science.

Let us first eliminate a possible error in interpretation. The rapid damping of the oscillations of the cork pendulum and their isochronous nature might suggest that the lead, which covers a greater distance than the cork in the same period, must also have a greater capacity for speed, in which case Galileo's entire explanation of alterations in speed would have to be rejected.[44] However, it is simple to show that the cork can, in fact, move more swiftly than the lead. Thus if we draw the cork pendulum aside through an arc of thirty degrees, and the lead pendulum through one of only three degrees, the isochrony of pendular vibrations ensures that the cork and the lead complete their swings in the same time and also that the motion of the cork, which traverses an arc of sixty degrees while the lead traverses an arc of only six degrees, will be "proportionally swifter."[45] The pendulum experiment thus in no way restores the link between the speed of descent and the weight of the moving bodies.

But if we look more closely at our two pendulums, we can also observe that, "having pulled aside the pendulum of lead through an arc of, say, fifty degrees, and set it free, it swings beyond the perpendicular almost fifty degrees, thus describing an arc of nearly one hundred degrees; on the return swing it describes a somewhat smaller arc; and after a large number of such vibrations it finally comes to rest. Each

[42]Ibid.
[43]Ibid., p. 129.
[44]Ibid.
[45]Ibid.

vibration, whether of ninety, fifty, ten, or four degrees occupies the same time: accordingly, the speed of the moving body keeps on diminishing since in equal intervals of time, it traverses arcs that grow smaller and smaller. Precisely the same things happen with a pendulum of cork suspended by a string of equal length, except that a smaller number of oscillations is required to bring it to rest, since on account of its lightness it is less able to overcome the resistance of the air; nevertheless the vibrations, whether large or small, are all performed in time intervals that are not only equal among themselves but also equal to the period of the lead pendulum."[46] The only difference between the motions of the two pendulums is therefore the more rapid damping of the vibrations of the cork, and this suggests that the latter is, so to speak, brought to move in increasingly smaller arcs during swings whose period remains quite unchanged. Now, if it can be shown that this phenomenon is due solely to friction, the basic idea presiding over Galileo's theory of speed differences would have been fully vindicated. In fact, the isochrony of pendular oscillations helps us to do just that. Thus, if we compare oscillations of equal amplitude by the cork and the lead either at the beginning or at the end of their motions, we find that they are accomplished in equal times, that is, with identical speeds, no matter whether the arc described measures a hundred, eighty, sixty, twenty, or ten degrees. Hence the reason why the amplitude of the cork's swing is reduced more swiftly than that of the lead cannot possibly lie in the "naturally" slower speed of the cork—an assumption that is incompatible with the isochrony of pendular oscillations—but lies only in the fact that, being "less apt, in view of its lightness, to overcome the resistance of the air," the cork is forced by the latter to traverse increasingly smaller arcs. There is, moreover, not the slightest doubt that if the retarding action of the medium could have been eliminated, no difference at all would have been detected between the motions of the two bodies; both of them would have traversed equal distances in the same overall period. In short, the pendulum experiment proved that "the gravity inherent in various falling bodies has nothing to do with the difference of speed observed among them,"[47] and thus fully justified Galileo's principle. Last but not least, it established that all bodies

---

[46]Ibid. [47]Ibid., p. 131.

whatsoever will fall with equal velocities through a vacuum.[48]

Crucial though it undoubtedly was, the pendulum experiment nevertheless failed to throw light on one important point, namely, whether our tacit assumption that the friction increases as the swiftness of the motion was, in fact, justified. The speeds of the cork and lead pendulums were much too small to settle this matter—only experiments involving high velocities could hope to do that. Is it then at least possible, in spite of this difficulty, to describe the principle? An earlier conclusion gives a clue. If the friction really increases with the speed of the motion, it follows that sooner or later the increment of speed added in equal successive times to the speed already acquired will be totally employed in overcoming the resistance of the medium; at this point the motion would no longer be accelerated but would become uniform. Now since this conclusion applies to all bodies in all media, we are entitled to assume not only that every freely falling body will have a maximum speed that it cannot possibly exceed but also that if this speed were increased by artificial means, it would quickly lose the excess.[49] Several experiments can be conceived to show that this is indeed what happens. Thus we can fire a gun, which, as we know, impresses upon a ball a speed far in excess of that associated with natural motions,[50] first downward from the top of a very high tower and then from an elevation of only four or six cubits. If we do so, we shall find that the first shot will make a smaller impression on the ground than the second—"clear evidence that the momentum of the ball fired from the top of the tower diminishes continually from the instant it leaves the muzzle until it reaches the ground."[51] But the increasing resistance of the medium to penetration could be demonstrated in yet

[48]Galileo, as the reader will have noted, assumed throughout his demonstration that the oscillations of his pendulums were isochronous, regardless of their amplitude of swing; this mistaken view in no way detracts from the general validity of his analysis, which merely stipulated that swings of the same amplitude on the part of pendulums of equal lengths must be isochronous, as in fact they are.

[49]*Discourses* III, p. 136.

[50]Galileo described as "supernatural" the speed of a ball fired from a musket or from a piece of ordnance; cf. *Discourses* IV, p. 278.

[51]*Discourses* III, p. 137. Galileo mentioned the same experiment during the Fourth Day (p. 279) but added that he never actually performed it. This was done most successfully by members of the Accademia del cimento, no doubt on Viviani's suggestion; for a summary of their experiment see A. Carugo and L. Geymonat's edition of the *Discourses*, p. 825.

another way, namely, by the "very likely" mechanical principle that "a heavy body falling from a height will, on reaching the ground, have acquired just as much momentum as was necessary to carry it to that height."[52] At least this would happen in a vacuum: for if our assumption about the resistance of the air is correct, we can easily predict that a shot fired vertically upward will not, on falling back, possess a "force of percussion" equal to that which it possessed just after the firing; "the resistance of the air would prevent the muzzle velocity from being equaled by a natural fall from rest at any height whatsoever."[53] While Galileo lacked the means to carry out this experiment, he defined its scope quite clearly so that, when it was eventually performed, the results were in full accord with his predictions.[54]

A further observation, however, was responsible for the last objection to Galileo's explanatory principle. Thus Simplicio, and later Sagredo, asked whether it was not a fact that "a cannonball falls more rapidly than a bird shot"[55] or that a stone in water will cover "in one pulse-beat" a distance that fine sand would not traverse in an hour.[56] Thus, while the first objection had been based on the alleged link between the speed of descent and the specific weight, the new one was based on the individual weights of two moving bodies. In fact, no new assumptions were needed to dispose of it—merely a closer look at the resistance that the medium put up to penetration and, above all, the realization that this resistance depends on the surface area of the moving body. Thus solid bodies, far from being smooth, are covered with "rugosities" that strike the air when the body is in motion and hence retard its speed, "and this they do so much the more in proportion as the surface is larger, which is the case of small bodies as compared with greater."[57] Nor is this the only way by which surfaces tend to retard the motions of smaller bodies. For bodies of the same specific weight and shape would only meet the same resistance on the part of the medium if their respective surfaces were proportional to their respective volumes; if, on the other hand, the ratio of the volumes decreased more

[52] *Discourses* III, p. 138.
[53] Ibid.
[54] Galileo's hypothesis can also be verified by comparing the theoretical height reached by a bullet fired vertically into the air with the height to which it actually rises—the difference between the two values then gives the measure of the air resistance. An experiment of type was performed at St. Petersburg at the beginning of the eighteenth century following the calculations of D. Bernoulli.
[55] *Discourses* III, p. 131.
[56] Ibid.
[57] Ibid., p. 132.

rapidly than that of the surfaces, it must follow that the smaller the body, the greater the resistance it meets. Now, if we take a cube two inches on a side, each face having an area of four square inches and the six sides a total area of twenty-four square inches, and saw it through three times so as to divide it into eight smaller cubes, each one inch on the side, so that the total surface of each cube will be six square inches instead of the twenty-four in the case of the larger cube. "It is evident therefore," Galileo pointed out, "that the surface of the little cube is only one-fourth that of the larger, namely, the ratio of six to twenty-four; but the volume of the solid cube itself is only one-eighth; the volume, and hence also the weight, diminishes much more rapidly than the surface."[58] It follows that "the resistance engendered by contact between the surface of the moving body and the medium increases more rapidly for smaller bodies than it does for bigger ones."[59] In other words, the resistance has a relatively greater effect on a grain of sand or a bird shot than on a stone or a cannonball, so that the medium once again provides the complete explanation of the observable differences in the speed of different bodies—a complementary proof of Galileo's assumption that the specific or absolute weight plays no direct part in the difference in speeds.

The problem we posed at the beginning of this chapter can now be

---

[58]Ibid., p. 134. "If we again divide the little cubes into eight others, we shall have for the total surface of each of these one and one-half square inches, which is one-sixteenth of the surface of the original cube; but its volume is only one sixty-fourth part." Galileo had previously used this example in his *Discourse on Bodies in Water* (Vol. IV, pp. 138–139). It should be noted that, though the surface plays an important role in friction phemonena, the *shape* as such does not influence the speed of a body in a given medium, as traditional physics had assumed. The *Discourse on Bodies in Water* was devoted almost entirely to the refutation of this idea: "The diversity of figures given to this or that solid cannot in any way be a cause of its absolute sinking or swimming. So that if the solid being formed, for example, into a spherical figure does sink or swim in the water, I say, that being formed into any figure, the same figure in the same water shall sink or swim: nor can its motion by the expansion or by other mutation or figure be impeded or taken away" (p. 87). In reality the motion is affected only by the specific weights of the moving body and of the medium. But in that case, how can we explain that a gold leaf will float on the surface of the water whereas a ball of the same metal will sink very rapidly? As opposed to the Peripatetics, who had seen the cause of this phenomenon in the viscosity of the liquid medium, Galileo contended that the air above it adhered to the leaf, thus reducing its specific weight below that of the water (pp. 97–99). The real cause of the phenomenon, as we now know, is the surface tension of the water. (For Galileo's observations, see also *Discourses* I, pp. 115 ff.)

[59]*Discourses* I, p. 134.

resolved. To geometrize the motion of heavy bodies, we had first to assume that the process of natural acceleration affects the fall of all mobiles in an identical manner; to prove the truth of this assumption, we had further to prove that any differences in speed that could actually be observed were not the result of differences in weight. That proof has now been adduced. The original speed of descent of a body moving in a given medium depends exclusively on the ratio of the specific weight of the medium to its own: the differences in speed are therefore the direct consequence of a retardation effect, not of the bodies' specific weights. But at the same time, geometrization had received an even more general justification. To account for speed differences, a theoretical situation was proposed which was identical for all bodies and to which the specific details could be added in each particular case. This situation is more than a simple fiction; it is, in fact, the case that would occur if the density of the medium became equal to zero. Proving that all differences in speed would disappear in a vacuum is thus tantamount to proving that the motion of heavy bodies can be analyzed in a purely rational way. And though no experimental verification of the laws and propositions had yet been adduced, the objection that the Third and Fourth Days of the *Discourses* resembled a mathematical hypothesis rather than a physical analysis had lost much of its weight.

There remained one more, but extremely difficult, problem. Let us return to Galileo's earlier assumption that the speed of descent varies as the specific weight; that assumption represents a clear dynamic statement, namely, that the specific weight is an adequate expression of the motor force by which the body is impelled downward. If, however, we accept the new explanation and assume that every body will fall through a vacuum with the same speed, we shall not only have robbed the specific weight of its power to influence changes in speed but have also introduced an important clue as to the true nature of the motive force on which natural motions depend: were it not for the medium, that force would have an identical effect on all bodies.[60] In other words, the idea of a force imparting upon bodies speeds varying as their densities had made way, at least implicitly, for that of a force whose effects are identical on all bodies, no matter what quantity of matter they contain;

[60] It should be recalled that Galileo also applied the term *gravità* to the natural motive force, as distinct from the body's specific weight.

if the action of the medium could be eliminated, all bodies would approach the center of the earth with the same speed. Hence one of the most important consequences of Newtonian dynamics was already implicit in Galileo's analysis of changes in speed. This immediately raises the question of whether, and to what extent, Galileo himself appreciated the dynamic significance of his analysis and, more important still, whether his known views on the causes of natural motion do indeed warrant so revolutionary a conclusion.

At first sight it seems impossible to answer these important questions in the affirmative. The first difficulty is that, no matter how carefully we examine Galileo's analysis, we are unable to discover any use of dynamic concepts. His argument was rather that, since observation shows that bodies of different specific weights tend to fall with speeds that are less differentiated, the rarer the medium, the original idea of a link between speed and specific weight cannot be maintained. The second phase of the explanation appealed no more to causal considerations than the first: to conclude that the differences in speed which are barely appreciable in a very rare medium would vanish altogether in one that was completely weightless meant extrapolating the results of an experimentally verifiable situation to one that was not, in other words, treating the vacuum not as a hypothetical case but as the simplest case open to physical analysis. Hence, though Galileo's method of exposition did not formally exclude certain dynamic notions, it reduced them to a supporting role. But this first difficulty was not the most important. How can we now offer a dynamic interpretation and a fortiori a dynamic justification of the idea that in a vacuum all bodies descend with the same speed if we still lack a clear and well-defined conception of inertia, mass, and the force-to-mass ratio? And if we claim that Galileo was able to do so, are we not reading him through Newtonian spectacles? Consider his conception of inertia, his failure to distinguish between mass and weight, and his rejection of the attractionist theory of gravity. In view of all this, how could he possibly have attached a precise dynamic sense to his analysis of speed differences? Would it therefore not be much wiser to keep to the letter of the *Discourses* instead of pursuing a line of inquiry that may well be based on nothing but hindsight?

This approach would certainly be the only valid one if the elimination of all direct links between the speed of descent and the specific weight had not forced itself on Galileo's attention in quite a different situation: if a smooth sphere is placed on an inclined plane and the inclination of that plane is gradually decreased, the downward speed of the sphere will decrease correspondingly, though its specific weight has remained completely unchanged. It was precisely to explain such changes in speed that Galileo introduced the new concept of "moment of descent," which, as we saw, served to express the action of the motive force directly, and no longer by the specific weight. More generally, the new concept implied that the centripetal force present in all bodies must be endowed not only with a gravific but also with an independent motor function responsible for the propensity for downward motion and also for the resistance to upward motion. Such, in brief, was the approach that Galileo first adopted in the *Mecaniche* and that, as he developed it in the *Dialogue*, led him to the enunciation of the principle of the conservation of uniform motion on a plane equidistant from the center.

Now there is an obvious similarity between the effect of the medium on freely falling bodies and that of the angle of inclination on bodies rolling down an inclined plane. Both change the speed of the downward motion in such a way that it is no longer possible to account for the change by direct reference to the specific weight. Hence, may we not take it that Galileo, struck by this obvious similarity, extended to the case of free fall the disassociation he had already introduced in the case of motion down an inclined plane? That disassociation was easily reconciled with the experimental facts: the rarer the medium, the less modified the moving body's specific weight. But since the ratio of the speeds never approaches that of the weights, the facts are much better described if we relate the natural motion to a motor function of the downward force inherent in all heavy bodies, that is, to a function quite distinct from the gravific function of that same force.[61] But experimentation not only highlights the independence of the ratio of the speeds from that of the specific weights in a very rare medium, but also shows that speed differences tend to disappear in it. Hence, we have good reason to suppose that all such differences will vanish altogether in a

[61]Certain expressions he used seem to confirm that Galileo appreciated this connection between natural free fall and motion down an inclined plane; thus he said that the *momento* of a bladder was "slight" when compared to that of a lead sphere (*Discourses* I, p. 119); by *momento* he probably referred to "moment of descent"; similarly on p. 128.

vacuum. Reformulating this conclusion in the terms of the distinction between the motor and gravific functions of the force inherent in all heavy bodies, we may therefore say without recourse to any new concepts that in a vacuum, in which weight has ceased to affect the speed of heavy bodies, the motor function of the centripetal force will impress an identical speed on all mobiles.

Our exposition, we repeat, was in no way meant to prove that Galileo might have based his analysis of the difference in velocities on dynamic considerations; a fortiori it in no way pretends to show that, in support of that analysis. he ever contended that a body is attracted to the center of the earth by a force proportional to its mass—in other words, that its gravitational mass is proportional to its inertial mass. On the other hand, we feel that our argument does establish two things: first of all that while the conceptual system of classical dynamics alone enables us to deduce that in a vacuum all bodies must descend with equal speeds, it is not the only one to accord with that proposition—Galileo's system was also capable of doing so, although in quite a different way. Moreover, the relative ease with which we were able to link the theory of the difference in speeds with the guiding ideas of Galilean dynamics makes it quite clear that even if the latter had not yet advanced to the heights of classical dynamics, it had left traditional dynamics a long way behind. In particular, the concept of moment of descent profoundly changed the causal analysis of natural motion; not only did it put an end to the identification of weight with the motive force but, by turning the latter into a distinct object of thought, it paved the way for the direct study of its *mode of action*, satisfying one of the most essential demands of the renewal of dynamics. This is borne out by the way in which Galileo toward the end of his life applied the science of uniformly accelerated motion to the description of the action of the natural motive force.

The Action of the Natural Motive Force

It does not seem difficult to decide whether Galileo did in fact have a "modern" grasp of the action of the natural motive force. On the one hand, he treated the free fall of heavy bodies as a case of uniformly accelerated motion; on the other hand, he believed that though the force responsible for the downward motion can be lessened by the lifting effect of the medium, it nevertheless conserves a constant value.

Now does it not follow directly from a combination of these two conclusions that a constant force produces, not a constant speed, but one that is uniformly accelerated? In other words, are we not entitled to treat the view presented in the *Discourses* as a clear anticipation of Newton's second law?[62]

However, we must not jump to hasty conclusions: all indications are that Galileo could not have failed to take a Newtonian view of the action of the natural motive force; unfortunately, however, the author of the *Discourses* did not write a single line that can be quoted in direct support of that interpretation. Nor is that all. For if Galileo had indeed assumed that a constant force engenders an accelerated motion, what better basis could he have found for his definition of naturally accelerated motion? However, not only did he fail to use this assumption, but the principle to which he appealed, that is, the principle of simplicity, was of a purely metaphysical order. Must we therefore agree with Duhem[63] that in this matter Galileo had not the least inkling of the classical approach? In fact, between Duhem's and the opposite view—that Galileo arrived at a correct and quite general appreciation of the action of a force—there is room for a less radical but far more positive interpretation, based on what views Galileo himself advanced in those rare passages in which he examined the special problem of the motion of heavy bodies from a frankly dynamic standpoint. If we use this approach, we shall arrive at two conclusions: (1) that Galileo was able, on one occasion at least, to link the changes in the motive force responsible for the downward motion of bodies on an inclined plane to a change in their rate of acceleration and (2) that, in one particular case, he succeeded in describing the effect of the continuous action of a force on a moving body as a sum of elementary impulses, each equal to the force in question.

The first conclusion represents the culmination of Galileo's work on the motion of heavy bodies on an inclined plane, begun in the *Mecaniche*. Because we have discussed the essential aspects of that work in the preceding pages, we need merely repeat that in order to account for the changes in speed accompanying changes in the inclination of

[62]That, incidentally, was Newton's own view; cf. his *Principia mathematica philosophiae naturalis*, Vol. I, p. 21, Cajori edition.

[63]Cf. footnote 73 below.

the plane, Galileo introduced the concept of moment of descent, thus distinguishing de facto the motor function of the force inherent in all bodies from its gravific function. In other words, the moment of descent served to represent the spontaneous propensity for downward motion displayed by every body along the vertical or along an inclined plane—a propensity that, moreover, is variable, since the moment of descent decreases from its maximum value along the vertical as the angle of inclination decreases, to vanish ultimately on a horizontal plane. The Third Day of the *Discourses* summed up these early conclusions as follows:

Since, then, I have your permission, let us first of all consider this notable fact, that the moments or speeds of one and the same moving body vary with the inclination of the plane. The speed reaches a maximum along a vertical direction, and for other directions diminishes as the plane diverges from the vertical. Therefore the *impeto*, ability, energy, or one might say the moment of descent of the moving body is diminished by the plane upon which it is supported and along which it rolls.

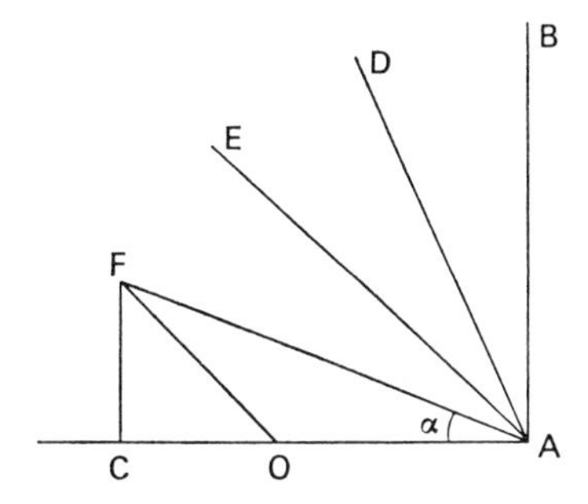

For the sake of greater clearness, erect the line *AB* perpendicular to the horizontal *AC*; next draw *AD*, *AE*, *AF*, etc., at different inclinations to the horizontal. Then I say that the highest and the total moment of descent of the falling body is along the vertical and is a maximum when it falls in that direction; the moment is less along *DA* and still less along *EA*, and even less yet along the more inclined plane *FA*. Finally along the horizontal plane the moment vanishes altogether; the body finds itself in a condition of indifference as to motion or rest, has no inherent tendency to move in any direction, and offers no resistance to being set in motion.[64]

Because it reveals the action of the motor force responsible for the natural motion of bodies, the moment of descent is indeed the dynamic

[64]*Discourses* III, p. 215.

factor to which the lesser or greater speed of a body's approach toward the center must be related. Let us add, finally, that the theory of the inclined plane allows a precise description of "the manner in which the moment varies with the inclination of the plane."[65] By a simple argument, but too long to be reproduced in full, Galileo established further that the moments of descent on planes of different inclination but of the same height are inversely proportional to the length of these planes; thus the moment along the plane *FA* will be to that along the plane *FO* as *FO* is to *FA*, or if we consider the plane *FA* and the perpendicular *FC*, the moment of descent along *FA* will be to the moment of descent along *FC* as *FC* is to *FA*.[66]

While staying with planes of different inclination but of the same height, let us now pass on from the moments of descent to the actual motions of heavy bodies on such planes. A proposition of the utmost importance sums up what, to Galileo, was their most remarkable quality: "The speeds acquired by one and the same body moving down planes of different inclinations are equal when the heights of these planes are equal."[67] Even though we are not told just how he arrived at this conclusion, we are justified to think that he had no difficulty in reconciling it with his analysis of uniformly accelerated motion. To do that, he had to meet just two conditions: to identify motion on an inclined plane with uniformly accelerated motion and then to treat the effect due to the inclination of the plane as a change in the rate of acceleration. Now there is no doubt that Galileo did meet these conditions: a scholium upon Theorem II explicitly extended the fundamental laws of uniformly accelerated motion to motion on an inclined plane;[68] moreover, in an earlier text dealing with the possible application to fluids of the laws governing the natural motion of heavy bodies, Galileo stated quite explicitly that the slope of an inclined plane has the power to change "the rate of acceleration" (*la ragione dell'accelerazione*).[69] These remarks, however, provide no more than a rough-and-ready justification of Galileo's conclusion; to prove it conclusively, a third condition had to be satisfied: to provide a precise evaluation of the "rate of change" that the inclined plane communicates to the rate

[65] Ibid., p. 216.

[66] Ibid., pp. 216–217. If α designates the angle between the inclined plane and the horizontal, we have, in the second case, moment along FA = moment along $FC \times \sin \alpha$.

[67] Ibid., p. 205.

[68] Ibid., p. 214.

[69] *Opere*, Vol. VI, p. 620.

of acceleration. Now, there was only one way to do that, namely, to use the ratio of the motor forces as determined on the inclined plane as a model for evaluating the ratio of the accelerations and to show that the proposition then follows as a matter of course. Short of this appeal to dynamics, there was no means of proving that the speeds acquired by one and the same body moving down planes of different inclinations are equal when the heights of the planes are equal. In fact, Galileo's own attitude bears out our interpretation to the full: in the first edition of the *Discourses* (1638), in which uniformly accelerated motions are discussed in purely kinematic terms, the proposition was presented as a principle corroborated by the evidence of the senses,[70] but when, following objections by Viviani, Galileo tried to revise his proof a year later, he made explicit use of dynamic ideas.[71] The argument that then helped him transform his principle into a theorem shows clearly that he had indeed established a link between the acceleration characteristic of a downward motion and the force responsible for it.

Thus if *AB* is an inclined plane of height *AC*, prove that a moving body will reach the horizontal *CB* with the same speed, no matter whether it descends down *AB* or down *AC*. We know that each of these motions is uniformly accelerated, which means that the increase in speed is proportional to the time of travel and that the distance traversed in one and the same time interval varies as the rate of acceleration (or, as Galileo put it, as the "moment of speed" which, being added in suc-

---

[70]*Discourses* III, pp. 205–207. For a discussion of that proposition as a principle, see the next section of this chapter.

[71]In a letter to Benedetto Castelli of December 3, 1639, Galileo explained the conditions under which he developed this proof: "The objections raised over the past few months by this young man, at present my guest and pupil, to the principle which I have adopted in my treatise on accelerated motion . . . have forced me to reconsider it in such manner . . . that I was finally led, to his great pleasure and my own, to discover a truth which, if I am not mistaken, is quite conclusive. . . . He immediately prepared a draft at my home because, being robbed of the sight of my eyes, I might well have become bogged down in figures and notations. Presented as a reminiscence by Salviati, in dialogue form, it is meant for insertion in a new edition of my *Discourses and Demonstrations*, immediately after the scholium upon the second proposition . . . as a theorem altogether essential for the consolidation of my science of motion" (Vol. XVIII, pp. 225 ff.). Viviani published this text in 1656 in the second edition of the *Discourses;* the text that Galileo actually dictated to Viviani appears in Volume VIII of the Edizione nationale, pp. 442–445. Several of Galileo's other correspondents had also raised objections to this "principle" and had asked him for irrefutable proof (cf. *Letter to Baliani*, August 1, 1639, Vol. XVIII, p. 78).

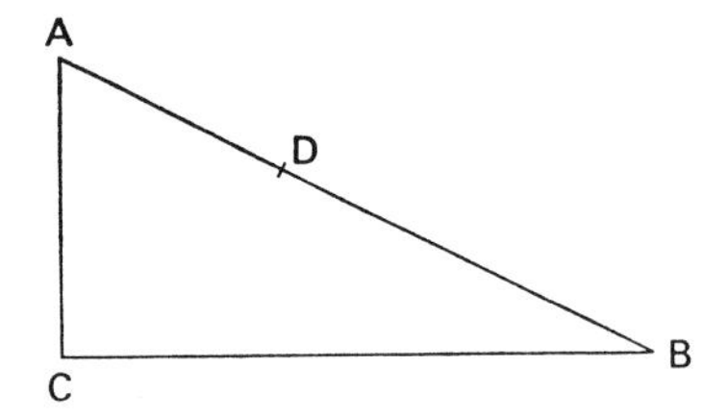

cessive and equal intervals of time to the speed acquired so far, determines the characteristics of every motion); we also know that the rate of acceleration is greater along AC than it is along AB; finally we know that the moment of descent down AC will be to that down AB as AB is to AC.

On the incline AB let us now measure off AD so that $AD/AC = AC/AB$, or $AC^2 = AD \times AB$. Now if we call M_1 the moment of descent down AC, and M_2 that down AB, we obtain $M_1/M_2 = AB/AC = (AB \times AC)/AC^2 = (AB \times AC)/(AD \times AB) = AC/AD$. "And therefore," Galileo concluded, "the body will traverse the space AD, along the incline AB, in the same time which it would occupy in falling the vertical distance AC (since the *momenti* are in the same ratio as these distances); also the speed at C is to that at D as the distance AC is to the distance AD. But, according to the definition of accelerated motion, the speed at B is to the speed of the same body at D as the time required to traverse AB is to the time required for AD; and, according to the last corollary of the second proposition, the time of passing through the distance AB bears to the time of passing through AD the same ratio as the distance AC to AD. Accordingly, the two speeds at B and C each bear to the speed at D the same ratio—that of the distances AC and AD; hence they are equal. This is the theorem which I set out to prove."[72]

The proof was therefore based on the fact that the distances AC and AD are traversed within identical time intervals, so that the speed attained at C is to that attained at D as AC is to AD. Now the motions along AC and AB are uniformly accelerated from rest, such that the

[72]*Discourses* III, p. 218. The reader will have noticed that Galileo proceeded by two distinct steps. In the first, he deduced the equation $V_C/V_D = AC/AD$ (where V_C is the velocity in C, and V_D that in D) from the ratio of the forces (*momenti*) to the distances traversed; in the second, starting from the definition of uniformly accelerated motion, he put $V_B/V_D = T_{AB}/T_{AD}$ (where V_B is the velocity in B, T_{AB} the time needed to traverse AB, and T_{AD} the time needed to traverse AD). Now, since $T_{AB}/T_{AD} = AC/AD$ (Corollary II of Theorem II), it follows that $V_B/V_D = AC/AD = V_C/V_D$.

speeds acquired after a given time interval are in the same ratio as the accelerations; hence, AC will be to AD as the acceleration through AC is to that through AB. That being the case—and this is the essential point—how did Galileo determine the distances traversed in equal times by a body moving along AC and AB, respectively? Quite simply by assuming that the ratio between the moments of descent, $M_1/M_2 = AC/AD$, is also the ratio between the distances that a body possessed successively of M_1 and M_2 would cover in the same interval along AC and AB. Now, since the distances traversed vary as the accelerations, it follows that when we deduce the ratio between them from the ratio between the moments of descent, we are, in fact, assuming that the natural motive force has an accelerative effect. In other words, it was solely the correlation of the moments of descent with the accelerations of the moving body along an inclined plane and along the perpendicular, thanks to its consequent determination of the ratio between the accelerations, that enabled Galileo to assert that the distances traversed in identical time intervals are equal to AD and AC, respectively, and hence to establish the equality of the speeds attained in B and C. The importance of this argument cannot be overstated: it is the only instance in the entire *Discourses* of a proposition on uniformly accelerated motion based directly on dynamic considerations. Near the end of his career Galileo thus discovered a means of deducing, with the aid of assumptions about the mode of action of the natural motive force, the most important propositions of his theory of motion. Had it come twenty years earlier, this discovery would have upset the entire balance of the *Discourses*; not only would it have enabled Galileo to substitute the link between the force and the acceleration for the principle of simplicity in his definition of naturally accelerated motion, but it would also have helped him to construct his proofs without reference to the medieval representation of speed. The result would have been a much more homogeneous theory of the motion of heavy bodies.

As it is, we know that by deducing the relative magnitude of the distances traversed by two bodies in differently accelerated motions in the same time from the ratio of their moments, Galileo was able to interpret, on one occasion at least, the action of the natural motive force in terms of its accelerative effects.[73] Hence, though he failed to state the general

[73] This is a conclusion that Duhem refuses to admit; a summary and discussion of his thesis will be given in Appendix 5.

principle that a constant force engenders a uniformly accelerated motion, Galileo had in any case broken with the traditional view that increases in speed call for corresponding increases in the strength of the motive force. For though he held that the moment of descent along the plane AB is smaller than that along the perpendicular AC, he nevertheless believed that both have a constant value and that both motions are uniformly accelerated, that the greater the force (that is, the moment of descent), the greater the acceleration, and conversely that increases in acceleration alone call for the application of increasingly powerful forces. For the first time in the history of mechanics, dynamic and kinematic precepts had been combined into a coherent whole, or, to put it in the traditional language, the analysis *quoad causas* had at last been fused with the analysis *quoad effectus*:[74] the moment of descent, on which the increase in speed depends in the sphere of causes, had been shown to correspond to the moment of speed or acceleration in the sphere of effects. Hence Galileo's final reflections on the moment of descent, the culmination of fifty years of research, may be said to have carried him to the threshold of classical dynamics.

However, Galileo's dynamic contributions had yet another basis, namely, his attempts, in 1638–1639, to come to grips with the force of percussion. As the reader will recall, he treated the latter as a dynamic magnitude, directly proportional to the mass (or more precisely to the weight) of the percussive body and to its speed.[75] However, a momentous event had occurred between his Paduan period and his final reflections on the subject, that is, the construction of a rigorous science of uniformly accelerated motions based, inter alia, on the elucidation of the nature of speed increases, a construction that opened the way for a reexamination of the entire problem. Thus natural percussion—the force with which a freely falling body strikes an obstacle—can be represented by the product of a constant factor (its weight) and a variable factor (the speed of its motion). Once the process underlying the variation or in-

[74]The recognition of the link between the characteristic continuity of speed increases and the fact that even the smallest force can, in the absence of friction, set even the largest body in motion, can, as we have seen, be considered a first fusion between the *quoad effectus* and *quoad causas* approaches; however, this fusion had no consequences and is therefore of purely historical interest.

[75]In other words, it could be expressed by some such formula as mv; see Appendix 3.

crease of that factor was known, it seemed reasonable to assume that the analysis of uniformly accelerated motion would throw light on the way in which a heavy body acquires a percussive force during its descent, thereby making it possible to advance from a purely descriptive account to a genuine explanation. One of the last texts Galileo devoted to the force of percussion was intended to do just that: his interpretation of the final "moment" characteristic of the force of percussion represented a substantially correct account of the way in which that force engenders its effect.

"The total moment of a heavy body during the act of percussion," Galileo wrote, "is nothing but the aggregate of an infinity of moments, each equal to a single elementary moment, be it internal and natural to the body itself (that is, due to the pressure which its absolute weight exerts while it rests on another, resistant, body) or be it an external and violent moment such as that of the motive force. Such moments accumulate throughout the body's motion by successive and equal additions from one instant to the next and conserve themselves in the body in precisely the same way as the speed in a freely falling body; for much as, in the infinitely many instants of however small a time, a heavy body can be seen to pass continuously through new and equal degrees of speed while conserving the degrees previously acquired, so also is there a conservation and composition from one instant to the next of those natural or violent moments that are impressed upon a body either by nature or by artificial means."[76]

This analysis is easily translated into the language of classical mechanics. Each of the elementary moments whose "aggregate" or summation constitutes the final moment of the percussive force is, in fact, identical with what Galileo had described elsewhere as the "moment of descent." As such it was determined first of all by the body's weight or, more simply, by its mass. As we saw, Galileo had meanwhile come

[76] *Opere*, Vol. VIII, p. 344. It was in 1638–1639 that Galileo dictated his last notes on the force of percussion to his pupil Marco Ambrogetti. The work, begun in dialogue form, was never completed, though Galileo on more than one occasion expressed his intention to extend the first Four Days of the *Discourses* with a Fifth Day devoted to the force of percussion (see *Letter to Louis Elzevir*, January 4, 1638, Vol. XVIII, p. 251; *Letter to Baliani*, August 1, 1639, Vol. XVIII, pp. 78 ff.). The original text was not published until 1718, by which time it was no longer of more than historical interest. The fragment from which we have been quoting appears among the notes Galileo added to the dialogue part.

to identify the moment of descent with the characteristic acceleration of a natural motion: hence it is not unreasonable to conclude that he treated each of the elementary moments whose "composition" engendered the percussive force of a body as a magnitude of the type $m\gamma$. As for the force of percussion, that is, the sum of the moments added "from one instant to the next," it was nothing other than the integral of the instantaneous effects as we have just defined them; it can therefore be expressed by some such formula as

$$\int_{t_0}^{t} m\gamma \, dt = \int_{t_0}^{t} m(dv/dt) \, dt = mv.$$

However, the reader does not have to accept this representation to appreciate the exceptional importance of Galileo's own remarks. Engendered by motion, the force of percussion becomes a magnitude whose final moment is due to the "composition of an infinity of elementary moments." For a given body, every one of these elementary moments is equal to the force that this body will manifest quite spontaneously; open to a statical evaluation (it is then measured by the balancing force), this force can thus, in the absence of obstacles, become also a source of motion, in which case it will represent the initial dynamic power whose continuous action confers upon the heavy body the final dynamic power, to which Galileo referred as the force of percussion. However, we still have to explain precisely how the continuous action of the first can engender the second. The successful analysis of uniformly accelerated motion provided an analogy, and it was Galileo's genius to use it when he assigned the same role to the elementary moment that he had done to that determinate amount of speed (or *momentum velocitatis*) whose addition in equal and successive time intervals holds the key to uniformly accelerated motion. "Such moments," he explained, "accumulate through the body's motion from one instant to the next by successive and equal additions and are conserved in the body in precisely the same way as the speed of a freely falling body increases." Now, in offering this definition, he was actually presenting a radically new interpretation of the mode of action of a force. Since the elementary moment, as the measure of the force impelling the heavy body downward, is repeated from one instant to the next, and since its addition to the moments previously acquired presides over the genesis of the force of percussion, it clearly appears that a motive force acting on a moving body communicates a continuous series of instantaneous impulsions to it: a uniformly increasing effect actually corresponds to the action of a constant force. Admittedly, since Galileo could not draw on the resources of infinitesimal calculus, he was unable to explain the precise mode of summation of these elementary impulsions; it is nevertheless true that on this point too his approach was already at one with that of classical dynamics. Analysis had previously shown that the action of a force must be evaluated in terms of the acceleration to which it gives rise; and his studies of percussion then added the idea that the continuous action of a force on a moving body is reflected in a summation of instantaneous impulsions, each of which is precisely equal to the magnitude of that force.[77]

We are now in a position to take stock of Galileo's contribution to dynamics. One idea dominated all his work on the motion of freely falling bodies and of bodies on inclined planes, namely, that all of them possessed an inherent force that impelled them toward the center of the earth. Two simultaneous functions had to be assigned to this force: a gravific function measuring the body's (specific or individual) weight at rest and a motor function presiding over its natural downward motion. Now, while it was not completely separable from the gravific function (a point to which we shall revert later), the motor function is by no means a simple appendage of the latter: in a vacuum, where it would be impossible to distinguish real from absolute weights, no differences in speed could be observed.[78]

From these principles, Galileo drew several remarkable conclusions. The centripetal force inherent in every body manifests itself, from a

[77]It should be added that the continuity of time greatly complicated the problem for Galileo. Being constant, time is comprised of a number of instants that can be as great as we like. In that case, since an elementary "moment" is added from one instant to the next, how can we avoid the conclusion that the force of percussion engendered by a motion must be infinite? Galileo never succeeded in solving this paradox, which highlights the incredible difficulties that seventeenth-century scientists had to overcome in the absence of infinitesimal calculus. Torricelli, too, failed to resolve this problem, as witness his second and third *Leçons académiques* (cf. Carugo and Geymonat's edition of the *Discourses*, pp. 863–864).

[78]Let us stress once again that Galileo himself never used the terms "gravific" and "motor" function; a single term, *gravità*, served him to describe the centripetal force present in all bodies, regardless of whether it was evaluated statically by means of the weight or whether it was considered as the cause of the motion.

dynamic point of view, as a moment of descent, reflecting the propensity of bodies to downward motion no less than their resistance to upward motion. But what happens on a plane equidistant from the center, on which both the propensity and the resistance to motion are, in the absence of friction, completely eliminated? This question was first examined in the *Mecaniche* and again during the Second Day of the *Dialogue*. According to the *Mecaniche*, the slightest force will always set even the largest body in motion; according to the *Dialogue*, if the least impulsion is given to a body on a horizontal plane, it will start to move and continue to do so indefinitely, unless an opposite force restores it to the state of rest. Finally, two of Galileo's conclusions may be called anticipations of classical dynamics: that the moment of descent, as the expression of the natural motive force, is proportional to the moment of speed (*momentum velocitatis*) on which the acceleration depends; and that a constant force engenders a uniformly increasing effect. Under these conditions, are we not entitled to claim that there was a direct filiation between the conceptual system of classical dynamics and Galileo's? In other words if we generalized Galileo's results by replacing the idea of an inherent force with the attractionist hypothesis, would we not be led directly to the propositions of Newtonian science?

This view, we believe, is untenable. For though Galileo's system did indeed agree with the idea that differences in speed do not spring from differences in weight, his failure to distinguish mass from weight[79] led him to paradoxical conclusions. On a plane equidistant from the center, on which bodies have lost all propensity and resistance to motion, the motor function of the force inherent in all bodies is provisionally suspended, so that the motion assumes all the characteristics of inertial motion. Yet the body has lost none of its weight; from this it follows that it can be as heavy as we like and yet be possessed of inertial motion; translated into classical language, Galileo's conception would thus signify that the weight must not be considered a force.[80] On the other hand, it was impossible to say that there was no relation between the body's weight and its propensity for motion. Thus we saw that the speed of freely falling bodies in media other than a vacuum is propor-

---

[79]That is, his treatment of weight as an intrinsic property of matter.

[80]We might also say that in order to arrive at the idea of inertia Galileo had to distinguish the weight from the motor force, a distinction that the concept of inertia itself, once generalized, would render superfluous.

tional to the ratio between the specific weight of the medium and that of the moving body; if this ratio is very small, the speed is almost identical with that which the body would possess in the vacuum; if it is close to unity, on the other hand, the speed becomes very small. Does that not suggest that the specific weight of the moving body tends to reinforce the body's propensity to motion since a very great weight allows the motor function of the inherent force to express itself to the full? Hence, even though the speed does not depend on the weight, we are entitled to ask whether the weight does not serve as an "instrument"[81] by which the motor force is helped to overcome the resistance of the medium. Thus we have the second paradox: while the weight sometimes intervenes in the determination of the behavior of moving bodies, at other times (and especially on a plane equidistant from the center) it suddenly plays no part in it at all. This paradoxical situation could be resolved only with the help of a clear distinction between mass and weight—it was because he characterized the former by its inertia and the latter as an effect of an attractive force proportional to the mass that Newton was able to explain why, in a vacuum, all bodies fall with the same speed or, better still, why their differences in weight are a prerequisite of their equal speeds; as a result any ambiguity as to the relationship between weight and motion disappeared, and the principle of inertia could at last be formulated correctly. Clearly, therefore, though Galileo succeeded in drawing conclusions that classical dynamics would fully endorse, he did so by means of a conceptual system that in no way foreshadowed that of classical dynamics. Exploiting the experiential data to the full, avoiding the pitfalls of a type of systematization that in the absence of a more probing analysis of motion (and notably of circular motion) would have been dangerously premature, Galileo could do no better than to erect a scaffolding, one that was bound to be removed just as soon as the new edifice had been completed.

But it would be wrong to end on so negative a note. For despite the shortcomings to which we have drawn attention, Galileo's last reflections also represented the culmination of an old project: the search for a lasting link between statics and dynamics. This search was begun in the *De Motu*, where, treating the propensity to downward motion as a

[81]Galileo used this telling expression during the Third Day of the *Discourses* (p. 119).

direct attribute of the specific weight, Galileo had tried to base the dynamics of natural motion on Archimedean hydrostatics. While no one can say precisely how long it took him to realize that this attempt was incompatible with experience,[82] we do know that the *Mecaniche* paved the way for what in some ways was the opposite approach. Thus, when he used *moment* to designate a magnitude we express by mv,[83] he expressly introduced virtual velocities into the determination of equilibrium states: two unequal bodies in a system of constraints will balance each other whenever their moments are equal, that is, whenever there is a strict compensation between the weight and the speed. From this dynamic definition of equilibrium, Galileo then tried to proceed to a description and evaluation of the force of percussion and of the centrifugal force. We know that his attempts were thwarted in both cases by his inability to conceptualize either of these forces in their specific dimensions,[84] a failure, moreover, that was quite predictable since as soon as the link with the static system is cut, the moment vanishes and with it the physical magnitude that was meant to fuse dynamics to statics.

Small wonder, therefore, that the final solution had to be sought along a third path. While the moment of a body suspended from a lever (or from a simple machine reducible to the lever) loses all meaning as soon as the body is disconnected from the lever, things are quite different with the moment of descent along an inclined plane. Being identical for all points of the plane, that moment measures both the force responsible for keeping the body in equilibrium and the magnitude of its propensity to downward motion once it has begun to move. However, a final step had to be taken before this concept, situated at the frontiers of statics and dynamics, could be used to forge a genuine passage from the one to the other, namely, the correlation of these general dynamic ideas with the science of uniformly accelerated motion. Galileo succeeded in doing just that in at least two ways. Our analysis of his proof that a body descending along inclined planes of different lengths but of the same height will reach the horizontal base with the same speed has shown first of all how he succeeded, on one occasion at least, in describing the action of a force in terms of its accelerative effects, that is, by

[82]It is possible that when he wrote the *Discourse on Bodies in Water* (1612), he still considered this attempt worthwhile.

[83]Cf. Chapter 3.

[84]For the force of percussion see Appendix 3; for the centrifugal force, see Chapter 5.

$m\gamma$ (which, of course, does not appear as such in the *Discourses*). Moreover, in his last reflections on the force of percussion, he once again used the example of uniformly accelerated motion to elucidate both the effect engendered by a constant force and also the precise way in which that effect is engendered. He admittedly left it at that, but there is no doubt that he had cleared the way for the fusion of statics with dynamics. Characterized dynamically by its coefficient of acceleration, the moment of descent that on an inclined plane is responsible for the motion is identical with the moment by which equilibrium conditions can be described from a static viewpoint.[85] Similarly, the moment, whose repetition from one instant to the next engenders the force of percussion in the form of elementary impulsions, can also be given a static interpretation: it can be used to represent the body's spontaneous "power" or "dead weight."[86] In short, though it was too closely bound up with a particular experimental situation to culminate in a general dynamics, Galileo's analysis of the natural motive force nevertheless paved the way for the union of statics and dynamics in a single conceptual system.[87]

---

[85]*Discourses* III, p. 216: "It is clear that the impelling force (*impeto*) acting on a body in descent is equal to the resistance or least force sufficient to hold it at rest."

[86]*Opere*, Vol. VIII, p. 325. (The term "dead weight" was subsequently adopted by Leibniz.

[87]Much as Leibniz was able, with the help of the infinitesimal calculus, to carry Galileo's analysis of uniformly accelerated motion to a successful conclusion, so his fusion of statics with dynamics may be considered a direct extension of Galileo's last studies (see M. Guéroult, *Dynamique et métaphysique leibniziennes*). In particular, Leibniz treated two heavy bodies in equilibrium as two "dead forces" whose combined effect is to offset the spontaneous propensity for motion of either; in that case, as Guéroult rightly pointed out, "the forces impede one another and exhaust themselves during each instant in which they are active" (p. 33); the result of such instantaneous actions that cannot, by definition, add their effects is the weight. Let us now pass on to a freely falling body. The same propensity for motion that in the last case exhausted itself in the weight now manifests its dynamic power during the first infinitesimal instant of the fall in the form of a *conatus*; and the constant accumulation of the latter gives rise to the *vis viva* (Leibniz's *potentia*) that the body acquires thanks precisely to its motion. And because Leibniz thus introduced a clear correlation between weight and *conatus*, he was able "to identify equilibrium with incipient motion, and weight with infinitesimal acceleration" (p. 80). The link between statics and dynamics was now complete; the latter had become a special case of the former. But the reader will also see that Leibniz's *conatus*, though more general in scope (the analysis was no longer limited to motion down an inclined plane), was a direct extension of Galileo's moment of descent; similarly, the way in which Leibniz's *potentia* was engendered by a summation of *conatus* bore a close resemblance to Galileo's treatment of the force of percussion as a summation of elementary moments.

The Status and Role of Galileo's Principles

Our analysis (Chapter 6 and the first two parts of the present chapter) has gone part of the way toward elucidating the conceptual basis of Galileo's approach to the motion of heavy bodies. The better to evaluate the soundness of that approach, we shall now examine its underlying principles.

If by a principle we understand a general expression, as a postulate, of some special property of these phenomena that are to be embraced by a theory, there is no doubt that Galileo indeed based his entire study of the motion of heavy bodies on a clearly determined principle, and explicitly so. Following soon after the definition of naturally accelerated motion and preceding the relevant theorems and propositions, this principle took the following form: "The degrees of speed acquired by one and the same body moving down planes of different inclinations are

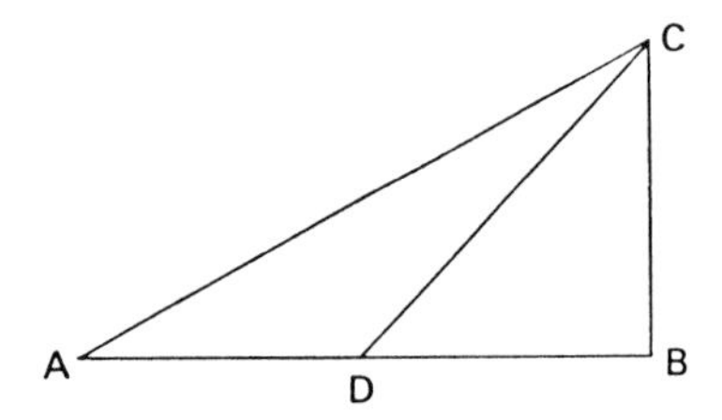

equal when the heights of these planes are equal."[88] Let us call the "height" of an inclined plane the perpendicular dropped from the upper extremity of the plane onto a horizontal line drawn through the lower extremity of that plane. Thus, if *AB* is the horizontal and if *CA* and *CD* are two inclined planes, Galileo referred to the perpendicular *CB* as the "height" of the planes *CA* and *CD*. Now, according to Galileo's principle, "the degrees of speed acquired by one and the same body descending along the planes *CA* and *CD* to the terminal points *A* and *D* are equal since the heights of these planes are the same." Moreover, "it must also be understood that the degree of speed is that which would be acquired by the same body falling from *C* to *B*."[89]

Of course, as Sagredo rightly remarked, this principle applied purely to ideal conditions, that is, to what happens in the absence of outside

[88] *Discourses* III, p. 205; the principle had previously been formulated in *Dialogue* I, p. 47, but without any justification. Descartes, as we saw, said that, in his opinion, the principle was not generally true (Adam and Tannery edition, Paris, 1897–1913, Vol. II, p. 386).

[89] *Discourses* III, p. 205.

resistances. However, though the experiment to which Galileo referred was a purely rational one, there was nevertheless no difficulty in reconciling it with sense experience; few principles can even be so easily justified or so strictly and satisfactorily harmonized with the observational data. Galileo's reply to Sagredo thus merits being quoted in full:

Your words are very plausible; but I hope by experiment to increase probability to an extent which shall be very short of a rigid explanation. Imagine this page to represent a vertical wall, with a nail driven into it; and from the nail let there be suspended a lead ball of one or two ounces by means of a fine vertical thread *AB*, say, from four to six feet long; on this wall draw a horizontal line *DC* at right angles to the vertical thread *AB*, which hangs about two finger breadths in front of the wall. Now bring the thread *AB* with the attached ball into the position *AC* and set it free; first it will be observed to descend along the arc *CBD*, to pass the point *B*, and to travel along the arc *BD* till it almost reaches the horizontal *CD*, a slight shortage being caused by the resistance of the air and the string; from this we may rightly infer that the ball in its descent through the arc *CB* acquired *impeto* on reaching *B*, which was just sufficient to carry it through a similar arc *BD* to the same height. Having repeated this experiment many times, let us now drive a nail into the wall close to the perpendicular *AB*, say, at *E* or *F*, so that it

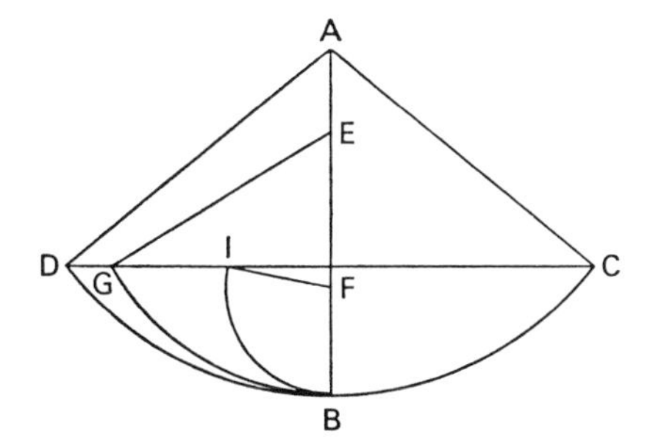

projects out some five or six finger breadths in order that the thread again carrying the ball through the arc *CB* may strike upon the nail *E* when the ball reaches *B*, and thus compel it to traverse the arc *BG* described about *E* as center. From this we can see what can be done by the same *impeto* which previously starting at the same point *B* carried the same body through the arc *BD* to the horizontal *CD*. Now, gentlemen, you will observe with pleasure that the ball swings to the point *G* in the horizontal, and you would see the same thing happen if the obstacle were placed at some lower point, say, at *F*, about which the ball would describe the arc *BI*, the rise of the ball always terminating exactly on the line *CD*. But when the nail is placed so low that the remainder of the thread below it will not reach to the height *CD* (which would happen if the nail were placed nearer *B* than to the intersection of *AB* with the horizontal *CD*), then the thread leaps over the nail and twists itself about it.

This experiment leaves no room for doubt as to the truth of our supposition; for since the two arcs *CB* and *DB* are equal and similarly placed, the *momento* acquired by the fall through the arc *CB* is the same as that gained by fall through the arc *DB;* but the *momento* acquired at *B*, owing to the fall through *CB*, is able to lift the same body through the arc *BD;* therefore the *momento* acquired in the fall *BD* is equal to that which lifts the same body through the same arc from *B* to *D;* so, in general, every *momento* acquired by fall through an arc is equal to that which can lift the same body through the same arc. But all these *momenti* which cause a rise through the arcs *BD*, *BG*, and *BI* are equal, since they are produced by the same *momento*, gained by fall through *CB*, as experiment shows. Therefore all the *momenti* gained by fall through the arcs *DB*, *GB*, *IB* are equal.[90]

Because it was "so reasonable that it ought to be conceded without question" and moreover because it was borne out by the most careful observation, the principle on which Galileo proposed to found his theory of the motion of heavy bodies thus seemed to provide all that can be expected of a scientific principle.

For all that, its role was not nearly as clear as its presentation might have suggested. To begin with, we are entitled to ask if Galileo himself treated it as a genuine principle, the more so as a careful reading of the discussions constituting the Third and Fourth Days of the *Discourses* shows that he introduced it during only two arguments, the first of which[91] he thought far from indispensable and the second of which merely took the form of a scholium upon a proposition; the principle was neither applied nor mentioned in connection with any proof. Moreover, a scholium that Galileo added to the second edition of the *Discourses* and that we have had occasion to discuss at some length was intended to do no less than adduce a "due and proper" proof of his alleged "principle." Now, this reversal, disconcerting though it may appear to be, was in fact quite logical: from the definition of naturally accelerated motion we know that speed increases with time, and that, as Galileo himself put it, the greater the inclination of the plane, the smaller "the rate of acceleration." From this it follows that it is by no means impossible to prove that a body moving down variously inclined planes of the same height will reach the horizontal base with one and the same speed. What

---

[90]Ibid., pp. 205–207.

[91]Theorem III; see Chapter 6 above.

we have here is not a principle but a simple theorem on naturally accelerated motion, namely, that the speed of a body on an inclined plane is determined exclusively by the height, not the length, of that plane.[92] Thus relegated to a single argument, ultimately subjected to a demonstration, this "principle" has all the appearances of a pseudoprinciple. Hence, when he tells us that in developing his mechanics he needed only this single assumption, it is difficult to avoid the conclusion that he was mistaken about both its meaning and its scope.[93]

This is certainly true if we keep to the letter of Galileo's formulation, but it is possible to show that in a different form, the principle did no more than reiterate one of the most fundamental propositions of mechanics. Thus, let us examine the only instance in which Galileo used it to develop an argument, namely, the scholium upon Proposition XXIII. In it, as the reader will recall, Galileo set out to prove that a body possessing a certain speed after a descent along a slope of a given length and continuing to move with the same speed along a horizontal plane will cover twice that length in the same time.[94] Now, since such a horizontal plane introduces neither acceleration nor retardation, it follows that uniform motion along it must be perpetual.[95] Let us next imagine that our moving body, upon reaching the lowest point of its original fall,

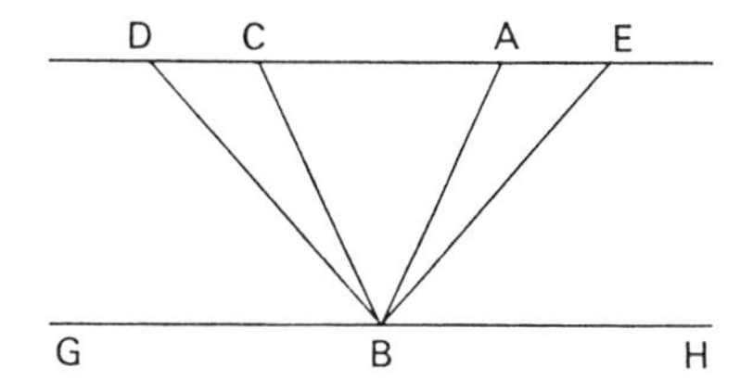

say, *B*, has its motion reflected along an ascending plane *BC* that makes the same angle with the horizontal line *GH* as the plane *AB*. On *BC* "there already exists a cause of retardation; for in any such plane this

[92]This was the view particularly of Christian Huyghens (*Horologium Oscillatorium*, 2nd part, Vol. XVIII, pp. 141–142).

[93]*Discourses* III, p. 205. Under the circumstances it may seem strange that we dwell on this principle at such great length. However, as the reader will see, it was in his attempts to prove it that Galileo succeeded in proving his real assumption, so that we may take it that the latter, as published in the second edition of *Discourses*, was the true principle of the Galilean theory of the motion of heavy bodies.

[94]See Chapter 4.

[95]*Discourses* III, p. 243.

same body is subject to natural acceleration downward. Accordingly we have here the superposition of two different states, namely, the velocity acquired during the preceding fall which, if acting alone, would carry the body to infinity at a uniform rate, and the velocity that results from a natural acceleration downward common to all bodies. It seems altogether reasonable, therefore, if we wish to trace the future history of a body that has descended along some inclined plane and has been deflected along some plane inclined upward, for us to assume that the maximum speed acquired during descent is permanently maintained during the ascent. In the ascent, however, there supervenes a natural inclination downward, namely, a motion which, starting from rest, is accelerated at the usual rate."[96] Let us now apply this general remark to the motion of a body descending along the downward sloping plane *AB* and then continuing its motion along the upward sloping plane *BC*. "At the very moment the body begins its ascent it is subjected, by its very nature, to the same influence that surrounded it during its descent from *A* along *AB;* it descends from rest under the same acceleration as that which was effective in *AB*, and it traverses, during an equal interval of time, the same distance along the second plane as it did along *AB;* it is clear that, by such a superposition upon the body of a uniform motion of ascent and an accelerated motion of descent, it will be carried along the plane *BC* as far as the point *C*, where the two velocities become equal."[97] From this we may deduce "logically" that "a body which descends along any inclined plane and continues its motion along a plane inclined upwards will, on account of the *impeto* acquired, ascend to an equal height above the horizontal; so that if the descent is along *AB* the body will be carried up the plane *BC* as far as the horizontal line *ACD*."[98] Can this conclusion, established for a particular case—that in which the plane *BC* makes the same angle with the horizontal *GH* as the plane *BA*—be extended to the general case of ascending and descending planes making any angle with the line *GH*, provided only that their elevation remains constant? Let the planes *EB* and *BD* make identical angles with the horizon. From our previous analysis it follows that its descent along *EB* will enable the moving body to rise to the point *D* along the slope. Moreover, from Galileo's principle of naturally accelerated motion it follows that, no matter whether the moving body descends along *AB* or

[96] Ibid.

[97] That is, equal to the speed of descent along *AB*; ibid., p. 244.

[98] Ibid.

along *EB*, it will arrive at *B* with the same speed. "Evidently then the body will be carried up *BD* whether the descent has been made along *AB* or along *EB*," the only difference being that "the time of ascent along *BD* is greater than that along *BC*."[99] In every case, the speed attained in *B* will therefore be such as to carry the moving body back to its original height.[100]

As we indicated, this argument involves two distinct steps: first, the demonstration that the speed (or *impeto*) attained on an inclined plane is such as to enable the moving body to ascend to the same height along an inclined plane of the same inclination; second, the extension of this proposition to all inclined planes of the same height of whatever slope. While the first step may seem obvious for purely symmetrical reasons, the second is not; it was, in fact, based on the principle which Galileo introduced in the Third Day. Now the true nature of the second step and the absurd consequences that would follow from its refutation bring out the true importance of Galileo's principle. Thus, if the principle were wrong and a heavy body could acquire, say, a greater speed along the plane *AB* than that which it acquires along the plane *EB*, it follows that if its motion were deflected onto a plane of smaller inclination, say, on to *BD*, the body would ascend to a height greater than that of its starting point. Conversely, if it reached the point *B* more quickly after descending down *EB* than after descending *AB*, then if its motion were deflected onto a plane of greater inclination, say, onto *BC*, it would rise to a greater height than that through which it had fallen. In either case, we should have to conclude that a body can be made to rise up by virtue of its own weight. But if we suppose, on the

[99] Ibid., p. 245.

[100] The assertion that the speed attained by a heavy body at the end of its descent will, in the absence of friction, invariably suffice to carry it back to the level from which it had started contains in it the germ of the principle of conservation of mechanical energy in an isolated system. Moreover, on p. 579 of Volume VIII of the Edizione nationale, Galileo is quoted as dealing with the same point more explicitly when he wrote that "the pendulum is, in a way, an inexhaustible store of force (*una conserva inesausta di virtù*), because it is capable of conserving indefinitely any *impeto* impressed upon it." The idea that the speed it attains after falling from a given height is always sufficient to carry the moving body back to the same height was hotly contested by most of Galileo's contemporaries. Galileo met their main objection—the enormous difference between the force needed to fire a bullet to a certain height and the force the latter acquires in returning from the same height—with the argument that the air offers much greater resistance to rapid motions than it does to slow; see *Letter to Baliani*, Vol. XVII, p. 77.

contrary, that a moving body will always reach the point *B* with the same speed no matter whether it has descended down *AB* or down *EB* (or, indeed, along the perpendicular), we can immediately dismiss as impossible a situation in such flagrant contradiction with "the fact that heavy bodies tend, not to rise, but to fall."[101] While it took the form of a proposition on naturally accelerated motion, Galileo's principle thus culminated in what constitutes the most fundamental insight of all mechanics, namely, that a heavy body cannot rise spontaneously. That our view of the matter is correct is moreover confirmed by the scholium published in the second edition of the *Discourses*, which, as we saw, was intended to prove Galileo's "principle." In a preliminary lemma, Galileo set out to determine the proportion in which the moment of descent, which has its greatest value along the vertical, decreases as the slope of the inclined plane is reduced. To that end—and this is the only point that interests us here—he explicitly mentioned the general idea underlying his principle: "For just as a heavy body or system of bodies cannot of itself move upward or recede from the common center toward which all heavy things tend, so it is impossible for any body of its own accord to assume any motion other than one which carried it nearer to the aforesaid common center."[102] Now this assertion may be considered a direct anticipation of Torricelli's famous axiom stating that two bodies connected together cannot move spontaneously unless their common center of gravity descends,[103] and it thus leaves us in no doubt as to the true significance of the "principle" around which the Third and Fourth Days of the *Discourses* revolved.

We are thus fully entitled to claim that its real purpose was to recall for the science of motion what may be described as the most primitive assumption possible about the behavior of heavy bodies. Though it failed to characterize motion in its specificity, Galileo's principle nevertheless introduced a proposition advanced with an absolute certainty and with which in no case the new science must conflict.[104] But Galileo

[101] E. Mach, *The Science of Mechanics*, p. 118.
[102] *Discourses* III, p. 215.
[103] P. Duhem, *Les origines de la statique*, Vol. II, p. 2. Though derived from Galileo's formulation, Toricelli's axiom nevertheless represented a clear advance over it, inasmuch as it dispensed with the common center of heavy bodies; cf. Chapter 3, footnote 151.
[104] It is interesting to note that when Christian Huyghens dealt with the motion of heavy bodies in the second part of his *Horologium oscillatorium*, he invariably assumed that a body's center of gravity cannot rise spontaneously. Proposition VI, in which he set out to prove that "the speeds acquired by

undoubtedly went much further still. For to assert that a body cannot move spontaneously unless its center of gravity approaches "the common center to which all heavy things tend" was tantamount to asserting that it is impossible to displace a body by whatever means, unless the motor work that is spent is equivalent to the resistant work; in other words, the proposition to which Galileo's principle can be reduced was also the basis of the theory of simple machines and hence of all statics. Thus when he was extending to mechanics the basic principle of statics, Galileo could well have thought that he was introducing into a nascent science the results of one that had long since been placed on solid foundations. True, the links between the two remained somewhat tenuous, for the idea that the center of gravity of bodies in motion or at rest cannot rise spontaneously is still unable by itself to fuse statics and dynamics into a single conceptual system; that fusion would come only with his final studies, in which he identified the "dead weight" with the moment of descent, and the latter with the moment of speed responsible for the increase in speed of accelerated motions. It nevertheless remains a fact that this "principle" helped to link the science of motion to the science of equilibrium, thus bringing the first the guarantee of the second. During this formative period, it seems highly probable that in Galileo's eyes this correspondence, if not a token of the truth, was at least a clear sign that his ideas were in accord with reason and experience.

Though they throw a great deal of light on the structure of the *Discourses*, these remarks tell us little about the actual principles that went into Galileo's construction of a new science of motion. Were there, in fact, such principles? Do the *Discourses* contain general postulates on the nature of motion or its properties? Both these questions can certainly

---

heavy bodies on descending along planes of different inclination are equal whenever the elevations of the planes are equal," is explicitly based on the "absurd" idea that a body can rise "to a height greater than that from which it has fallen," in other words, that its center of gravity can rise spontaneously (*Œuvres*, Vol. XVIII, p. 142). Huyghens described this idea as the "great principle of mechanics" (Vol. XVI, p. 21), and based his analysis of impact and his determination of centers of oscillation upon it. Thus he wrote in the fourth part of the *Horologium oscillatorium* that "if any number of weights begin to move under their own weight, their common center of gravity cannot rise to a height greater than that from which it began its motion" (Vol. XVIII, p. 246).

be answered in the affirmative, a reading of the *Discourses* showing unequivocally that Galileo based his demonstrations on two principles in the authentic sense of that term: that of the conservation of uniform horizontal motion and that of the composition of velocities, stating that when two motions combine, they in no way impede each other and can therefore be considered independently.[105]

To begin with, let us look at Galileo's formulation of the principle of conservation during the Third Day: "Furthermore we may remark that any velocity once imparted to a moving body will be rigidly maintained as long as the external causes of acceleration or retardation are removed, a condition that is found only on horizontal planes; for in the case of planes that slope downward there is already present a cause of acceleration, while on planes sloping upward there is retardation; from this it follows that motion along a horizontal plane is perpetual; for if the velocity is uniform, it cannot be diminished or slackened, much less destroyed."[106] But it was at the beginning of the Fourth Day that there appeared what from the modern point of view is by far the most important statement of the principle of conservation, one that is quite remarkable for its concision: "Imagine any particle projected along a horizontal plane without friction; then we know, from what has been more fully explained in the preceding pages, that this particle will move along this same plane with a motion that is uniform and perpetual, provided the plane has no limits."[107] In full accord with this definition, Theorem I on the motion of projectiles then provided the first instance of the use of the principle of conservation of uniform motion in a *geometrized argument.*

THEOREM I—PROPOSITION I

A projectile carried by a uniform horizontal motion compounded with a naturally accelerated vertical motion describes a path which is a semiparabola.

Let us imagine an elevated horizontal line or plane *ab* along which a body moves with uniform speed from *a* to *b*. Suppose this plane to end abruptly at *b;* at this point the body will, on account of its weight, acquire also a natural motion downward along the perpendicular *bn*. Draw the line *be* along the plane *ba* to represent the flow, or measure, of time; divide this line into a number of segments, *bc*, *cd*, *de*, representing equal

---

[105]For this reason we also speak of the "principle of the independence of motions."

[106]*Discourses* III, p. 243.

[107]*Discourses* IV, p. 268.

intervals of time; from the points *b*, *c*, *d*, *e*, let fall lines that are parallel to the perpendicular *bn*. On the first of these lay off any distance *ci*, on the second a distance four times as long, *df;* on the third, one nine times as long, *eh;* and so on, in proportion to the squares of *cb*, *db*, *eb*, or, we may say, in the squared ratio of these same lines. Accordingly we see that while the body moves from *b* to *c* with uniform speed, it also falls perpendicularly through the distance *ci*, and at the end of the time interval *bc* finds itself at the point *i*. In like manner, at the end of the time interval *bd*, which is the double of *bc*, the vertical fall will be four times the first distance *ci;* for it has been shown in a previous discussion

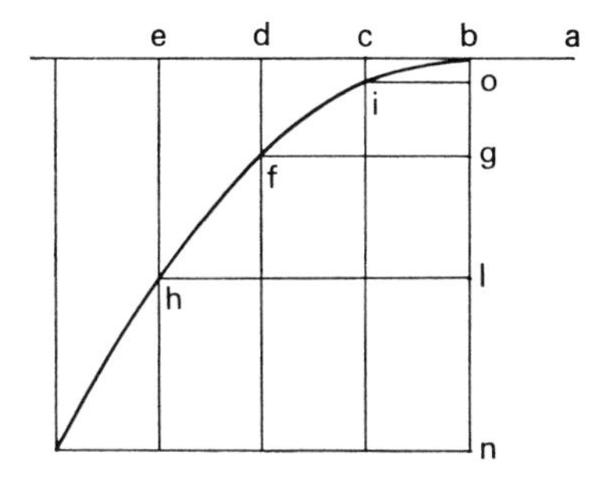

that the distance traversed by a freely falling body varies as the square of the time; in like manner the space *eh* traversed during the time *be* will be nine times *ci;* thus it is evident that the distances *eh*, *df*, *ci*, will be to one another as the squares of the lines *be*, *bd*, *bc*. Now from the points *i*, *f*, *h* draw the straight lines *io*, *fg*, *hl* parallel to *be;* these lines *hl*, *fg*, *io* are equal to *eb*, *db*, and *ch*, respectively; so also are the lines *bo*, *bg*, *bl*, respectively, equal to *ci*, *df*, and *eh*. The square of *hl* is to that of *fg* as the line *lb* is to *bg;* and the square of *fg* is to that of *io* as *gb* is to *bo;* therefore the points *i*, *f*, *h*, lie on one and the same parabola. In like manner it may be shown that if we take equal time intervals of any size whatever and if we imagine the particle to be carried by a similar compound motion, the positions of this particle at the ends of these time intervals will lie on one and the same parabola. Q.E.D.[108]

Galileo's approach to the principle of the composition of velocities marked a similar advance. On the Fourth Day of the *Discourses*, he declared his intention to demonstrate in a rigid manner all those properties that belong to a body whose motion is compounded of two other motions,"[109] and went on to a general analysis of the manner in which two velocities can be combined. The case of two uniform velocities was treated in Theorem II:

---

[108]Ibid., pp. 269 and 272–273.

[109]Ibid., p. 268.

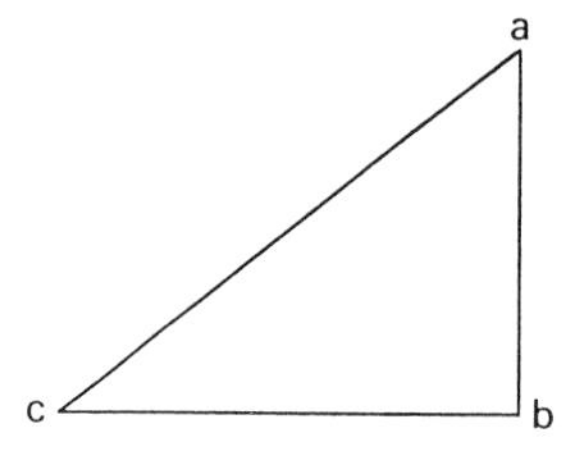

THEOREM II—PROPOSITION II

When the motion of a body is the resultant of two uniform motions, one horizontal, the other perpendicular, the square of the resultant moment of translation (*momentum lationis*) is equal to the sum of the squares of the two component *momenti*.

Let us imagine any body impelled by two uniform motions, and let *ab* represent the vertical displacement while *bc* represents the displacement which in the same time interval takes place in a horizontal direction. If then the distances *ab* and *bc* are traversed during the same time interval with uniform motions, the corresponding *momenti* of translation will be to each other as the distances *ab* and *bc* are to each other; but the body impelled by these two motions describes the diagonal *ac;* its translational moment is proportional to *ac*. Also the square of *ac* is equal to the sum of the squares of *ab* and *bc*. Hence, the square of the resultant moment is equal to the sum of the squares of the two moments *ab* and *bc*. Q.E.D.[110]

An example may help to clarify this analysis.

Let us imagine a body to move along the vertical *ab* with a uniform *impeto* of 3, and on reaching *b* to move toward *c* with a speed and *impeto* of 4, so that during the same time interval it will traverse 3 cubits along the vertical and 4 along the horizontal. But a particle that moves with the resultant velocity will in the same time traverse the diagonal *ac*, whose length is not 7 cubits—the sum of *ab* (3) and *bc* (4)—but 5, which is *in potenza* equal to the sum of 3 and 4, that is, the squares of 3 and 4 when added make 25, which is the square of *ac*, and is equal to the sum of the squares of *ab* and *bc*. Hence *ac* is represented by the side—or we may say the root—of a square whose area is 25, namely, 5.

As a fixed and certain rule for obtaining the *impeto* that results from two uniform *impeti*, one vertical, the other horizontal, we have therefore the following: take the square of each, add these together, and extract the square root of the sum, which will be the *impeto* resulting from the two.[111]

[110]Ibid., p. 280.

[111]Ibid., pp. 288–289. We have quoted this passage in full not only because of its bearing on the composition of velocities but also to show to what extent Galileo used speed and *impeto* as interchangeable terms—a habit that, as we saw at the beginning of Chapter 6, reflects his treatment of speed as a quasi-physical magnitude coextensive with motion.

But Galileo did not merely examine the case of uniform compound speeds. With the same facility (except for the vocabulary) he also determined the speed of a projectile (whose compound motion has a non-uniform, that is, accelerated, component) on each point of its trajectory. His procedure, which we shall summarize in Appendix 6 because it is too long to be examined here, proves to what high degree of accuracy and subtlety he had succeeded in carrying his analysis.

We are therefore not exaggerating the modern character of Galileo's approach when we claim that in the *Discourses* he formulated two principles in what were substantially classical terms and which, as such, played a crucial part in the geometrization of mechanics.

It is all the more astonishing that he should have failed to place these two principles at the head of his *Discourses*. Would the strict science he claimed to have constructed not have had to begin with the most general assumptions?[112] In this respect he was far from the ideal he professed on more than one occasion,[113] namely, the imitation of Euclid. But let us look more attentively at the manner in which he introduced his two principles. The principle of conservation appears in the scholium upon Proposition XXIII (Third Day), but only after it has been used in a proof that is invalid without it. Similarly, the principle of the composition of velocities was introduced explicitly to determine the *impeto* of a projectile on each point of its trajectory, but again only after it had been used implicitly in Theorem I. In other words, though they defined the general properties of motion and hence presided over all the relevant theories and propositions, the principles of conservation and composition made their entry in the form of special statements (commentaries upon a scholium, one theorem among others) designed to meet special needs and in any case long after they were first applied. Hence, though the assertions that uniform motion on a horizontal plane is perpetual and that compound motions do not impede one another did indeed serve Galileo as principles of demonstration, they had not yet attained the status of genuine principles. How can we explain this paradoxical situation, of which the ambiguity of Theorem II is perhaps the best example?[114] What prevented Galileo, while giving a near-correct

[112]The first to show clearly what principles were needed for the geometrization of the science of motion of heavy bodies was Christian Huyghens (cf. *Œuvres*, Vol. XVIII, p. 124).

[113]For example, at the end of the Third Day of the *Discourses* (p. 267).

[114]Despite its title, that theorem adduced no proof; at best it provided a description.

formulation of the two essential principles of classical mechanics, from presenting them as such? We think that a clear answer is possible in the case of the principle of conservation, and that it can then be extended to cover the case of the principle of composition.

Let us quickly recall Galileo's theory of gravity. It is endowed, as we saw, with two chief characteristics: its immanence—it is within the body itself that it exercises its gravific and motor functions—and its tendency to impel all heavy bodies toward the center of the earth. It was from these two characteristics that Galileo was able to deduce the idea of the conservation of uniform motion. Thus when a heavy body falls freely toward the center of the earth, its gravity will communicate to it a certain moment of descent, and the latter in turn will produce a speed that increases as the time. If that motion toward the center is replaced by an upward motion, the role of the gravity will be immediately reversed: transformed into a source of resistance, it will gradually lower the speed of the moving body and finally cause it to return toward the center. However, it is reasonable to assume that on a plane that neither approaches nor recedes from the center the acquired speed will neither increase nor diminish, since any cause of acceleration and retardation will have disappeared. Reasoning by an approximation to the limit, Galileo then arrived at the idea that uniform motion is conserved indefinitely. Now this analysis, though not a novelty, reminds us of two things: first that inertial motions call for a supporting plane that prevents the mobile from approaching or fleeing the common center of heavy bodies; second, that only a plane equidistant from the center, that is, a spherical surface, is capable of playing this role. Since Galileo never changed his views on gravity, we have good reason to think that in the *Discourses*, as in the *Dialogue*, the principle of conservation is always restricted, from a theoretical point of view,[115] to the case of uniform motions on planes equidistant from the center.

Let us look once again at the passage in which Galileo defined his principle of conservation. It states quite explicitly that inertial motions call for the presence of a plane: "Imagine any particle projected along a horizontal plane without friction; then we know, from what has been more fully explained in the preceding pages, that this particle will move

---

[115]Which, moreover, he recalled during the Fourth Day (pp. 283–284).

along the same plane with a motion which is uniform and perpetual, provided the plane has no limits."[116] Theorem I was no less explicit. It dealt first of all with the uniform component of projectile motion: "Let us imagine an elevated horizontal line or plane *ab* along which a body moves with uniform speed from *a* to *b* . . ."; and it went on to the precise cause of projectile motion: "Suppose this plane to end abruptly at *b;* at this point the body will, on account of its weight, acquire also a natural motion downward along the perpendicular *bn.*"[117] In the *Discourses*, as in the earlier works, a plane was therefore necessary to offset the inherent tendency of all heavy bodies to approach the center and hence to ensure the conservation of their motion. However, and this was the great innovation of the *Discourses*, that plane had ceased to be a spherical surface.

Thus while he continued to attribute to the supporting plane the elimination of "every cause of acceleration or retardation," as his theory of gravity demanded, Galileo saw fit in his *Discourses* to abandon the second consequence of that theory, namely, that this plane must also be spherical. The reason for that transformation must doubtless be sought in the demands of geometrization: it was essential to assume that the inertial element in the compound motion of projectiles must be uniform and rectilinear if projectile motion was to be treated as a direct extension of naturally accelerated motion; that assumption alone enabled Galileo, with the mathematical resources at his disposal, to construct a geometrical expression of projectile motion and to determine its most fundamental properties. To appreciate this point, we can do no better than turn back to those pages of the *Dialogue* in which Galileo tried to determine the trajectory of a body in free fall on a rotating earth; applying then the principle of conservation in its original form, he did not even succeed in formulating the problem in geometrical terms, and he finally abandoned all attempts to combine uniform circular with uniformly accelerated rectilinear motion. A simplification, which was as well a rationalization, was therefore needed before a general science of the motion of heavy bodies could be constructed.

But it is also clear that this simplification or rationalization was devoid of any physical justification. For as long as gravity is described as an inherent force tending to carry bodies toward their common

---

[116]*Discourses* IV, p. 268.

[117]Ibid., p. 272.

center, inertial motion can only be confined to a spherical surface; on a horizontal plane, that is, on a plane tangent to any point on the earth's surface, a body must necessarily recede from the center toward which gravity impels it, and can never conserve its speed. Moreover, Galileo fully appreciated this fact, for having established the parabolic form of projectile motion in the abstract, he went on to question the legitimacy of his assumptions. And though his appeal to "geometrical license" (as employed by Archimedes in the quadrature of the parabola)[118] and to the fact that projectile motions occur over distances so small in comparison with the earth's radius that their inertial uniform motion becomes indistinguishable from rectilinear motion[119] undoubtedly bridged the gap between physical intuition and the demands of geometrization, there is no doubt that the rectilinear inertia Galileo postulated and used in his study of projectile motion was lacking in theoretical, if not in practical, foundations.[120]

This conclusion can be justified in yet another way. In his analysis of the motion of heavy bodies, Galileo introduced an abstract medium lacking in most of the elements that keep distorting motion in natural media. The chief characteristics of that medium, essential to anyone interested in developing mathematical physics,[121] were not difficult to define. All friction resulting from the uneven surfaces of bodies will have disappeared in it; polished spheres with a perfectly smooth surface will roll on perfectly smooth planes, so that the speed acquired will depend exclusively on the duration of the motion. The resistance characteristic of material media and notably of the air will have been eliminated as

[118]Ibid., p. 274.

[119]Ibid., p. 275.

[120]This view may not have been shared by classical physicists, most of whom held that Galileo had given the principle of inertia its full generality. For this, they had the authority of Newton, who in his *Principia* (Vol. I, p. 21, Cajori edition) acknowledged his debt to Galileo for the first law of motion—the law of conservation of uniform rectilinear motion. It was Emil Wohlwill who first cast doubt on Newton's opinion by pointing out that Galileo had never gone beyond the idea of an inertial motion on a spherical surface (cf. "Die Entdeckung des Beharrungsgesetzes" in *Zeitschrift für Völkerpsychologie*, Vol. XV, pp. 287 ff.). Both Mach and Cassirer prefer the traditional interpretation to Wohlwill's, but A. Koyré, after a review of the diverse opinions, comes down squarely in favor of the latter (cf. *Études galiléennes*, III, pp. 50 ff.).

[121]*Discourses* IV, p. 276: "Of these properties of weight, of velocity, and also of form, infinite in number, it is not possible to give any exact description; hence in order to handle this matter in a scientific way, it is necessary to cut loose from these difficulties; and having discovered and demonstrated the theorems, in the case of no resistance, to use them and apply them with such limitations as experience will teach."

well; moreover, since no physical basis is needed for such motion, the new medium could readily be identified with a vacuum, devoid of physical properties, and hence treated as a limiting concept or a necessary abstraction. Moreover, the vacuum, being both homogeneous and of indefinite extent, is indifferent to divisions into qualitatively distinct regions, and therefore quite different from the ideal medium Galileo had described in his *De Motu*. But, purified and abstract though it was, the new medium was not yet a rational entity. First of all, gravity as a downward tendency continued to be an inalienable attribute of heavy bodies, and though the specific weight had ceased to be a measure of the natural motive force, the idea that the most general properties of motion could be based on matter lacking in gravity held no appeal for Galileo. But, above all, it was the further stipulation of a common center toward which all bodies tend that most seriously encumbered his mechanical reasoning. Here again the role he attributed to that center had lost a great deal of its former importance, as witness his choice of time instead of distance when explaining the increase in speed associated with naturally accelerated motion. It remains a fact, however, that Galileo's medium continued to have a center; this forced him to employ expressions with an oddly traditional flavor. Thus, as late as 1630 he still described the motion of heavy bodies as the acquisition of a "better state," since by such motion each body approaches the center toward which "its nature as a heavy body impels it."[122] Similarly, the use throughout the *Discourses* of such expressions as "natural motion" served to highlight the theoretical and practical importance Galileo continued to attach to the common center of heavy bodies. Though it subjected motion to mathematical reason, Galileo's mechanics still contained limitations alien to a fully rational science. The fact that Torricelli, who could look upon the new science of motion as an accomplished fact, encountered the same problems and solved them in almost the same terms clearly shows that the reasons for Galileo's failure to proceed to a complete mathematicization must be sought above all in the conceptual system he handed down to his pupils.

Like Galileo, Torricelli treated gravity as a force residing within bodies and responsible, inter alia, for projectile motion: "Because its gravity acts from inside," he wrote, "the moving body will immediately

---

[122] *Letter to Raffaello Staccoli*, January 16, 1630, Vol. VI, p. 637.

begin to deviate from the direction of the throw, and since this deviation keeps increasing, the body will describe a certain curve."[123] Torricelli thus came up against Galileo's problem: how to justify a form of mechanics, or more particularly a science of motion, that could explain the real behavior of moving bodies only by such assumptions as the conservation of rectilinear motion, when the very nature of such bodies made such conservation a theoretical impossibility. Under these circumstances, how could the geometrical analysis of the motion of heavy bodies pretend to reflect the physical truth? Like Galileo, Torricelli, who examined and amplified what few indications he could cull from the Fourth Day of the *Discourses*, spent a great deal of effort on an attempt to resolve this basic contradiction.

To that end he went back to the simpler case of statics. Ever since Archimedes, he noted, statics had been based on two suppositions which, from a strictly physical point of view, were quite arbitrary: ". . . first that surfaces lacking in weight are nevertheless conceived as being ponderous, and second that the strings by which magnitudes are attached to a balance are assumed to keep an even distance from each other, when in fact they converge toward the center of the earth."[124] What are we to make of these two assumptions? According to Torricelli, the first could be justified on purely geometrical grounds; because geometrical objects exist solely "by definition and in the intellect," it is no less absurd to attribute weight to geometrical bodies than it is to endow them with "a center, a perimeter, a surface, solidity, and so on."[125] As for the equidistance of the threads, Torricelli produced the following argument: Imagine, he wrote, that our balance, although material, were transported into the highest regions above the sun's orbit; in that case, the strings, if still converging toward the center of the earth, will have become almost equidistant.[126] Now let us go further still and move our balance "beyond the stellar balance of the firmament," in other words, at an infinite distance; this time the threads will no longer converge but be perfectly parallel. Hence, our physical balance will have become quite indistinguishable from Archimedes' geometrical balance, so that the same mathematical formula will apply to both. If, finally,

---

[123]E. Torricelli, *Opera geometrica* (Florence, 1644); and *De Motu projectorum* II, p. 156, as quoted in Koyré, *Études galiléennes* III, pp. 138 ff.

[124]Koyré, *Études galiléennes*, p. 140.

[125]Ibid.

[126]Ibid., p. 142.

we restore the balance to our regions, the equidistance of the threads will disappear, but "the proportion of the figures already demonstrated will not be destroyed."[127]

Torricelli's artifice was certainly ingenious; not only did it enable him to offer a convincing plea in favor of geometrical abstraction but it also allowed him to justify it while remaining in the real world. The case of statics is, however, relatively simple, and Galileo himself had been able to cope with it even in his youth.[128] But when he went on to consider the science of motion, was Torricelli still able to devise an artifice to eliminate the constraints imposed by his theory of gravity? The answer, strictly speaking, is that he did not, as witness the following application of the principle of conservation to the analysis of projectile motion: "Let the moving body be projected from a point A in any direction AB above the horizon. It is clear that without the pull of gravity, the moving body will proceed with a rectilinear and uniform motion following the direction of the line AB."[129] In constructing his theory of motion, Torricelli thus found himself in the same uncomfortable position as Galileo: on the one hand, only gravity could explain the curvilinear nature of projectile motion; on the other hand, his geometrical analysis forced him to introduce the physically meaningless fiction of a motion impervious to the action of gravity. In short, as part of his geometrical analysis he both used and discarded the action of gravity on a given body. Like Galileo's, Torricelli's geometrization of projectile motion was thus based on postulates that were both necessary and yet impossible to justify, so that there appeared an unbridgeable gulf between geometrized mechanics and the real world that this mechanics was meant to represent. Hence, though both Galileo and Torricelli endeavored to build a rational medium for the science of mechanics, neither granted it full autonomy. It was left to Newton to reconcile the geometrical abstraction with the physical reality, but by then the question of gravity, which had so sorely tried Galileo and Torricelli, had ceased to pose a problem. For, having defined matter by its mass, Newton had no need to justify the principle of inertia with the help of a more or less ambiguous fiction; the tendency of bodies to persevere in a state of rest or uniform motion in a straight line had become the normal physical behavior of bodies considered in isolation.

---

[127] Ibid., p. 143.

[128] Although not nearly as elegantly as Torricelli (cf. Chapter 3).

[129] Koyré, *Études galiléennes*, p. 138.

Thus, in Newton's theory of gravity, unlike in those of Galileo and Torricelli, free fall was treated as a violent motion, the gravitational force impelling a body toward the center of the earth having become the attribute of an external factor, namely, the attraction exerted upon the moving body by the mass of the terrestrial globe. Newton thus brought the demands of geometrization into perfect harmony with the physical properties of bodies, and the world of mathematical mechanics with the real world.[130]

If we now return to the *Discourses*, we have no difficulty in grasping why Galileo failed to offer a theoretical justification of the conservation of uniform motion in a straight line. During the Fourth Day, he formulated the principle of inertia and applied it in a substantially correct way; the abstract medium to which he referred was such nevertheless that the principle could not be attached to any possible phenomenon. However fruitful his solution was from a geometrical point of view, it had perforce to remain a borderline supposition—one that was too incompatible with the prevailing conception of gravity to be granted permanent domicile. The reason why the principle of conservation does not appear at the head of the *Discourses* is therefore simply that the form in which Galileo would have been forced to present it could play no useful part in his analysis of projectile motion. This fully explains why the idea of inertial motion in a straight line had to be smuggled in during the general exposition, and even then only when it appeared indispensable; in short, even though it served as a principle in Galileo's demonstrations, it cannot be granted the real status of a principle.

The last two chapters permit us to draw several conclusions about the specific nature and the scope of Galileo's geometrization of the motion

[130]The reader may wonder why, in retracing the history of the principle of inertia from Galileo to Newton, we have failed to mention Gassendi or Descartes, who were the first to formulate that principle correctly (cf. ibid., pp. 144 ff.), and who succeeded precisely because they considered gravity as an effect and not as an inherent force. However, neither Gassendi nor Descartes developed mathematical mechanics in the manner of Galileo, Torricelli, or Newton; in particular, their interpretation of gravity, while based on the assumption that a body shielded from all forces might well continue indefinitely in uniform motion in a straight line, did not go hand in hand with an analysis of the mode of action of a force. Important though their contributions (that of Descartes in particular) undoubtedly were, they must therefore be considered marginal to the main current that flowed from Galileo and Torricelli to Huyghens and Newton.

of heavy bodies. Let us deal with the specific nature first. It is easy to show that Galileo's conceptual system was not yet that of classical mechanics. Thus, though his definition of naturally accelerated motion was basically correct, it played no part in the body of his exposition; lacking the resources of the infinitesimal calculus, he was forced to prove his first theorems with the help of fourteenth-century concepts. In the sphere of dynamics the split was no less marked. His distinction between the gravific and the motor function of gravity enabled Galileo to frame the principle of the conservation of uniform motion, first in its circular form and then, thanks to an ultimate geometrical abstraction, in its rectilinear form. Similarly, he also concluded that gravity, as a motive force, must impress the same speed on all bodies in a vacuum, and there is good reason to suppose that at the end of his life he had come to identify natural acceleration with the action of gravity. However, it is certain that the system within which he made these advances remained not only preclassical, but it was also quite incapable of leading him directly to classical mechanics; we need merely recall that in Galileo's world inertial motions concerned heavy bodies, not simple masses. Considered as a whole, Galileo's science of the motion of heavy bodies thus had a paradoxical aspect: while his theorems were continuous with Newton's, the underlying concepts were not. Of a piece with classical mechanics with respect to its results, Galilean mechanics was nevertheless based on a totally different conceptual system.

A further series of conclusions can be drawn concerning the scope of Galileo's geometrization. Here his achievements were most impressive. He succeeded in constructing a kinematics in which naturally accelerated motion was transformed into an intelligible object, lending itself, like a mathematical object, to the construction of an indefinite number of propositions by purely demonstrative means; in revealing the crucial part of the medium in the difference in speeds, he established the soundness of a general and abstract science, by definition independent from the nature of bodies. His dynamic analysis was no less remarkable: the correlation of acceleration with the moment of descent, on the one hand, and of the moment of descent with the "dead weight," on the other, heralded the advent of a science in which the passage from kinematics to dynamics and from dynamics to statics no longer presents any difficulties. Admittedly, to be perfect, geometrization would have to meet other conditions as well; in particular, it would have to be in full accord with the deductive ideal of which Euclid and Archimedes had supplied the model. That the *Discourses* erred in this respect, no one would care to deny; in particular, the repeated use of the method of indivisibles detracted from the rigor of Galileo's general exposition. Similarly, the manner in which he introduced his principles was most unsatisfactory from a deductive point of view: the proposition on which he ostensibly based his entire analysis played only a minor role in the rest of his argument and at best may be said to have furnished the new science with the guarantee of statics. As for his "true" principles, that is, the postulates concerning the conservation of uniform rectilinear motion and the composition of velocities, these were announced in the middle of the *Discourses* and, as it were, independently of the theory of gravity. It would be wrong, however, to exaggerate the importance of these shortcomings. True, the *Discourses* did not have the formal perfection of the works of Huyghens and Newton, yet its theorems were nevertheless arranged in an authentic deductive order. Certain demonstrations had doubtful foundations, but Galileo's determination to follow a demonstrative path was so great that it led him almost automatically to the correct formulation of the principle of inertia. Hence we are doing no more than justice to Galileo if, after having drawn attention to the preclassical character of most of his concepts, we grant that his inspiration and general procedure were on a level with those of Newtonian mechanics. That he should have succeeded with an inadequate conceptual system to construct a science of motion which by and large was of a piece with its classical counterpart is but another proof of his genius; Galileo may unreservedly be proclaimed the father of modern mechanics.

IV

8 Reason and Reality

Three contributions may be said to summarize Galileo's role in the formation of classical mechanics: the construction of a model of the universe that set cosmology on a new path; the translation into the language of mechanics of the traditional arguments against the earth's diurnal motion; and the creation of a geometrized science of the motion of heavy bodies. Moreover, this great reorganization of the contents of science went hand in hand with a transformation of the very basis of physical analysis. Eschewing a priori speculation no less than pure description, Galileo set himself the task of elaborating a conceptual system in which rational necessity took the place of physical causality; as the clearest expression of that necessity, geometry became the language of scientific research, it was transformed from a technical aid into the master key to the door of experience. The reason, therefore, why no scientific problem was ever the same again as it had been before Galileo tackled it lay largely in his redefinition of scientific intelligibility and in the means by which he achieved it: only a new explanatory ideal and an unprecedented skill in combining reason with observation could have changed natural philosophy in so radical a way. No wonder then that, as we read his works, we are struck above all by the remarkable way in which he impressed the features of classical science upon a 2000-year-old picture of scientific rationality.

Now whatever differences persisted between Galileo's cosmology and his theory of the motion of heavy bodies, far from complicating the study of that new rationality, bring out its full import. True, the *Discourses*, by subjecting motion to the law of number and by introducing geometry into the description of nature, may be said to have provided a more direct justification of the Galilean revolution than did the cosmological works; yet in them, too, Galileo, forced as he was to engage in ceaseless polemics to defend his astronomical conclusions and moreover impeded by the lack of a rigorous language, made a no less determined effort to define the aims and methods of natural philosophy in the clearest possible way. Such works as the *Letters on Sunspots*, the *Saggiatore*, and above all the *Dialogue*, all underpinned the new scientific rationality. Hence, if we combine and coordinate Galileo's cosmological contribution with that of the *Discourses*, we shall obtain the clearest possible idea of his methodology.

Return to Reality

Like every true advance, Galilean science sprang from a rejection—the rejection of the traditional approach and of its explanatory ideal. A discussion of the motives responsible for this rejection not only will lead us to a clear diagnosis of the ailments of the old science but will also reveal some of the most characteristic features of Galileo's remedy.

The chief maxim of Peripatetic physics was never to oppose the evidence of the senses; there is thus no better way to appreciate its explanatory power than to examine its ability to assimilate facts. However, the value of a science lies less in its interpretation of known and generally accepted facts than in its ability to cope with new and unexpected phenomena. And when it came to these, the attitude of the traditionalists was altogether deplorable, as Galileo showed, not without irony. There was, first of all, their bland rejection of everything new, which generally took the form of naïve astonishment: how could the ancients possibly have been unaware of what the moderns claimed they had discovered? Then there was the no less naïve belief that, by turning their backs on the new discoveries, they could also escape from the consequences. Galileo mentions the case of a famous philosopher who, in order to protect his library from contagion, removed Gilbert's book from its shelves, "starting back at the first hint of a new proposition like a horse at its shadow."[1] Finally, and more subtly, there was the deliberate attempt to "neutralize" the recent discoveries in one way or another: Were not the so-called sunspots so many small stars that revolved about the sun and fused from time to time? And what was the moon's earthlight if not the scattering of sunlight by a very dense crown of ether? In fact, the entire works of Aristotle had become as constant and incorruptible as his Heavens. This explains why the corruptibility of the latter should have proved a matter of such great importance to Galileo—introducing generation and corruption into celestial bodies meant celebrating the triumph of observation over the dead letter of the texts.

The principle of authority—the ultimate line of defense of those afraid of the new theories—was thus the unavoidable complement of their refusal to bow to experience. Unable to muster the facts, the traditionalists had perforce to rely on books, and in so doing they opened

---

[1]*Dialogue* III, p. 427.

the door to all sorts of tyrannical practices: "And what is more revolting in a public dispute, when someone is dealing with demonstrable conclusions, than to hear him interrupt it by a text (often written to some quite different purpose) thrown into his teeth by an opponent?"[2] And the odious goes hand in hand with the ridiculous, as witness the case of the Peripatetic physician who, having observed during a dissection that the nerves start from the brain and not from the heart, declared ingenuously that he would have been quite ready to grant it all, had Aristotle not asserted the contrary;[3] or the case of the lecturer from a well-known academy who, upon hearing the telescope described but not yet having seen it, said that the invention was taken from Aristotle. For had Aristotle not declared that it was possible to see the stars during daytime from the bottom of a very deep well?[4] However, the pernicious character of the principle of authority did not so much manifest itself in the farfetched utterances of its upholders as in their complete rejection of the scientific method and its ideals. For creative thought based on observation it substituted the art of stringing disparate texts together; for research it offered mere booklearning. Simplicio gave an excellent summary of their technique when he argued that, since all knowledge is contained in the works of Aristotle, it is enough to have his sayings constantly before the mind and to "combine this passage with that, collecting together one text here and another very distant from it." To this Sagredo made the following, most judicious, reply: "What you and the other great philosophers will do with Aristotle's texts, I shall do with the verses of Virgil and Ovid, making centos of them and explaining by means of these all the affairs of men and the secrets of nature."[5] In short, the Peripatetics were like "certain capricious painters who occasionally constrain themselves, for sport, to represent a human face or something else by throwing together now some agricultural implements, again some fruits, or perhaps the flowers of this or that season." Such bizarre behavior, so long as it is proposed in jest, is both pretty and pleasant, but who would ever consider it the best method of imitating nature?[6]

---

[2]*Dialogue* II, pp. 138–139.
[3]Ibid., p. 134.
[4]Ibid., p. 135.
[5]Ibid., pp. 134–135. Cf. *Letter to Liceti*, January 1641, Vol. XVIII, p. 294.
[6]*Third Letter on Sunspots*, Vol. V, pp. 190–191; cf. *Dialogue* II, p. 135. The painter Galileo had in mind was Archimboldo. Galileo launched even more vitriolic attacks on the Peripatetic method, for instance, when he wrote: "Oh, the inexpressible baseness of ab-

When it is thus reduced to the art of compiling references, science becomes indistinguishable from literary criticism or the study of law and loses its most essential characteristics: contact with reality and the separation of truth from falsehood by experimental verification. "If what we are discussing were a point of law or of the humanities, in which neither truth nor false exist, one might trust in subtlety of mind, readiness of tongue, and the greater experience of the writers, and expect him who excelled in those things to make his reasoning most plausible, and one might judge it to be the best. But in the natural sciences, whose conclusions are true and necessary and have nothing to do with human will, one must take care not to place oneself in the defense of error; for here a thousand Demosthenes and a thousand Aristotles would be left in the lurch by every mediocre wit who happened to hit upon the truth for himself. Therefore, Simplicio, give up this idea and this hope of yours that there may be men so much more learned, erudite, and well-read than the rest of us as to be able to make that which is false become true in defiance of nature."[7]

Because it ran away from the facts and relied exclusively on the principle of authority, traditional science suffered from a mortal disease exceedingly simple to diagnose: its symptoms are a total incapacity to abandon concepts that have lost their usefulness and hence to fit the theory to the experimental facts. But while the symptoms of the disease are obvious, its etiology has still to be explained; when we say that the Peripatetics, by taking refuge in their master's text, had abandoned "the honorable title of philosopher" for that of "doctor of history,"[8] we

ject minds! To make themselves slaves willingly; to accept decrees as inviolable; to place themselves under obligation and to call themselves persuaded and convinced by arguments that are so 'powerful' and 'clearly conclusive' that they themselves cannot tell the purpose for which they were written, or what conclusion they served to prove! But let us call it greater madness that among themselves they are even in doubt whether this very author held to the affirmative or the negative side. Now what is this but to make an oracle out of a log of wood and run to it for answers; to fear it, revere it, and adore it?" (*Dialogue* II, p. 138). Of course, the results are no better when the principle of authority is brandished by theologians; Galileo expanded at length on the relations between science and theology in his letter to the Grand Duchess Christina (Vol. V, especially pp. 324 ff. and pp. 341 ff.).

[7] *Dialogue* I, p. 78. In his letter to the Grand Duchess Christina (1615) Galileo had previously remarked that a mathematician cannot be ordered about like a lawyer or a merchant, "for demonstrated conclusions about things in nature or in the heavens cannot be changed with the same facility as opinions about what is or is not lawful in a contract, bargain, or bill of exchange" (Vol. V, p. 326).

[8] *Dialogue* II, p. 139.

are stating no more than the truth but have not yet adduced the causes. Why, in fact, did a philosophy whose avowed rational aims were hardly less cogent than those of modern science and which had originally tried to keep as close to experience as it possibly could, nevertheless prove so unable or unwilling to face up to any kind of new idea? Must the true reasons for this failure perhaps be sought in the Aristotelian explanatory ideal rather than in the traditional concepts?

Now that ideal is easily summed up in a single formula: the interpretation of natural effects—the relative position of bodies, their apparent qualities, and any changes in them—in terms of essences or principles considered as causes; in other words, the assumption that explaining an effect is tantamount to accounting for its existence on physical grounds. From a theoretical point of view it is hard to conceive of a more perfect ideal, since on satisfying it we should be in possession of absolutely certain knowledge. But, unless it is to remain an idle dream, this ideal demands that we have some insight into the natural order of essences or principles; if that demand cannot be fully met we may easily mistake for essences or principles such abstractions as merely reflect the world of our perceptions. This is best illustrated by Aristotle's "proof," based on the theory of natural motions, first of the difference in essence between the earth and celestial bodies, and then, in a second step, of the nongenerability, incorruptibility, and immutability of the latter.

Aristotle, as the reader will remember, began by assuming the existence of three simple motions: rectilinear upward motion away from the center, rectilinear downward motion toward the center, and circular motion about the center. On the assumption that each element must be associated with a particular motion, he next introduced the two primary terrestrial elements, earth and fire (to which he later added water and air), corresponding to simple upward and downward motions in a straight line, and another element, the ether, corresponding to circular motion and in fact providing its physical basis. All these motions had distinct attributes. Thus while rectilinear motion was discontinuous and indefinite because it took place from one terminus to another and was therefore characteristic of imperfect bodies subject to generation and corruption, circular motion, which was continuous and changeless because parts of the trajectory described in equal time were identical, could be associated only with immutable bodies, that is, with bodies that were nongenerable and incorruptible. Galileo had no difficulty in showing that this argument was based on a series of preconceptions that, to say the least, were completely arbitrary. Thus it was an arbitrary assumption to allow of only two types of natural motions, namely, the rectilinear and the circular, when the very criterion by which they were chosen (the precise superposition of successive fragments) also turns spiral motion into a simple form of natural motion. Another preconception was the unwarranted assumption that the universe could have only one center, when it could equally well be said to have a hundred or even a thousand centers associated with as many circular and rectilinear motions. And why discard out of hand the possibility that simple bodies may be immobile by nature once nature has been defined as the internal principle of movement and rest? More generally, why base the number of elements on the number of simple motions, when simple motion is no guarantee of a body's simplicity? After all, a compound body moves downward with the same natural motion as the earth, which is supposed to be a simple body. One could quote many further examples, all tending to show that the essential difference between the earth and celestial bodies, let alone the nongenerability, incorruptibility, and immutability of the latter, could not possibly be established with Aristotle's own arguments. In fact, the only interesting problem was the one set by the weakness of the whole thesis. How could so many thinkers have failed for so long to appreciate the logical fallacy on which their division of the world into two quantitatively distinct regions was based? The only possible explanation was that they endorsed Aristotle's "proof" because, in the final analysis, they considered it an endorsement by reason of two of the most elementary and hence the most persistent data of sense experience, namely, that terrestrial bodies move in a straight line while celestial bodies revolve uniformly about the earth. As for the nongenerability and incorruptibility of celestial bodies, they followed directly from the fact that while generation and corruption were common phenomena on the earth, "throughout all times, according to the records handed down by the ancients, we find no trace of change either in the whole of the outermost heaven or in any one of its proper parts."[9] In short, the essences of the Peripatetics were so many abstractions from sensible experience: when they thought in terms of essences, they were really appointing perception the

[9]*On the Heavens* I, 3, 270 b 11–16. Cf. *Dialogue* I, p. 72.

supreme guide of all knowledge. A remarkable passage in the *Third Letter on Sunspots* underlines this conclusion and at the same time shows the vanity of all scientific explanations based on essences:

For in our speculating we either seek to penetrate the true and internal essence of natural substances or else content ourselves with a knowledge of some of their properties. The former I hold to be as impossible an undertaking with regard to the closest elemental substances as with more remote celestial things. The substances composing the earth and the moon seem to me to be equally unknown, as do those of our elemental clouds and of sunspots. I do not see that in comprehending substances near at hand we have any advantage except copious detail; all the things among which men wander remain equally unknown, and we pass by things both near and far with very little or no real acquisition of knowledge. When I ask what the substance of clouds may be and am told that it is a moist vapor, I shall wish to know in turn what vapor is. Peradventure I shall be told that it is water, which when attenuated by heat is resolved into vapor. Equally curious about what water is, I shall then seek to find that out, ultimately learning that it is this fluid body which runs in our rivers and which we constantly handle. But this final information about water is no more intimate than what I knew about clouds in the first place; it is merely closer at hand and dependent upon more of the senses. In the same way I know no more about the true essences of earth or fire than about those of the moon or sun, for that knowledge is withheld from us, and is not to be understood until we reach the state of blessedness.[10]

Because we lack the superhuman intuitive power which alone would allow us to come to grips with essences, or at least with what we designate by that term, we run the risk of merely repeating in another language the data of sense experience. This may explain why traditional science was so rigid—for if "essences" are simple reflections of observations, will the least progress in experience not only show up their inadequacies but also reveal their true origins? Hence, the old system must be preserved in toto at all costs, which means the rejection of all new discoveries.[11]

[10] *Third Letter on Sunspots*, Vol. V, pp. 187–188.

[11] Cf. *Dialogue* I, p. 81, and the following lines in Galileo's *Letter to Castelli* of November 30, 1610 (Vol. X, pp. 503–504): "I almost laughed when you said these observations would convince even the most obdurate. Do you not realize that men capable of reasoning and concerned with the truth would already have been convinced by my other demonstrations, and that these obstinate men, desirous only of the applause of the stupid and narrow-minded crowd, would not even heed the evidence of the stars if the latter came down to earth and addressed them? Let us therefore be content with increasing knowledge for the great satisfaction we ourselves derive from it, and abandon all hope of persuading the multitude or of gaining the acclaim of the bookish philosophers."

However, not only did the traditional explanatory ideal turn science into a mere reflection of sense experience; by explaining natural phenomena in terms of essences or principles, it also attempted to adduce physical causes for their occurrence. Now it is easy to show that this approach constitutes an even greater source of confusion, as witness the Peripatetic attempt to discover the *principles* of natural and violent motions. To that end they asked themselves what causes must be assigned to natural and violent motions, or rather what joint motor was responsible for their existence. In the case of spontaneous natural motions the answer was simple: they must result from an internal property of bodies—let us say, their weight. The case of violent motions, however, which involves the intervention of external agents, was far more complicated. Must the motor be placed in the medium, or must we rather follow Buridan and invoke an impetus, that is, a provisional motor quality that the agent impresses upon the moving body before he releases it? Neither solution, as we saw, was ever adopted without reservations—not surprisingly when we consider that one and the same cause could engender both natural and violent motions. Thus imagine a hole drilled from one of the earth's poles to the other; a stone dropped into this hole would travel spontaneously toward the center "by its natural and intrinsic principle." But after it reached the center, its speed will be such as to cause it to move on further. Now, although this motion is violent and preternatural, to what other principle can it be referred than to the natural and intrinsic one that previously drove the stone toward the center? "Thus you see," Galileo concluded, "how a movable body may be moved with contrary motions by the same internal principle."[12] But the very fact that violent and natural motions can be attributed to one and the same cause shows that neither has been properly explained, so that making the search for causes the aim of natural philosophy means hampering the very formulation of scientific problems.[13] We can now appreciate the real nature of the explanations elaborated by traditional science: by invoking essences, entities, and principles that did not even lend themselves to clear definition, it

[12]*Dialogue* II, p. 263. Galileo had previously (p. 47) compared this situation with that of a pendulum that, having been displaced from the perpendicular (its state of rest) and then set free, will fall back toward the perpendicular and go the same distance beyond it.

[13]Only one contribution of traditional science is free from this taint, namely, the theory of *latitudines motus*, and this precisely because it was not framed in causal terms.

substituted mere words for true explanations. Thus when Simplicio claimed that the cause of downward motion was well known, since "every body is aware that it is gravity," Salviati rightly retorted in a famous passage: "You are wrong, Simplicio; what you ought to say is that everyone knows that it is called 'gravity.' What I am asking you for is not the name of the thing but its essence, of which essence you know not a bit more than you know about the essence of whatever moves the stars around. I except the name which has been attached to it and which has been made a familiar household word by the continual experience that we have of it daily. But we do not really understand what principle or what force it is that moves stones downward, any more than we understand what moves them upward after they leave the thrower's hand, or what moves the moon around."[14] A comparison proposed by Sagredo serves to round off this discussion: "Now this method of philosophizing seems to me to have great sympathy with a certain manner of painting used by a friend of mine. He would write on the canvas with chalk: 'This is where I'll have the fountain with Diana and her nymphs; here some greyhounds; here a hunter with a stag's head. The rest is a field, a forest, and hillocks.' He left everything else to be filled in with color by a painter, and with this he was satisfied that he himself had painted the story of Actaeon, not having contributed anything of his own except the title."[15]

---

[14] *Dialogue* II, pp. 260–261. Galileo amused himself on several occasions ridiculing the special role of words in traditional science, for instance, when he had Simplicio raise the following objection to Salviati's explanation, based on an experiment, of why an armature adds to the power of a magnet: "Truly, I think that Salviati's eloquence has so clearly explained the cause of this effect that the most mediocre mind, however unscientific, would be persuaded. But we who restrict ourselves to philosophical terminology reduce the cause of this and other similar effects to sympathy, which is a certain agreement and mutual desire that arise between things that are similar in quality among themselves, just as, on the other hand, that hatred and enmity through which other things naturally fly apart and abhor each other is called by us antipathy" (ibid., III, p. 436). For another example, this time concerning the luminosity of celestial bodies, see ibid., I, p. 99.

[15] *Dialogue* III, p. 436. In a later passage, Galileo leveled a similar dart at those who claimed that the tides were caused by the attraction of the moon: "But that concept is completely repugnant to my mind; for seeing how this movement of the oceans is a local and a sensible one, made in an immense bulk of water, I cannot bring myself to give credence to such causes as lights, warm temperatures, predominances of occult qualities, and similar idle imaginings. These are so far from being actual or possible causes of the tides that the very contrary is true. The tides are the cause of them, that is, make them occur to mentalities better equipped for loquacity and ostentation than for

The subordination of science to perception and the confusion inherent in causal explanations were the two chief factors responsible for the impotence of the traditional method. But it is possible to offer an even more radical critique and to show that its explanatory method doomed traditional science to self-destruction, as it were. This appears most patently from the Aristotelian's reaction to Galileo's lunar researches. In their view, the moon, being incorruptible and immutable, had to have a perfect and uniform surface, and its substance, being condensed from celestial matter, had to be homogeneous. Hence, when the telescope showed beyond the least doubt that the lunar surface was mountainous and could be divided into regions with distinct properties, these philosophers found themselves in a terrible quandary. From their polemics with Galileo we know how some of them tried to "save the appearances." First of all, they "leveled" the moon's "cavities and prominences"[16] with the help of a crystal envelope, thus restoring its perfect sphericity.[17] As for the other problem, the juxtaposition of extremely bright and dark zones in one and the same lunar region, they had but one possible escape: to assume that the moon was composed of two substances with distinct properties. In short, they needed no less than three new "essences" to account for Galileo's discoveries, or rather they needed four, since they had never formally abandoned the old celestial substance.[18] In other words, the use of essences in scientific explanation not only led the champions of traditional science to make clear the links between these essences and perception but also forced them to go back on their own words. The assumption by which certain philosophers tried to banish sunspots from the heavens—that the so-called sunspots were so many asteroids that kept fusing and separating as they revolved about the sun—had much the same consequences. Since fifty or more such bodies were needed to engender the observed phenom-

reflections upon and investigations into the most hidden works of nature" (pp. 470 f.). Not even Kepler was spared: despite the great esteem in which Galileo held him, he was openly reproached for attributing the tides to "the moon's dominion over the water," thus reintroducing into natural philosophy "occult properties and such puerilities" (p. 486). For Galileo's own tidal theory, see Appendix 4. It goes without saying that he considered final causes even more meaningless than efficient causes; to Galileo, finality was simply a naïve form of anthropocentrism; he criticized it in *Dialogue* I, pp. 85–86; III, pp. 394 ff., and 397.

[16] *Siderius Nuncius*, Vol. III_1, pp. 59–60.

[17] Cf. *Letter to Marcus Welser*, February 1611, Vol. XI, pp. 38 ff.; *Letter to Gallenzone Gallenzoni*, July 16, 1611, Vol. XI, pp. 142 ff.

[18] *Letter to Gallenzone Gallenzoni*, p. 143.

ena and since, moreover, such fusions could not possibly result from the regular motions of which the asteroids were said to be possessed, the only alternative was to endow them with "tumultuous and deformed motions"—in other words, to introduce permanent disorder into nature and thereby to abandon the call for regularity and harmony on which traditional cosmology had always prided itself. Irretrievably lost if it accepted the recent discoveries, Peripatetic physics was equally doomed if it tried to absorb them. As "a permanently discordant organ,"[19] it could therefore do no better than seek a reprieve in a complete denial of the facts, under the protective shelter of the principle of authority.

This breakdown of traditional thought convinced Galileo that a return to observation was the only means of freeing science from its age-old shackles. What principles must guide scientific reason in its attempts to come to grips with reality? To what constraints must it submit itself to build up meaningful statements? What is the precise sphere to which it can legitimately be applied? These questions, which define the context in which Galileo was determined to construct his natural philosophy, also hold the key to his explanatory ideal.

The evil it was meant to remedy prescribed its own cure: since the principle of authority was the clearest symptom of the ultimate impotence of traditional science, that principle had to be banished once and for all. The *Saggiatore* summed it up most sharply: "Sarsi says he does not wish to be numbered among those who affront the sages by disbelieving and contradicting them. I say I do not wish to be counted an ignoramus and an ingrate toward nature and toward God; for if they have given me my senses and my reason, why should I defer such great gifts to the errors of some man? Why should I believe blindly and stupidly what I am told to believe, and subject the freedom of my intellect to someone else who is just as liable to error as I am?"[20] "In philosophy," Galileo wrote to Castelli in 1639, "doubt is the father of invention. It opens the way for the discovery of the truth."[21] However, the rejection of the Peripatetic principle of authority did not constitute an end in itself, nor did it signify that Galileo recognized no authority of any kind. For though human opinion does not enter into the study

[19]*First Letter on Sunspots*, Vol. V, p. 113.
[20]*Saggiatore*, Vol. VI, p. 341; cf. pp. 232, 337, and 339–340.
[21]*Letter to B. Castelli*, December 3, 1639, Vol. XVIII, p. 125.

of nature, natural philosophy cannot deny one authority without denying itself, namely, that of direct experience. "All human reason must be placed second to direct experience," Galileo wrote in his *Second Letter on Sunspots*; "hence they will philosophize better who give assent to propositions which depend upon manifest observations than they who persist in opinions repugnant to the senses and supported only by probable reasons."[22] Now what meaning could the phrase "propositions which depend upon manifest observations," possibly have if not to express the fact that "nature has no obligation to men, has passed no agreement with them,"[23] or that "the force of human authority on the effects of nature is nil, for nature is deaf and unyielding to our vain desires."?[24] The physicist must persuade himself of the absolute objectivity of nature, "the faithful executrix of God's orders,"[25] and as such "unable to sin against the laws imposed upon her."[26] If he does that, he will cease arguing as if "bodies only came into existence when we began to discover and know them,"[27] or, which comes to the same thing, as if nature had first "constructed the brains of men and then ordered things accordingly."[28] "But I should think rather," Galileo concluded, "that nature first made things in her own way and then made human reason skillful enough to be able to understand, but only by hard work, some part of her secrets."[29] To substitute the authority of the facts for human authority, to accommodate one's ideas to things, not things to one's ideas,[30] such was indubitably the first principle on which reason must rely in its relations with reality.

That this was no mere profession of methodological faith, but that Galileo did indeed restore the continuous dialogue between reason and reality, which he rightly accused traditional science of having neglected, is amply borne out by his cosmological works no less than by his writings on mechanics. We need merely to recall the manner in which he established the material identity of the earth and celestial bodies dur-

[22] *Second Letter on Sunspots*, Vol. V, p. 139.
[23] *Letter to Grienberger*, September 1, 1611, Vol. XI, p. 192.
[24] *Saggiatore*, p. 337; cf. *Third Letter on Sunspots*, Vol. V, p. 218; also *Letter to Ingoli*, Vol. VI, p. 538.
[25] *Letter to the Grand Duchess Christina*, Vol. V, p. 316.
[26] *Letter to Castelli*, December 21, 1613, Vol. V, p. 283.
[27] *Letter to Monsignor Dini*, May 21, 1611, Vol. XI, p. 108.
[28] *Dialogue* II, p. 289.
[29] Ibid.
[30] *Dialogue* I, p. 120; cf. *First Letter on Sunspots*, Vol. V, p. 97: "For names and attributes must be accommodated to the essence of things, and not the essence to the names, since things came first and names afterwards."

ing the First Day of the *Dialogue* by reference to the lunar mountains, sunspots, the phases of Venus, and the moon's earthlight to appreciate that his scientific reflections were based on experimental studies of a scope his predecessors never even suspected. And, though it was less spectacular, the role that fell to observation in his mechanical studies was no less obvious or fruitful. Take, for example, his treatment of the difference in speeds of freely falling bodies. When he first examined this problem in the *De Motu*, he already challenged Aristotle's conclusions but had not yet openly abandoned the traditional approach; despite interesting remarks on the effects of the medium, his argument was still conducted from a causal point of view, and the speed differences observed *in concerto* were attributed simply to differences in the specific weights of the bodies concerned. In the *Discourses*, however, speed had lost its links with weight, and a truly novel relationship between experience and scientific reason appeared. Two bodies of different specific weights were released in a relatively dense medium and their speed differences noted; next, the same two bodies were released in a medium of very low density and the differences in speed noted once again. A comparison showed that the differences had decreased considerably and that in air (the rarest medium known under normal conditions) enormous distances were needed for any appreciable speed differences to appear. Admittedly, these observations were guided throughout by a process of reasoning which alone made them meaningful, and the final hypothesis (that all bodies fall with equal speed in a vacuum) was much more than a simple statement of fact. For all that the initial datum which guided the entire development of the new theory of the difference in speeds (namely, the variation in speed differences when the specific weight of the moving body is "combined" with that of the medium) was explicitly suggested by experiment, which alone enabled Galileo to discard the traditional solution once and for all. A return to the facts and active and constant attention to experiment were thus the true sources of this analysis and of the justification it contributes to the geometrized science of motion.

In a very similar manner an unexpected technical observation was greatly responsible for another of Galileo's discoveries. He had noted that shipbuilders employ stocks, scaffolding, and bracing of disproportionately larger dimensions for launching a big vessel than they do for a small one built of the same material, because otherwise the larger ship might fall apart under its own heavy weight.[31] This fact, Galileo found extremely disconcerting, because the larger vessels were so constructed that their parts bore to another the same ratio as those of the smaller one, and also because "mechanics has its foundation in geometry, where mere size is not determinant" ("the properties of circles, triangles, cylinders, cones, and other solids do not change with their size").[32] Moreover, experience itself shows that "imperfections in the material, even those which are great enough to invalidate the clearest mathematical proof, are not sufficient to explain the deviations observed between machines in the concrete and in the abstract."[33] The only reasonable solution was to accept that relations of geometrical proportionality do not fully determine the resistance of bodies and then, by continuing to rely on the evidence of our senses, to decide what other relations may obtain between the resistance of solids to breaking stresses, on the one hand, and their length, diameter, shape, etc., on the other. In other words, though observation as such does not provide the full explanation, its evidence alone, particularly if it runs counter to the predictions of established science, can reveal the true state of affairs in all its complexity and hence can help us elaborate an appropriate hypothesis. "Therefore, Sagredo, you would do well to change the opinion which you or many other students of mechanics have entertained concerning the ability of machines and structures to resist external disturbances, thinking that when they are built of the same material and maintain the same ratio between parts, they are able equally, or rather proportionally, to resist or yield to such disturbances and blows. For we can demonstrate by geometry that the large machine is not proportionately stronger than the small. Finally, we may say that for every machine and structure, whether artificial or natural, there is set a necessary limit beyond which neither art nor nature can pass; it is here understood, of course, that the material is the same and the proportion preserved."[34]

It is by an examination of the conditions under which Galileo's return to observation was effected that we can best appreciate its full importance. For the kind of observation on which Galileo relied and

---

[31]*Discourses* I, p. 50. Galileo himself explained that he had learned this during his frequent visits from Padua to the Venice arsenal.

[32]Ibid.

[33]Ibid., p. 51.

[34]Ibid.

which he consistently proclaimed as the true basis of the new science was quite unlike its traditional counterpart. Quite unlike it, first of all, because the enormous extension of the old frontiers—notably in astronomy, thanks to the invention of the telescope—introduced a sharp break between observation as it had been understood and practiced before Galileo and observation as it would be understood and practiced after him. Not that the idea of improving observation with the help of instruments was a Galilean innovation; astronomers had made a habit of working with various instruments ever since antiquity, and more recently it was thanks to astrolabes and quadrants that Tycho Brahe had been able to develop positional astronomy to a degree of accuracy that Galileo himself could not surpass.[35] This remark, however, in no way detracts from Galileo's originality; Tycho and his predecessors used instruments exclusively to support and refine the evidence of the senses, so that their science could never transcend the limits of sense perception. Galileo's telescope, by contrast, swept away the frontiers of the traditional world, revealing an unsuspected universe high up in the skies, bringing "thirty or forty times nearer" what had been up to then a supremely regular, but somewhat unreal, machine, and converting celestial bodies into true physical objects. Hence, when it opened a previously closed window to observation, the telescope not merely refined the evidence of the senses but, endowing man with a "superior and better sense than natural and common sense,"[36] renewed the very content of sensible knowledge and brought it the promise of a vast extension.[37] In other words, it held out the hope that the most crucial problems of science could be henceforth solved by experiment so that, when he offered his cosmology as an alternative to the traditional one, Galileo could rightly claim that, unlike all his anti-Peripatetic precursors, he was at last doing more than substituting one a priori discourse for another.

However, the telescope had an even more important significance for Galileo, since by restoring observation to its rightful place it also revealed in the sharpest possible way the organic links by which science is bound to experimentation. What the discoveries of 1610, 1611, and

[35]Even though Tycho never succeeded in eliminating the irradiation phenomenon, when, as Galileo noted, he could simply have used a cardboard tube with a tiny opening; *Dialogue* III, p. 364.

[36]This expression is used in *Dialogue* III, p. 355.

[37]One example of that extension is the construction of the microscope almost immediately afterward; cf. Vol. XII, pp. 199, 201, and 208.

1612 made clear beyond the least doubt was that scientific theories depend intimately no less on the genius of their authors than on the experiential knowledge by which they are supported. Every theory is first of all the theory of our experience, the expression of a certain state of our knowledge of reality. Did Galileo himself take this view? In other words, did he consider differences in method secondary to that most fundamental difference which is a renewal of objective experience? The answer is provided by his general attitude toward Aristotle, whom he was always careful to distinguish from his disciples: "Now, in order that we may harvest some fruit from the unexpected marvels that have remained hidden until this age of ours, it will be well if in the future we once again lend an ear to those wise philosophers whose opinion of the celestial substance differed from Aristotle's. He himself would not have departed far from their view if his knowledge had included our present sensory evidence, since he not only admitted manifest experience among the methods of forming conclusions about physical problems but went so far as to give it first place. So when he argued the immutability of the heavens from the fact that no alteration had been seen in them throughout the ages, it may be believed that, had his eyes shown him what is now evident to us, he would have adopted the very opinion to which we are led by these remarkable discoveries. I should even think that, in making the celestial material alterable, I contradict the doctrine of Aristotle much less than do those people who still want to keep the sky unalterable; for I am sure that he never took its inalterability to be as certain as the fact that all human reasoning must be placed second to direct experience."[38] In contrasting the scientific spirit of Aristotle with the dead letter of his philosophy, Galileo thus intended to show that, since the objective base of knowledge had changed, it was essential to reexamine the entire conceptual structure of science—a truth that has long since become a commonplace but proves that Galileo's assertions on experience were no mere oratory. At the same time that he constructed the modern image of the world, Galileo was also

---

[38]*Second Letter on Sunspots*, Vol. V, pp. 138–139. Inasmuch as it drew a distinction between Aristotle and his modern sectarian disciples, this passage was of course highly polemical. Nevertheless it certainly expressed a sincerely held view, and one reiterated on many occasions, namely, that, were Aristotle still alive, he would most certainly be . . . a Galilean; cf. *Dialogue* I, pp. 80–81; *Letter of August 14, 1612*, Vol. XI, p. 422; *Letter to Prince Leopold*, May 1640, Vol. VIII, p. 542; and *Letter to Liceti*, September 15, 1640, Vol. XVIII, pp. 248–250.

the first to realize that a priori dogmas are always the counterpart of a sterilized experience.

Completely renewed in its content, observation became the object of a new method of organization and construction whose importance cannot be overestimated. For it is not enough to recall, or even to prove, the irreplaceable role of experiment; we must also allow it to play that role, that is, we must make homogeneous to rational knowledge what at first is no more than the simple evidence of our senses. Now this aspect is the more important in that Galileo never disguised the uncertainty of astronomical observations, even when made with the telescope. "There are many occasions," he wrote in May 1640, "on which our senses when first apprehending the facts may be in error and must therefore be corrected with the help of a well-conducted argument (*mediante l'aiuto del retto discorso razionale*)."[39] The same idea is reflected in the ridicule with which in the *Saggiatore* he greeted Sarsi's claim that perception alone can decide whether or not comets produce their own light: "I confess that I do not possess such a perfect faculty of discrimination. I am more like the monkey that firmly believed he saw another monkey in a mirror, and the image seemed so real and alive to him that he discovered his error only after running behind the glass several times to catch the other monkey. Assuming that what Sarsi sees in his mirror is not a true and real man at all, but just an image like those which the rest of us see there, I should like to know the visual differences by which he so readily distinguishes the real from the spurious."[40] Only a stable and well-conducted experiment in which the essential has been carefully distinguished from the contingent can supply natural philosophy with the support it needs. How can this be achieved?

A rational construction by which observation can be organized and developed systematically offers one solution. This was precisely the approach that had enabled Galileo to demonstrate so convincingly that sunspots must be contiguous with the sun's body. Let us recall the general approach he used. Having looked at the sun over a continuous period, Galileo was able to identify two particular spots near the edge of the sun's rim, which he called *A* and *B*. Daily drawings then showed that the two spots kept moving across the solar disk until they had reached the other side in a little less than two weeks. He also deter-

[39] *Letter to Prince Leopold*, Vol. VIII, p. 511. See also the critique of common sense in *Dialogue* II, pp. 280 ff.

[40] *Saggiatore*, Vol. VI, pp. 277 f.

mined by careful measurements that the apparent distance between them was much smaller when they were near the edges of the disk than when they were near the center of the sun. So much for the direct experiential evidence. Now, the changes in the shape and position of the sunspots and the relations to which they point can also be made to yield further information on their nature and especially on whether or not they are contiguous to the sun's surface. How do we elicit this additional information, which, though immanent in experience, is not directly accessible to us?

We can do so with the help of geometry. The procedure used by Galileo was as simple as it was convincing. Let us represent the sun by the hemisphere *CDE*, which, because of its distance from the earth, appears in the form of a plane surface. On July 1, 1611, the two spots *A* and *B* occupied the positions *F* and *I*, and their apparent separation was given by the chord *HL*. Four days later, on July 7, the two spots, which had meanwhile moved toward the right, were seen to be equidistant from *G*, the center of the hemisphere, and Galileo noted that

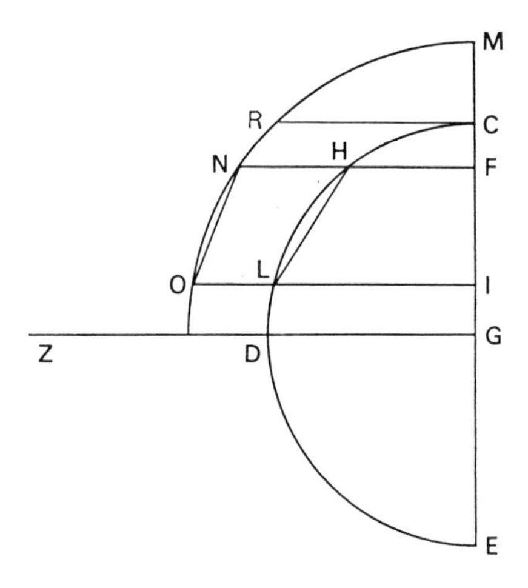

the increase in their apparent distance was in precisely the same ratio as *FI* was to *HL*; from this he concluded that the spots must be contiguous to the sun's surface. For let us suppose that the spots moved across a sphere *MNO* whose radius is 5 percent longer than that of the sun; if *FI* still designates the apparent distance of the two spots, *NO* will be their real distance, and that distance must be appreciably smaller than HL. Hence, when the two spots are equidistant from the center of the sun, their apparent distance will not increase in the proportion FI/HL but in the much smaller proportion FI/NO. However, observation shows that nothing of the kind happens, so that we may conclude with perfect certainty that the spots are indeed part of, or very close to, the solar surface.

Even though this proof lacked the solidity we have come to expect of a scientific demonstration, it was truly remarkable, for it showed how geometry can be used to read quite unsuspected messages into observational data. And once the latter are supported by an appropriate rational structure, they can also reveal previously hidden relationships and in so doing transform them into genuine topics of science. At the same time, the importance of geometry itself is confirmed and enlarged; no less essential for the analysis and refinement of observation than for the framing of hypotheses, it has thus become part and parcel of the very structure of science. By raising it to that role Galileo showed that when the resources of mathematics are suitably deployed, no hidden or occult powers are needed to gain full understanding of reality.[41]

However, mathematics was not the only tool Galileo employed, for he also showed that instruments or special experimental arrangements could be deployed in such a way as to organize experimental findings in a rational manner and hence to remove them from the sphere of mere opinions to which philosophical tradition had relegated them. The main problem, as Galileo realized full well, was to devise appropriate measurement techniques, because measurement alone can save experience from the uncertainty of individual impressions, thereby lending it

[41]A similar illustration of Galileo's mathematical approach can be found in his *Third Letter on Sunspots*, Vol. V, pp. 213 ff. We have chosen the case of sunspots because geometry played so clear a part in their analysis, but we might equally well have considered Galileo's use of the pendulum in his analysis of speed alterations, because here too he relied not simply on crude observations but on the rational interpretation and organization of the observed data. Thus, if we cause a cork and a lead pendulum of identical lengths to vibrate with identical amplitudes, we shall find that the vibrations of the cork pendulum are damped much more swiftly than those of the lead one, which might suggest that the cork had a smaller natural speed. One of the consequences of the isochrony of pendular oscillations is that all oscillations of the same amplitude are of equal durations from which it follows that the faster retardation of the cork pendulum can be due only to the action of the medium. Hence, it was indeed his rational organization of the observational data and not the simple evidence of his senses that guided Galileo's thought in this case no less than in his analysis of sunspots.

that stability and objectivity without which science can never be said to be firmly rooted in reality. Galileo's determination to introduce measurement into experiment was expressed in many spheres of his scientific endeavor. Thus in 1612, anxious to perfect his astronomical observations, he fitted his telescope with what may be called a rudimentary micrometer.[42] Thanks to the greater accuracy of the resulting estimates not only of the dimensions but also of the apparent elongations of the satellites of Jupiter, he was able to construct periodic tables of their motions; so observational data could be fully utilized and definitely integrated into the body of science.[43] But the areas in which Galileo obtained his most spectacular results were by no means the only ones relevant to our present discussion; the same desire to raise experience from the purely perceptive to the scientific stage also went into his construction of the first instrument designed to measure temperature differences in two solids or liquids.[44] A like significance also attaches to his experimental attempts, carefully described in the *Discourses*, to measure the weight of air and to determine, with the help of the telescope, whether or not light is propagated instantaneously,[45] and also to the work on magnets which he pursued so assiduously over many long months during his stay in Padua and in which, rather than concern himself with the nature of their attractive powers, he established some of their general properties—for instance, to what extent a magnet's strength can be multiplied by an armature[46]—by measurement and calculation. There is little point in citing further examples

---

[42]Cf. Vol. III_1, p. 446.

[43]Cf. Vol. III_2 (2nd part). In particular, he hoped these tables would improve the determination of the position of ships at sea; cf. *Letter to Lorenzo Realio*, June 1637, Vol. XVII, pp. 96 ff. In the same letter Galileo also explained how the telescope could be stabilized to offset a ship's vibrations and described an accurate time-measuring device based on the isochrony of pendular oscillations.

[44]This was certainly the first attempt to measure intensive magnitudes; but since Galileo never succeeded in perfecting a satisfactory temperature scale, his instrument was more in the nature of a thermoscope than of a thermometer. It is mentioned in several of his letters; see particularly Vol. XI, p. 506, and Vol. XII, p. 139; for its description, see Viviani's *Raconto istorica della vita del Sig. Galileo Galilei*, Vol. XVII, p. 377.

[45]*Discourses* I; for the weight of air see pp. 123–124; for the propagation of light, see pp. 87–88.

[46]These experiments are reported in *Dialogue* III, pp. 433-436. Galileo applied what would later be called the methods of elimination and combined variation. His Paduan correspondence shows that he took great and constant care to explain his experimental methods and data in the fullest possible detail, as witness particularly his letters of 1602–1604 to G. F. Sagredo, whom he later immortalized in the *Dialogue* and the *Discourses* (cf. Vol. X, passim).

or dwelling at greater length on the pains Galileo took to describe his observational techniques;[47] our remarks will have shown conclusively that his many attempts to organize experiments with the help of geometry and measurement were precisely what modern scientists mean when they speak of the establishment of scientific facts. Not only did Galileo realize that science is nothing if it is not systematic observation; he also knew that observation can be useful only if it yields geometrically and quantitatively determinable data. Transformed into an integral part of science, observation thus ceased to be a vague guarantee of the truth of theories developed without it, to become the indispensable basis of scientific explanation.

Explanation and Its Validation: Galileo's Rationalism

By restoring and strengthening the normal and fruitful links between reason and reality, Galileo undoubtedly set science on the right road. But that was only a beginning, for just as a scientific explanation is not a description of causes, so it cannot be identified with observation, however refined by geometry or measurement. And truly the first difference is one of difficulty, since as the *Saggiatore* pointed out, while chance often plays a large part in observation, it completely disappears from explanations based on the "work of reasoning."[48] An even more important difference between explanation and observation is one of level. "In all the effects of nature," Galileo explained, "experience informs me about the *an sit*, but tells me nothing about the *quomodo*;"[49] describing phenomena and making their production or development intelligible are two operations as irreducible to each other as they are complementary. Take, for instance, the resistance solid bodies offer to fracture. Observation tells us that this resistance does not increase with the size of the bodies—indeed, that if we keep increasing the various

---

[47]For instance, when analyzing the effects of irradiation on astronomical observations, or when describing his search for a device that would show up sunspots without obstructing the view (cf. *Second Letter on Sunspots*, Vol. V, pp. 136–137). In the same vein, we could also mention his careful definition of the principles that should be applied to the observations of the nova of 1572 (*Dialogue* III, pp. 305–318) or his discussion in the same section of the *Dialogue* of the best means for detecting the parallactic displacement of the fixed stars, which turned out to be the very method by which Bessel succeeded, in 1838, to make the first parallax measurement for one such star—61 Cygni.

[48]*Saggiatore*, Vol. VI, pp. 258–259.

[49]*Letter to Liceti*, June 23, 1640, Vol. XVIII, p. 208.

parts of a machine in the same proportion, we shall quickly reach the point where their very dimensions will cause them to break down. But that is all it can tell us. Observation can never adduce reasons for this apparent paradox or explain the relationship between the resistance of a machine and the dimensions of its parts. In short, observation helps us pose problems, but their solution demands the kind of intellectual initiative that no inspection of the facts, however scrupulous, can ever hope to replace.

That being the case, what is really an explanation?

A look at a problem raised by the apparent motion of sunspots and its methodological solution will take us some of the way toward the answer. The reader will recall the main conclusions about the nature of sunspots at which Galileo arrived in the summer of 1612: sunspots are of variable dimensions, occur in regular formations, cross the surface of the sun in approximately two weeks, and sometimes make a brief reappearance after the same interval. Moreover, an attentive analysis of their displacement shows that they must be contiguous to the sun's surface and that they probably participate in its rotation, accomplished in a little less than a month; finally, their changes in shape and the fact that they divide or recombine before disappearing suggest that they must be generable and corruptible. To these early discoveries, to which we have been confining our attention, Galileo added another one: in the course of their displacement, the spots invariably describe rectilinear and slightly inclined trajectories—a fact that could be explained, so it seemed, by "the obliquity of the horizon."[50] Soon afterward, however, a new series of observations was to lend this last discovery an unexpected significance. As he followed the motion of an easily identifiable sunspot day after day, Galileo noticed that its passage was no longer in a straight inclined line but in a somewhat bent one, whose extremities were on the same level on both sides of the solar disk; other observations, on other spots, showed that he could not have been mistaken.[51] What did all this mean? What unknown factor saw to it that a spot would describe a straight line on some occasions and a curve on others? No established astronomical theory could account for this strange fact, but Galileo quickly realized that if just one supplementary assumption was added to the Copernican hypothesis, it would be

[50]That is, to the ecliptic. See *First Letter on Sunspots* V, pp. 15–16.

[51]*Dialogue* III, pp. 373–374.

possible not only to explain this phenomenon but also to predict the future course of any one spot.

Let *O* be the center of the ecliptic and also of the solar globe, whose distance from us is so great that we can perceive only one-half of it at any one time (see Figure I). Hence, if we describe the circle *ABCD* around *O*, it will cut off that hemisphere of the sun which we can see from that which is hidden.[52] Moreover, since our eyes, like the center of the earth, are in the plane of the ecliptic (in which the center of the

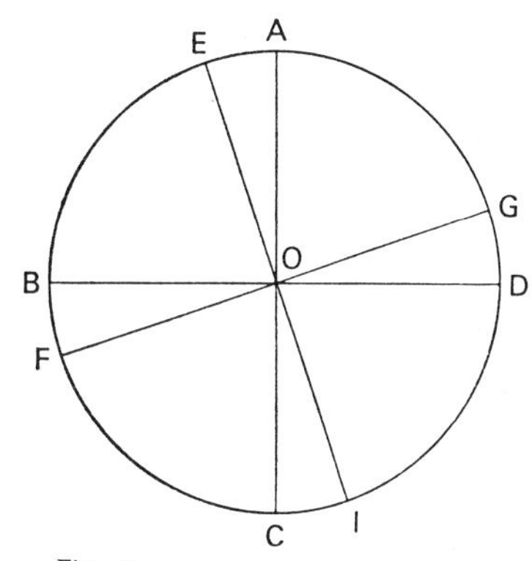

Fig. I

sun lies likewise), it follows that, if we imagine the solar body as cut by the plane of the ecliptic, the section will appear to us as a straight line. Let this be *BOD*, and assume it to be perpendicular to *AOC*, that is, to the axis of the ecliptic and of the earth's annual motion. Now it is certain that if the sun revolved about *AOC*, spots appearing on either side of *B* and moving toward *D* would appear to the terrestrial observer to be describing straight lines parallel to the plane of the ecliptic, and this they would do all year round. Let us now assume instead that the sun revolves, not about this axis *AOC*, but about the tilted axis *EOI*, and let this be fixed and perpetually pointing toward the same part of the firmament. In that case, the greatest distance a spot can traverse will be represented by the diameter *FOG*, at right angles to the axis *EOI*, and a terrestrial observer will notice "extraordinary changes" in its motion.

For assume the earth to be at such a point on the ecliptic that the

[52]Ibid., p. 376.

solar hemisphere visible to us is bounded by the circle *ABCD* which, passing through the poles *A* and *C* (as it always does), also passes through the poles *E* and *I*. The great circle whose diameter is *FG* is (by construction) at right angles to the circle *ABCD*, as is the ray reaching our eyes from the center *O*. "Therefore the same ray falls in the plane of the circle whose diameter is *FG* and whose circumference therefore appears to us as a straight line and is the same as *FG*."[53] Hence, whenever a spot is at *F* and is carried by the sun's rotation, it will mark on the surface of the sun the circumference of a circle, which appears to us as a straight line: its passage will seem straight and tilted upward and so will the motions of any other spots that describe smaller circles in the same revolution, because all of these are parallel to the great circle, *FG*. Six months later, when the earth will have run through half of its orbit and will be situated opposite that solar hemisphere which is now hidden from us, the boundary of the part seen by us will still be the same circle *ABCD* passing through the poles *E* and *I*, and the spots will again appear to describe straight lines. The only perceptible difference is that this time the path of the spots will descend toward *F* on the right.

The argument shows that, when Galileo deduced from his earliest observations that the spots follow a rectilinear path, he was not the victim of an illusion; it also explains why, on closer examination, he observed "extraordinary changes" in the apparent movements of the sunspots: they can describe straight lines only when the circumference bounding the visible part of the sun passes through the poles *E* and *I*, and this happens no more than twice a year.[54] But let us suppose the earth to be one quadrant removed from its present place. A terrestrial observer would still observe only half the sun, that is, the hemisphere *ABCD*, and *AC* would still appear to be "the axis through which the plane of our meridian would pass."[55] However, by virtue of the motion of the earth, the sun's axis of rotation *EI* will temporarily occupy the same plane as *AC*, with its upper pole *E* turned toward us and its lower pole *I* falling in the hidden hemisphere. A simple glance at Figure II shows that, in that case, the greatest circle a spot can describe, namely, *BFDG*, will no longer appear as a straight line but as a curve, with the convex part toward the bottom; it is obvious that the

[53]Ibid.

[54]Ibid., p. 377.

[55]Ibid.

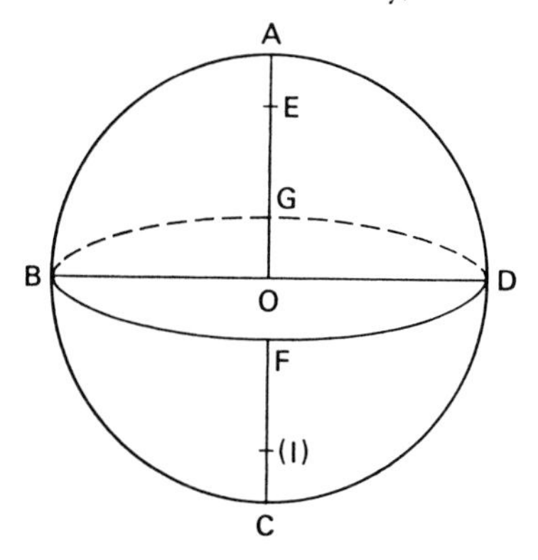

Fig. II

same will hold for all lesser circles parallel to the great circle *BFDG.*[56] Thus the transformation of rectilinear into curvilinear trajectories becomes perfectly intelligible; to explain the observed changes, we need merely assume that the sun's axis of rotation is oblique to the plane of the ecliptic, and then trace out the earth's annual motion in our mind's eye.

This example clearly illustrates the nature and scope of explanation in Galilean science. The physicist's first task is to define a *rational system* —that is, a set of principles and concepts providing the framework in which explanations can legitimately be sought: in the present case this framework was the Copernican doctrine together with the additional assumption that the sun's axis of rotation is oblique to the plane of the ecliptic. Under these conditions explaining means nothing else but accounting for the observed phenomena from the principles and concepts constituting the rational system employed, and this can be done by the elaboration of a model designed to express the phenomena under investigation within the rational system. Galileo's own procedure provides an excellent example of how a model can serve to transform the bald description of phenomena into a coherent whole governed by general principles and ideas, that is, to transform simple matters of fact into objects of interpretation, thus allowing what Galileo himself has called the transition from the *an sit* to the *quomodo*. But our example also illustrates two other points of the utmost importance. Let

[56]Ibid. Moreover, this is the only position in which the points marking the entrance and exit of the spots will be "balanced," that is, appear on the same level of the solar disk.

us take a closer look at Galileo's procedure. He defined a rational system and constructed a model appropriate to it; then exploiting the latter to the full, he showed that the observed phenomena must follow as a matter of course. Explanation thus became an act of rational genesis, namely, the demonstration that the observed data are a necessary consequence of a purely intellectual construction. Nor is that all. In the situation illustrated by Figure II, a sunspot appearing in *B* and moving toward *D* was describing a curve with the convex part below. Six months later, on the other hand, the lower pole *I* will be turned toward us, while the upper pole *E* will have become invisible from the earth: it is then predictable that the path of a sunspot traveling from *B* to *D* would still be curved, while the convex part of its trajectory would now

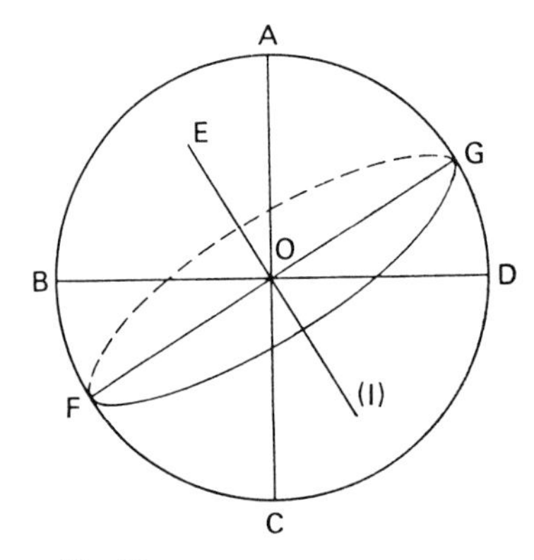

Fig. III

point toward the top. We can also imagine a situation in which the upper pole *E* is visible, though lying in a plane different from that of the axis *AC* (see Figure III): a spot traveling from *F* to *G* would then describe a rising curve; six months later it would still describe a curve but this time one that would slope slightly downward.[57] Observations fully confirmed all these changes which Galileo had deduced from his model: "Continuing to make very careful observations for many, many months, and noting with consummate accuracy the displacement of various spots at different times of the year, we found the results to accord exactly with the predictions."[58] Explaining therefore means more than seizing upon the phenomena; it also means revealing the intelli-

[57]Ibid., p. 378.

[58]Ibid., p. 379.

gible process governing their production, and in such a way that reason, by exercising its deductive powers, can anticipate the actual course of events. Starting from known facts and adducing their genesis in rational terms, explanation finds its true meaning in the anticipation of facts that have yet to be discovered.

Though our example elucidates the general structure of explanatory theories, it was nevertheless taken from an area in which Galileo, lacking an adequate conceptual framework, remained incapable of deploying the resources of mathematics to the full; even though mathematics played a certain part in the construction of his model, it chiefly served to add greater precision to his exposition, without yet constituting the very language of scientific research. In order to depict Galileo's principle of explanation in its most exalted form, we must therefore complete our analysis with another example, this one taken from the *Discourses*.

Consider a perfectly smooth spherical body being rolled along a plane surface of fixed dimensions; its motion will at first be uniform and horizontal, but as soon as it leaves the supporting surface it will become more complex and reproduce exactly a projection. Could its precise trajectory be determined? Galileo thought it could, with the help of three assumptions: the principle of conservation of uniform motion on a horizontal surface, the principle of the composition of motions, and the theory of naturally accelerated motion.[59] Let us now consider a surface *ab* on which a projectile is released with a certain speed; it is clear that once it has arrived at *b*, it will acquire a natural motion downward. Moreover, it follows from our first assumption that it will also conserve its uniform motion, which must therefore be combined with its free fall. Now we know the laws governing naturally accelerated motion, and quite especially the one stating that the distances traversed increase as the square of the time. Let us therefore extend *ab* by the equal and successive segments *bc*, *cd*, and *de*, representing equal intervals of time and also the horizontal distance that the moving body would have traversed in the course of one of these intervals if the plane had not ended abruptly in *b*. From *b*, *c*, and *d* let us now drop perpendiculars; at the end of the first interval of time, the projectile will have reached *i*, not *c*, having meanwhile

[59]These three assumptions are explicitly mentioned immediately after the proof of Theorem I on projectile motion; *Discourses* IV, p. 273.

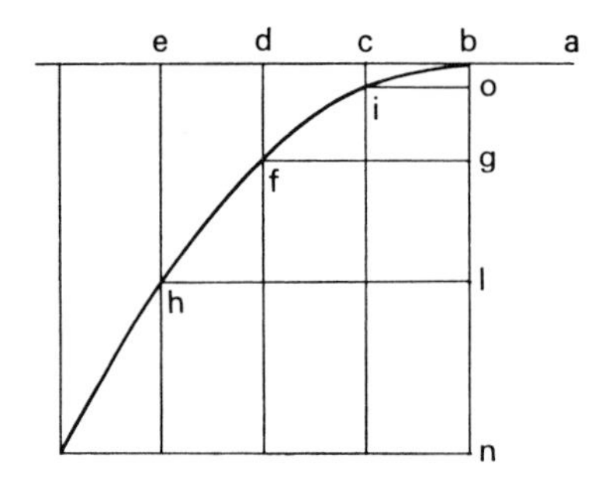

descended through *ci*. At the end of the second time interval, *cd*, it will have descended through *df*, and that distance (according to Theorem II, on naturally accelerated motion) is the quadruple of *ci;* similarly, after the third time interval the projectile will have reached *h*, *eh* representing nine times the distance of *ci*. Thus the points *b*, *i*, *f*, and *h* will mark successive positions of the projectile, and the curve joining these points will represent the trajectory. At first sight there seems to be no great difference between this explanatory method and that which enabled Galileo to account for the apparent motion of sunspots—in both cases the phenomena had to be expressed within a chosen rational system for the precise purpose of obtaining an intelligible genesis of their production; in both cases too the expression and the genesis were carried out thanks to a model. However, despite the formal similarity between the two situations, the *Discourses* introduced a completely new element. To make this clear, let us compare the two models: while the first was above all a mechanical model, admittedly constructed with the help of geometry, but one whose explanatory basis was the motion of the earth about the sun, the second was a truly mathematical model, because the expression it provided of the phenomena in the chosen rational system could at once be subjected to mathematical reasoning and hence benefit from all its resources. Let us now return to our figure and draw the lines *io*, *fg*, and *hl* parallel to *be*. Is it possible to define the nature of the curve joining the points *b*, *i*, *f*, and *h*? "The lines *hl*, *fg*, *io* are equal to *eb*, *db*, and *cb*, respectively; so also are the lines *bo*, *bg*, *bl* respectively equal to *ci*, *df*, and *eh*. The square of *hl* is to that of *fg* as the line *lb* is to *bg*; and the square of *fg* is to that of *io* as *bg* is to *bo*."[60] Now Apollonius

[60]Ibid.

had shown that a parabola is a curve with precisely these properties:[61] hence, a parabola helps us represent the motion of projectiles within our chosen rational system. The Galilean approach to physical explanation may thus be said to have found its ultimate expression in the *Discourses*. In the case of sunspots, as we saw, he equated explanation with the expression of a phenomenon or of a set of phenomena in a rational system and with the help of a model. In the *Discourses*, he went much further than that: here an explanation became the transformation, with the help of an appropriate construction, of a physical fact into a mathematical problem, followed by its analysis and solution in the light of previously established mathematical truths. Hence, though Galileo did not fundamentally change his explanatory ideal as he passed on from his cosmological studies to the *Discourses*, it was only in the latter that this ideal came into its own and that physical research was set on the road it has been following ever since.

But though the transformation of physical into mathematical problems thus became the true task of natural philosophy, it could perform that task only under certain conditions. Two of those are quite obvious: the ability to draw on conceptualizations thanks to which mathematical arguments can be applied to natural phenomena without deforming them, and the availability of the requisite mathematical tools. That these are indeed indispensable conditions may readily be gathered from the fact that it was an inadequate conceptualization that faulted Galileo's centrifugal theory during the Second Day of the *Dialogue*, and that it was the lack of the right mathematical techniques that during the Third Day of the *Discourses*[62] impeded his determination of the swiftest line of descent between two points. However, these two conditions are not sufficient. To explain, as we have said, means transforming a physical problem into a mathematical one and then solving it with the help of established mathematical truths; in other words, mathematizing means understanding the rational justification needed by every physical proposition on the model of mathematical demonstrations. Though this fact may seem self-evident today, it certainly was not so at the beginning of the seventeenth century, for its acceptance pre-

[61]Ibid., pp. 270–271.

[62]*Discourses* III, p. 264. Galileo concluded that this line must be a circle. The correct solution demanded that one be able to treat the problem of a function of a function, and to determine the value of the variable so that the depending function will be a minimum (calculus of variations). In 1696, Jean Bernoulli showed that the required curve was a cycloid.

supposes that all differences between physical and mathematical intelligibility have been eliminated. When Galileo rejected the traditional causal ideal, he was already suggesting that nothing opposed the fulfillment of this third condition; several arguments in the *Dialogue*, all the more convincing in that they were not yet couched in the language of geometry, show clearly that this is actually the case.

Take, for instance, the proof that if the earth did indeed spin east on its own axis every twenty-four hours, then a stone dropped from the top of a tower would not be deflected toward the west. In it, Galileo contended that if the entire earth were covered by an ocean, it must follow from the principle of conservation that, in the absence of friction, a ship, once set in motion, would forever keep circling the globe. From the same principle it also followed that all the objects carried along by the ship and having received the same initial impulsion will equally participate in the same inertial circular motion, characterized by the same power of indefinite conservation. That being the case, let us assume that a stone on top of a mast leaves its support and begins to descend. Because an inertial motion identical to that of the ship has been impressed on it, the stone will readily accompany the vessel; keeping straight beneath the top of the mast, it will hit the same point on the deck that it would have hit had the ship been at rest. Let us next replace our ship with a tower on a rotating earth: the mechanical situation being identical with the last, a rock dropped from that tower will also keep to a strictly vertical line of descent. The principle of conservation thus leads to the conclusion that the vertical fall of heavy bodies is in no way altered by the rotation of the earth.[63]

Despite its simplicity, Galileo's argument was extremely revealing. How did he really establish the soundness of his proposition? Taking the principle of conservation for granted, he applied it first of all to a case in which the earth's diurnal motion, whose possibility he wished to

---

[63]Other examples of a similar type: the argument that enabled Galileo to refute Tycho Brahe's objection, and that by which he established that the speed of a naturally accelerated motion increases in a continuous manner. It should be noted that Mach described these arguments as *Gedankenexperimente* (thought experiments; cf. E. Mach, *Erkenntnis und Irrtum*, pp. 183 ff.), that he contended they played a crucial role in the formation of classical science and, moreover, that he used them to justify his own empiricist interpretation. In so doing, he forgot that traditional science was perfectly familiar with this type of argument, though it did not attach the same importance to it as Galileo was to do (see footnote 65 below).

establish, could not possibly affect the issue:[64] on a ship rounding the globe with an inertial motion that is shared by all the objects it carries along there could be no changes in the relative positions of the objects and their support. Now the case of a tower carried along by the earth's diurnal motion is in all respects *identical* with that of the ship; a stone dropped from the top of that tower will therefore behave in precisely the same way as one dropped from the mast of a moving ship; conserving the motion impressed upon it by the earth, it will follow the motion of the tower and will continue to fall in a vertical line, just as it would had the earth been perfectly immobile. No causal considerations and no factor involving the nature of bodies had to be introduced. At no point, for example, did Galileo speak of a natural circular motion inherent in the stone as the efficient cause of the descent. He merely introduced a principle, that is, a general statement on motion and then tried to discover whether, having admitted that principle, it followed as a matter of course that heavy bodies keep to the vertical, no matter whether the earth is in motion or at rest; the rational justification presiding over his explanation was therefore precisely the demonstration that the proposition to be proved was a necessary consequence of, and implied by, the rational system employed. Now that implication differed in no way from a purely mathematical one: both involve the reduction of a statement to one or several previously established propositions with the help of successive identifications. The concordance of scientific with mathematical intelligibility had at last been fully established.[65]

---

[64] Thus avoiding the risk of engaging in a circular argument.

[65] Aristotle, too, as we mentioned earlier, was not averse to thought experiments; a good example was his rejection of the vacuum on the grounds that it would lead to a host of absurd consequences (infinite or instantaneous motions; the descent of all bodies with the same speed, and so on; cf. *Physics*, IV, 8). However, Aristotelian thought experiments had purely negative ends; they served exclusively to reject certain possibilities, while Galileo used them to introduce positive scientific statements. This transformation of the status of thought experiments was highly significant and proved in yet another way that scientific rationality had joined hands with mathematical rationality. In this connection it is interesting to compare Galileo's approach with the analytical method known and practiced by geometers ever since antiquity. In Book VII of his *Collection*, Pappus defined it as follows: "In analysis, we assume that which is sought as if it were already done, and we inquire what it is from which this results, and again what is the antecedent cause of the latter, and so on, until by retracing our steps we come upon something already known or belonging to the class of first principles, and such a method we call analysis as being solution backwards (*ἀνάπαλιν λύσιν*)" (cf. T. E. Heath, *A History of Greek Mathematics*, Vol. II, p. 400). In the argument just examined, Galileo

Moreover, throughout his work Galileo emphasized the innumerable advantages accruing from the identification of physical with mathematical demonstrations. Thus the *Dialogue* presented mathematics as the most perfect knowledge to which man can aspire—knowledge so perfect, in fact, that it can be compared intensively if not extensively with divine understanding—"extensively, that is, with regard to the multitude of intelligibles, which are infinite, the human understanding is as nothing even if it understands a thousand propositions; for a thousand in relation to infinity is zero. But taking man's understanding intensively, insofar as this term denotes understanding some proposition perfectly, I say that the human intellect does understand some of them perfectly and thus in these it has as much absolute certainty as Nature itself has. Of such are the mathematical sciences alone; that is, geometry and arithmetic, in which the Divine intellect indeed knows infinitely more propositions, since it knows all. But with regard to those few which the human intellect does understand, I believe that its knowledge equals the Divine in objective certainty, for here it succeeds in understanding necessity, beyond which there can be no greater sureness."[66] This certainty founded on necessity explains why a proof conducted *more mathematico* provides the strictest possible and also the most indispensable test of physical speculations. If this is granted, then no subterfuge, no amount of rhetoric, can serve to preserve false or doubtful propositions by the side of true ones. "To a person who wants to convince others of something which, if not false, is at least very questionable, it is a great advantage to be able to use probable arguments, conjectures, examples, analogies, and other sophisms, and to fortify himself further with unimpeachable texts, entrenching himself behind the authority of philosophers, scientists, rhetoricians, and historians.

similarly "retraced his steps" from a certain assertion to "something already known" thus proving the soundness of his assertion. This procedure is also described in a passage of the *Dialogue* (I, p. 75) in which Galileo, dealing with the demonstrative sciences remarked that the "analytical method" allows one to hit upon some proposition previously demonstrated or to arrive at "some axiomatic principle." We know that the analytical method and its natural complement, the "compositive" or "synthetic" method, were the subject of many discussions in the sixteenth century (notably on the part of Zabarella), and it is likely that Galileo was familiar with them (see J. H. Randall, "The Development of Scientific Method in the School of Padua," *Journal of the History of Ideas*, Vol. I, pp. 177–206), but it is difficult to tell what influence these confused discussions might have had on Galileo's thought.

[66]*Dialogue* I, pp. 129–130. Moreover, while human reasoning proceeds by steps from one conclusion to another, Divine understanding is based on direct intuition.

To reduce oneself to the rigor of geometrical demonstrations is too dangerous an experiment for anyone who does not thoroughly know how to manage these, for just as there is no middle ground between truth and falsity in physical things, so in rigorous proofs one must either establish one's point beyond any doubt or else beg the question inexcusably, and there is no chance of keeping one's feet by invoking limitations, distinctions, verbal distortions, or other fireworks."[67] A theory that passes the test of geometrization will thus vindicate the dictum that "to contradict geometry is to fly in the face of the truth."[68] For the same reason, geometrical proofs alone help science in meeting this deductive ideal that Aristotle, hard though he tried, could never reach. In that sense, Galileo's critique of traditional physics was two-edged: on the one hand, it showed up the Peripatetics' failure to gather the very principles on which their arguments were ostensibly based, a failure that often prevented them from appreciating the nonconclusive nature of their demonstrations;[69] on the other hand, it highlighted their neglect of a strict demonstrative path, often putting "the proof of a proposition among texts that seem to deal with other things."[70] With geometrization, however, all these disadvantages disappear: a strict dividing line is drawn between assumptions and conclusions, and each new assertion must, on pain of being dismissed as valueless, be the object of a complete demonstration. As a result, every proposition, no less than the whole edifice, acquires that coherence and inevitability that are the chief characteristics of all scientific thought.

However, identifying scientific proofs with mathematical proofs had more than purely formal advantages—there is nothing to rival the power of geometry to pose and resolve problems in a clear and concise manner—for instance, in demonstrating that the friction effect of the medium is the greater, the smaller the dimensions of the moving body. Moreover, the language of geometry is an unequaled means of conferring a general validity on particular conclusions; thus the parabolic form of the trajectory described by a projectile was established after an analysis of the special case of a body moving along a horizontal surface that is cut off abruptly, so that the inertial motion of the body becomes coupled to a naturally accelerated downward motion. Can

---

[67]*Saggiatore*, Vol. VI, p. 296.
[68]Ibid., p. 214.
[69]For example, when they appealed to the theory of contraries to "prove" the essential difference between the earth and the heavens.
[70]*Dialogue* II, p. 134.

this theorem be extended to the far more common case of upward projection? In order to answer that question in the affirmative, we need merely observe that the change in the direction of the inertial motion (which is now oblique and no longer parallel to the horizon) does not require any additional geometrical assumptions—hence the same construction applies, and with it the same proof. Successful geometrization bestows upon physical arguments a general validity that is both immediately apparent and completely convincing.[71] Finally, it is undeniable, as we have already noted, that by substituting geometrical demonstrations for causal explanations Galileo not only bestowed on science all the advantages of mathematical language but also vastly increased its analytical potential: science could take full advantage of all the resources of mathematical thought and convert the knowledge mathematicians had gathered for their own use into a powerful means of expressing and interpreting physical phenomena. For once the appropriate conceptualization had been adopted, the whole body of mathematical constructions, propositions, and relations were made freely available to science, and opened up new vistas that were as rewarding as they were wide.

Finally, nothing could have been more symbolic of this general transformation than the way in which Galileo reversed the traditional roles of logic and mathematics. Without doubt, the frequently heard claim that Aristotelian physics was totally ignorant of mathematics is false; we need merely recall the pains Aristotle took to reconcile his conception of the continuum with that of the geometers, carefully eschewing any assertions that ran counter to their principles. For all that, mathematics played a very limited part in Aristotle's work; it helped determine the coherence or incoherence[72] of certain theories, but was never

---

[71]In the first edition of the *Discourses* Galileo took the general validity of Theorem I for granted. It was only in a note to the second edition that he made explicit mention of its generality, explaining that geometrically the case of a bullet fired upward is equivalent to the case considered in his original proof (Vol. VIII, pp. 446–447). Now, remarkably enough, that generalization involved the conservation of uniform motion not only on a horizontal plane but also on a horizontal plane *inclined in any direction*, which was clearly incompatible with Galileo's own theory of gravity. This constitutes further proof that it was indeed the inner demands of geometrization that led him to the threshold of the correct formulation of the principle of conservation.

[72]Thus in *On the Heavens* (III, 4, 303 a 20–22) Aristotle rejected Democritus' atomism on the pretext (among others) that in accepting it "one puts oneself in contradiction with the principles of geometry"; cf. ibid., III, 1, 301 a 11 ff.; III, 7 306 a 27 ff.

thought to preside over their proof. Only a scientific syllogism (the syllogism of the διοτι) was considered capable of providing valid explanations, because it alone could reveal the causes of natural phenomena. Galileo, for his part, reversed the traditional roles and restricted logic to the part previously assigned to mathematics, namely to weigh the coherence of an argument or make out the nonacceptability of a conclusion; as such it was a valuable aid but of a purely critical type and hence quite unconnected with the actual construction of proofs.[73] "Logic, it appears to me," Galileo wrote in the *Discourses*, "teaches us how to test the conclusiveness of any argument or demonstration already discovered and completed; but I do not believe that it teaches us to discover correct arguments and demonstrations."[74] In fact, it is mathematics alone that can provide "stimulation to discovery" or usher in a truly demonstrative science; the art of proof, Galileo wrote elsewhere, is acquired by "the reading of books filled with demonstrations, and these are exclusively mathematical works, not logical ones."[75] Thus not only was Galileo the first to construct a geometrized theory of motion but from the very outset he also rejected the idea of two distinct forms of intelligibility, one proper to the physicist, the other to the geometer. "I can hear my adversaries scoff that it is one thing to treat nature as a physicist and another to treat it as a mathematician; that geometers should stick to their reveries and not interfere in philosophical matters, in which truth is quite other than it is in mathematics, as if there was more than one truth, as if geometry were prejudicial to the development of true philosophy, as if it were impossible to be a philosopher and a geometer all at once, and as if he who knows geometry cannot know physics as well or argue as a physicist about physical problems."[76] In short, once explanation has become the ideational reconstruction of natural phenomena and its justification the demonstration of its rational necessity, the norms of mathematical intelligibility can in no way be distinguished from those of physical intelligibility.

---

[73]For examples of this use of logic, see *Saggiatore* VI, pp. 241 and 263 ff.; *Dialogue* III, p. 379; *Letter to Prince Leopold*, May 1640, Vol. VIII, pp. 530–531; *Letter to Liceti*, August 25, 1640, Vol. XVIII, p. 234, etc. Admittedly, Galileo also employed mathematics for critical purposes, as witness the following reply by Salviati to Simplicio in *Dialogue* III, p. 423: "I endorse the policy of these Peripatetics of yours in dissuading their disciples from the study of geometry, since there is no art better suited for the disclosure of their fallacies."

[74]*Discourses* II, p. 175.

[75]*Dialogue* I, p. 60.

[76]Vol. IV, p. 49; fragment from the *Discorso interno alle cose che stanno in su l'acqua*, 1612.

Let us now try to determine the scope and also the limits of this penetration of mathematics into natural philosophy. Once he had demonstrated that the theorems on uniformly accelerated motion were correct, Galileo asked himself whether his "definition of accelerated motion fits the essence of naturally accelerated motion."[77] To prove that it did, he had to take a final step, namely, to demonstrate by experiment that there is a strict correlation between the laws he had constructed in the abstract and those governing the actual fall of heavy bodies. This was a step from which no "true man of science" can escape, "for this is the custom—and properly so—in those sciences where mathematical demonstrations are applied to natural phenomena, as is seen in the cases of perspective, astronomy, mechanics, music, and others where the principles once established by well-chosen experiments become the foundations of the entire superstructure."[78]

To what extent did Galilean science take full advantage of these "well-chosen experiments"?

At first sight it seems difficult to deny that Galileo adduced the clearest and most accurate verification of his laws. Thus the passage in *Discourses* III in which he described his verification of the square law does not seem unworthy of a modern treatise. In it, Salviati tried to counter the objections of all those who claimed that Galileo's experiments did not justify the "conclusions reached."

So far as experiments go, they have not been neglected by the author,[79] and often in his company I have attempted in the following manner to assure myself that the acceleration actually experienced by falling bodies is that which we have described.

A piece of wooden molding or scantling, about seventeen cubits long, half a cubit wide, and three finger breadths thick was taken; on its edge was cut a channel a little more than one finger in breadth; having made this groove very straight, smooth, and polished, and having lined it with parchment, also as smooth and polished as possible, we rolled along it a hard, smooth, and very round bronze ball. Having placed this board in a sloping position, by lifting one end some one or two cubits above the other, we rolled the ball, as I was just saying, along the channel, noting, in a manner presently to be described, the time required to make the descent. We repeated this experiment more than once in order to measure the time with an accuracy such that the deviation between two observations never exceeded one-tenth of a pulsebeat. Having performed this operation and having assured ourselves of

[77] *Discourses* III, p. 197.
[78] Ibid., p. 212.
[79] That is, by Galileo himself.

its reliability, we now rolled the ball only one-quarter the length of the channel; and having measured the time of its descent, we found it precisely one-half of the former. Next we tried other distances, comparing the time for the whole length with that for the half, or with that for two-thirds, or three-fourths, or indeed for any fraction; in such experiments, repeated a full hundred times, we always found that the spaces traversed were to each other as the squares of the times, and this was true for all inclinations of the plane, that is, of the channel, along which we rolled the ball. We also observed that the times of descent, for various inclinations of the plane, bore to one another precisely that ratio which, as we shall see later, the author had predicted and demonstrated for them.

For the measurement of time we employed a large vessel of water placed in an elevated position; to the bottom of this vessel was soldered a pipe of small diameter giving a thin jet of water, which we collected in a small glass during the time of each descent, whether for the whole length of the channel or for a part of its length; the water thus collected was weighed after each descent on a very accurate balance; the differences and ratios of these weights gave us the differences and ratios of the times, and this with such accuracy that although the operation was repeated many, many times, there was no appreciable discrepancy in the results.[80]

Can one ask for any better verification or for one more carefully conducted? And on reading so lucid a passage, are we not bound to agree with all those scientific historians who until the end of the nineteenth century proclaimed Galileo the founder of the modern experimental method? Actually, matters are not nearly as straightforward, and it is with good reason that modern critics have drawn attention to some glaring flaws in Galileo's procedure. Thus a closer look at the way in which he conducted his experiment, and especially at his time-measuring device, will quickly convince us that his approach was extremely rudimentary. How, for instance, could he assume that a vessel from which a pipe drew a thin jet of water could serve as a reliable clock, that the flow of water could be started and stopped to coincide precisely with the ball's descent down the inclined plane? If we remember further that the weight of the water collected had to be determined with the utmost accuracy before the distances traversed could be compared with the time elapsed, we cannot help concluding that results obtained under such poor conditions had little chance of proving anything at all—according to Paul Tannery,[81] they could not even serve to

[80] *Discourses* III, pp. 212–213.

[81] Paul Tannery, "Galilée et les principes de la dynamique," in *Mémoires scientifiques*, Vol. VI, especially pp. 407–408.

decide between Galileo's own law and such closely related formulas as that of Baliani, according to which the distances varied as the series of positive integers. In fact, as Koyré has put it, Galileo still lived in "the world of approximations;"[82] the construction of a precision clock demanded an effort that Galileo, despite his great technical skills (we need only think of the astronomical telescope) and his theoretical acumen (the discovery of the isochrony of pendular oscillations) obviously found too great. This explains the makeshift arrangement he was forced to use.[83] Indeed, just as he was unable to verify his conclusions, so he was also prevented from applying his laws; this would have called for a prior determination of the natural constants involved in the motion and especially of g, which Galileo was unable to provide. Thus, when discussing a problem whose solution called for an approximate evaluation of the natural acceleration experienced by heavy bodies,[84] Galileo concluded from experiments, which he did not even bother to describe, that a freely falling body would traverse a hundred cubits in eight seconds, thus making g eight cubits per second per second, that is, barely half its true value in the vicinity of the earth.[85] In other words, he was no more able to confirm his formula by successful daily applications than he could by direct experiment.[86] Does all this, and especially

[82]Alexandre Koyré, "Du monde de l'à peu près à l'univers de la précision," in *Critique*, September 1948 (No. 28) pp. 815 ff.

[83]Galileo did in fact begin the construction of a clock based on the isochrony of pendular oscillations but was unable to finish it (the clock was perfected by his son, Vicenzio, seven years after Galileo's death). Moreover, in a letter to Baliani he described a method of measuring sidereal time with the help of a pendulum (letter of August 1, 1639, Vol. XVIII, pp. 76–77), which, though of some theoretical importance, proved of small practical value. It was left to Huyghens to construct the first reliable timepiece: the cycloidal pendulum, based on his discovery that the cycloid is a brachistochronic as well as a tautochronic curve.

[84]*Dialogue* II, pp. 245 ff. The problem was to compute the time it would take a ball to fall from the moon to the earth; in accordance with his view that gravity was a force residing within heavy bodies, Galileo assumed that the acceleration was constant throughout the ball's descent.

[85]The Florentine cubit measuring 0.573 meters, Galileo arrived at $g = 4.584$ m/sec^2 when its true value is 9.81 m/sec^2; cf. *Dialogue* II, p. 249. A. Koyré has retraced the history of attempts to measure g from Galileo to Huyghens in *Une expérience de mesure*, republished in *Études d'histoire de la pensée scientifique*, pp. 253 ff.

[86]Cf. A. Koyré: "Thus modern science started under highly paradoxical circumstances. It adopted accuracy as its principle; it proclaimed that reality was geometrical and hence strictly measurable; and it was in possession of formulas that allowed it to deduce and calculate the position of a moving body at each point of its trajectory and at each moment of its travel and yet was incapable of doing anything about it because it could neither determine an

the unavoidably inaccurate nature of his conclusions, not suggest that Galileo failed to attach nearly as great importance to experiment as he always claimed he did? Did he not rather consider that any proposition (such as his "square law") that could be justified by strictly rational arguments had a very high probability of being true, that is, in accordance with reality? In brief, if Galileo accepted as crucial any experiments that were manifestly inconclusive, was not the motive his firmly rooted belief in the necessary correspondence between natural and rational necessity?

This is borne out further by the importance he attached to the principle of simplicity, particularly in his cosmological studies. We saw how the idea that nature has the "habit and custom to employ only those means which are most common, simple, and easy"[87] suggested to him that the speed of a falling body varies as the time of descent. The role of the principle on that occasion was, however, purely formal: it helped elaborate a definition but did not establish its physical truth. In his cosmological studies, on the other hand, the principle of simplicity became a superior criterion of deciding between physical truth and falsehood. To illustrate this point, let us look again at the apparent motion of sunspots. We saw that, in order to account for their variation, Galileo simply combined the Copernican doctrine with the supplementary assumption that the axis of the sun was oblique to the ecliptic; all observed phenomena immediately fell into place, and it became possible to predict the subsequent course of events with great accuracy. Now, once we have a model that can anticipate experience so well, may we not assume that it must be valid? That it reflects reality most faithfully? But let us now pose the problem in logical terms. It is true that by means of Galileo's hypothesis all the apparent motions of the sunspots become intelligible, but before we can conclude, *ex converso*, that the hypothesis itself is true—which is in any case not permissible on purely logical grounds—we should first of all have to show that "such peculiarities ought not also be seen in a sun moving along the ecliptic by inhabitants of an earth stationary in its center."[88] And this we simply cannot do since, as we shall see, the Ptolemaic doctrine can

instant nor measure a speed. And without these determinations the laws of the new physics remained empty and abstract." From "Un experimentum au XVIIe siècle: la détermination de *g*," *Congrès international de philosophie des sciences*, Vol. VIII (Paris, 1949), p. 197.

[87]*Discourses* III, p. 197.

[88]*Dialogue* III, pp. 379–380.

account for the "peculiarities" just as well as the heliocentric hypothesis can. Must we then give up the attempt to prove the physical truth of the model and consider it a mere artifice whose sole purpose is to "save the phenomena"? Galileo did not think so, but, being unable to decide the issue by direct appeal to facts, he tried to reach the solution along another path.

Let us adopt the Ptolemaic view and try to account for all the variations in the motion and shape of the sunspots in terms of the sun's own motion. How many distinct motions should we then have to introduce? First of all, a rotational motion, as the only possible explanation of why the spots, whose contiguity to the sun's body has previously been established, traverses the solar disk from east to west in just under a month.[89] But we also need a second motion, for if we assume the sun's axis of rotation to be perpendicular to the plane of the ecliptic, it follows that the "passage of the spots would appear to us to be made in straight lines, and parallel to the ecliptic" whereas, as we saw, "their courses appear for the most part to be made along curved lines."[90] We should have to assume further that the tilt of this axis is not fixed and "facing continually toward the same point of the universe" but that it changes direction from one moment to the next. For if the obliquity were always pointed in the same direction, the paths of the spots would never change their appearances "whether they were straight or curved, bent up or down, ascended or descended."[91] To obviate all these disadvantages, we should therefore have to assume that the sun's axis of rotation had a circular motion of its own parallel to the axis of the ecliptic.[92] In that case each of the sun's poles would have to describe a circle whose radius would correspond to the tilt of the axis of rotation.[93] Finally, it is obvious that, if the rotational motion was completed in just under a month, the second motion must have a period of one year, because this is how long it takes the celestial appearances to recur. In short, to explain the observed irregularities in the behavior of a sunspot on the assumption that the earth is at rest at the center of the ecliptic, we must attribute to the sun "two motions about its own center and on different axes." All these assumptions are very difficult, indeed almost impossible, the more so because to these two motions we must also add the sun's annual motion and the diurnal motion by which it describes

---

[89]Ibid., p. 380.
[90]Ibid., pp. 380–381.
[91]Ibid.
[92]Ibid.
[93]Ibid.

"a circle parallel to the equinoctial plane" once a day.[94] Though theoretically feasible, the Ptolemaic solution is thus vastly more complicated than the Copernican. Hence, "if these four motions, so incongruous with each other yet necessarily all attributable to the single body of the sun, could be reduced to a single and very simple one, the sun being assigned one inalterable axis, and if, with no innovations in the movements assigned by so many other observations to the terrestrial globe, one could still easily preserve the many peculiar appearances in the movements of the solar spots, then really it seems to me that this decision could not be rejected."[95]

Nor did Galileo consider this appeal to the principle of simplicity a mere expedient. For if the purpose of an explanation is the ideational re-creation of phenomena, the *form* of that re-creation, which is a rational creation, cannot be without importance: of two explanations that agree equally well with experience, the simpler must be the rationally superior. Hence, the greater simplicity of a theory must be clear proof of its closer agreement with reality. For all that, the principle of simplicity is a metaphysical rather than a physical postulate: though demanded by reason and explicitly by mathematical reason, it is, strictly speaking, no more than a prejudice when applied to experience. If we thus make it the touchstone of the merit of a given theory, we are willy-nilly turning reason into a yardstick of reality. Indeed, we might go further still, since from certain remarks repeated with emphasis it would appear that to Galileo the very idea of a possible conflict between reason and experience was quite inconceivable. A few pages back we retraced the argument from which he concluded that the vertical fall of heavy bodies was in no way affected by the earth's diurnal motion. That argument was based in essence on the assumption that a tower on a moving earth was comparable to a stone on a moving ship. But had Galileo, in fact, tried to verify by experiment whether the stone actually landed at the foot of the mast from which it had been dropped? In his *Letter to Ingoli* of 1624, in which he first brought up this argument, he made it perfectly clear that he attached no more than secondary importance to such investigations: "And in this instance, I was twice as good a philosopher as all those who have compounded error with lies, pretending they had seen the experiment, for I actually performed it, once natural reason (*il natural discorso*) had fully convinced

---

[94] Ibid., p. 382.

[95] Ibid.

me that the effect must follow, as indeed it did."[96] Eight years later, in the *Dialogue*, experimental verification had become quite superfluous. Thus when Simplicio asked Salviati if he had checked his conclusions, he was told that no experiment was needed—"the effect will follow as I tell you because it must happen that way."[97] Here the difference between rational and experimental justification has shrunk to such a point that we are entitled to ask what useful role, if any, the latter continued to play in Galileo's work.

On the basis of the examples we have been quoting, Tannery and Koyré have both answered this question in the negative. Koyré, in particular, who has dealt with the matter more systematically, felt convinced that, as far as Galileo was concerned, "experience was never the essential key to the recognition of truth."[98] Inaccurate and generally subsidiary to reason, Galileo's experiments were primarily "thought experiments" and served more as illustrations than as genuine verifications.[99] Galileo's rationalism thus embraced an a priori element that, in many respects, bore a close resemblance to that of the great classical philosophers; as far as Galileo was concerned, "good physics is conducted a priori."[100] Moreover, Koyré has tried to uncover the philosophic roots of this a priori element: according to him, Galileo based his views sincerely and explicitly on Plato's theory of knowledge. There are many good reasons for this interpretation, not least the typically Platonic form of many passages in the *Dialogue*. Thus on the Second Day, when Simplicio failed to see that the motion of a body on a plane equidistant from the center, that is, on a spherical surface, would be conserved indefinitely, Salviati conducted an interrogation that led his interlocutor step by step to the discovery of the truth "in the light of natural reason."[101] Moreover, Koyré believes that the Platonic influence

---

[96]*Letter to Ingoli*, Vol. VI, p. 545.

[97]*Dialogue* II, p. 171. Galileo's refutation of Tycho's objection to the earth's diurnal motion was also based on reasoning rather than on experiment.

[98]A. Koyré, *Études galiléennes* III, p. 67.

[99]Koyré even went so far as to claim that in the *Dialogue* the true champion of experiment was Simplicio, the Aristotelian. Ibid.

[100]Ibid. This was also the view of Mersenne: "I doubt," he wrote, "whether Galileo ever performed experiments on the inclined plane, because he never spoke of them and because the proportion he mentions is often contradicted by experimental evidence." (See *Harmonie universelle* (Paris, 1636), Vol. I, p. 112).

[101]*Dialogue*, pp. 171 ff. Koyré described these passages, in which Galileo adopts Socrates' maieutic process of examination, as the "experimental proof" of Galileo's Platonism; *Études galiléennes*, III, p. 128.

was no less obvious in Galileo's general inspiration, as witness his repeated adulatory references to Plato as the stalwart champion of geometry in science, that is, of the very method that Galileo, in contrast to Aristotle, held to be the only fruitful approach. "Who argued more justly," Galileo asked one of his Peripatetic adversaries, "Plato, who held that philosophy could not be learned without mathematics, or Aristotle, who scoffed at Plato for delving too far into geometry?"[102] And there is a great deal of further evidence that Galileo embraced the Platonic conception of truth. Thus, during the Second Day of the *Dialogue*, when Simplicio claimed that there could be no possible refutation of Ptolemy's argument that on a moving earth bodies would be flung out into space, Salviati made the following reply: "The unraveling depends upon some data well known and believed by you just as much as by me, but because they do not strike you, you do not see the solution. Without teaching them to you then, since you already know them, I shall cause you to resolve the objection by merely recalling them."[103] Arguing in this way, was Salviati not leaning toward Plato's opinion that *nostrum scire sit quoddam reminisci*?[104] And much as he accepted Plato's view of recollection, so there would seem to be little doubt that Galileo also accepted the Platonic theory of the acquisition of true knowledge. "To Galileo," Koyré concluded, "human understanding is possessed *ab initio* of 'clear and distinct' notions, whose very clarity guarantees their *truth*; man needs only to turn inward to discover in 'his memory' the foundations of his knowledge of reality: the alphabet of the mathematical language spoken by nature as created by God."[105] Thus whereas the modern scientist holds that the links between theory and experiment are constant and, as it were, organic, Galileo could be said to have taken the view that reason guided by geometry can advance unaided to the full comprehension of reality.

This interpretation, at least in the extreme form Koyré has given it, strikes us as rather exaggerated. Clear though Galileo's declarations in

[102]Vol. VII, p. 744. Cf. the following fragment of a projected title: "Wherein the usefulness of mathematics in the study of natural proportion may be gathered from numerous examples, all demonstrating the impossibility of philosophizing correctly without the aid of geometry, in full accord with Plato's just affirmation." (Vol. VIII, pp. 613–614); see also *Discourses* I, p. 135.

[103]*Dialogue* II, p. 217; cf. ibid., p. 183.

[104]See also this note in an undated text: ". . . while showing with the help of the following analysis how right Plato was to argue that science is but the recollection of familiar and hence evident proportions." (Vol. VIII, p. 598).

[105]*Études galiléennes* III, p. 126.

favor of Plato undoubtedly were, they cannot be treated in isolation from the circumstances in which he was placed, quite especially when he wrote the *Dialogue*. The very nature of his subject and the approach he decided to adopt were such that he knew full well how hostile the reaction of philosophers and theologians would be; hence it was only a measure of elementary prudence that, when dealing with adversaries to whom the principle of authority was sacrosanct, he should have sought the protection of a tradition that, though less powerful than the Peripatetic, nevertheless enjoyed universal respect, the more so as Platonism, thanks to its long association with Christianity, was clear of the least taint of suspected heresy.[106] Other factors also intervened. Thus, as Koyré himself has shown, the clash between Plato and Aristotle on the very question of the application of mathematics to physical science was a subject of keen discussion that cropped up almost as a matter of course in philosophical disputes.[107] Buonamico and Mazzoni, in particular, two men whom Galileo had met in Pisa, had deliberated at length on this theme, and it is certain that Galileo was familiar with their writings. The *Dialogue* more particularly echoed these discussions and on several occasions used them to refute Aristotelian objections to the idea of a mathematical science of nature.[108] Under these circumstances, vaunting one's adherence to the Platonic doctrine meant participating in a current debate at least as much as expressing one's blind adherence to the letter of that doctrine. All in all, therefore, Galileo's advocacy of Platonic ideas is certainly more ambiguous than one might be tempted to think at first sight.

There is an even more convincing reason why it is quite wrong to claim that Galileo adopted a Platonic explanatory ideal. For before we can claim that he accepted the idea of a priori elements of knowledge, we must first prove that he treated his own principles of explanation as so many innate objects of pure intellectual intuition. Now neither the systematic way in which he adduced these principles nor the importance he invariably attached to experience in their formulation entitles us to argue that he did. Take, for example, the principle of mechanical relativity whose construction in the *Dialogue* preceded

[106]This was not true of the Atomists; Galileo, in fact, never quoted any of them.

[107]*Études galiléennes* III, pp. 117 ff; cf. "Galileo and Plato," *The Journal of the History of Ideas* (October 1943), pp. 400–428.

[108]For example, *Dialogue* II, pp. 229 ff. We shall return to this problem.

that of the principle of conservation. What do we find? Galileo began with an appeal to a datum of direct experience, namely, that, in describing the motion of an object, it makes no difference whether we suppose that the object itself is in motion and the observer at rest, or vice versa. The general idea of relativity having been introduced, Galileo proceeded to his principle in two steps: The first was to show that for a system of bodies in uniform motion there would be no difference between the state of motion and that of rest. Thus let us imagine we are in a ship, and more particularly in its hold where the goods are placed; we shall easily agree, if the motion is actually uniform, that it will have no effects since it in no way alters the relative positions of the objects.[109] But understanding the principle of mechanical relativity means also understanding that in the hold, where there are no external reference points, an observer can never determine whether the ship is in motion or at rest. Similarly, a man resting on the deck and looking at the top of the mast will find that the mast follows the same pitch and toss of the ship, so that he does not have to shift his head to keep his gaze permanently fixed on the mast. Indeed, "if I had aimed a musket, I would never have to move it a hair's breadth to keep it aimed."[110] These remarks show better than anything else that, even while constructing his principle, Galileo never lost touch with direct experience. Does this mean that his principle was a simple form of experimental induction? To adopt that view would be as mistaken as to overlook the role of concrete intuition altogether. There is not the slightest doubt that a genuine rational intention—the generalization of the idea of relativity—presided over Galileo's successive appeals to observation as well as over their coordination, and that it alone led him to a proposition of such great scientific importance. The reason why his principle was of general validity was thus chiefly that it reflected the intimate association of reason with experience.

The case of the principle of conservation was no different. Here experience took the form of research with inclined planes and informed every step in the argument leading up to the formulation of that principle. It was thanks to the inclined plane that Galileo first of all introduced his distinction between the motor and gravific functions of the natural force inherent in all bodies—a distinction without which the

---

[109]*Dialogue* II, p. 142.

[110]Ibid., p. 274.

principle of conservation could never have been conceived. The inclined plane next suggested to Galileo that the motor function of that force was responsible for the body's propensity no less than its resistance to motion. Finally, by taking his study of inclined planes to its ultimate conclusion, Galileo discovered that, in one case at least, it was possible to eliminate both the propensity and also the resistance to motion, and hence to assert that even the smallest force could set the heaviest body in motion. A final abstraction then led him to the idea of inertial motion. Once again it was by taking maximum advantage of the experimental data that he constructed his principle. However, the rational contribution was no less plain. Thus when, passing from planes of increasingly smaller inclinations to the horizontal plane, he asserted that on the latter even the smallest force would invariably set a body in motion, his use of trajectories of increasingly smaller inclinations was merely a means of depicting before the mind's eye an authentic mathematical argument, namely, the method of approximation to limits. Similarly, when from the elimination of any propensity of resistance to motion on a surface equidistant from the center he deduced the indefinite conservation of motion, it was indeed the principle of sufficient reason and not some vague intuition that guided his conclusions. All this helps us in making an accurate assessment of the status of Galileo's principles. Rooted as they were in experiments that reason not only helped to coordinate but varied to suit its own needs, these principles were neither forcibly imposed on reality nor simply induced from observation: they were rational constructions, and as such served the physicist as guidelines in the effective interpretation of natural phenomena. Nothing in the manner in which Galileo introduced his principles thus suggests that he might have considered them innate elements of human knowledge or, a fortiori, that he considered man to "be in possession of the true principles of the nature of the physical world prior to any experience,"[111] as any faithful disciple of Plato would undoubtedly have done.[112]

---

[111] *Études galiléennes* III, p. 67. On the same page Koyré grants that the term "innate" was not part of Galileo's vocabulary.

[112] Koyré (*Études galiléennes* II, p. 17) was not the only one to assert that Galileo was a Platonist. The same view has also been put forward inter alia, by E. Cassirer, *Das Erkenntnisproblem in der Philosophie und Wissenschaft der neueren Zeit* (Berlin, 1911), I, p. 389; E. A. Burtt, *The Metaphysics of Sir Isaac Newton* (London, 1925), pp. 71 ff. and L. Olschki, *Galilei und seine Zeit* (Halle, 1927), pp.

All in all, therefore, the a priorist interpretation seems to have little substance. A closer look at the claim in the *Dialogue* that there is no point in turning to experience for confirmation that a stone dropped from the mast of a moving ship will not be deflected toward the stern will show that this apparently irrefutable proof of Galileo's Platonism is not nearly as convincing as Koyré seems to think. What, in fact, was Galileo trying to prove? Precisely that, once we have accepted the principle of conservation, we cannot possibly assume that the stone would be deflected—in other words, that the absence of any deflection is an analytical consequence of the principle of conservation. Now the conditions under which he had established the principle were such as to convince Galileo that it must be in perfect agreement with both nature and reason. And what applied to the principle itself applied equally well to propositions directly deduced from it; since all were covered by the same guarantee, there was no need to subject them to special verification. Hence, when Galileo wrote that he was sure, even without experience, that "the effect would happen as I tell you, because it must happen that way," he was simply underlining the half-rational, half-experimental status of his principles and not expressing an a priorist view of scientific truths.

Moreover, though Galileo's experimental techniques were not altogether beyond reproach, it would be flying in the face of the facts to deny that he gathered and assimilated experiential data with the utmost care and subtlety. Take again the case of the resistance of materials to fracture. His entire theory was based on the precise observation that large machines are not proportionally stronger than small ones. Now this was a completely unsuspected experiential fact, not an a priori assumption based on geometrical conclusions. Experimentation also played a crucial role in his formulation and verification of several fundamental ideas of mechanics. In this area, apart from the inclined-plane experiments, we need only recall Galileo's analysis of changes in the speed of freely falling bodies: the way in which he varied the specific weights of moving bodies and media simultaneously, no less than his use of the pendulum to prove that in relatively rare media friction

164–174, all of whom fell victim to the myth that Plato was a mathematical physicist in embryo. Strangely enough, all these writers deduced Galileo's Platonism from the "spirit" of his work, and none of them seems to have asked whether Galileo himself presented his principles as elements of innate knowledge or as rational construction closely linked to experience.

alone is responsible for the observed speed alterations, bears witness to a remarkable facility of combining reason with experience—so remarkable, in fact, that it plainly foreshadowed the approach of Newtonian science. Finally, it is impossible to side unreservedly with Tannery and Koyré when they object to the experiment by which Galileo tried in *Discourses* III to justify his square law. For though his procedure left much to be desired, it remains a fact that, by substituting motion down an inclined plane for naturally accelerated motion, he succeeded in opening the phenomenon of free fall to both observation and measurement. And what better means was there of verifying the square law? Would he have evinced so acute an experimental sense if the experiment itself had struck him as a mere formality?[113] Hence, it is by no means idle to wonder whether the view put forward by Tannery and Koyré was not perhaps the result of a somewhat overhasty extrapolation, namely, that because Galileo's experiments are no longer convincing today (and this is true), experiment cannot possibly have played any part in the genesis of Galilean science. At the very least, this interpretation goes far beyond what the premises, that is, the content of the work itself, allow us to assert with any degree of certainty.

Must we therefore conclude that the thesis of Tannery and Koyré is completely false? Nothing could be more mistaken, since in our view that thesis represents an important step in the correct appreciation of Galileo's method. For just as it is wrong to underestimate the place in this method of the interrogation of experience, so it is equally wrong to equate it with the explanatory procedure of classical science. For one thing, no theorem in the *Discourses* was induced by experiment; every one was obtained with the help of principles that, though framed in close contact with the facts, were nevertheless based chiefly on reason. Galileo's use of the principle of simplicity was by no means arbitrary;

---

[113]Let us note, however, that in causing bodies to roll rather than slide down an inclined plane, Galileo modified their natural acceleration (that observed in free fall) in a manner that he was still quite incapable of evaluating. Thus, if the plane has the inclination α and the body (for example, a cylinder) has the radius r, the mass is m and the moment of inertia about the center of gravity Ig, then, if the body slides down an inclined plane its natural acceleration will in the absence of friction be reduced in the ratio $g/g \sin \alpha$; however, if it rolls down the plane its acceleration will be reduced in the ratio $g/\{(g \sin \alpha)/[1 + (Ig/mr^2)]\}$: the greater the moment of inertia, the slower the descent. However, since the speed still increases as the time, Galileo's basic experimental idea still applied.

indeed it reflected his clear faith in the rational structure of reality. Nor must we overlook the many passages in which he placed experimental proof on a par with experience. For example, during the Second Day of the *Dialogue* he declared quite bluntly that "one single experience or conclusive demonstration suffices to dash a hundred thousand other probable arguments."[114] Hence, despite our many reservations, we must give full credit to Tannery and Koyré for the stress they have laid on the fervent rationalism with which Galilean science is suffused and for having been the first to try to grasp it in its proper context and full significance. Their mistake was rather that they based their interpretation on a single fact, namely, the great distance that in the area of experimentation still separates Galileo from the classical physicists. It was on this rather slender basis that they (and especially Koyré) felt entitled to replace the century-old view that Galileo's ideal was one of pure experimentation (which, in fact, it was not) with that of an a priorist ideal and to justify this step with the very doubtful hypothesis of Galileo's adherence to the Platonic doctrine. In our view, they would have done much better, had they, after first emphasizing the keen rationalism to which Galileo's experimental practice bore clear witness, gone on to characterize this rationalism from within Galileo's own doctrine instead of judging, or rather prejudging, it with the help of extraneous norms. To do so, moreover, they would have had to meet but a single demand: to make a deliberate and methodical attempt to reconstruct Galileo's own explanatory procedure. By thus restoring Galileo's rationalism to its true role—to inspire and guide research—they would have been able not only to demonstrate its originality but also to determine its limitations. We believe that this originality and these limitations can be clearly determined by an attentive examination of the following two questions:

First, was the representation of reality that Galileo tried to construct side by side with his mechanics in such perfect agreement with mathematical reason as to persuade him that natural order was identical with rational order?

Second, were the conditions under which he introduced his geom-

[114]*Dialogue* II, p. 148. Galileo reiterated this view on many occasions, such as in his *Letter to the Grand Duchess Christina*, Vol. V, p. 316, and in *Discourses* III, p. 200: "But without depending upon the above experiment, which is doubtless very conclusive, it seems to me that it ought not to be difficult to establish this fact by reasoning alone."

etrization not such as to ensure that experimentation was transformed into an essential element of all scientific research?

A rapid recall of the main arguments by which traditional science claimed that mathematical reason lacked the capacity to define the principles of physics will bring out the full importance of Galileo's contribution. For Aristotle, physics was the science of changes in nature, including displacements of bodies, transformations of quality and quantity, generation and corruption. Principles and concepts were supposed to serve as representations of reality in such a way that they not only helped to render the general phenomenon of change intelligible (how could there be changes?) but also lent it meaning (why were there changes?). We know that Aristotle solved these problems by introducing the concepts of actual and potential existence, and of form and matter; forced by the very nature of matter to exist potentially before coming into actual existence, forms must necessarily be subject to changes, which, although transitory phenomena by definition,[115] nevertheless play an indispensable role in any true philosophy of nature. Now that which form and matter can accomplish, mathematical entities, *in which nothing calls for change*, obviously cannot: any attempt to construct physics with the help of mathematical concepts is tantamount to robbing changes of their physical foundations.[116] All this follows quite clearly if we agree with Aristotle's views about the origins of mathematical concepts. According to Book II of the *Physics*, the fundamental concepts of geometry are volumes, areas, lines, and points. Being neither essences capable of existing independently of experience nor simple logical constructions, these concepts must refer to such properties of physical bodies as the geometers originally "abstracted" from their physical context.[117] Not subject to the contingencies that invariably accompany the presence of matter, these properties may be studied in a purely theoretical way, that is, in terms of rational necessity. But, that being the case, how could mathematics possibly render the least service to natural science?[118] How could concepts that simply refer to some

[115]At least in the sublunary world.
[116]This was the critique Aristotle leveled at Plato's *Timaeus*: cf. *On the Heavens* III, 8.
[117]"The mathematician, too, occupies himself with these things, but [only] inasmuch as they are limits of natural bodies." *Physics* II, 2, 193 b 31–32; cf. 194 a 9–10: "Although geometry investigates lines, it ignores their physical aspects."
[118]Except in a critical sense.

external attributes of reality they themselves have idealized ever lead to an understanding of that reality as a whole? When we seek in natural effects "the same necessity as in mathematical demonstrations,"[119] are we not forgetting the basic fact that "these mathematical subtleties do very well in the abstract, but they do not work out when applied to sensible and physical matters"?[120] This can be proved with the help of a simple example. Thus if we take a plane and sphere made of any material, can we truly assert that they will touch in a point like a geometrical sphere and a geometrical plane? Far from doing that, the material sphere will be so heavy as to cause the plane to bend around it, even if it were perfectly smooth, which, by the very nature of things, it can never be.[121] Because it is as distinct from geometrical shapes as what changes is from what is motionless, as the contingent is from the essential, or the irregular from the regular, matter is completely beyond the reach of mathematical reason.

That Galileo had ceased to pay heed to this kind of argument must by now be quite obvious. Once motion had become a state, there was no longer the least need to base it on the properties of matter, which was given a new status in turn. From a source of potential existence matter had turned into an inalterable part of reality, comparable in all respects to an "eternal and necessary" property; as such it was open to mathematical analysis, and the physicist had no reason for maintaining that his concepts were totally distinct from those of the geometer. The First Day of the *Discourses* dispelled any remaining doubts on this point: "Since I assume matter to be unchangeable and always the same, it is clear that we are no less able to treat this constant and invariable property in a rigid manner than if it belonged to simple and pure mathematics."[122] Moreover, were physical shapes really so unlike mathematical forms that it made no sense to treat them alike? Was it true, for example, that two tangent material spheres must necessarily touch over an extended surface? In fact, even if we could ignore the pressure of the weight, the spheres would have to be perfectly smooth or such that the

---

[119]*Dialogue* I, p. 38.

[120]*Dialogue* II, p. 233; cf. *Discourses* I, p. 96: "The arguments and demonstrations which you have advanced are mathematical, abstract, and far removed from concrete matters; and I do not believe that when applied to the physical and natural world these laws will hold." Galileo put these words into the mouth of Simplicio, the Aristotelian.

[121]*Dialogue* II, pp. 229–233. In his *Metaphysics* (B, 2, 998 a 2), Aristotle states that the same example had been used by Protagoras to confound the geometers.

[122]*Discourses* I, p. 51.

convexity of one fits perfectly into the concavity of the other. Now these conditions are practically excluded by the very irregularity of matter; thus, if we are careful not to crush the two spheres together, we can make certain that they will touch at just one or two points, instead of over an extended surface.[123] Moreover, the problems posed by the application of geometry to natural bodies are in no way different from those posed by the application of arithmetic to counting and weighing. Just as a scrupulous merchant must take certain precautions —for example, discounting the weight of his containers—so "the mathematical scientist (*filosofo geometra*), when he wants to recognize in the concrete the effects which he has proved in the abstract, must deduct the material hindrances."[124] If he does so, he will find that the laws of geometry apply as well to the behavior of natural bodies as numerical computations apply to practical trade. "The errors, then, lie, not in the abstractness or concreteness, not in geometry or physics, but in a calculator who does not know how to make a true accounting."[125] The traditional position had thus been completely reversed. To Aristotle and his disciples mathematical concepts were so many abstractions from the sensible world and hence incapable of elucidating reality. The "mathematical scientist" by contrast, firmly convinced as he is that "what happens in the concrete happens the same way in the abstract,"[126] tends to treat natural bodies as complex geometrical forms. In his view it is perfectly legitimate to proceed from the abstract to the concrete and hence to arrive at a representation of reality that is in full accord with the demands of mathematical understanding. "Philosophy," Galileo explained in a famous passage, "is written in this grand book, the universe, which stands continually open to our gaze. But the book cannot be understood unless one first learns to comprehend the language and read the letters in which it is composed. It is written in the language of mathematics, and its characters are triangles, circles, and other geometric figures, without which it is humanly impossible to understand a single word of it; without these one wanders about in a dark labyrinth."[127]

---

[123] *Dialogue* II, pp. 234–235.
[124] Ibid., p. 234. Example: Galileo's analysis of the diversification of speeds in free fall.
[125] Ibid.
[126] Ibid., p. 235.
[127] *Saggiatore*, Vol. VI, p. 232. Cf. *Letter to Liceti*, January 1641, Vol. XVIII, p. 295: "I hold that the book of philosophy is that same book which is continuously kept open before our eyes: but as it is written in characters other than our alphabet, it cannot be read by all: the characters best suited to its reading are triangles, circles, spheres, cones, pyramids, and other mathematical figures."

But how can we construct a theory of matter based on mathematics? It is not enough to define matter as an "eternal and necessary" affection; everyday experience tells us that it is also possessed of a host of qualities or properties of which some—color, sound, smell, and so on—are not directly accessible to geometrical analysis. What is the mathematical scientist to make of these? A crucial passage in the *Saggiatore*, setting out the classical distinction between primary and secondary qualities, was to supply the answer:

> Now I say that whenever I conceive any material or corporeal substance, I immediately feel the need to think of it as bounded, and as having this or that shape; as being large or small in relation to other things, and in some specific place at any given time; as being in motion or at rest; and touching or not touching some other body; and as being one in number or few, or many. From these conditions I cannot separate such a substance by any stretch of my imagination. But that it must be white or red, bitter or sweet, noisy or silent, and of sweet or foul odor, my mind does not feel compelled to bring in as necessary accompaniments. Without the senses as our guides reason or imagination unaided would probably never arrive at qualities like these. Hence I think that tastes, odors, colors, and so on, are no more than mere names so far as the object in which we place them is concerned, and that they reside only in the consciousness. Hence if the living creatures were removed, all these qualities would be wiped away and annihilated. I may be able to make my notion clearer by means of some examples. I move my hand first over a marble statue and then over a living man. As to the effect flowing from my hand, this is the same with regard to both objects and my hand; it consists of the primary phenomena of motion and touch, for which we have no further names. But the live body which receives these operations feels different sensations according to the various places touched. When touched upon the soles of the feet, for example, or under the knee or armpit, it feels in addition to the common sensation of touch a sensation on which we have imposed a special name, tickling. This sensation belongs to us and not to the hand. Anyone would make a serious error if he said that the hand, in addition to the properties of moving and touching, possessed another faculty of tickling, as if tickling were a phenomenon that resided in the hand that tickled. A piece of paper or a feather drawn lightly over any part of our bodies performs intrinsically the same operations of moving and touching, but by touching the eyes, the nose, or the upper lip it excites in us an almost intolerable titillation, even though elsewhere it is scarcely felt. This titillation belongs entirely to us and not to the feather; if the live and sensitive body were removed, it would remain no more than a mere word. I believe that no more solid an existence belongs to many qualities which we have come to attribute to physical bodies: tastes, odors, colors, and many more.[128]

---

[128]*Saggiatore*, Vol. VI, pp. 347–349.

On the basis of this distinction, Galileo felt free to offer a squarely mechanistic explanation of certain secondary qualities. Thus taste and smell were apparently due to the penetration of small corpuscles into the upper part of the tongue and the nostrils.[129] Similarly, the various qualities we associate with sound—sweet, strident, deep, and so on—result from the impact of waves, transmitted by the air, on our eardrums.[130] But it was probably in connection with heat that Galileo expressed his mechanistic ideal most forcefully. In particular, he completely rejected the idea that heat was a "prime quality" and could be passed on as such from one body to the next by contact or radiation:

> Having shown that many sensations which are supposed to be qualities residing in external objects have no real existence save in us, and outside ourselves are mere names, I now say that I am inclined to believe heat to be of this character. Those materials which produce heat in us and make us feel warmth, which are known by the general name of "fire," would then be a multitude of minute particles having certain shapes and moving with certain velocities. Meeting with our bodies, they penetrate by means of their extreme subtlety, and their touch as felt by us when they pass through our substance is the sensation we call "heat." This is pleasant or unpleasant, according to the greater or smaller speed of these particles as they go pricking and penetrating; pleasant when this assists our necessary transpiration and obnoxious when it causes too great a separation and dissolution of our substance. The operation of fire by means of its particles is merely that in moving it penetrates all bodies, causing their speedy or slow dissolution in proportion to the number and velocity of the fire corpuscles and the density or tenuity of the bodies. Many materials are such that in their decomposition the greater part of them passes over into additional tiny corpuscles, and this dissolution continues so long as these continue to

[129]Ibid., p. 349.

[130]Ibid. This idea was developed further at the end of *Discourses* I, where Galileo tried to show that the agreeable or disagreeable nature of certain musical chords depends on the regular or irregular impact of the vibrations produced by the notes on our ear. "The unpleasant sensation produced by the latter [dissonances] arises, I think, from the discordant vibrations of two different tones which strike the ear out of time. Especially harsh is the dissonance between notes whose frequencies are incommensurable; such a case occurs when one has two strings in unison and sounds one of them open, together with a part of the other which bears the same ratio to its whole length as the side of the square bears to the diagonal; this yields a dissonance similar to the augmented fourth or diminished fifth. Agreeable consonances are pairs of tones which strike the ear with a certain regularity; this regularity consists in the fact that the pulses delivered by the two tones, in the same interval of time, shall be commensurable in number, so as not to keep the eardrum in perpetual torment, bending in two different directions in order to yield to the ever-discordant impulses" (*Discourses* I, pp. 146–147).

meet with further matter capable of being so resolved. I do not believe that in addition to shape, number, motion, penetration, and touch there is any other quality in fire corresponding to "heat"; this belongs so intimately to us that when the live body is taken away, heat becomes no more than a simple name.[131]

Shape, size, and contact, location in space and time: it is then clear that these "primary qualities" are all open to measurement and therefore susceptible to mathematical understanding. If we add that motion, the other essential quality of matter, is also subject to numerical laws, we see that there is little left to obstruct a mathematical representation of reality *ex parte rerum*. The faith of the "mathematical scientist" who thinks that matter and its properties can be fully embraced by mathematical concepts thus seemed fully vindicated. Now there was at least one problem Galileo tried to solve while adhering as closely as possible to his mechanistic-*cum*-mathematical ideal, and we can find no better and more precise test of his confidence in the rational character of reality.

The problem was the nature of the cohesive force thanks to which bodies, instead of dissolving into a cloud of particles, exist as fixed and durable entities. This property, though most obvious in solids (which resist attempts to fracture or penetrate them), is present in liquids as well. Thus, when water flows in a very thin stream over a great enough height, it does not dissolve into vapor but descends in droplets of relatively large volume; like the parts of a solid, the parts of a liquid are therefore held together by a cohesive force. Can we explain this force with the help of purely rational concepts? Galileo based his answer on an analysis of the resistance to penetration by certain liquids. Once again the starting point was an argument with the Aristotelians, who on the authority of a passage in the *De Caelo*[132] maintained, in opposition to Archimedes, that the motion of a solid in a liquid cannot be determined by reference to the ratio of their specific weights alone. An ebony ball, for example, will sink in water, while a thin strip of the same material will float on the surface, provided that it is placed there lightly enough.[133] Similarly, balls of gold or lead will sink immediately, while leaves of the same metal will float. According to these Aristotelians, a single explanation could account for all these disconcerting phenomena,

[131] *Saggiatore*, pp. 350–351.
[132] *On the Heavens* IV, 6.
[133] *Discorso intorno alle cose che stanno in su l'acqua*, 1612, Vol. IV, p. 90.

namely, that very thin sheets of ebony, gold, or lead are too light to overcome the cohesive force by which the upper parts of the water are held together.[134]

Against this theory, which accepted the existence of a cohesive force as a prime and qualitatively irreducible datum, Galileo mustered a whole series of arguments. Let us begin with the logical ones. Thus, if we say that the upper parts of a liquid have a certain degree of tenacity, there is no reason why the same tenacity should not be granted to the lower parts as well. But in that case a thin plate of gold or ebony would have to float not just on the surface but on whatever level of the water we choose to immerse it, "which is false."[135] And even if the cohesive force was limited to the upper parts of the liquid, would the plate not sink if we agitated the liquid or transferred the lower parts to the top by some other means before placing the plate in it?[136] Moreover, the Aristotelian theory was not merely lacking in coherence; it was also incompatible with experience, which shows us that no body, however fine and light, is incapable of moving through water. Thus muddy waters, if they are allowed to settle long enough, will become perfectly pure and clean.[137] And what are we to make of a tenacity or cohesion that even the tiniest grains of sand have no difficulty in surmounting? The conclusion is therefore perfectly clear: since it is impossible to discover or even to imagine a force, however minute, to which the resistance of water against division and penetration is not inferior, that force must of necessity be nothing.[138]

Two concepts, as free as possible of the taint of qualitative intuition, enabled Galileo to develop his own views on the cohesion of bodies and to offer an explanation of the differences between liquids and solids in this respect. He asked himself first of all in what manner the constituent parts of a body could be combined, and found that they could be

---

[134]Ibid., pp. 102–103. In fact, the Aristotelian hypothesis involved the assumption that the motion of a solid in a liquid is determined by its shape: the reason why a thin strip of ebony cannot break the "tenacity" of the upper parts of the water is precisely that the large surface of the latter allows the tenacity to express itself to the full. Galileo's *Discorso* was largely devoted to the refutation of this assumption; see footnote 58 of Chapter 7.

[135]Ibid., p. 103.

[136]Ibid., p. 107.

[137]Ibid., p. 103.

[138]Ibid., pp. 104–105. The resistance the sea puts up to the motion of a ship in no way weakens this argument—it depends, not on the cohesion of the water, but on the swiftness with which the adjacent parts of the water must be pushed aside by the ship's prow.

joined in two distinct ways: continuously, no part having an extremity of its own, and contiguously, each part having its own extremity and merely touching the rest.[139] On the basis of this definition he distinguished two fundamentally different types of penetration: that characteristic of continuous bodies and calling for a true division of parts, and that of contiguous bodies calling for a mere repulsion of parts. And since liquids are clearly repulsed rather than divided by penetration, it seems reasonable to suppose that their continuity is purely apparent and that they are, in fact, made up of a host of contiguous elements. This hypothesis explains in particular why even "the minutest and lightest particle" can always cut a path through even the thickest layers of water. "Were we able to contemplate the nature of water and other fluids, perhaps we should discover the constitution of their parts to be such not only that they do not oppose division but that they have nothing in them to be divided: so that the resistance that is observed in bodies moving through the water is like that which we meet when we pass through a great throng of people—they may impede our progress, not by any difficulty in the division, for none of the persons of which the crowd is composed are divided, but only in the moving of these persons sideways who were already divided and disjoined." Moreover, the contrast between continuity and contiguity also throws some light on the nature of solid bodies. Because they are, in fact, continuous, that is, the extremities of their parts are joined and not merely touching, they can be penetrated only by an act of division, which no matter how minute their parts, will have to be repeated ad infinitum. The very fact that only a change of state, the transformation of the solid into the liquid, can eliminate the intrinsic resistance of solids to penetration struck Galileo as an experimental confirmation of his hypothesis. Thus, if we melt a lump of metal, are we not merely resolving it into its "last and least particles" so that not only will any resistance to division have been eliminated but there will be nothing left to divide? And while it is exceedingly difficult to understand what actually happens during this process, there is good reason to suppose that such ultimate division is due to the action of a fire whose "most tenuous" parts succeed in insinuating themselves into the solid body and hence cause them to separate.

[139]Ibid., p. 106. Galileo took this distinction from Aristotle (*Physics* V, 3 227 a 6 ff.), who had established it on geometrical grounds.

Whatever its true importance, there is little doubt that this explanation of the cohesion of bodies was in perfect keeping with a rationalist and even with a mechanistic ideal—like the substantial forms and secondary qualities, the forces of cohesion had been banished from the scientific description of matter, so much so that anyone who had read no Galileo other than the *Saggiatore* or the *Discourse on Bodies in Water* might easily think that Galileo's conception of matter differed little from that of the traditional Atomists. In both all matter was said to be composed of particles whose sole attributes were form and motion. However, this would be true only if Galileo had declared himself satisfied with his first assumptions. But far from doing that, he repeatedly stressed their shortcomings. In particular, he realized that they failed to explain quite ordinary phenomena—for instance, why, when we submerge a body and then gently draw it out again, "we see the water follow it and rise notably above its surface, before it separates from it."[140] To explain this phenomenon with the assertion that, whenever bodies are joined together so closely that no air can enter between them, they tend to stay together is merely substituting one difficulty for another, explaining one unknown with a second.[141]

Solids presented Galileo with still more formidable problems than liquids. Thus he thought it paradoxical that the latter should be considered discontinuous down to their last particles, and the former completely continuous. Could continuity alone explain why it takes such great force to break cylinders made of wood or other solid coherent materials, or why different solids put up different resistance to fracture?[142] In fact, the *Discourse on Bodies in Water* fails to define or even to prove the existence of "the binding material which holds together the parts of solids so that they can scarcely be separated."[143] For that reason alone, we can take it that it was no more than a rough draft of a treatise Galileo intended to perfect in due course. This he did twenty-six years later, at the very beginning of the *Discourses*, incidentally providing his readers with a striking illustration of how mathematical reason alone is unable to construct an adequate picture of reality.[144]

It was the unexpected introduction of a new concept, that of the

---

[140]Ibid., p. 102.
[141]Ibid., p. 103.
[142]*Discourses* p. 55.
[143]Ibid., p. 56.
[144]In what follows we shall omit any further references to the inner cohesion of liquids; Galileo did not change his views on them substantially, refusing to the end to attribute their relative coherence to "any internal tenacity (that is, viscosity) acting between their particles" (*Discourses* I, p. 115). It should

vacuum, which transformed the entire analysis at one stroke. With it, Galileo resuscitated one of the most typical themes of traditional philosophy—"that much-talked-of repugnance which nature exhibits toward a vacuum."[145] He tried, first of all, to show that this very repugnance provided a satisfactory explanation of several phenomena otherwise highly disconcerting. Thus, "if you take two highly polished and smooth plates of marble, metal, or glass and place them face to face, one will slide over the other with the greatest of ease, showing conclusively that there is nothing of a viscous nature between them. But when you attempt to separate them and keep them at a constant distance apart, you find that the plates exhibit such a resistance to separation that the upper one would carry the lower one with it and keep it lifted indefinitely even when the latter is big and heavy."[146] Suction pumps posed a similar problem: though water, being a heavy body, tends to fall spontaneously, any fountain maker knows that a suction pump provided with a long enough pipe will raise water to a height of eighteen cubits.[147] Now both phenomena involve the presence of a vacuum, as witness the eventual separation of the plates or the breaking up of the column of water; this suggests very strongly that it is nature's abhorrence of the vacuum which is responsible for the resistance to separation in both cases. Does not the plate experiment, in particular, suggest that nature has an aversion for empty space "even during the brief moment required for the outside air to rush in and fill up the region between the two plates"?[148] And since nature never runs counter to itself, its aversion to the vacuum must surely be considered one of the chief causes of the cohesion of materials.[149]

---

however be noted that some of his pupils, especially Borelli, thought that viscosity provided the only possible explanation of the known effects (Borelli, *De Motu animalium* (Rome, 1680), I, p. 454). Actually, Borelli's solution was no better than Galileo's; the phenomena under consideration (the spherical form of water droplets, the flotation of gold leaves, and so on) were not fully explained until Laplace introduced the concept of surface tension.

[145] *Discourses* I, p. 59.

[146] Ibid.

[147] Ibid., pp. 63–64.

[148] Ibid., p. 59.

[149] Cf. ibid. "This resistance which is exhibited between the two plates is doubtless likewise present between the parts of a solid and enters, at least in part, as a concomitant cause of their coherence." Moreover, Galileo was persuaded to think that the part played by nature's horror of the vacuum in the cohesion of bodies could be isolated and measured. Thus if we suppose that it is that horror alone which maintains a column of water in a suction pump up to a height of eighteen cubits, then we need merely weigh the water contained in a tube of that height to obtain the value of the resistance of a vacuum contained in a cylinder of any solid material having a bore of the same diameter (p. 64).

However, despite its importance, nature's abhorrence of the vacuum could not yet explain all the phenomena Galileo was investigating—for instance, why different bodies put up different resistances to fracture. Moreover, the vacuum had always been presented as a negative or *a contrario* cause, whereas the adherence of plates or the ability of water to rise up to a height of eighteen cubits were positive phenomena calling for positive causes.[150] Could these be adduced without reverting to the "tenacity" of the Peripatetics? Galileo not only believed that they could but felt that there was no need to introduce new causes. Taking up an idea that probably went back to Hero of Alexandria,[151] he thought that the ultimate cause of the cohesion and particularly of the resistance of bodies to fracture might well lie in the presence of a very great number of small vacua spread "between the tiniest particles of the solids." Acting as so many attractive forces, negligible when considered by themselves but immensely strong when combined, these small intervening vacua would readily hold the constituent parts of matter together.[152] However imperfect its formulation still was,[153] this hypothesis had numerous advantages. First of all, it enabled Galileo to offer a simple explanation of changes in state and especially of why certain solids turn into liquids and persist in that state for some time before reverting to the solid state.

Sometimes when I have observed how fire winds its way in between the most minute particles of this or that metal and, even though these are solidly cemented together, tears them apart and separates them, and when I have observed that, on removing the fire, these particles reunite with the same tenacity as at first without any loss of quantity in the case of gold and with little loss in the case of other metals, even though these parts have been separated for a long while, I have thought that the explanation might lie in the fact that the extremely fine particles of fire, penetrating the slender pores of the metal (too small even to admit the finest particles of air or of many other fluids), would fill the small intervening vacua and would set free these small particles

[150] Ibid., p. 60.

[151] *Heronis Alexandrini opera quae supersunt*, Vol. I, *Pneumatica*, A, pp. 5–7 (Teubner, Leipzig, 1899).

[152] *Discourses* I, p. 66: "And who knows but there may be other extremely minute vacua which affect the smallest particle so that the cause which binds together the contiguous parts is throughout of the same vintage?" We are simply trying to follow Galileo's own trend of thought rather than lend it greater coherence than it actually possessed; it should, however, be added that Galileo's minute vacua were infinitely small (a point to which we shall be returning).

[153] Ibid., p. 57. As Galileo himself readily admitted.

from the attraction which these same vacua exert upon them and which prevent their separation. Thus the particles are able to move freely so that the mass becomes fluid and remains so as long as the particles of fire remain inside; but if they depart and leave the former vacua then the original attraction returns and the parts are again cemented together.[154]

More generally, Galileo believed that the hypothesis of small intervening vacua provided a far better explanation than the simple contrast between continuity to contiguity of the most essential differences between solids and liquids:

When I take a hard substance such as stone or metal and when I reduce it by means of a hammer or fine file to the most minute and impalpable powder, it is clear that its finest particles, although when taken one by one are, on account of their smallness, imperceptible to our sight and touch, are nevertheless finite in size, possess shape and capability of being counted. It is also true that when once heaped up they remain in a heap; and if an excavation be made within limits, the cavity will remain and the surrounding particles will not rush in to fill it; if shaken, the particles come to rest immediately after the external disturbing agent is removed; the same effects are observed in all piles of larger and larger particles of any shape, even if spherical, as is the case with piles of millet, wheat, lead shot, and every other material. But if we attempt to discover such properties in water, we do not find them; for when once heaped up, it immediately flattens out unless held up by some vessel or other external retaining body; when hollowed out, it quickly rushes in to fill the cavity; and when disturbed, it fluctuates for a long time and sends out its waves through great distances.[155]

Now, what simpler way is there of explaining all these facts than by the assumption that the vacua present in all solid bodies prevent the total separation of their constituent parts, no matter how long we continue to divide them? And that it is the absence of such vacua from water and all other liquids which ensures the separation of the elementary particles?

The exquisite transparency of water also favors this view; for the most transparent crystal when broken and ground and reduced to powder loses its transparency; the finer the grinding, the greater the loss; but in the case of water where the attrition is of the highest degree we have extreme transparency. Gold and silver when pulverized with acids more finely than is possible with any file still remain powders and do not become fluids until the finest particles of fire or the rays of the sun dissolve them, as I think, into their ultimate, indivisible, and infinitely small components.[156]

---

[154] Ibid., pp. 66 f.

[155] Ibid., pp. 85–86.

[156] Ibid.

Galileo thus believed not only that the idea of intervening vacua provided a clearer explanation of the cohesion of bodies and of their resistance to fracture but that it also led to a better grasp of matter in its unity, no less than in the diversity of its states.[157]

It is hardly necessary to stress the profound transformation Galileo's theory had undergone during the twenty years that separated the *Discourse on Bodies in Water* from the *Discourses*. To the principles of continuity and contiguity, both perfectly open to, and homogenous with, mathematical reason, the later work added the idea of intervening vacua, which was anything but that.[158] Similarly, the mechanistic ideal of the *Saggiatore* and the *Discourse on Bodies in Water* suffered a severe setback: for the first, and in fact the only, time, Galileo felt compelled to introduce the concept of attraction, that is, what he himself had referred to as an occult force.[159] In other words, he had no inhibitions about using concepts of disparate origin, provided only they struck him as being capable of leading to a better grasp of reality. Not only is an unsatisfactory theory better than no theory at all, but it is sometimes (as in the case under discussion) preferable to a mathematically more significant, but ineffectual, one; despite his rationalism, Galileo thus never lost sight of the complexity of the facts. However, the changes he introduced with the *Discourses* appeared at the best in his justification of the hypothesis of intervening vacua. For its nonmathematical character did not mean that there was no need to establish its general coherence, and more especially its coherence with mathematical reason; though mathematical reason had ceased to be the sole and direct basis of the new concepts, it nevertheless retained its explanatory function, and it is by the latter that a theory stands or falls. How, and under what conditions, can mathematical reason perform this task?

Let us return to intervening vacua. What dimensions should we assign to them? If they are finite, then, however minute they may be, it follows that all matter must be penetrable, an idea Galileo considered

[157]However, let us note that, though the presence of small interstitial vacua provided a positive cause for the cohesion of bodies, it in no way explained the other difficulty mentioned by Galileo, namely, the observed differences in the resistance of solids to fracture.

[158]And this despite the fact that it obviated recourse to a gluey or viscous substance cementing the parts of a body together. That substance would not really have solved the problem, since any difficulty that arises with respect to the cementation of the parts of the body itself will also arise with regard to the parts of the glue; ibid., pp. 59 and 65.

[159]*Dialogue* IV, pp. 470 f.

quite absurd. But because matter is continuous and hence infinitely divisible, we may think of it as being made up of an infinite number of infinitely small particles (*di infiniti atomi non quanti*) connected by an infinite number of infinitely small vacua (*vacui infiniti non quanti*).[160] Now this interpretation was in perfect accord with Galileo's theory of indivisibles, the geometrical significance of which we examined earlier; in other words, the truth of the hypothesis of intervening vacua hinges on our ability to show, without introducing logical contradictions, that all continuous bodies comprise an infinite number of infinitely small parts—in other words, that intervening vacua have the same rational status as indivisibles.

As the reader will remember, there were several reasons why Galileo did not reject the theory of indivisibles out of hand despite its paradoxes. One of these paradoxes was that, once we construct the continuum with the help of indivisibles, we cannot say that one line is greater than another, since all comprise an infinite number of constituent parts. However, our ordinary system of integers is no less paradoxical, since there are as many square and cube numbers as there are integers. It also seemed paradoxical to maintain that continuous magnitudes are composed of an infinite number of indivisible parts, when no division, even if continued ad infinitum will ever succeed in reducing such a magnitude to its parts. However, this paradox arises only if we place our imagination above our reason. For reason tells us that every continuous magnitude can be subjected to an infinite process of division, and such division can be performed only if the continuum is indeed made up of an infinite number of parts. Moreover, if these parts were not lacking in magnitude, all continuous magnitudes would perforce have to be infinite. And it was precisely because of their resemblance to integers and because indivisibles alone provide a valid explanation of the divisibility of the continuum that Galileo accepted them in the first place. But this simply means that he thought the theory of indivisibles useful; to demonstrate its truth he would also have had to show that the underlying concepts were perfectly clear and unambiguous. Now the concept of infinity from which the concept of indivisibles is inseparable is anything but unambiguous, so much so that its use in an argument raises difficulties "that lie far beyond our grasp."[161] In particular, as we

---

[160] *Discourses* I, p. 72.

[161] Ibid., p. 89.

saw, it challenges the long-established distinction between the whole and its parts and makes it impossible to account for the manifest differences between continuous magnitudes considered *sub specie quantitatis*.[162] Most perplexing of all are the qualitative changes that occur as we pass from the finite to the infinite. Take a line *AB* and let the point *C* divide it into two unequal parts. It can easily be shown that "if pairs of lines be drawn from each of the terminal points *A* and *B*, and if the ratio between the lengths of these lines is the same as that between *AC* and *CB*, their points of intersection will all lie on the circumference of one and the same circle."[163] Let us now bring *C* closer to *O*, the center of *AB*: the circumference of the circle on which the lines drawn

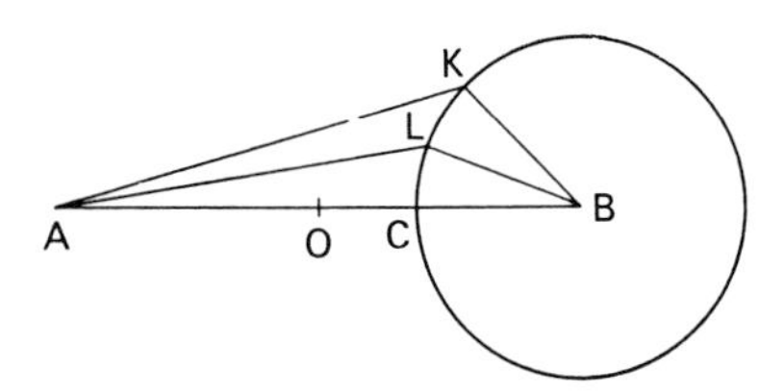

from *A* and *B* intersect will become bigger and bigger, so much so that when *C* is in the immediate vicinity of *O*, the circle becomes "larger than the celestial equator."[164] Finally, if *C* coincides with *O*, we shall obtain a circle "larger than the largest of the others, a circle which is therefore infinite," or rather, because no figure can be both infinite and determined, our locus becomes "a straight line drawn perpendicular to *BA* from the point *O*" and extending to infinity without ever turning. As we pass from the finite to the infinite, the circle therefore ceases to exist as such and experiences a "metamorphosis,"[165] that is, is transformed into a straight line. Under these conditions it would certainly be wrong to claim that the theory of indivisibles, with the infinite dogging its every step, had been justified; Galileo had tried to show that reason could accept it without introducing contradictions: at no moment had he suppressed this fundamental fact that, in using that theory, reason is willy-nilly driven to conclusions contrary to its very nature. Hence Galileo's stress on the conjectural character of his analysis: the infinite is an "abyss" that our mind cannot grasp and on which it can never hope to pronounce with certainty.[166]

---

[162]Ibid., pp. 78–79.
[163]Ibid., p. 83; proof on pp. 89–91.
[164]Ibid., p. 84.
[165]Ibid., p. 85.
[166]Ibid., p. 90. Same remark on pp. 73, 76, 77, and 83.

Let us now return to our hypothesis of intervening vacua. The mere fact that he introduced it was already a sign of Galileo's inability to offer a purely mechanistic explanation of the cohesion of bodies. But did its affinity with the theory of indivisibles at least help establish its rational acceptability? Our analysis of the concept of indivisibles has shown that it did not: before we can say that the hypothesis of intervening vacua is fully supported by reason, we must first prove that reason is capable of grasping infinity, and this, by definition, it cannot do. If, instead, we consider that the hypothesis was simply the most plausible explanation of the forces of cohesion and of the resistance of bodies to fracture that Galileo could offer, then we must also grant that in treating it as such he admitted, if not a radical split between reason and reality, at least a certain incapacity of reason to grasp reality in its full complexity. In other words, we must hold that part of reality is forever closed to human reason; to grasp it, we should have to possess far greater mental resources than we normally do—resources of a kind that God alone can bring in to play. In short, Galileo's rationalism in no way reflected an a priori and dogmatic faith in the complete coincidence of the rational with the natural order of things. Not only did he make persistent efforts to modify and improve the concept to which he looked for a coherent representation of matter, but in the end he was forced to admit that in its dealings with reality reason comes up against an irreducible fact—the infinite—which, though we may grasp it with the help of analogies, we can never hope to understand in its full meaning. Hence it was from within that Galileo's rationalism was forced to take stock of its limitations, that it came to appreciate its true scope and thus armed itself against an oversimplified interpretation of reality.

Galileo also justified the distinction between reason and reality with semiphilosophical, theological arguments. Thus on July 16, 1611, he wrote to Gallenzoni:

> Of the proportions holding between quantities, some strike me as being more perfect and others less so; the more perfect are those obtaining between proximate numbers, for instance, the double, triple, and sesquialter proportions, and so on; the less perfect are those obtaining between more remote prime numbers, such as the proportions 11/7, 17/13, 53/37, and so on; the imperfect finally are those obtaining between incommensurable quantities. These we can neither explain nor even name. In these circumstances, if we had to organize and arrange to the best of our ability and in accordance with perfect proportions the differences between the principal motions of the celestial spheres, I believe that we should have to rely on proportions of the first type, which

are the most rational; God, on the other hand, not bothering about symmetries that man can understand, has ordered these motions with the help of proportions that are not only incommensurable and irrational but totally inaccessible to our intelligence. . . . If the most famous architect would have had to distribute the multitude of fixed stars through the great vault, I believe he would have arranged them in elegant squares, hexagons, and octagons, fitting the largest between the medium-sized and the smallest, using familiar ratios on the assumption that these would provide him with the best proportions; God, however, by apparently scattering them at random, impresses us as having arranged them without heeding any rules or any demands of symmetry and elegance.[167]

The same sentiments were reiterated in a letter to Ingoli, to whom Galileo wrote in 1624: "And who can guarantee that the motions of the planets are not incommensurable among themselves, thus calling for endless corrections, seeing that we cannot reason about them unless we treat them as commensurables?"[168] We do not have to read between the lines of these texts to appreciate their true importance. First of all, they stress the specificity of the natural order of things—indeed, the irrationality that it often seems to evince.[169] But at the same time they make it clear that this irrationality does not so much reside in nature itself as in human understanding. When God created the world, He did so as a geometer. Of this Galileo was fully convinced. But God's infinity shows that His mathematics transcends all human mathematics. Perfect though the latter may be, it is a mere fragment of the mathematics that went into the construction of the world and that alone can render it fully intelligible. Now it is this contrast between human and divine reason, between finite and infinite mathematics, that may explain why reality, though accessible to our reason, so often eludes it. But in that case, is it not true to say that experience alone can make up for the comparative weakness of our mathematical reason, or at least help it, by means of familiar relations and proportions, to grasp the particular relations and proportions that natural phenomena obey?

An examination of the simplifications without which, as he admitted himself, Galileo could not have undertaken the geometrization of the

---

[167]*Letter to Gallenzone Gallenzoni*, July 16, 1611, Vol. XI, pp. 149–150.

[168]*Letter to Ingoli*, Vol. VI, p. 534.

[169]Or that there is not a single effect in nature, even the least that exists, such that the most ingenious theorists can arrive at a complete understanding of it; cf. *Dialogue* I, pp. 126–127.

motion of heavy bodies and especially of the motion of projectiles will not only confirm this interpretation but also clarify it. As we know, when he constructed his mathematical model of projectile motion, he was forced to make a number of assumptions that from a strictly physical point of view were nothing but simplifications. The Fourth Day of the *Discourses* dwelled on this point at some length. First of all, he had to assume that it was possible to ignore the tendency of all heavy bodies to fall toward the center of the earth.[170] Next he had to transform the spherical plane equidistant from the center, which alone is compatible with the conservation of uniform motion, into a horizontal plane tangent to the earth's surface and on which, strictly speaking, all motions are accelerated or decelerated. Finally—and this was perhaps the most important simplification—he had to assume that the retarding role of the medium could be ignored or, as he himself put it, that, contrary to observation, it did not "destroy the uniformity of the horizontal motion or change the law of acceleration of falling bodies." As we saw, all these simplifications were so many prerequisites of successful geometrization. Without the first and the second, the curve representing the trajectory of projectiles would cease to be a parabola, and the construction of an alternative curve raised insoluble mathematical difficulties. As for the third simplification, we need only recall that the resistance of the medium tends to reduce all naturally accelerated motions to uniform motions to appreciate that in its absence Galileo's whole theory of naturally accelerated motion would have become meaningless. In short, Galilean science was based essentially on the substitution of a simpler, ideal world for the very complex world in which we really live. But when we admit that, are we not also admitting that geometrization must necessarily lead us into error? "I grant," Galileo wrote, "that these conclusions proved in the abstract will be different when applied in the concrete and will be fallacious to this extent, that neither will the horizontal motion be uniform, nor the natural acceleration be in the ratio assumed, nor the path of the projectile a parabola, etc."[171] To what extent then are the simplifications to which Galileo was forced to resort scientifically acceptable?

The legitimacy of the first two simplifications is easily established because they merely introduce into mechanics what had long since

---

[170]*Discourses* IV, p. 274.

[171]Ibid.

been a commonplace in statics, namely, that the influence of a point at an extremely great distance from the system under investigation can be safely ignored. Thus Archimedes took it for granted that "the beam of a balance or steelyard is a straight line, every point of which is equidistant from the common center of all heavy bodies, and that the cords by which heavy bodies are suspended are parallel to each other."[172] Similarly, a distance of four miles—the maximum range of artillery—is insignificant compared with that which separates the gun from the center of the earth, so that we shall not be erring too gravely if, having treated this length as a straight line, we transform the spherical plane on which a uniform motion is normally conserved into a horizontal plane.[173] Moreover, since the paths of projectiles terminate on the surface of the earth, only "very slight changes" can take place in their parabolic figure, which, Galileo conceded, would be greatly altered if they terminated at the center of the earth. The third simplification, the elimination of the resistance of the medium, proved much more difficult to justify, so much so that at first sight it seemed to make geometrical truth incompatible with physical truth. Not only does the resistance of the medium gradually slow down the acceleration of natural motions, but it also ensures that no uniform horizontal motion can continue indefinitely; moreover, as Galileo remarked, the perturbations arising from the resistance of the medium are so complex—they depend on the form, weight, and speed of projectiles alike—that it is quite impossible to describe them accurately.[174] However, observations also show that experimental conditions can be so designed as to reduce the perturbations to a minimum. Thus, if we take two balls of the same size but with one weighing ten or twelve times as much as the other (for example, one of lead and one of cork) and drop them from a height of, say, 200 cubits, experiment shows that they will reach the earth with a slight difference in speed, from which it follows that retardation caused by the air is small, for otherwise its greater weight would give the lead ball a considerable advantage.[175] Nor is it even certain that the resistance of the air is very much greater for a rapidly moving body than it is for one moving slowly (unless the difference in speed is enormous); indeed, the pendulum suggests that the opposite may be the case:

[172]Ibid.

[173]Ibid., p. 275.

[174]Ibid., p. 276.

[175]Ibid., pp. 276–277.

Attach to two threads of equal length—say, four or five yards—two equal leaden balls and suspend them from the ceiling; now pull them aside from the perpendicular, the one through 80 or more degrees, the other through not more than 4 or 5 degrees; so that, when set free, the one falls, passes through the perpendicular, and describes large but slowly decreasing arcs of 160, 150, 140 degrees, etc.; the other swinging through small and also slowly diminishing arcs of 10, 8, 6 degrees, etc. In the first place, it must be remarked that one pendulum passes through its arcs of 180, 160 degrees, etc. in the same time that the other takes to swing through its 10, 8 degrees, etc., from which it follows that the speed of the first ball is 16 and 18 times greater than that of the second. Accordingly, if the air offered more resistance to the high speed than to the low, the frequency of vibration in the large arcs of 180, 160 degrees, etc., ought to be less than in the small arcs of 10, 8, 4 degrees, etc., and even less than in arcs of 2, or 1 degrees; but this prediction is not verified by experiment; because if two persons start to count the vibrations, the one the large, the other the small, they will discover that after counting tens and even hundreds they will not differ by even a single vibration—not even by a fraction of one.[176]

Hence, in respect to the tendency of moving bodies to approach the center of the earth, no less than to the resistance of the medium, it seems most likely that "the errors, neglecting those which are accidental, in the results which we are about to demonstrate are small in the case of our machines where the velocities applied are mostly very great and the distances negligible in comparison with the semidiameter of the earth or one of its great circles."[177]

From these remarks we can proceed to a more accurate assessment of the scope and role of experimentation in Galilean science. We have seen that, provided that observations are confined to relatively slow motions over short enough distances and are far enough from the center of the earth, Galileo's simplifications do not lead to any serious split between geometrical theory and the actual course of events. In other words, if we keep within the framework to which our technical means (*artifizii nostri*) confine us,[178] we have a very good chance to correlate the laws to the facts. But when we say this, are we not also saying that an experiment can tell us only what happens under clearly defined conditions which, though they make the experiment possible in the first place (and for that very reason), also and inexorably limit its scope? And does it not mean further that other experiments conducted under quite different conditions are needed to establish the general

---

[176]Ibid., p. 277.
[177]Ibid., p. 278.
[178]Ibid.

validity of a law? Take the experiment by which Galileo, in the Third Day of the *Discourses*, tried to verify the fundamental law of naturally accelerated motion. Having cut a groove in a piece of wooden molding and having smoothed, polished, and lined it with parchment also as smooth as possible, he rolled a hard, smooth, and perfectly round ball along it over measured distances, and timed each descent with a water clock. From the weight and quantity of water collected during each descent he then established that, in practice no less than in theory, the distances traversed bore to one another the same ratios as the squares of the times. But was this experiment really and fully conclusive? Did Galileo not attach greater importance to it than the circumstances justified? More particularly, did he immediately grant it a general validity, or did he rather think that it must be corroborated by further experiments conducted under different conditions? We can answer this question quite unequivocally, for already on the First Day of the *Discourses* Galileo made it perfectly clear that his geometric law held only for bodies descending at a relatively slow pace; as their speed increases, an ever-increasing part of the acceleration had to be devoted to surmounting the resistance of the medium, until there came a point when the speed and the resistance were in balance and the accelerated motion was transformed into a uniform motion. But it was in his letter to Carcavy of June 1637 that Galileo expressed his views on the relevance of his experiment most plainly. Having noted that "no sensible difference has ever been detected in experiments conducted on the surface of the earth from heights and over distances within our reach,"[179] he went on to deliberate at some length on the scope of such verifications. Would the same results be obtained in the immediate vicinity of the center of the earth? In particular, would the distances traversed still bear to one another the double ratio of the times of descent, or would new physical factors, undetectable on the surface of the earth, intervene in such a manner that the mathematical laws of uniformly accelerated motion would cease to apply to naturally accelerated motion? Only a further experiment could settle this issue. Let two observers, one on the highest possible site and the other at the lowest, take two strings of equal length and, suspending two identical balls, set them in motion at the same instant and count the number of oscillations over the same,

---

[179]*Letter to Carcavy*, June 5, 1637, Vol. XVII, p. 91.

relatively long, time interval: they will always count the same number of beats—irrefutable proof that every swing of one takes as long as every swing of the other.[180] But we know that what applies to motions on arcs of circles applies equally to motions along the chords subtending these arcs. Hence it seems highly probable that at very great distances from the surface of the earth, two motions on two inclined and parallel planes will also take the same time; this conclusion greatly enhances the validity of the geometrical law. Aware of the simplifications he had been forced to introduce with his first experiment, Galileo thus admitted quite freely that other experiments were needed to confirm it; in other words, he fully realized that experimental conclusions hold only within clearly defined limits. This adds to rather than detracts from the importance of experimental verifications, for it is only by repeated experiments that we can hope to show that the abstract model elaborated by the physicist is not a mathematical construction without a counterpart in nature,[181] and more particularly to what domains and under what circumstances the experimental evidence is applicable. No matter what their shortcomings, Galileo's experiments were unquestionably the indispensable complements of his geometrization. Though his faith in science never flagged, his manner of analyzing and discussing the exigencies of geometrization nevertheless shows that we should be distorting his thought if we claimed he had a naïve and dogmatic faith in the spontaneous concordance of reason and reality.

We can now take stock of Galileo's rationalism, using his conception of scientific explanation as our guideline. To explain, according to Galileo, meant above all to proceed from a certain number of principles and concepts and with the help of a model to an intelligible reproduction of the phenomena under investigation. This definition highlights two characteristic traits of Galileo's rationalism, above all, the almost complete reduction of physical to rational necessity: once the causal idea of Aristotelian physics had been abandoned, explanation had the sole task of establishing an implicative relationship between the facts, as derived from a model, and the guiding principles of reason. At the same time, simplicity became an important physical criterion:

---

[180]Ibid.

[181]A possibility Galileo mentioned explicitly on several occasions: cf. *Discourses* III, p. 197; *Letter to Carcavy*, pp. 90–91; and *Letter to Baliani*, January 7, 1639, Vol. XVIII, pp. 12–13.

nature follows the simplest path, that is, the one which permits, on the side of reason, the simplest deductions.[182] However, explanation does not merely tend to provide a rational reconstruction of phenomena; its true aim is to turn every physical problem into a mathematical one and thereby to improve for its analysis the existing mathematical science. Now this aim helped to impress some quite special features on Galileo's rationalism. On the most general level, it led him to assume that reality, far from being irreducible to mathematical reason, is in basic accord with it, as witness his dictum that *Deus posuit omnia in numero, pondere et mensura*,[183] which not only summed up his faith but incidentally drew attention to the distinction between primary and secondary qualities. But Galileo did not leave it at that; rather, under the influence of geometrization, he came increasingly to conceive of physical truth as being shaped in the image of mathematical truth. This is reflected, inter alia, in his view that the Ptolemaic and Copernican hypotheses were two contradictory, not simply contrary, interpretations of the World,[184] and more profoundly still in his yes-or-no conception of the role of experiments. To Galileo, the natural order was perfectly determined, and the object of physics was precisely to make its laws explicit. "*Ex parte rei*," he observed in the *Saggiatore*, "there is no mean between the true and the false."[185] A physical law can thus only be true or false, and this truth or falsity is always relative to a particular region of the world. "The deliberations of nature," he wrote elsewhere, "are perfect, unambiguous and necessary; our opinions have nothing to do with them nor, a fortiori, arguments that are only probable: for all arguments concerning nature are either correct and true or else incorrect and false. . . . To pretend that the truth is so deeply

[182]Newton expressed much the same sentiment when he wrote: "For nature is pleased with simplicity and affects not the superfluous pomp of causes"; *Principia mathematica philosophiae naturalis* (Philosophic Library, New York, 1962), p. 324.

[183]Vol. IV, p. 52.

[184]Cf. *Considerazioni sopra l'opinione copernicana*, Vol. V, p. 356. We know that Galileo's identification of physical with mathematical truth misled Duhem into thinking that Galileo did not appreciate the true nature of scientific knowledge, in contrast to the philosophers and theologians of his day who maintained that the Ptolemaic and Copernican hypotheses were equivalent. In fact, Duhem has failed to appreciate that Galileo's approach stemmed from a completely new conception of scientific rationality, the very rationality whose triumph Newton was to ensure fifty years later. Cf. my article "Galilée et le problème de l'équivalence des hypothèses," in *Revue d'histoire des sciences et leur applications*, Vol. 17, no. 4. (Oct.-Dec. 1964), pp. 305 ff.

[185]*Saggiatore*, p. 296.

hidden from us and that it is hard to distinguish it from falsehood is quite preposterous: the truth remains hidden only while we have nothing but false opinions and doubtful speculations; but hardly has truth made its appearance than its light will dispel the dark shadows."[186] This is indeed an attitude typical of a "mathematical philosopher": just as a mathematical proof is either correct or incorrect, so an argument concerning nature is either entirely true or entirely false. The idea of approximate knowledge in the modern sense, that is, knowledge open to indefinite improvement, and the belief that experiments, even if negative, can provide crucial hints about the necessary reconstruction of a theory, thereby playing a paramount role in its development, were alien to Galilean science, which was above all a child of reason.

But clear and incisive though Galileo's rationalism undoubtedly was, it can be judged only in its proper context. One of the most remarkable aspects of the latter was Galileo's constant determination to question experience and to link it as closely as possible to his principles of explanation. If he sometimes accepted verifications that strike us as being far too rough-and-ready, he was also the first to combine geometry with observation and to grasp the crucial importance of measurement. Similarly, we should be distorting his thought if we claimed that he was not aware of the specificity of the natural order. Though he was convinced of its intelligibility, and though he tried to reduce it as far as possible to mathematical reason, he was nevertheless aware that the agreement between reason and reality was not complete. In particular, before reason could be proclaimed the only competent legislator on nature, would it not have to embrace infinity and master its paradoxes and difficulties? Analyzing the conditions appropriate to geometrization in the *Discourses*, Galileo gave a most lucid definition of the status of scientific laws and propositions: though they can be verified satisfactorily within the limits of our technical resources, they must of necessity be based on simplifications and hence must be so many interpretations of reality.

Finally, just like the immediate context of Galileo's work, his rationalism cannot be truly appreciated without regard to its historical background. Traditional philosophy had placed the philosopher above the physicist and more generally above anyone who tried to construct

---

[186] Vol. IV, p. 24.

explanations by discussion or direct examination of the phenomena; this was perhaps most clearly reflected in Geminus's views regarding the dependence of astronomy on the philosophy of essences, a relationship in which the student of nature had no true freedom to choose his assumptions and in which philosophical intelligibility was prized greatly above scientific intelligibility.[187] Hence, when Galileo insisted that mathematical reason was capable of embracing reality, when he claimed that rational necessity was identical with natural necessity, and when he raised simplicity into a cornerstone of scientific explanation, he made a clear break with the past, deliberately replacing the scientist-philosopher with the mathematical scientist. In particular, when he introduced the central ideas of his rationalist approach, he acted as a physicist determined to discover for himself which are the indispensable assumptions of his own science. To consider this attitude the expression of an a priorist ideal means shutting one's eyes to Galileo's revolutionary contribution and forgetting that no science can take shape without a certain number of postulates that confront reality as so many a priori assumptions. What Galileo's rationalism meant, in the final analysis, is that every scientist has the right to be his own philosopher.

---

[187]Cf. Chapter 1, and my article "Galilée et le problème de l'équivalence des hypothèses," p. 312.

Conclusion

The most general and most obvious conclusion to be drawn from our analysis is that experimentation played a crucial part in the formation of classical mechanics. It was observation and observation alone that led to the unification of heaven and earth and hence to a universal mechanics; it was thanks to the systematic analysis and development of the experiential facts that some of the most sterile ideas of traditional thought were discarded, among them the identification of weight with a motor force (inclined-plane experiments), or the assumption that heavy bodies have intrinsic differences in speed (pendulum experiments, and the study of the motion of different bodies in different media). Does this mean that experimentation alone explains the rise of the new science, that once Aristotelian physics had been rejected, it was a more attentive interpretation of facts which led science from Galileo to Newton within two generations? While there can be no doubt about the preeminent role of experiment, a methodical study of Galileo's work shows quite unequivocally that this question cannot be answered in the affirmative.

Three theories or methods played a decisive part in the formation of classical science: Copernican astronomy, the medieval theory of *latitudines*, and the mathematical approach of Archimedes. Copernicanism, though he embraced it fervently, also set Galileo a difficult problem: how to justify the idea of an orderly world in which motion, not rest, was the normal state of the earth. That justification he sought on two levels at once—first of all, on that of the general assumptions or premises of the Copernican doctrine. His examination of the conditions under which alone Copernicanism could hold sway persuaded Galileo that the cosmological premises of Aristotle's *De Caelo*, which constitute the true foundations of traditional mechanics, must be replaced with others in better accordance with his own views on motion. At the same time, an analysis of the problems posed by the earth's diurnal motion led him to outline some of the most significant concepts of classical mechanics—for instance, when during the Second Day of the Dialogue he introduced in turn the idea of an inertial system, the principle of the conservation of uniform motion, and the principle of the composition of motions. Copernicanism had created a new theoretical situation, and it was largely the attempt to defend it that helped classical science to come into its own. However, the Copernican influence was not the only one to make itself felt. For much as he was heir to the Copernican tradition, so Galileo also received the theory of *latitudines*, that is, the medieval attempt to represent motion as a process in space and time by means of a *quoad effectus* analysis. Not only did he appreciate the heuristic value of the concepts used by the Mertonians and the Parisians but he also relied on these very concepts in his construction of a geometrized science of motion, so much so that the treatment of speed as an intensive magnitude continued to inspire his investigations to the end of his life. Now this second influence cannot be separated from that of Archimedes. Thus even at the time when he wrote the *De Motu* he looked to Archimedean hydrostatics for support in his struggle against Aristotelian dynamics. But above all it was Archimedes' successful construction of statics with the help of a geometrical model that persuaded Galileo to transform a physical phenomenon, naturally accelerated motion, into a mathematical object, that is, into one whose properties could be established by deduction: seen in this new light, the concepts underlying the theory of *latitudines* could at last prove their fruitfulness and play an important part in the construction of the new science.

This fusion of currents had several important consequences. To begin with, the widespread belief that the formation of classical science involved a resuscitation of Platonic and Atomistic ideas had to be radically revised. All the evidence suggests that the doctrines employed by seventeenth-century scientists can be divided into two distinct groups: those that served to justify the adoption of a new explanatory ideal or to elaborate a general representation of matter in place of the Aristotelian; and those that actually introduced new ideas and concepts. Though the importance of the former, which provided the philosophy of nature informing classical science with some of its most characteristic elements, cannot be denied, it is nevertheless certain that their role cannot be compared with that of the second group. No Platonic or Atomistic ideas ever entered into the latter. Thus although Galileo used Plato as an ally against the Peripatetics, he borrowed no concepts or methodological ideas from him; the explanatory ideal informing both the *Dialogue* and the *Discourses* was as different from the speculations of the *Timaeus* as it was from those of Aristotle's *Physics* or *De Caelo*. As for the Atomists, though Galileo discussed some of their views, he did so more on philosophical than on scientific grounds; it is impossible to discover the slightest link between Atomistic ideas and the geometrization of the motion of heavy bodies. Quite different was the role of the Copernican doctrine, of the medieval theory of *latitudines*, and of the Archimedean current; they helped Galileo to demolish the traditional cosmology, to conceive of motion as a process in space and time, and to think mathematically about physical problems, thus providing him not only with valid anti-Peripatetic arguments and a general philosophy of nature but also with the theoretical tools he needed to help modern physics to its first successes. Though classical science undoubtedly marked a partial return to pre-Aristotelian rationalist ideas, it was thanks chiefly to the triple heritage of Archimedes, of the fourteenth-century nominalists, and of Copernicus that Galileo could carry science over the threshold of a new era. Nor can we ignore the general epistomological consequence of these influences. For the very fact that the doctrines we have mentioned played as great a part in the formation of classical mechanics as did observation itself shows that this science was born not only of a renewed respect for the facts but also of reflection on earlier theories. Not that Galileo simply fused disparate currents—a view whose fallacy we think we have demonstrated in this book; for all that, it is impossible to maintain that he made a clear break with the past, basing his own science solely on the direct interpretation of experience. All science is somehow born of science, that is, of earlier theoretical systems, even if it uses them for quite different and often quite unsuspected ends.

However, an analysis of Galileo's contribution in its historical singularity helps us to better appreciate the true sources of classical mechanics and shows that Galileo's work represents a clear stage in the development of mechanics, a stage that cannot be fully appreciated by those who consider it a mere chapter of Newton's great work. We have had several occasions to stress the untenability of that view. To begin with, Galileo's contribution to classical mechanics cannot be evaluated on a single plane. His original reflections on motion were based on two sets of problems that were still clearly distinct: those posed by the justification of the Copernican doctrine and those posed by the construction of a geometrized theory of motion. The results obtained in the two spheres may not have been totally dissimilar, but neither can they be combined into a coherent body of propositions: the principles of conservation and composition of the *Dialogue* are not identical with their counterpart in the *Discourses*, and the science of motion played no effective part in Galileo's refutation of the traditional objections to the Copernican doctrine. If further proof of this duality is necessary, we need merely point to the absence of all attempts on Galileo's part to discuss the motions of the celestial bodies in mechanical terms. Not that he had no opinions on the matter: various passages in the *Dialogue* suggest that he tended to attribute these motions to magnetic phenomena, such as when he explained that the moon faces the earth constantly with one surface "as if drawn by a magnetic force."[1] At no time, however, did he attempt to construct a celestial mechanics in the manner of Kepler. The example of Borelli, who, thirty years later and still lacking a satisfactory theory of circular motion or concept of mass, tried unsuccessfully to offer a mechanical explanation of the planetary orbits, shows what Galileo could have done but omitted to do because he realized that his resources were not adequate to the task.[2] But the lack of cohesion of his mechanical ideas is not the only reason why we must grant the specificity of Galileo's work and hence reject the view that classical science was a continuous whole, of which Newton's system was merely the culmination. In this respect again the *Discourses* show quite unequivocally that identical results do not necessarily spring from identical concepts. Thus while the definition of naturally accelerated motion needed only the advent of infinitesimal calculus to reveal its full fruitfulness, the case of the concept of gravity was quite different and so, quite generally, was the case of the concepts Galileo employed when he adopted a dynamic approach. We have only to think of his notion of moment of descent: helping to define the motor function of gravity as such and showing that matter may in certain cases be indifferent as to motion or rest, that concept led him to the principle of conservation and enabled him, toward the end of his life, to interpret the action of the natural motor force in terms of accelerations. In Newton's conceptual system, as a matter of fact, there was no room for the moment of descent: let us say only that by endowing heavy bodies with an inertial motion, it made impossible the treatment of weight as a force. Galilean science had certainly ceased to be traditional science, but it was still a far cry from classical science: the fairest assessment is perhaps that it was the first stage of that science, and that it needed a conceptual metamorphosis

---

[1]*Dialogue* I, p. 91. For similar expressions, see *Dialogue* III, pp. 425 ff; and *Letter to Castelli*, December 21, 1613, Vol. V, p. 288.

[2]For Borelli's celestial mechanics, see A. Koyré: *La révolution astronomique*, pp. 467 ff.

before the Newtonian stage could be reached. To the extent that it coordinated Galileo's contributions, the *Principia Mathematica* may indeed be called the culmination of the *Dialogue* and the *Discourses*, though not their simple extension.

But imperfect though it may have been, Galilean science had many positive aspects as well. Precisely because it was the first stage of classical mechanics, Galilean mechanics provides us with an outstanding illustration of the conditions under which modern science was born, and with the help of what methods. In the *Dialogue* and the *Discourses* we are, in fact, privileged spectators of the gradual presentation of all the postulates and assumptions on which the new science would be based, of postulates and assumptions that, once their success was proved, appeared as direct expressions of the nature of things when, in fact, they were no more than happy anticipations of the course of events. Better than anyone else, Galileo thus lets us see to what extent the principles of modern science, far from being a negation of philosophy, are rooted in philosophical procedures. To hold that mathematical reason is capable of embracing reality, to assume that rational necessity is akin to natural necessity, to turn simplicity into a touchstone of scientific explanation is not to introduce so many ostensive definitions based on the evidence of our senses, but rather to choose a metaphysical platform. In thus taking us to the true sources of classical science, Galileo shows us not only that this science sprang from the substitution of one explanatory ideal for another but also that it was a wager that, despite its successes, was one of the greatest gambles of reason. There is no better protection against the ever-present temptation of positivism or pragmatism than a return to Galileo.

Appendixes

Appendix 1
Additional Remarks on Oresme's Theory of *Configurationes Qualitatum*

The following remarks are meant to supplement the arguments presented in Chapter 2:

1. While Oresme found it relatively easy to describe uniform variations in speed (uniformly difform motions), difform variations, even of a uniform type, presented him with very difficult problems. This was because increases in speed may be associated with decreasing accelerations, and decreasing speeds with increasing decelerations.

To describe such variations, he had to resort to a remarkable artifice: he considered the "line of intension or summit line" (*linea intensionis seu linea summitatis*),[1] that is, the line describing the upper limit of his diagrams. On these, a uniform variation had a straight summit line, while a difform variation had a curved summit line. That curve could be concave or convex and, as Oresme added, either circular or non-circular (that is, it could be rational or irrational). A concave summit line might, for example, represent a decreasing acceleration, a convex line an increasing acceleration. Altogether there would thus be four types of possible configurations representing the *intensio* and *remissio* of a speed subject to uniformly difform variations.[2] Difformly difform variations, on the other hand, demand an infinite number of summit lines.

2. Duhem suggested that Oresme's use of summit lines might be considered an anticipation of analytic geometry.[3] This view is allegedly supported by various passages of Oresme's work, and particularly by the one in which he defined a uniformly difform quality as being such that "when any three points [of extension] are taken, that proportion of the distance between the first and the second to the distance between the second and the third is the same as the proportion of the excess in intensity of the first over the second to the excess of the second over the third."[4] Duhem considers this statement equivalent to the modern definition that "the intensity varies with the extension in such a way that

---

[1] *Tractatus* I, 13. As quoted in M. Clagett, *The Science of Mechanics in the Middle Ages*, p. 348.

[2] P. Duhem, *Système du monde*, Vol. VII, p. 546.

[3] Clagett put it more cautiously when he said that the summit line is comparable to a curve in modern analytic geometry, *Science of Mechanics*, p. 341.

[4] *Tractatus* I, 11; as quoted in Clagett, *Science of Mechanics*, p. 352.

we can represent it by a straight line inclined to the axis of the longitudes or abscissae"; or to the algebraic expression: "given any three points M_1, M_2, and M_3, of which x_1, x_2, and x_3 are the longitudes or abscissae and y_1, y_2, and y_3 the latitudes or ordinates, we always have $(x_1 - x_2)/(x_2 - x_3) = (y_1 - y_2)/(y_2 - y_3)$."[5]

However, before we can call Oresme the first to offer an analytic description, by means of a line, of the successive values of a dependent variable (y) considered a function of an independent variable (x), we must first show that the relevant passage in his text contains an explicit reference to the summit line. Now, he simply reiterated the standard definition of a quality subject to uniform variations; that is, he based his analysis on the quality considered as a whole. Admittedly, he went on to show that a geometrical diagram can express this quality, but he again based this remark on the diagram as a whole, not on its summit line.[6] In fact, Oresme used his summit line exclusively to classify the possible modes of variations associated with particularly difficult cases. However, when it came to the actual analysis of the process of variation, it was always the whole figure Oresme examined, not the summit line. All that mattered was the form and the *overall* magnitude of the resulting figures, and it was on these two properties that all of Oresme's arguments were wholly based.

3. The *configurationes* by which Oresme represented variations in speed were the simplest cases considered by him. They helped him either to analyze a "linear quality" (*qualitas linearis*) whose base (the magnitude taken in *longitudo*) could be represented by a straight line, or to study the distribution of an intensive quality (for instance, heat, or even speed) along a line traced out inside the body. But Oresme was also familiar with more complex cases; thus he tried to discover how the intensity of a quality varies over a whole area instead of over a straight line. This he called the analysis of "superficial qualities"; in it the *latitudines* erected from different points of the surface make up three-dimensional figures, which he believed were to the *qualitas superficialis* what the two-dimensional figures were to the *qualitas linearis*.[7] Without a fourth dimension it is, of course, impossible to represent the *configurationes* expressing vari-

---

[5]Duhem, *Système du monde*, Vol. VII, p. 548.

[6]*Tractatus* I, 11; as quoted in Clagett, *Science of Mechanics*, p. 352.

[7]For Oresme's text, see A. Maier, *Zwei Grundprobleme der scholastischen Naturphilosophie*, pp. 100 ff.

ations in the intensity of a quality over the subject as a whole, but Oresme believed that this failure could be remedied by the division of the subject into an infinite number of surfaces, each of which could be analyzed separately. It goes without saying that this was a purely theoretical suggestion; Oresme confined his practical studies to variations in linear qualities, speed being chief among them.

4. However, he assigned yet another role to his *configurationes intensionum*, one that was deeply rooted in traditional science.[8] Thus he believed that the *configuratio* resulting from the analysis of the intensity of a quality either *quoad subjectum* or *quoad tempus* revealed not only the distribution of that quality over the body or its variation in time but also its nature or its mode of action. He accordingly spoke of rectangular or triangular "heat" and assumed that the heat represented by a sharp *configuratio* had a mode of action quite different from that represented by a nonsharp *configuratio*, just as a sharp tool does not produce the same effect as a blunt one. Moreover, he believed that a *configuratio* can also reveal how a body reacts to external influences: the more irregular the *configuratio*, the more responsive the body. In other words, as A. Maier has stressed, he characterized material bodies by the possible "forms" of their qualities, much as the Atomists characterized them by the nature and arrangement of their atoms. This is further proof that, despite their very important contributions, fourteenth-century writers preserved what was basically a qualitative approach to reality.

---

[8]This aspect is discussed in A. Maier's *Die Vorläufer Galileis im XIV. Jahrhundert*, pp. 125 ff.; and in "La Doctrine de Nicole Oresme sur les configurationes intensionum," *Revue des sciences philosophiques et théologiques* 32 (January–April 1948).

Appendix 2
Bradwardine's Law

As expounded in Bradwardine's *Tractatus* of 1328,[9] his law represents the first serious modification of Aristotelian dynamics by a fourteenth-century writer.

Although Bradwardine did not question Aristotle's choice of factors (namely, force and resistance or F and R), he posed the problem of dynamic explanation in a much more precise manner. Thus he was concerned not merely with determining the general dependence of the speed on F and R but also with discovering the formula expressing "the proportion of speeds in motion" (*proportionem velocitatum in motibus*). In other words, he asked himself how, given the variation in the speed of a motion, we can express the correlated variation in the ratio F/R responsible for it. In what way must the ratio F/R be changed so that the speed is doubled, tripled, and so on? Conversely, what changes will lead to a variation in the *virtus motiva* responsible for the speed of a moving body?

Aristotle's answer to this problem (namely, that $V = F/R$) was quite unacceptable. To begin with (and Bradwardine could not have been the first to note this fact), it does not follow from Aristotle's formula that the speed must be zero when $F = R$; rather, when $F/R = 1$, the corresponding velocity is of degree 1 and hence equivalent to rest. Bradwardine added four more precise objections. In the first place, the Aristotelian formula was insufficient, since, when applied to the comparison of two speeds, it demands that either the motor force F or the moving body R remain constant, so that the formula cannot be applied to cases in which F and R vary simultaneously.[10] The second and much more crucial objection was that, according to Aristotle's formula, "any motive force must be of infinite capacity." Thus let a force F communicate a speed V to a moving body; if the mass of the body was doubled or quadrupled, the same force would then communicate to the body a speed of $V/2$ or $V/4$, and so on, ad infinitum, from which it follows that "any mobile could be moved by any mover," an absurd consequence but one that follows necessarily from Aristotle's law. Finally, experience itself bears witness against Aristotle: if a man moves a certain weight at

---

[9] Thomas Bradwardine, *Tractatus de proportionum seu de proportionibus velocitatum in motibus*.

[10] Ibid., pp. 96–98.

a certain speed, two men will obviously move it at a speed that will be much more than twice as great; by arguing the contrary, Aristotle was patently in conflict with the facts.[11] From this, Bradwardine concluded that "the proportion of the speeds of motion does not vary in accordance with the proportion of the resistances nor, if the resistance remains constant, does it vary in accordance with the proportion of the movers."[12]

He accordingly proposed to use a formula quite different from Aristotle's,[13] a formula based on the view that, far from depending on simple variations of the force or the resistance taken in isolation, "the proportion of the speeds of motions varies in accordance with the proportion of the power of the mover to the power of the thing moved."[14] In other words, a double speed must correspond to a "double proportion" of F to R, a triple speed to a triple proportion, and so on. The reader will remember that doubling or tripling a proportion does not mean multiplying it by 2 or 3, but it means raising it to the power 2 or 3. Hence, Bradwardine's law states that if the quotient F/R is squared, the speed is doubled; if the quotient is cubed, the speed is trebled, and so on. Similarly, the square or cube root of F/R corresponds to half or a third the original speed. As Maier has noted, the modern equivalent of Bradwardine's law is a logarithmic expression of the type $v = \log_a F/R$ where $a = F_1/R_1$, F_1 and R_1 being the initial values of F and R.[15] Unlike Aristotle, to whom speeds increased directly as the ratio F/R, Bradwardine's law thus affirms that speeds increase or decrease as the logarithms of the quotient of that ratio.[16]

Did this formula obviate the difficulties Aristotle encountered? It is clear, first of all, that in trying to obtain a speed two, three, or four times smaller than a given speed v, we no longer run the risk of ending up with a ratio in which F is smaller than R; thus if a decrease in the speed by a factor of 2, 3, or 4 can be expressed by the square, cube, or fourth root of the ratio F/R, that is, by a fractional exponent, the root corresponding to that exponent will always be a positive number, and Bradwardine's law thus enables us to obtain, without fear of contra-

[11]Ibid., p. 98.
[12]Ibid., p. 100.
[13]After having criticized three other alternatives; cf. *Tractatus*, pp. 86–94 and 104–110.
[14]Ibid., p. 110.
[15]Maier, *Die Vorläufer Galileis*, pp. 91–92.
[16]In fact, if a body descends with unit degree of speed while the ratio F/R is, say, 3/1, then for it to descend with two degrees of speed, F/R must be $(3/1)^2 = 9/1$ (and not 6/1, as Aristotle believed); in other words, $2v = (F/R)^2$, or, more generally, $nv = (F/R)^n$; thus, $v = \log (F/R)$.

diction, an infinite variation in speed. It can also be shown that his law is in far better agreement with experience than Aristotle's.

Moreover, the acclaim with which it was greeted during the fourteenth century also bears witness to its superiority. The Mertonians, needless to say, adopted it unanimously, as did the Parisians and later the whole of Europe: Buridan, Oresme, and Albert of Saxony all accepted Bradwardine's interpretation and, like him, believed that speed comparisons cannot be based on the successive values of each of the terms of the ratio F/R but rather must reflect the proportions between the successive values of the ratio considered as a whole.[17] Bradwardine's law was quickly extended to apply to changes of all kinds, so much so that it is generally true to say that, whenever variations in intensive magnitudes could be determined by the quotient of two other magnitudes, fourteenth-century writers invariably had recourse to Bradwardine's formula.[18] Finally, Bradwardine's law was undoubtedly responsible for the renewed interest in proportions evinced by fourteenth-century writers, and especially for their investigation of all cases in which a ratio A/B was found to be proportional to a ratio C/D, characterized by a fractional exponent, such that $(C/D)^{m/n}$. This was obviously the approach Oresme used in his *De proportionibus proportionum* (c. 1360), a treatise that set out to give a systematic description of complex proportions or of "proportions of proportions." "To encourage the studious," Oresme explained, "it is useful to say a few words about those proportions of proportions whose study is an inestimable aid to the analysis not only of the proportions of motions but also of the most secret and arduous problems of philosophy."[19]

However, valuable though it undoubtedly was, the importance of Bradwardine's law must not be overestimated. Unlike the theory of *latitudines* or the *impetus* concept, it played no part in the formation of classical mechanics, and it was quite a different tradition on which Galileo based his youthful attempts to improve Peripatetic dynamics (cf. Chapter 3). The fact that he ignored Bradwardine's law can perhaps be explained by its purely theoretical character, by the complications his functional equations (medieval writers were not familiar with logarithms), and by the absence of a standard measurement defining to

---

[17]Maier, *Die Vorläufer Galileis*, pp. 98–104.
[18]Ibid., pp. 97–98.
[19]E. Grant, "Nicole Oresme and his De proportionibus proportionum," in *Isis* 165 (September 1960), p. 297.

what ratio of F/R a speed of degree 1 must be assigned.[20] Moreover, Bradwardine's formula had no repercussions on the traditional viewpoint. Admittedly, scientists began to consider the ratio F/R as a whole, but they were handicapped by the fact that, just as Bradwardine never abandoned the idea of resistant forces, so he never considered that a speed could vary without a simultaneous variation in the fundamental dynamic ratio. Despite his reformulation, the analysis *quoad causas* thus remained quite distinct from the analysis *quoad effectus*, and even though an increase in speed was known to call for an increase in the dynamic ratio, there was no common analytical language to link the concepts employed in the two spheres. Because it conserved the traditional division and failed to suggest any possible application of the new theory of *latitudines* to the description of the natural motion of heavy bodies, Bradwardine's law cannot be said to have extended the frontiers of Peripatetic dynamics.

---

[20]Maier, *Die Vorläufer Galileis*, p. 93.

Appendix 3
On the Force of Percussion

Galileo left two texts on the force of percussion: the one he wrote in 1598–1600 was published by Favaro as an appendix to the *Mecaniche* (in Vol. II of the Edizione nazionale); the other, which appeared in dialogue form in about 1638, may be considered an appendix to the *Discourses*.

1. In his first essay, Galileo introduced the idea that was to preside over his later study of the force of percussion: if we hit a nail with a hammer, we drive it forward, but if we simply place the hammer on the nail the nail will not budge.[21] Why the difference? Since the weight of the hammer is constant, the only possible explanation is the speed with which the hammer has been brought down.

This suggested to Galileo that percussion must be governed by the general laws of statics. Take a lever with unequal arms; we know that a small body placed on it will balance a much heavier body whenever the virtual velocities of the two bodies are in the inverse ratio of their weights.[22] The effect of the small body (its *momento*) is therefore equal to the product of its weight and the speed of its displacement, and so is that of the heavier body. But the effect of a percussive body on striking another is also a function of its weight and its speed. This suggests that if we apply the general law of equilibrium to the case of percussion, we shall be able not only to describe it in physical terms but also to determine its effects.

Let us designate the moment of the striking body as the product of its weight and its velocity, that is, $W_1 \times V_1$; similarly, let us call the moment of the struck body the product of its weight and the speed with which it is driven forward, that is, $W_2 \times V_2$.

Galileo's solution was to assume that the two moments must be equal, and that if the effect lasts the *same time* for the striking and struck bodies, then the ratio of the distance through which the latter advances under the impact to the distance the former covers as a result of its *impeto* is inversely proportional to the ratio of their respective weights:[23] in other words, percussion effects can be expressed by a formula of the type $S_2/S_1 = W_1/W_2$. Hence, if we know three of the variables, we can

[21] *Opere*, Vol. II, p. 188.
[22] Ibid., p. 189.
[23] Ibid.

easily determine the fourth. To use Galileo's own example, if $W_1 = 4$, $W_2 = 4000$, and if $S_1 = 10$ paces, then it follows that the struck body will be driven forward 1/100 of a pace.[24]

These remarks prove two things. First of all, they show that Galileo believed that percussion effects are best explained on the analogy of simple machines, that is, by treating them as a special case in statics.[25] But at the same time he apparently considered the percussive force of a body a mere function of its weight, or rather as a multiplication of its weight due in some way to its speed at the moment of impact.[26] Hence, he held that there was no natural distinction between percussion and pressure; percussion was merely an artificially augmented form of pressure. Theoretically it is always possible to obtain the same effect with percussion as with the application of a large enough weight.

2. In the text written in 1638, Galileo did not appreciably change his original conclusions—the central argument was still the similarity between the "operation of percussion" and the "operation of mechanical engines."[27] The final proposition of the second study gave a precise definition of the relationship between the force of percussion and pressure (henceforth called the "dead weight"): "If the percussive effect of a given weight dropped from a fixed height is to displace a body of even resistance over a certain distance, and if the same effect can be obtained by a certain quantity of dead weight acting by simple pressure, I say that, if the same weight falling on a body of double the resistance displaces it through a distance equal to, say, half that through which it displaced the former, then, to produce this second displacement by pressure, the dead weight previously used would not be sufficient; we should need one twice as great; and similarly for all other proportions: the smaller the displacement caused by the same percussive body, the greater will be the quantity of dead weight needed to produce the same result."[28]

However, though Galileo's central idea had remained basically unchanged, the later text nevertheless marked an advance over the earlier, inasmuch as Galileo twice referred to the somewhat unsatisfactory character of his original solution.

---

[24]Ibid.

[25]Galileo, as the reader will recall, tried to do the same with the centrifugal force; cf. Chapter 5.

[26]This conclusion was more or less identical with that offered in the *Mechanical Problems* (Question XVII) of the Peripatetic School, 853 b 14 ff.

[27]*Opere*, Vol. VIII, p. 329.

[28]Ibid., p. 340.

Consider a stake on which a ram weighing 100 pounds is dropped from a height of four cubits; the stake will be driven forward, say, 4 inches. Let us now try to determine what dead weight, acting by pressure alone, would produce the same effect: the answer is, say, 1000 pounds. This suggests that a weight of 100 pounds traveling with a speed acquired by a fall through four cubits has "a force and an energy" equivalent to the force and energy of a dead weight of 1000 pounds, so that the "virtue of the speed alone accounts for a pressure of 900 pounds in the form of the dead weight." However, let us now drop the ram on the stake a second time: it will clearly drive it forward less than the original distance, say 2 inches. Replace the dead weight of 1000 pounds on the stake. Experiment shows that it will produce no effect at all and that a much greater weight is needed to drive the stake another 2 inches forward. If we drop the ram a third time, it will drive the stake forward over a still smaller distance, and an even larger dead weight is needed to obtain the same effect. "Now which of these dead weights," Galileo asked, "must we take as our measure of the force of percussion, a force that, for its part, never changes?"[29]

This was, however, only the first of a series of difficulties. We saw that each successive blow of the ram drives the stake forward a little less than the preceding one. Thus if the second blow drives the stake forward 2 inches, the third will only drive it forward 1½ inches, the fourth 1 inch, the fifth ½ inch, and so on: the distance thus keeps decreasing, but, unless the resistance of the ground is "infinite," each blow of the ram will drive the stake forward a fraction further. Now, the dead weight needed to produce the same result must be increased very much more rapidly, so much so that, very soon, we should need an "immensely great" weight to produce the least effect; hence, if we equate the force of percussion with pressure, do we not turn it into an infinite magnitude? Admittedly, this consequence, too, has its parallel in the lever, for if we extend the arm of a balance far enough, the smallest weight will balance even the largest.[30] For all that, Galileo never succeeded in resolving the (inevitable) paradox that the dead weight must be increased to infinity while the force of percussion remains constant.[31]

---

[29]Ibid., p. 326–328.
[30]Ibid., pp. 328–330.
[31]He came up against the same problem when dealing with percussion in another context; see Chapter 7, footnote 77.

In short, Galileo's study of the force of percussion ended in failure. Ingenious though some of the experiments he devised may have been, he never succeeded in conceptualizing the force of percussion as such. The main cause was, of course, that he did not appreciate that the force of percussion depends, not on the weight, but on the product of the mass and the speed of the percussing body. "Today we would say," Mach remarked, "that the force of the impact, or the momentum mv is a magnitude of quite another dimension than the pressure p. The dimension of the former is mlt^{-1} and that of the second mlt^{-2}. In fact, the ratio of the pressure to the impulsion resulting from the impact is comparable to the ratio of a line to an area."[32] By treating them as magnitudes of the same type, Galileo was bound to end up with a paradoxical situation; that he appreciated this fact is yet another proof of his genius.

---

[32]Ernst Mach, *The Science of Mechanics*, p. 267.

Appendix 4
Did Galileo Consider the Moving Earth an Inertial System?

The *Dialogue* gives two contradictory answers to this question, on the Second and Fourth Days, respectively.

1. Second Day

The answer is in the affirmative. Having examined the traditional arguments against the earth's diurnal motion, Galileo thought he could dismiss them unreservedly. The first three of these (the perpendicular descent of heavy bodies, the trajectory of cannonballs fired westward and eastward, and the flight of birds) he rejected on the basis of the principle of the conservation of uniform circular motion: since such motion tends neither to separate bodies from, nor to bring them nearer to, the common center, its conservation disposes of all the objections that a moving earth would inevitably introduce deflections.[33] Similarly, though by arguments that must strike the modern reader as ineffective, he also disposed of the objection that all freestanding objects would be flung into space. And though he dealt specifically with the earth's diurnal motion, Galileo's arguments applied equally well to the earth's annual motion: on a moving earth, bodies behave in precisely the same way as they would on an immobile earth.

2. Fourth Day

While the Second and also the Third Days simply demolished all the common objections to the Copernican doctrine, the Fourth Day tried to adduce positive proof of its soundness. To that end, Galileo considered two remarkably regular phenomena on the earth's surface: the tides and the trade winds.

THE TIDES (*Dialogue* IV, pp. 442 ff.)

Galileo set out to prove that the daily ebb and tide of the oceans can be explained only on the assumption that the earth is in double motion. That proof was expressly based on the principle of conservation. Consider a very large vessel filled with water and in uniform motion. If the

---

[33]Cf. Chapter 5.

vessel is suddenly speeded up, the water will retain a "part of its slowness" and hence "fall somewhat behind while becoming accustomed to the new *impeto*, remaining toward the back end [of the vessel], where it would rise somewhat." Conversely, if the vessel is suddenly retarded, the water (being contained in the vessel but not firmly adhering to it) will retain part of the old *impeto* and "run toward the forward end, where it would necessarily rise."[34] Galileo believed that if the Copernican doctrine was true, then the oceans must behave like the water in the vessel.

Describe about the center *A* the circumference of the earth's annual orbit *BC*, and around the point *B* describe the smaller circle *DEFG*,

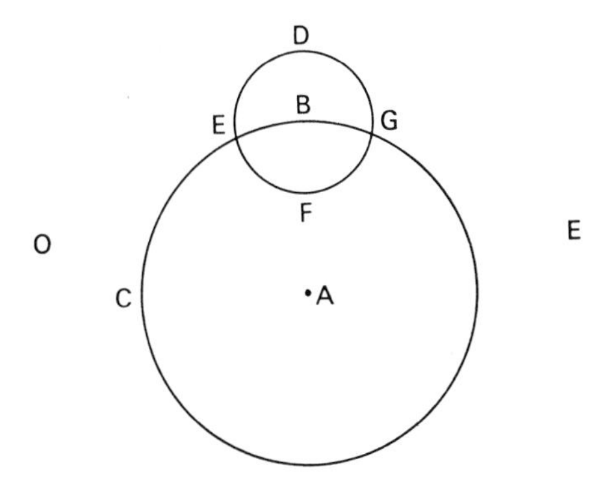

representing the terrestrial globe. Let us suppose that its center *B* runs along the whole circumference of this orbit from west to east, that is, from *B* toward *C*; and let us suppose further that the terrestrial globe turns about its own center *B* from west to east, in the order of the points *D*, *E*, *F*, *G*, during a period of twenty-four hours. "From the composition of these two motions, each of them in itself uniform, I say that there results a difform motion in the parts of the earth." To appreciate this fact, we need merely note that when a circle revolves about its own center, every part of it must move with contrary motions at different times. Thus, while the parts of the circumference around the point *D* will move toward the left, that is, toward *E*, the parts around the point *F* will move toward the right, that is, toward *G*; moreover, once the parts around *D* have reached the vicinity of *F*, their

[34]*Dialogue* IV, p. 450.

motion will be contrary to what it originally was around *D*.[35] Hence, if we combine this "contrariety in the motion of the parts of the terrestrial surface" with the earth's annual motion, we shall obtain "an absolute motion of the parts of the surface which is at one time very much accelerated and at another retarded by the same amount." This becomes obvious when we consider first the parts around *D*, whose absolute motion will be very swift, resulting as it does from two motions in the same direction, that is, toward the left. The first of these is part of the annual motion "common to all parts of the globe"; the other is that of the point *D* which is also carried to the left by the diurnal rotation, so that "in this case the diurnal motion increases and accelerates the annual motion." From *D*, let us now pass on to the parts around *F*. While the annual motion still carries them to the left, the diurnal motion now carries them to the right, so that to obtain their "absolute motion," this time we must subtract the diurnal motion from the annual one. Finally, at *E* and *G*, where the diurnal motion carries the bodies neither toward the left nor toward the right, the absolute motion is equal to the annual one. Thus, subjected to acceleration and retardation every twenty-four hours, the ocean waters will run forward and backward in their basins to produce the tides. "Now if it is true (as indeed proved by experience) that the acceleration and retardation of the motion of a vessel makes the contained water run back and forth along its length, and rise and fall at its extremities, then who will make any trouble about granting that such an effect may—or, rather, must—take place in the ocean waters? For their basins are subjected to just such alterations; especially those that extend from west to east, in which direction the movement of these basins is made."[36]

Quite obviously this explanation of tidal phenomena ran counter to the arguments advanced in the rest of the *Dialogue*.[37] Thus, while the

[35]Ibid., p. 452.
[36]Ibid., p. 453.
[37]In fact, it failed to offer a direct solution of all of the problems posed by the tides. Thus it implied that there can be only one high and one low tide every twenty-four hours, instead of the two we normally observe (ibid., p. 458). This apparent anomaly could, according to Galileo, be easily explained by reference to the weight of the ocean waters. For if liquids are indeed impelled forward or backward in their vessels by changes in the speed of the original motion, then their own weight will impress upon them a "reciprocal" motion in the opposite direction (p. 454). The same happens with the oceans—the real rhythm of the tides is due to the "reciprocal oscillation" of the water (p. 458). However, Galileo also noted that the twelve-hourly cycle, though the most common, was not the only possible one—the size and depth of the basins could change the tidal rhythm quite considerably (pp. 459–460).

entire Second Day was devoted to the proof that the diurnal rotation cannot cause the least perturbation, that nothing would happen on a moving earth that would not also happen on an earth at rest, Galileo clearly abandoned the idea that the earth was an inertial system when he attributed the tides to the double motion of the earth on the Fourth Day. Moreover, why should the oceans alone reflect the changes in speed experienced by every part of the earth once a day? If Galileo's explanation of the tides were correct, would it not follow that all objects not rigidly bound to the earth's surface ought to move forward and backward every twenty-four hours? In short, he failed to see that the Fourth Day of the *Dialogue* was in conflict with the Second, in other words, that the principle of conservation cannot serve to refute the traditional objections to the earth's diurnal motion and explain the tides at the same time.

THE TRADE WINDS (*Dialogue* IV, pp. 463 ff.)
From sailors Galileo had learned that perpetual currents of air blow between the subtropics and the equatorial belt, and that these currents occur so constantly throughout the year that they could not be ordinary winds. Could they, too, be caused by the earth's diurnal motion? If we compare the behavior of heavy and light bodies (Galileo meant dense and rare bodies), we find that the latter are much easier to set in motion than the former, but conversely that the former, once set in motion, are much more likely to conserve their impressed motion than the latter. This being so, let us examine the behavior of the lightest bodies known, namely, the air that surrounds the terrestrial globe. Above the continents, the mountains, acting as so many movers, exert some pressure on the air and force it to follow the earth's diurnal rotation; however, where the earth's surface has large flat spaces, such as on the oceans, the "reason for the surrounding air to obey entirely the seizure of the terrestrial rotation would be partly removed."[38] Under these conditions, it is only to be expected that above the oceans, and particularly above those placed where the terrestrial rotation has the greatest linear velocity (that is, between the equator and the tropics), the relative immobility of the air should be reflected in the presence of a constant easterly wind. The trade winds, thanks to which ships can make for the West Indies with such ease but find it hard to sail in the opposite direc-

---

[38]Ibid., pp. 463–464.

tion, may therefore be considered a direct consequence of the earth's diurnal motion, and, moreover, to provide crucial evidence in its favor.[39]

Unlike the tides, the trade winds do constitute a proof of the earth's diurnal rotation. However, not only were the reasons adduced by Galileo incorrect,[40] but they were once again incompatible with the conclusions reached during the Second Day. In particular, in order to interpret the trade winds as being a consequence of the earth's diurnal motion, Galileo was compelled to change all his previous ideas about the behavior of air. Thus, while he argued during the Second Day that the air, like all terrestrial bodies, has the power of conserving the uniform circular motion impressed upon it by the earth's diurnal motion and that it is therefore possessed of the same inertia as solids and liquids, during the Fourth Day he suddenly, and without feeling the least need to explain his change of approach, transformed the air into a body whose rarity was such that even an infinitesimally small force sufficed to set it in motion—into a body, moreover, that was "quite incapable of conserving this motion once the motor has ceased to act."[41] It was only thanks to the direct impulsion it received from "the roughnesses of the earth's surface" that the air could follow the earth's rotation;[42] in other words, the power to conserve motion varies as the density of the moving body. Now this interpretation ran counter, at least in part, to some of the arguments by which Galileo had tried to refute the traditional objections during the Second Day, for instance, when citing the inertia of the air to dismiss those based on the flight of birds or the motion of the clouds.[43]

In respect to the trade winds no less than to the tides, Galileo's arguments were therefore fundamentally mistaken. However—and this fact cannot be stressed often enough—their shortcomings and even their contradictions in no way detract from the inherent importance of the *Dialogue* or that of its contribution to classical mechanics. In both cases Galileo came up against problems whose solution had to await further

---

[39]Ibid., p. 465.

[40]Very simply put, the trade winds are actually produced by masses of cold air moving from the polar region toward the equator. Because the poles have a smaller tangential velocity than the tropics, the trade winds are deflected westward in the northern hemisphere and especially so between the two tropics. In the summer, they make their presence felt as far north as 30°.

[41]Ibid., p. 463.

[42]It seems that Galileo took this idea from Copernicus' *De revolutionibus orbium coelestium* I, 8, p. 93 (A. Koyré, ed.).

[43]*Dialogue* II, pp. 169, 209–210, and 278–279.

progress in that science. It should also not be forgotten that the Fourth Day of the *Dialogue*, in which Galileo tried to cement his defense of the Copernican doctrine with the most spectacular demonstrations he could think of, was not entirely of a piece with the other Days. Its arguments can therefore be split off from the rest of the work (as, in fact, they were by Galileo's successors) without detracting from the value of the remainder.

Appendix 5
Did Galieo Change His Belief that Forces Are Proportional to Speeds?

In his "De l'accélération produite par une force constante" (On the acceleration due to a constant force),[44] Duhem suggested that Galileo at no time characterized the action of forces in terms of acceleration, not even in the text that he added to the second edition of the *Discourses* (see Chapter 7).

He based this view on Galileo's own proof, which relied on the idea that the speed attained by a body on reaching the point C (see the accompanying diagram) is to the speed it attains on reaching point D in the same time as the distance AC is to the distance AD. However, the ratio AC/AD is also the ratio of the body's moment of descent down AC to its moment of descent down AB, so that the whole argument rests on the fact that the moment of descent is proportional to the distance traversed. "The principle assumed by Galileo," Duhem explained, "is therefore the following: two forces acting on one and the same mobile starting from rest will in equal times make it describe distances that are in the same ratio as the magnitudes of these forces."[45] This principle, Duhem continued, is admittedly a corollary of the modern principle that "two forces communicate accelerations to one and the same moving body that are to each other as the magnitudes of these forces . . . but it so happens that it is also a corollary of the fundamental axiom of Peripatetic dynamics. . . . The reason, therefore, why Galileo was able to formulate the law governing the descent of a body down an inclined plane so accurately was that the Aristotelian postulate he used was, by a happy coincidence, also a consequence of the dynamics we

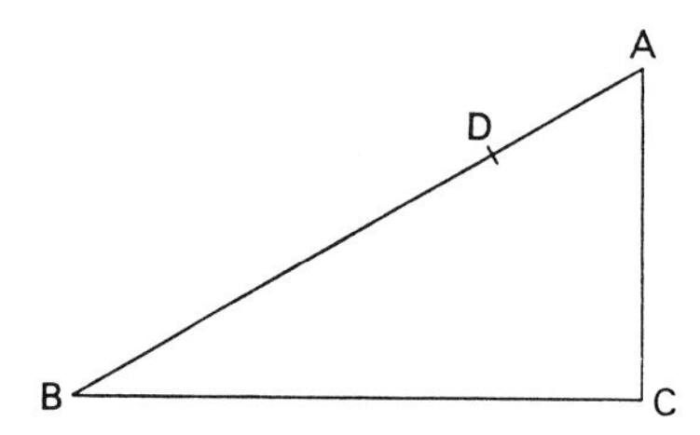

[44]*Proceeding of the Second International Congress of Philosophy* (Geneva, 1904), pp. 861 ff. Duhem expressed the same view in his *Origines de la statique*, Vol. I, pp. 260–261.

employ today."[45] Hence Duhem's conclusion: "Galileo always assumed and proclaimed the great principle of Peripatetic dynamics, namely, the proportionality of the force to the velocity it engenders;"[46] at no time did he even "suspect the fundamental law of modern dynamics."[47]

What are we to make of this argument? To begin with, it should be noted that, strictly speaking, the central idea of Duhem's critique proves nothing at all: if it is true that the proportionality of the distances *AC* and *AD* to the moments of descent can be deduced from Peripatetic dynamics, it does not follow in the least that Galileo should have failed to see that the motor action of the moments of descent is linked to the acceleration of the moving body down the plane *AB* or down the perpendicular *AC*. But, above all, Duhem overlooked the fact that Galileo had expressly described motions down an inclined plane as being uniformly accelerated. Hence, the only reason why a body descends more swiftly down the vertical than it does down an inclined plane is that it experiences a greater acceleration or, to use the language of the *Discourses*, that the moment of speed added, from one instant to the next, to the acquired velocity (on which alone the uniform increase in speed depends) is greater in the first case than in the second. In other words, Galileo could not have failed to realize that the speeds acquired down *AD* and *AC* must have the same ratio as the acceleration acquired by the moving body either on descending along the inclined plane *AD* or on falling freely along *AC*; his remark that an inclined plane serves to change "the ratio of the accelerations" could have meant nothing less.[48] We are therefore fully entitled to think that in the text he dictated to Viviani in 1639 Galileo did indeed interpret the motor action of a force in terms of the resulting acceleration. However, so great was Duhem's antipathy to Galileo that he attributed this interpretation to Scaliger and Benedetti.[49] Now the passages he quoted from the works of these two men prove the precise opposite: far from characterizing the effects of a force, as Galileo did with his concept of moment of descent, their concept of impetus merely served to explain the increase in the motor force of gravity, so that they attributed changes in speed to a simultaneous increase in the motor force. The solutions of Scaliger and Benedetti were therefore on a par with that offered by Buridan.[50]

---

[45]Duhem, *Proceeding*, pp. 900–901.
[46]Ibid., p. 888.
[47]Ibid., p. 901.
[48]Cf. Chapter 7.
[49]Duhem, *Proceeding*, pp. 884–885.
[50]Cf. Chapter 2.

Appendix 6
The Determination of the *Impeto* (or Speed) of a Projectile at Every Point on Its Trajectory

Let bc be a parabola representing the motion compounded of a uniform horizontal motion along cd and a uniformly accelerated motion down the vertical, bd. The problem is to determine the *impeto*, or speed, of a projectile at each point in its parabolic path.

1. Galileo's solution was based on four propositions or remarks that we shall recall in brief:

A property of the parabola. Draw the parabola cb and produce its axis db upwards. If $ba = bd$, it is possible to show that the straight line joining the points a and c will be tangent to the parabola at c (*Discourses*

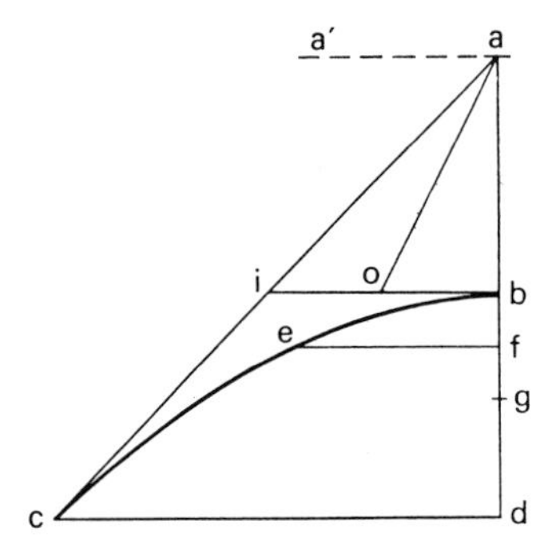

IV, p. 271). Again, if bi is drawn parallel to the base cd of the parabola, it is clear that $bi = cd/2$ and that, in the particular case of $cd = da$, $bi = ba = bd$.

The principle of the composition of velocities. Consider the triangle abi, and imagine that two uniform motions, one along ab and the other along aa' drawn parallel to bi, are impressed simultaneously on a moving body located in a; then if the body covers a distance equal to ab along ab in the same time that it covers a distance aa' equal to bi along aa', it follows from Theorem II of projectile motion that in order to obtain the magnitude of the real speed, we need merely draw the diagonal joining the points a and i, that is, the diagonal joining the extremities of the segments proportional to the respective speeds of the two motions (ibid., p. 280).

The ratio of the successive speeds attained by a body in uniformly accelerated downward motion. On the same figure, let *bd* be the vertical distance traversed by a body starting from rest at *b*. Mark on *bd* any point *f*; if V_f is the speed of the body at *f* and V_d its speed at *d*, the problem is to determine the value of V_d/V_f.

Since the speed increases as the time, the ratio of the speeds at *d* and *f* will be as the ratio of the times needed to traverse *bd* and *bf* or as T_{bd}/T_{bf}. Construct *bg*, the mean proportional between *bf* and *bd* (that is, $bg = \sqrt{bd \times bf}$). From a proposition established previously (Corollary II to Theorem II on uniformly accelerated motion), we have $T_{bd}/T_{bf} = bg/bf$; thus $V_d/V_f = T_{bd}/T_{bf} = bg/bf$ (ibid., pp. 281–282).

How can the speed of the uniform horizontal and of the uniformly accelerated vertical components of a projectile motion be depicted on a single diagram? Let us return to our figure. The projectile will describe the parabola *bc*; the uniformly accelerated component of its motion will be represented by the vertical *bd*, and the speed acquired during this motion will have a fixed value; hence it can be represented by *bd*. The uniform horizontal component of the projectile motion, by contrast, can assume an infinite number of theoretical values, and to represent these, we must extend the axis *bd* of the parabola upward. It is clear that we can always mark a point *a* on this axis such that a moving body falling freely from *a* to *b* will reach *b* with a speed precisely equal to the uniform (inertial) component of its motion. This remark supplies the necessary artifice. We can, in fact, represent the uniform inertial component of the projectile motion by a fixed distance, such as *ab*, and we shall call this distance the "sublimity" (of the parabola) (ibid., pp. 282–283).

2. *Solution of the problem* (ibid., pp. 284–285).

Galileo considered the speed attained by the body at the end of its trajectory and again at any point along that trajectory.

Let *bec* be a semiparabola whose amplitude is *cd* and whose height is *db*, which latter extended upward will cut the tangent of the parabola *ca* in *a*. Through the vertex draw the horizontal line *bi* parallel to *cd*; if $cd = da$, then $bi = ba = bd$.

Let *ab* be the time of descent along *ab* and also the value of the speed acquired in *b* at the end of the descent; that speed will represent the uniform horizontal component (inertial speed) of the projectile motion. Now from the scholium upon Proposition XXIII (Third Day) we know that a body moving at a uniform rate along a horizontal plane will traverse the distance $2ab = cd$ in the time ab (duration of descent from a to b).

Let us now combine the downward motion from b to d with this horizontal motion; since $bd = ab$, the downward motion will take the same time as the uniform horizontal motion through cd. As a result of the combination of these two motions, the projectile will describe the parabola bc, and its real speed on reaching the terminal point c will be the resultant of the speed ab of the uniform horizontal motion and the speed acquired on free descent down bd. We have just seen that these two speeds are equal; hence, bi will represent the speed acquired during the descent down bd, and we need merely join i to a to obtain the magnitude of the speed resulting from the composition of these two elementary speeds, that is, the real speed of the moving body upon reaching the point c.

Let us now consider any point e along the parabola and try to determine the magnitude of the real speed attained by the moving body on reaching that point. The initial component of this speed will not have changed, and we can still represent it by ba. The problem, therefore, is to measure the other component, that is, that due to the fall from b to f. Choose a point g such that bg is the mean proportional between bd and bf, that is, such that $bg = \sqrt{bd \times bf}$. Since bd $(= ab)$ measures the time of free fall from b to d and hence the magnitude of the speed acquired at d, it follows that bg measures the time of descent from b to f, and thus also the magnitude of the speed acquired at f. If we now measure off $bo = bg$ on bi and join ao, then the diagonal ao will give us the magnitude of the real speed of the projectile at the point e of its trajectory.

Admittedly, this analysis was confined to a particular case, for Galileo assumed throughout his argument that $cd = ad$. Still, the manner in which he employed the resources of traditional geometry was altogether remarkable. Projectile motion was turned into an object of mathematical reasoning; in particular, Galileo showed that it is possible to resolve the real speed of projectile motion into its two components at any chosen moment, and hence to evaluate its true magnitude. It was therefore not simply in his definition but throughout his analysis that Galileo asserted the dominance of reason over motion. There is perhaps no better example of the powerful impetus he gave to the science of mechanics.

Bibliography

All Galileo references are to *Opere di Galileo Galilei*, Edizione nazionale, 20 vols., published by A. Favaro from 1890 to 1909 (Florence).
For the best English versions of the major works the reader is referred to the following translations:
On Motion and *On Mechanics*, translated by I. E. Drabkin and Stillman Drake, University of Wisconsin Press, Madison, 1960.
Discoveries and Opinions of Galileo (comprising the *Siderius Nuncius*, the *Letters on Sunspots* and excerpts from the *Saggiatore*), translated by Stillman Drake, Doubleday Anchor Books, New York, 1957.
A Dialogue Concerning the Two Chief World Systems, translated by Stillman Drake, University of California Press, Berkeley, 1962.
Dialogues Concerning Two New Sciences (the *Discourses*), translated by Henry Crew and Alfonso de Salvio, Northwestern University Press, Evanston, Ill., 1914.

D'Abro, A.
The Rise of the New Physics, 2 vols., Dover Publications, New York, 1951.

Archimedes
The Works of Archimedes, edited by T. L. Heath, Dover Books, New York, 1897.

Aristotle
Physics, University of Nebraska Press, Lincoln, 1961.
On the Heavens, Loeb Classical Library, London, 1939.
On Generation and Corruption, Clarendon Press, Oxford, 1930.
Metaphysics, Loeb Classical Library, London, 1935.
Minor Works, Loeb Classical Library, London, 1955.

Armitage, A.
Copernicus, A. S. Barnes, New York, 1962.

Bachelard, G.
Le nouvel esprit scientifique, Presses Universitaires de France, Paris, 1949.
La formation de l'esprit scientifique, J. Vrin, Paris, 1957.

Banfi, A.
Galileo Galilei, Casa editrice ambrosiana, Milan, 1949.

Belaval, Y.
Leibniz, critique de Descartes, Gallimard, Paris, 1960.

Boffito, G.
Bibliografia galileiana, 1896–1940, Rome, 1940.

Borel, E.
La mécanique et la gravitation universelle, A. Michel, Paris, 1942.

Boutroux, P.
Les principes de l'analyse mathématique, Exposé historique et critique, 2 vols., Paris, 1914.

Boyer, Carl B.
The Concepts of the Calculus: A critical and historical discussion of the derivative and the integral, Hafner Publications, New York, 1949.

Bradwardine, T.
Tractatus de proportionum seu de proportionibus velocitatum in motibus, edited by H. Lamar Crosby, University of Wisconsin Press, Madison, 1955.

Brunschwicg, L.
Les étapes de la philosophie mathématique, Alcan, Paris, 1922.

Burtt, E. A.
The Metaphysical Foundations of Modern Physical Science, K. Paul, Trench, Trubner and Co., London, 1925.

Carugo, A., and Geymonat, L.
Discorsi e dimostrazioni matematiche intorno a due nuove scienze, new edition with introduction and historical notes, Turin, 1958.

Clagett, M.
The Science of Mechanics in the Middle Ages, University of Wisconsin Press, Madison, 1959.

Copernicus
De revolutionibus orbium coelestium, Book I, edited by A. Koyré, Paris, 1934.

Crombie, A. C.
Augustine to Galileo, Falcon Press, London, 1952.

Descartes, R.
Œuvres, published by Ch. Adam and P. Tannery, 13 vols., Paris, 1897–1913.

Dijksterhuis, E. J.
Archimedes, E. Munksgaard, Copenhagen, 1956.

Drake, S.
Discoveries and Opinions of Galileo, Doubleday Anchor Books, New York, 1957.
Galileo Studies, University of Michigan Press, Ann Arbor, 1970.

Dreyer, J. L. E.
A History of Astronomy from Thales to Kepler, Dover Books, New York, 1953.

Dugas, R.
Histoire de la mécanique, Éditions Dunod, Paris, 1950.
La mécanique au XVIIe siècle, Éditions Dunod, Paris, 1954.

Duhem, P.
"De l'accélération produite par une force constante," in *Extraits des comptes rendus du deuxième congrès international de philosophie*, Geneva, 1904.
Études sur Léonard de Vinci, 3 vols., A. Hermann, Paris, 1905–1913.
Les origines de la statique, 2 vols., A. Hermann, Paris, 1905–1906.
Le mouvement absolu et le mouvement relatif, Montligeon, 1907.
Essai sur la théorie physique de Platon à Galilée, A. Hermann, Paris, 1908.
Le système du monde, Histoire des doctrines cosmologiques de Platon à Copernic, 10 vols., A. Hermann, Paris, 1954–1959.

Euclid
The Elements, edited by T. L. Heath, 3 vols., Dover Books, New York, 1956.

Favaro, A.
Galileo Galilei e lo studio di Padova, Sucessori Le Monnier, Florence, 1883.
Cronologia galileiana, Padua, 1892.
Bibliografia galileiana (1568–1895), Tip. dei Fratelli Bensini, Rome, Florence, 1896.
Galileo Galilei, Modena, 1910.

Fraenkel, A. A.
Abstract Set Theory, Amsterdam, 1966.

Gandillac, M. de
La philosophie de Nicolas de Cuse, Éditions Montaigne, Paris, 1941.

Geymonat, L.
Galileo Galilei, Turin, 1957.

Giaccomelli, R.
Galileo Galilei giovane e il suo "De Motu," Domus Galileiana, Pisa, 1949.

Gilbert, W.
De Magnete, Dover Books, New York, 1958.

Gilson, E.
History of Christian Philosophy in the Middle Ages, Random House, New York, 1955.

Goldbeck, E.
Der Mensch und sein Weltbild, Quelle und Meyer, Leipzig, 1925.

Guéroult, M.
Dynamique et métaphysique leibniziennes, Les Belles Lettres, Paris, 1934.

Hall, A. R.
Ballistics in the Seventeenth Century: A Study in the relations of science and war, Cambridge, England, 1952.

Heath, T. L.
A History of Greek Mathematics, 2 vols., The Clarendon Press, Oxford, 1921.

Huyghens, C.
Œuvres complètes, 22 vols., M. Nijhoff, The Hague, 1888–1920.

Kahn, C.
Anaximander and the Origins of Greek Cosmology, Columbia University Press, New York, 1960.

Koyré, A.
Études galiléennes, 3 vols., A. Hermann, Paris, 1939.
"Du Monde de l'à peu près à l'univers de la précision," *Critique* 2, 1948.
"Le Vide et l'espace infini au XIVe siècle," in *Archives d'histoire doctrinale et litteraire du Moyen Age*, 1949, pp. 45–91.
"A Documentary History of the Problem of Fall from Kepler to Newton," in *Transactions of the American Philosophical Society*, New Series, Vol. 45, Part 4, Philadelphia, October 1955.

La révolution astronomique, A. Hermann, Paris, 1961.
Du monde clos à l'univers infini, Paris, 1962.
Études d'histoire de la pensée scientifique, Librairie A. Colin and Presses Universitaires de France, Paris, 1966.

Mach, E.
The Science of Mechanics, translated by T. J. McCormack, The Open Court Publishing Co., La Salle, Ill., 1942.
Erkenntnis und Irrtum, J. A. Barth, Leipzig, 1906.

Maier, A.
Die Vorläufer Galileis im XIV. Jahrhundert, Edizioni di storia e letteratura, Rome, 1949.
An der Grenze von Scholastik und Naturwissenschaft, Edizioni di storia e letteratura, Rome, 1952.
Metaphysische Hintergründe der Spätscholastischen Naturphilosophie, Edizioni di storia e letteratura, Rome, 1955.
Zwei Grundprobleme der scholastischen Naturphilosophie: Das Problem der intensiven Grössen, Die Impetustheorie, Edizioni di storia e letteratura, Rome, 1957.
Zwischen Philosophie und Mechanik, Edizioni di storia e letteratura, Rome, 1958.

Marcolongo, R.
"La Meccanica di Leonardo da Vinci," *Atti della Reale Accademia delle scienze fisiche e matematiche*, XIX, Stabilimento industrie editoriali meridionali, Naples, 1933.
"La Meccanica di Galileo," in *Nel terzo centenario della morte di Galileo Galilei*, Milan, 1942.
Lo sviluppo della meccanica sino ai discepoli di Galileo, New Accad. Lincei, 13, 2, 1919.

Michalsky, D.
"La physique nouvelle et les différents courants philosophiques au XIVe siècle," *Bulletin international de l'académie polonaise des sciences et des lettres*, Cracow, 1927.

Michel, P. H.
La Cosmologie de G. Bruno, Paris, 1962.

Mieli, A.
"Il tricentenario dei Discorsi e dimostrazioni matematiche, di Galileo Galilei," *Archeion* 21, 1938.

Milhaud, G.
Les philosophes géomètres de la Grèce, F. Alcan, Paris, 1900.

Montucla, J. E.
Histoire des mathématiques, 4 vols., H. Agasse, Paris, 1799–1802.

Moody, E. A., and Clagett, M.
The Medieval Science of Weights, University of Wisconsin Press, Madison, 1952.

Moreau, J.
Aristote et son École, Presses Universitaires de France, Paris, 1962.

Namer, E.
Galileo, R. M. McBride & Co., New York, 1931.

Newton, Isaac
Principia mathematica philosophiae naturalis, 2 vols., Cajori edition, University of California Press, Berkeley, 1962.

Olschki, L.
Galilei und seine Zeit, M. Niemeyer, Halle, 1927.

Perrin, J.
Les éléments de la physique, Paris, 1946.

Ronchi, V.
Galileo Galilei e il cannochiale, Udine, 1942.

Rosen, E.
The Naming of the Telescope, H. Schumann, New York, 1947.

Santillana, G. de
The Crime of Galileo, University of Chicago Press, Chicago, 1955.

Tannery, P.
"Galilée et les principes de la dynamique," *Mémoires Scientifiques*, Vol. IV, Paris, 1926.

Taton, R.
A General History of the Sciences, translated by A. J. Pomerans, Basic Books, New York, 1963–1966.

Taylor, F. S.
Galileo and the Freedom of Thought, Watts and Co., London, 1938.

Vailati, G.
Scritti, Barth, Leipzig-Florence, 1911.

Vuillemin, J.
Mathématiques et métaphysique chez Descartes, Paris, 1960.

Wohlwill, E.
"Die Entdeckung des Beharrungsgesetzes," *Zeitschrift für die Völkerpsychologie und Sprachenwissenschaft*, Nos. 14 and 15, 1884.
Galileo Galilei und sein Kampf für die Copernicanische Lehre, L. Voss, Hamburg, 1909.

Zeuthen, H. G.
Histoire des mathématiques dans l'Antiquité et le Moyen Age, Gauthier-Villars, Paris, 1902.

Index

Accelerated motion, 72, 96–97, 324–345. *See also* Naturally accelerated motion
Albert of Saxony, 62, 93, 104n, 105, 110, 128, 130, 139n, 285–286
Albertus Magnus. *See* Albert of Saxony
Anaximander, 12, 13, 31n
Antiphon, 35
Antonini, Daniello, 287n
Aquinas, St. Thomas, 66, 87, 128n
Archimedes, xiii, 37, 44, 59, 147, 157–158, 262, 450, 457
  and current, 141, 161–162, 457
  hydrostatics of, 124–125, 271, 358, 437, 458
  influence on Galileo, 61, 118, 121, 173
  statics of, 271, 376, 379
Aristotle, 61, 112, 116. *See also* Peripatetics
  on being, 2–7, 22–23
  on celestial motion, 100
  on circular motion, 15, 26–28, 30, 49–51, 52, 184, 187
  on compound motion, 9–10, 130, 227
  on continuity of motion, 39, 50, 282
  on geocentrism, 31, 186–187
  on gravity, 112, 217
  and indivisibles, 314, 315n
  on local motion, 1–60 passim, 217–219
  motion doctrines disputed, 61–62, 103, 120–121, 125–126
  on natural motion, 7–8, 14–15, 17–18, 48, 52, 53–55, 58, 121–126, 182–187
  on nonbeing, 4–5, 6
  and ordered cosmos doctrine, 12–21, 208–212, 457–458
  on privation, 6–7, 8
  on rectilinear motion, 15, 25–28, 30, 49–50, 52, 184
  and scientific method, 387–388, 392–393, 413n, 415
  and theory of contraries, 182, 187–189
  on time, 41–42, 291–292
  on use of mathematics, 31, 416–417
  on violent motion, 48, 52, 53–55, 57–58, 91–95
  on weight, 17–21, 58, 59, 121–123
Atomists, 14, 17, 18, 24, 189, 440, 458–459
Attractionist theory, 343
Avempace (Ibn Bajja), 109, 110
Averroës, 103, 109
Averroists, 107

Baliani, Giovanni Batista, 420
Barberini, Maffeo, xix
Being, 2–7, 22–23, 217–219
Bellarmine, Cardinal, xvii, xviii
Benedetti, Giovanbattista, 98n, 102n, 118, 119, 126n, 130n, 165n
Bernouilli, Jean, 164, 411n
Bessel, Friedrich Wilhelm, 403n
Blasius of Parma, 116n, 143, 154
Borelli, Giovanni Alfonso, 248, 249n, 264n, 441, 460
Borrius, 128
Bradwardine, Thomas, 61, 62n, 63, 65, 72n, 107, 108, 110, 112, 113, 282, 310, 468–471
Brahe, Tycho, 198n, 210, 213–214, 245, 253, 397, 412n
  on earth's diurnal motion, 224, 225–226, 232, 424n
Bruegger, G. G., xvi
Bruno, Giordano, 102n, 189, 210, 249n, 258
Buonamico, 97n, 116, 128n, 130n, 426
Buridan, Jean, 61, 62n, 91n, 108n, 115n, 126, 127, 130n
  impetus theory of, 92–98, 101, 102–103, 130, 132, 139, 250, 390
  on motion in void, 104–105, 110

Capra, Baldassare, xv
Campanella, Tomasso, xviii
Carcavy, 452
Cardano, Geronimo, 139n
Cassirer, Ernst, 374n
Castelli, Benedetto, xvii, 193n, 218, 349n, 393n
Cavalieri, Francesco Bonaventura, 280n, 314n
Celestial motion, 48
  Aristotle on, 27, 29–32
  versus local motion, 179, 182, 186–189, 199
  medieval treatment of, 99–111
Center of gravity, 143–147, 155, 173n, 360–367
Centrifugal force, 172n, 235–244
Centripetal force, 237, 355–356
Circular motion, 211–215, 237, 243, 246–247
  Aristotle on, 15, 26–28, 30, 49–50, 52, 184, 187
Clagett, Marshall, 61n, 84n, 89n, 148n

Clavius, Father Christopher, xvi, 192n
Cohen, I. B., 219n
Cohesive force, 437–448
Colombe, Lodovica delle, xvi
Comets, 198, 213–215
Commandinus, 147
Composition of motions. *See* Compound motion
Compound motion, 206–212
 Aristotle on, 9–10, 17, 45, 114, 218
 Galileo on, 217–218, 259–261, 264, 266
 Galileo's principle of, assessed, 369–378
Conservation of motion, 100–101, 102–103, 171–172, 180, 220–221, 231–234, 247–252, 254, 257–259, 367–369
Continuity of motion, 39, 50, 114–115, 279–284
Copernicanism, 176–267, passim, 407
Copernicus, Nicolaus, xiv, 31, 115, 211n, 219, 221, 237–238, 250n, 457–460
Costabel, 147n
Current, 141, 160–161, 164–165

Del Monte, Guido Ubaldo. *See* Guido Ubaldo del Monte
Democritus, 14, 35–36, 310n
Descartes, René, 94, 164, 280n, 308n, 360n, 378n
De Soto, Domingo, 99, 102n, 292n
De Thienis, Gaetano, 116n
Difform motion. *See* Nonuniform motion
Displacement, 141, 149–150, 162, 164
Diurnal motion. *See* Earth's diurnal motion
Drake, Stillman, 120n
Duhem, Pierre, 31n, 58, 61, 84, 88, 89n, 90, 104n, 109n, 114, 311n
 on Buridan, 93–94, 100–101, 102n, 103
 on de Soto, 292n
 on Galileo, 142, 346, 454n
 on Jordanus, 148n
 on Leonardo, 154
 on Oresme, 106
Dumbleton, John, 61, 62n, 84
Duns Scotus, John, 64, 88, 104, 107, 108, 109n
Durand of St. Pourcain, 87–88, 90

Earth's diurnal motion, 179, 224–267, 343, 412–413, 424n
Eclipses, 202
Eleatics, 4, 7, 9
 on local motion, 34
Elements theory, 21–34, 139, 179
 and Aristotle's motion doctrine, 182, 186, 187–188
Empedocles, 12, 24
Equilibrium, 143, 149, 155–156, 358
Equivalence principle, 162–163, 164
Euclid, xiv, 37, 118, 308n, 311, 379
Eudoxus, 44, 311

Fabri, Honoratius, 102n
Fabricus, Johann, 193n
Favaro, A., 116n, 287n
Fermal, Pierre de, 264n
Fraenkel, A. A., 313n
Franciscans, 109n
Franciscus de Marchia, 88, 91n, 100
Franciscus de Mayronis, 88, 104
Free fall. *See* Natural motion
Friction. *See* Resistance

Galilei, Galileo. *See* Galileo
Galilei, Vicenzio, 420n
Galileo, xiii–xxiii, 61, 68n, 82n, 111, 115, 118, 121, 285
 on Aristotle, 26, 398
 Aristotle's influence on, 116, 117
 on center of gravity, 147
 center of gravity principle assessed, 360–367
 on compound motion, 206–212, 217–218, 259–261, 264, 266, 409–411
 compound motion principle assessed, 369–378
 on conservation of motion, 171–172, 220–221, 231
 conservation principle assessed, 367–369
 on continuity of motion, 46, 279–284
 De Motu analyzed, 120–141
 on elements theory, 139
 and equivalence principle, 162–163, 164
 on geocentrism, 265–267
 on gravity, 11, 165–172, 255–259, 324–345
 heavy body principles assessed, 360–380
 on horizontal conservation principle, 409–411
 on impetus, 94, 128, 130, 132, 279
 on inertial motion, 244–250, 257, 260, 266
 on intervening vacua, 442–444, 447
 and law of chords, 120
 Le Mecaniche analyzed, 141–172
 on lever theory, 156–158, 161, 162, 166–167

Galileo (continued)
- on mechanical relativity, 230–231, 426–427
- on moon surface, 190–193
- and motion along inclined plane, 141, 165–172, 272, 273, 307–308, 346–352
- on motion versus rest, 256–257
- on motion in vacuum, 136–139
- and naturally accelerated motion, 80, 180n, 272, 276–279, 285–298, 345, 348–349, 350n, 352–353, 358–359, 409–411
- and natural motion, 122–126, 131–137, 345–360
- and natural motive force, 345–360
- on pendulum's motion, 120, 138n
- and percussive force, 172n, 352–355, 358–359
- on Peripatetics, 384–385, 386, 390–391, 392–393, 417n
- on planet luminosity, 199–201
- and projectile motion, 126–133, 135–136, 272, 281, 409–411
- refutes Aristotle's motion doctrine, 121–126, 184–189, 206–212
- on relative motion, 33, 226–231
- on resistance, 356–357, 450–453
- scientific method of, 173–174, 189–190, 193–199, 383, 389, 393–395, 396, 399–456 passim
- on solar rotation, 257–259
- square law of, 180, 272, 286–287, 289–290, 298–299, 307
- on sunspots, 193–199, 203–204, 399–401, 421–423
- and unification of dynamics, 124–125, 128n
- use of mathematics by, 61, 409–418, 434–437, 454–455
- on volume, 134–135

Gallenzoni, Gallenzone, 447
Gassendi, Pierre, 378n
Geocentrism, 31, 116–117, 186–187, 206–208, 265–267. *See also* Earth's diurnal motion
Geometry, 31, 147, 172, 174, 383, 415–416. *See also* Archimedes; Euclid
- Galileo's use of, 139, 173–174, 238–244, 262–264, 271–380, 400–401
- Greek, 118, 259, 262, 264
- medieval use of, 76–79

Gerard of Brussels, 86
Gilbert, W., 229, 252, 291n, 384
Giovanni de Fontana, 116n
Grassi, Orazio, xviii, 213–214, 399
Gravity, 92–93, 109–110, 208–209, 235–244, 324–345. *See also* Positional gravity; Specific gravity; Specific weight
- Galileo on, 11, 154, 165–172, 213
- motor function of, 95–98, 250–252, 256

Gregory de St. Vincent, 43n
Gregory of Rimini, 311n
Gualdo, Paolo, 198
Guérolt, M., 309n
Guidicci, Mario, xviii, 214n
Guido Ubaldo del Monte, xiii, 118, 120, 143, 147, 155n, 165n, 171n, 180

Heath, T. L., 37
Heliocentrism. *See* Earth's diurnal motion
Henry of Ghent, 87, 107
Heracleides of Pontus, 32
Hero of Alexandria, 442
Heytesbury, William, 61, 62n, 67n, 70n, 71, 81, 84, 294, 306
Hipparchus, 132
Huyghens, Christian, 244, 317n, 319, 363n, 366n, 380, 420n

Impetus theory, 92–98, 101, 111, 128, 130, 132, 279. *See also* Moment; Moment of descent
Incommensurability, 37–38
Indivisibles, 35–36, 39–40, 79, 310–323 passim, 445–446. *See also* Intervening vacua
Inertial motion, 17, 94–95, 172, 221
- Galileo on, 244–250, 257–260, 266, 415–416

Infinitesimals, 35, 37
Ingoli, Father F., xviii, 448
Instantaneous motion, 42, 69–72, 136–137, 295–296, 308
Intervening vacua, 442–444, 447. *See also* Indivisibles
Irradiation effect, 206
Irrationals, 34, 36, 37, 38

Johannes de Marchanova, 116n
John of Basoles, 88, 89
John of Jandun, 107n
Jordanus Nemorarius, 141, 148–149, 152–154, 155n, 170n
Jupiter
- Galileo's observation of, 402
- motion of, 206
- satellites of, xvi, xvii, 192n

Kepler, Johannes, xiv, xvi, 79, 192n, 210n, 252, 253–257, 258n, 460
and ordered cosmos doctrine, 209, 210n
Koyré, A., 210n, 219n, 253–255, 287n, 291n, 314n, 374n, 420, 424–426, 428n, 430–431

Latitude (*latitudo*) and intensive magnitudes, 66–67
Latitudines concept, 69–70, 140, 221–222, 294, 295–297, 457–459
Law of distances, 81–82, 84–85
Law of mean degree, 80–81, 84, 86, 113
Leibniz, G. W., 309n, 359n
Leonardo da Vinci, 154, 201n
Lever theory, 141–143, 144–146, 148–151, 152n, 156–158, 161, 162, 166–167
Local motion, 10, 64–65, 108, 285
agency of, 21–23, 48–49
Aristotelian constraints removed, 179, 180–181
Aristotle on, 1–60 passim, 217–219
and change, 86–87
continuous nature of, 34–48
ontological perspectives of, 11, 20–21, 45, 48, 60, 216–223
Locher (philosopher), 247–248
Lorenzi, Antonio, xv
Lorini, Niccolo, xvii

Mach, Ernst, 145–146, 157, 374n, 412n
Maestlin, 201n
Maier, Anneliese L., 55, 61, 64–65, 78, 90n, 95n, 102n, 104, 311n
Mars, 206
Marsilius of Inghen, 62, 65
Mathematics, 31–32, 77, 119, 440–441
Aristotle's use of, 416–417, 425
Galileo's use of, 61, 140, 177, 266, 409–418, 425, 431–437
medieval use of, 61, 70–76, 89
Maurolicus, 147
Mazzoni, Jacopo, xiv, 426
Mechanical relativity, 228, 230–231, 426–427
Medici, Antonio de', xv
Medieval physics, 61–117, 151–154, 221–222, 292–297, 307, 458. *See also* Mertonians; Parisians
Melissus, 12
Mersenne, Marin, xxi, 164n, 263n, 424n
Mertonians, 68–69, 72–79, 80, 89–91. *See also* Medieval physics; Parisians
Michel, P. H., 210n
Moment, Galileo on, 154–155, 157, 158, 160, 168
Moment of descent, Galileo introduction of, 166, 167n, 168–169, 171, 172n, 344, 345, 347–352, 356
Momentum. *See* Moment
Moody, E. A., 148n
Moon, 180, 190–193, 200–203, 204, 205n
Motion along inclined plane, 141, 151–154, 165–172, 255, 259, 272, 273, 307–308, 346–352
Motion in vacuum, 16, 18, 103–111, 136–139, 174
Motive force. *See* Natural motive force
Movement of a movement. *See* Compound motion

Naturally accelerated motion, 180n, 272, 324–345, 348–349, 350n, 352–353, 358–359
Galileo's analysis of, 276–279, 285–298, 298–323, 452–453, 458, 460
Natural motion, 29, 95–99, 179, 213, 272, 307–308
Aristotle on, 14–15, 17–18, 48, 52, 53–55, 58
and vacuum, 174
Natural motive force, 91–99, 342n, 344, 345–360, 427–428
Natural places doctrine, 19, 20–21, 222
Newton, Sir Isaac, 106, 107, 170, 245, 255, 265, 320, 337, 346, 379, 380, 454n, 459–460
on gravity, 377–378
Nicholas of Cusa, 189, 210n
Nifo, Agostino, 102n
Nonbeing, 4–5, 6
Nonuniform motion, 23, 56, 66–72, 79–91, 113, 245
Novae, 198

Ockham, William of, 89, 109
Ordered cosmos, 24, 47, 182–183. *See also* Aristotle
Oresme, Nicole, 62, 65, 85, 86n, 90, 112, 115, 116, 130, 222, 306, 465–467
on acceleration, 68–71, 294
and continuity of motion, 282–283
geometric innovations of, 76–79
influence on Galileo, 126, 301–302
on local motion, 104–106
on motion in vacuum, 106–107, 108, 110
on nonuniform motion, 81, 82, 83
Oxford. *See* Mertonians

Padua, 116
Panofsky, M., 212n
Pappus, 152, 155n, 171n
Parisians, 76–79, 104–105, 127
Parmenides, 3, 12–13
Pascal, Blaise, 311n
Paul of Venice, 166n
Percussive force, 172n, 352–355, 358–359
Peripatetics, 58, 59, 141, 162, 249–250, 386–388, 390
 and cohesive force, 437–438
 mathematical limitations of, 85, 116–117
Perpetual motion, 147n
Philiponus, John, 91n, 110, 127, 128n
Philolaus, 12
Piccolomini, Alessandro, 98n, 102n
Planets, 199–200
Plato, 17–18, 21, 24, 135, 424–426
Platonism, 9, 424–426, 427–428, 431, 458–459
Plutarch, 189
Positional gravity, 151–152, 170n
Potential being, 7, 8, 22–23
Projectile motion, 118, 139n, 272, 281, 307–308, 448–451
Protagoras, 433n
Ptolemy, xiv, 30, 117, 210, 226
 doctrine applied to sunspots, 421–423
 on earth's diurnal motion, 425
Pythagoreans, 34

Qurra, Thabit ibn, 142n

Rectilinear motion, 15, 25–28, 30, 49–50, 52, 184, 216n, 218–219
 and ordered cosmos, 211
Relative motion, 33, 218, 226–231, 244–247
Resistance, 46–47, 55, 57, 93, 95, 121, 327–345, 356–357, 449–453
 Aristotle on, 17, 20–21, 27–28, 41, 121–123
 Galileo on, 130–131
 Kepler on, 255
 and projectile motion, 127
Richard of Middleton, 107
Rothmann, 229

Sagredo, Francesco, xviii
Sagredo, G. F., 402n
St. Vincent, Gregory de, 43n
Salviati, Filippo, xvii
Sarpi, Paolo, 180, 286
Sarsi, Lothario (Father Grassi), xviii, 399
Saturn, xvii
Scaliger, Julius Caesar, 98n, 102n
Scheiner, Father Christopher, xvii, 193n
Scotists, 65, 88–90, 104, 117, 296. *See also* Duns Scotus
Solar rotation, 257–259. *See also* Sunspots
Specific gravity, 59, 122–126, 134–139. *See also* Gravity; Specific weight; Weight
Specific weight, 171, 324–345. *See also* Gravity; Specific gravity; Weight
Speed, as intensive magnitude, 278, 283, 295–296, 298, 301–302, 305, 308–309, 323. *See also* Velocity
Square law, 180, 272, 286–287, 289–290, 298–299, 307
Stevin, Simon, 147n
Sunspots, 193–199, 203–204, 257–258, 399–401, 408, 421–423
Swineshead, Richard, 61, 62n, 63n, 71, 72–73, 74–75, 80, 82, 84, 93n

Tannery, Paul, 37, 419, 424, 430–431
Tartaglia, Niccolò, 118, 139n, 154, 165
Telescope, xv, 190, 199
 and earth's diurnal motion, 224
 impact of, 397–398
Telesio, Bernadino, 102n
Tempier, Stephen, 103, 104, 108, 222
Terrestrial motion. *See* Local motion
Thabit ibn Qurra, 142n
Theaetetus, 37–38
Theodorus, 34, 37–38
Torricelli, Evangelista, 164n, 366, 375–378

Ubaldo del Monte, Guido. *See* Guido Ubaldo del Monte
Uniformly accelerated motion. *See* Naturally accelerated motion
Uniform motion, 50–51, 52, 56, 58, 231, 244–247, 271–272, 276
University of Coïmbra, 102n

Vacuum. *See* Motion in vacuum
Vailati, G., 142, 149, 152n
Valerio, Luca, xv, 147, 180
Velocity, 46, 121–140, 162
 Aristotle on, 23, 47, 55–58
 as intensive magnitude, 63–79, 111, 113

Venus, xvi, 199–200, 206
Vinci, Leonardo da, 154, 201n
Vinta, Belisario, 180
Violent motion, 48, 52, 53–55, 57–58, 91–95
Viviani, xxiii, 339n, 349

Weight, 17–21, 58, 59, 121–123, 356–357. *See also* Gravity; Specific gravity; Specific weight
Welser, Marcus, xvii, 194
William of Ockham, 89, 109
William of Ware, 88
Wohlwill, Emil, 172n, 374n

Xenophanes, 12

Zabarella, 414n
Zeno, 34, 36–37, 38, 43–44, 56, 282, 310–311, 312
Zeuthen, H. G., 37

WITHDRAWN
FROM
COLLECTION

FORDHAM
UNIVERSITY
LIBRARIES